MOBILE WiMAX

WIRELESS NETWORKS AND MOBILE COMMUNICATIONS

Series Editor: Yan Zhang

AUERBACH PUBLICATIONS

www.auerbach-publications.com
To Order Call: 1-800-272-7737 • Fax: 1-800-374-3401
E-mail: orders@crcpress.com

MOBILE WiMAX

Toward Broadband Wireless Metropolitan Area Networks

Edited by Yan Zhang ◆ Hsiao-Hwa Chen

Auerbach Publications
Taylor & Francis Group
New York   London

CRC Press is an imprint of the
Taylor & Francis Group, an **informa** business

Auerbach Publications
Taylor & Francis Group
6000 Broken Sound Parkway NW, Suite 300
Boca Raton, FL 33487-2742

International Standard Book Number-10: 0-8493-2624-9 (Hardcover)
International Standard Book Number-13: 978-0-8493-2624-0 (Hardcover)

Library of Congress Cataloging-in-Publication Data

Mobile WiMAX : toward broadband wireless metropolitan area networks / editors, Yan Zhang and
    Hsiao-Hwa Chen.
       p. cm.
    Includes bibliographical references and index.
    ISBN 978-0-8493-2624-0 (alk. paper)
    1. Wireless metropolitan area networks. 2. IEEE 802.16 (Standard) I. Zhang, Yan, 1977- II. Chen,
Hsiao-Hwa.

TK5105.85.W5685 2007
004.67--dc22
                                                    2007008436

Visit the Taylor & Francis Web site at
http://www.taylorandfrancis.com

and the Auerbach Web site at
http://www.auerbach-publications.com

Contents

PART III Security, Systems, and Policies

Preface

Wireless metropolitan area networks (WirelessMANs) is emerging as a promising broadband wireless access (BWA) technology to provide high-speed, high bandwidth efficiency and high-capacity multimedia services for residential as well as enterprise applications. It is observed that WirelessMAN (e.g., WiMAX) is even regarded as a 4G technology. For the success of the WirelessMANs, international standardization organizations are very actively specifying the standards IEEE 802.16, ETSI HiperMAN and Korea WiBro. The WiMAX forum has been working hard to make its efforts to ensure the compatibility between these technologies. WirelessMANs provide two working modes: point-to-multipoint (PMP) and mesh networking. Two optional architectures have also been designed to enable their flexible integration, quick deployment, easy maintenance, great scalability and resilient services. To achieve these goals, considerable efforts in both academia and industry communities have been made to address the various problems in all layers of protocol stack, i.e., physical, MAC, network and application layers. Efficient signal processing techniques, network architecture, secure protocols, programmable configuration, and flexible system design are needed to guarantee that the system will meet the all requirements.

This book concentrates on the basic concepts, recent advances, and latest standard specifications in the WirelessMANs with the emphasis on IEEE 802.16-based WiMAX. The book is comprised of 17 chapters, the topics of which comprehensively cover almost all essential issues in WirelessMANs. The subjects are explored in different issues and scenarios, including OFDM/OFMDA-based physical layer, RF and circuit, MIMO, signal processing algorithms, capacity, multimedia capability, QoS, medium access control (MAC), mobility management, handoff, dynamic channel assignment, cross-layer optimization and security protocols, etc. The book aims to provide readers with an all-in-one reference containing all aspects of the technical, practical, economic, and policy issues in WirelessMANs.

The first chapter introduces the basics of WirelessMAN. The other 16 chapters in this book are organized in three parts:

- Part I: RF, Signal Processing, and MIMO
- Part II: Protocol Issues
- Part III: Security, Systems, and Policies

Chapter 1 identifies the challenges and possible solutions in the physical layer. Chapter 2 discusses the challenges in aspects of RF and circuitry design. Chapter 3 presents the OFDMA-based physical layer from a signal processing point of view. Chapter 4 and Chapter 5 explore the advantages of employing MIMO technology in WiMAX. Part II concentrates on the protocol issues, which include MAC, QoS in PMP and mesh networks, cross-layer optimization, mobility management, handoff in heterogeneous networks, energy management, and link adaptation mechanism. Part III is focused on the issues of security, economy and system capacity in the WiMAX.

This book has the following salient features:

- Provides a comprehensive reference for WirelessMANs
- Studies different layers of protocol stack for WiMAX
- Identifies basic concepts, techniques, advanced topics and future directions
- Offers many illustrative figures that enable easy understanding
- Details the particular techniques for performance improvement of WiMAX

The book serves as a useful reference for students, educators, faculties, telecom service providers, research strategists, scientists, researchers, and engineers in the field of wireless networks and mobile communications.

We would like to acknowledge the effort and time invested by the contributors for their excellent work. All of them are extremely professional and cooperative. Our thanks also go to the anonymous chapter reviewers who provided invaluable comments and suggestions that helped to significantly improve the whole text. Special thanks go to Richard O'Hanley and Jessica Vakili of Taylor & Francis Group for their support, patience, and professionalism during the whole publication process of this book. Last but not least, a special thanks also goes to our families and friends for their constant encouragement, patience, and understanding throughout this book project.

Yan Zhang and Hsiao-Hwa Chen

About the Editors

Yan Zhang received a PhD degree from the School of Electrical & Electronics Engineering, Nanyang Technological University, Singapore. From August 2004 to May 2006, he worked with National Institute of Information and Communications Technology (NICT) Singapore. Since August 2006, he has worked at the Simula Research Laboratory, Norway.

Dr. Zhang is on the editorial board of the *International Journal of Network Security*; he is currently serving as book series editor for the book series "Wireless Networks and Mobile Communications" (Auerbach Publications, CRC Press, Taylor & Francis Group). He is serving as co-editor for several books: *Resource, Mobility and Security Management in Wireless Networks and Mobile Communications; Wireless Mesh Networking: Architectures, Protocols and Standards; Millimeter-Wave Technology in Wireless PAN, LAN and MAN; Distributed Antenna Systems: Open Architecture for Future Wireless Communications; Security in Wireless Mesh Networks; Mobile WiMAX: Toward Broadband Wireless Metropolitan Area Networks; Wireless Quality-of-Service: Techniques, Standards and Applications; Broadband Mobile Multimedia: Techniques and Applications; Internet of Things: From RFID to the Next-Generation Pervasive Networked Systems; Unlicensed Mobile Access Technology: Protocols, Architectures, Security, Standards and Applications; Cooperative Wireless Communications; WiMAX Network Planning and Optimization; RFID Security: Techniques, Protocols and System-On-Chip Design; Autonomic Computing and Networking; Security in RFID and Sensor Networks; Handbook of Research on Wireless Security*.

He serves or has served as Industrial Co-Chair for MobiHoc 2008, Program Co-Chair for UIC-08, General Co-Chair for CoNET 2007, General Co-Chair for WAMSNet 2007, Workshop Co-Chair FGCN 2007, Program Vice Co-Chair for IEEE ISM 2007, Publicity Co-Chair for UIC-07, Publication Chair for IEEE ISWCS 2007, Program Co-Chair for IEEE PCAC'07, Special Track Co-Chair for "Mobility and Resource Management in Wireless/Mobile Networks" in ITNG 2007, Special Session Co-organizer for "Wireless Mesh Networks" in PDCS 2006, and a member of the Technical Program Committee for numerous international conferences including CCNC, AINA, GLOBECOM, ISWCS, and ICC. He received the Best Paper Award and Outstanding Service Award as Symposium Chair in the IEEE 21st International Conference on Advanced Information Networking and Applications (AINA-07).

Dr. Zhang's research interests include resource, mobility, energy, and security management in wireless networks and mobile computing. He is a member of IEEE and IEEE ComSoc.

Hsiao-Hwa Chen is currently a full Professor in the Institute of Communications Engineering, National Sun Yat-Sen University, Taiwan. He received BSc and MSc degrees with the highest honor from Zhejiang University, China, and a PhD degree from the University of Oulu, Finland, in 1982, 1985, and 1990, respectively, all in electrical engineering. He worked with the Academy of Finland as a Research Associate from 1991 to 1993 and the National University of Singapore as a Lecturer and then a Senior Lecturer from 1992 to 1997. He joined the Department of Electrical Engineering, National Chung Hsing University, Taiwan, as an Associate Professor in 1997 and was promoted to a full Professor in 2000. In 2001 he joined National Sun Yat-Sen University, Taiwan, as a founding Director of the Institute of Communications Engineering of the University. Under his strong leadership the institute was ranked second in the country in terms of SCI journal publications and National Science Council funding per faculty member in 2004. In particular, National Sun Yat-Sen University was ranked first in the world in terms of the number of SCI journal publications in wireless LANs research papers during 2004 to mid-2005, according to a Research Report (www.onr.navy.mil/scitech/special/354/technowatch/textmine.asp) released by the Office of Naval Research, U.S.A. He was a visiting Professor to the Department of Electrical Engineering, University of Kaiserslautern, Germany, in 1999, the Institute of Applied Physics, Tsukuba University, Japan, in 2000, the Institute of Experimental Mathematics, University of Essen, Germany in 2002 (under DFG Fellowship), the Chinese University of Hong Kong in 2004, and the City University of Hong Kong in 2007.

Dr. Chen's current research interests include wireless networking, MIMO systems, next generation CDMA technologies, information security, and Beyond 3G wireless communications. He is a recipient of numerous research and teaching awards from the National Science Council, the Ministry of Education and other professional groups in Taiwan. He has authored or co-authored over 200 technical papers in major international journals and conferences, five books, and several book chapters in the areas of communications, including *Next Generation Wireless Systems and Networks*, published by John Wiley in 2005. He has been an active volunteer for IEEE various technical activities for over 15 years. Currently, he is serving as the Chair of IEEE Communications Society Radio Communications Committee. He served or is serving as symposium chair/co-chair of many major IEEE conferences, including IEEE VTC 2003 Fall, IEEE ICC 2004, IEEE Globecom 2004, IEEE ICC 2005, IEEE Globecom 2005, IEEE ICC 2006, IEEE Globecom 2006, IEEE ICC 2007, and IEEE WCNC 2007.

He served or is serving as Editorial Board Member or/and Guest Editor of *IEEE Communications Letters, IEEE Communications Magazine, IEEE Wireless Communications Magazine, IEEE JSAC, IEEE Network Magazine, IEEE Transactions on Wireless Communications,* and *IEEE Vehicular Technology Magazine.* He is serving as the Chief Editor (Asia and Pacific) for Wiley's *Wireless Communications and Mobile Computing (WCMC) Journal* and Wiley's *International Journal of Communication Systems.*

His original work in CDMA wireless networks, digital communications, and radar systems has resulted in five U.S. patents, two Finnish patents, three Taiwanese patents, and two Chinese patents, some of which have been licensed to industry for commercial applications. He is also an adjunct Professor of Zhejiang University, China, and Shanghai Jiao Tung University, China.

Contributors

Fatih Alagöz
Department of Computer Engineering
Bogazici University
Istanbul, Turkey

Michel Barbeau
School of Computer Science
Carleton University
Ontario, Canada

Jalel Ben-Othman
PRiSM Laboratory
Versailles Saint-Quentin-en-Yvelines
    University
Versailles, France

Balvinder Bisla
Intel Corporation
Santa Clara, California, U.S.A.

Jiannong Cao
Department of Computing
Hong Kong Polytechnic University
Hong Kong, China

Hsiao-Hwa Chen
Institute of Communications Engineering
National Sun Yat-Sen University
Kaohsiung City, Taiwan

Yifan Chen
Nanyang Technological University
Singapore

You-Lin Chen
Department of Computer Science
National Chaio Tung University
Hsinchu, Taiwan

Dong-Ho Cho
Department of Electrical Engineering
KAIST
Daejeon, South Korea

Yang-seok Choi
Intel Corporation
Hillsboro, Oregon, U.S.A.

Alexi Davydov
Intel Corporation
Hillsboro, Oregon, U.S.A.

Marc Emmelmann
Telecommunication Networks Group
Technische Universität Berlin
Berlin, Germany

Xiaopeng Fan
Department of Computing
Hong Kong Polytechnic University
Hong Kong, China

Sahar Ghazal
PRiSM Laboratory
Versailles Saint-Quentin-en-Yvelines
    University
Versailles, France

Gurkan Gür
Department of Computer Engineering
Bogazici University
Istanbul, Turkey

Heikki Hämmäinen
Networking Laboratory
Department of Electrical
    and Communications Engineering
Helsinki University of Technology
TKK, Finland

Kun-Chien Hung
Department of Electronics Engineering and
    Center for Telecommunications Research
National Chiao Tung University
Hsinchu, Taiwan

Dimitris Katsianis
National and Kapodistrian University
    of Athens
Athens, Greece

Mehmet S. Kuran
Department of Computer Engineering
Bogazici University
Istanbul, Turkey

Taesoo Kwon
Communication & Connectivity Lab
Samsung Advanced Institute
    of Technology
South Korea

Steven Y. Lai
Department of Computing
Hong Kong Polytechnic University
Hong Kong, China

Christine Laurendeau
School of Computer Science
Carleton University
Ontario, Canada

Youn-Tai Lee
Networks and Multimedia Institute
Institute for Information Industry
Taipei, Taiwan

Hongxiang Li
Department of Electrical Engineering
University of Washington
Seattle, Washington, U.S.A.

David W. Lin
Department of Electronics Engineering
    and Center for Telecommunications
    Research
National Chiao Tung University
Hsinchu, Taiwan

Hui Liu
Department of Electrical Engineering
University of Washington
Seattle, Washington, U.S.A.

Kanchei Loa
Networks and Multimedia Institute
Institute for Information Industry
Taipei, Taiwan

Jianhua Ma
Hosei University
Tokyo, Japan

Maode Ma
School of Electrical and Electronic
    Engineering
Nanyang Technological University
Singapore

Gabriel-Miro Muntean
Performance Engineering Lab
UCD School of Electronic Engineering
Dublin City University
Dublin, Ireland

John Murphy
Performance Engineering Lab
UCD School of Computer Science
    and Informatics
University College Dublin
Dublin, Ireland

Olga Ormond
Performance Engineering Lab
UCD School of Computer Science
    and Informatics
University College Dublin
Dublin, Ireland

Apostolos Papathanassiou
Intel Corporation
Hillsboro, Oregon, U.S.A.

Berthold Rathke
Telecommunication Networks
    Group
Technische Universität Berlin
Berlin, Germany

Theodoros Rokkas
National and Kapodistrian University
    of Athens
Athens, Greece

Atul Salvekar
Intel Corporation
Hillsboro, Oregon, U.S.A.

Timo Smura
Networking Laboratory
Department of Electrical
    and Communications Engineering
Helsinki University of Technology
TKK, Finland

Roshni Srinivasan
Intel Corporation
Hillsboro, Oregon, U.S.A.

Shiao-Li Taso
Department of Computer Science
National Chaio Tung University
Hsinchu, Taiwan

Shailender Timiri
Intel Corporation
Hillsboro, Oregon, U.S.A.

Tuna Tugcu
Department of Computer Engineering
Bogazici University
Istanbul, Turkey

Hung-Yu Wei
Department of Electrical Engineering
National Taiwan University
Taipei, Taiwan

Adam Wolisz
Telecommunication Networks
    Group
Technische Universität Berlin
Berlin, Germany

Laurence T. Yang
St. Francis Xavier University
Nova Scotia, Canada

Hujun Yin
Intel Corporation
Hillsboro, Oregon, U.S.A.

Yan Zhang
Simula Research Laboratory
Lysaker, Norway

Jun Zheng
Department of Computer Science
Queens College
City University of New York
New York, New York, U.S.A.

Yuan Zheng
Department of Computing
Hong Kong Polytechnic University
Hong Kong, China

Chapter 1

IEEE 802.16-Based WirelessMAN

Sahar Ghazal and Jalel Ben-Othman

Contents

The Institute of Electrical and Electronics Engineers (IEEE) 802.16 standard is a real revolution in wireless metropolitan area networks (WirelessMANs) that enables high-speed access to data, video, and voice services. Worldwide Interoperability for Microwave Access (WiMAX) is the industry name given to the 802.16-2004 amendment by the vendor interoperability organization. The standard supports point-to-multipoint (PMP) as well as mesh mode. In the PMP mode, multiple subscriber stations (SSs) are connected to one base station (BS). The access channel from the BS to the SS is called the downlink (DL) channel, and the one from the SS to the BS is called the uplink (UL) channel. Two duplexing techniques are used: frequency division duplex (FDD), where uplink and downlink operate on separate frequency channels and sometimes simultaneously, and time division duplex (TDD), where uplink and downlink share the same frequency band. IEEE 802.16 defines connection-oriented media access control (MAC). The physical (PHY) layer specification works in both license bands and license-exempt bands, and thus covers a range of 2 to 66 GHz. IEEE 802.16 is designed to support quality of service (QoS) mainly through the differentiation and classification of four types of service flows: unsolicited grant service (UGS), real-time polling service (rtPS), non-real-time polling service (nrtPS), and best effort (BE). Uplink scheduling is supported by the standard for only UGS flows. In this chapter, the QoS specified by the IEEE 802.16 standard is detailed and some previous works on scheduling uplink service flows are discussed. To support mobility, the IEEE has defined the IEEE 802.16e amendment, which is also known as mobile WiMAX. Battery life and handover are essential issues in managing mobility between subnets in the same network domain (micromobility) and between two different network domains (macromobility).

1.1   Introduction

IEEE 802.16 is mainly aimed at providing broadband wireless access (BWA) and thus it may be considered as an attractive alternative solution to wired broadband technologies like digital subscriber line (xDSL) and cable modem access. Its main advantage is fast deployment which results in cost savings. Such installation can be beneficial in very crowded geographical areas like cities and in rural areas where there is no wired infrastructure [1, 2]. The IEEE 802.16 standard provides network access to buildings through external antennas connected to radio BSs. The frequency band supported by the standard covers 2 to 66 GHz. In theory, the IEEE 802.16 standard, known also as WiMAX, is capable of covering a range of 50 km with a bit rate of 75 Mb/s. However, in the real world, the rate obtained from WiMAX is about 12 Mb/s with a range of 20 km. The Intel WiMAX solution for fixed access operates in the licensed 2.5 GHz and 3.5 GHz bands and the license-exempt 5.8 GHz band [3]. Research for a new standard of WirelessMAN dates to early 1998. In 2001 the 802.16 standard was finally approved by the IEEE. A brief history of the standard is presented here:

- IEEE 802.16-2001: IEEE 802.16 was formally approved by the IEEE in 2001 [1]. It is worth mentioning that many basic ideas of 802.16 were based on the Data Over Cable Service Interface Specification (DOCSIS). This is mainly due to the similarities between the hybrid fiber-coaxial (HFC) cable environment and the BWA environment. 802.16 utilizes the 10 to 66 GHz frequency spectrum and thus is suitable for line-of-sight (LOS) applications. With its short waves, the standard is not useful for residential settings because of the non-line-of-sight (NLOS) characteristics caused by rooftops and trees.
- IEEE 802.16a-2003: This extension of the 802.16 standard covers fixed BWA in the licensed and unlicensed spectrum from 2 to 11 GHz. The entire standard along with the latest amendments was published on April 1, 2003 [3]. This amendment was mainly developed for NLOS applications and thus it is a practical solution for the last-mile problem of transmission where obstacles like trees and buildings are present. IEEE 802.16a supports PMP network topology and optional mesh topology BWA. It specifies three air interfaces: single-carrier modulation,

256-point transform orthogonal frequency division multiplexing (OFDM), and 2048-point transform orthogonal frequency division multiple access (OFDMA). The IEEE 802.16a standard specifies channel sizes ranging from 1.75 to 20 MHz. This protocol supports low latency applications such as voice and video.

■ IEEE 802.16-2004: This standard revises and consolidates 802.16-2001, 802.16a-2003, and 802.16c-2002 [1]. WiMAX technology based on the 802.16-2004 standard is rapidly proving itself as a technology that will play a key role in fixed broadband WirelessMANs. The MAC layer supports mainly PMP architecture and mesh topology as an option. The standard is specified to support fixed wireless networks. The mobile version of 802.16 is known as mobile WiMAX or 802.16e.

■ IEEE 802.16e (mobile WiMAX): This is the mobile version of the 802.16 standard. This new amendment aims at maintaining mobile clients connected to a MAN while moving around. It supports portable devices from mobile smart-phones and personal digital assistants (PDAs) to notebook and laptop computers. IEEE 802.16e works in the 2.3 GHz and 2.5 GHz frequency bands [2].

In broadband wireless communications, QoS is still an important criteria. The WiMAX standard is designed to provide QoS through classification of different types of connections as well as scheduling. The standard supports scheduling only for fixed-size real-time service flows. The admission control and scheduling of both variable-size real-time and non-real-time connections are not considered in the standard. Thus WiMAX QoS is still an open field of research and development for both constructors and academic researchers. The standard should also maintain connections for mobile users and guarantee a certain level of QoS. IEEE 802.16e is designed specifically to support mobility.

The rest of this chapter is organized as follows: Section 1.2 presents the IEEE 802.16 MAC and Section 1.3 presents physical layers. In Section 1.4, QoS and mobility issues are discussed, concentrating on QoS scheduling methods and mechanisms proposed by some authors. We conclude in Section 1.5.

1.2 IEEE 802.16 MAC Layer

The MAC layer is a common interface that interprets data between the lower physical layer and the upper data link layer. In IEEE 802.16, the MAC layer is designed mainly to support the PMP architecture with a central BS controlling the SSs connected to it. The 802.16 MAC protocol is connection oriented. Upon entering the network, each SS creates one or more connections over which data are transmitted to and from the BS. The application must establish a connection with the BS as well as associated service flows. The BS assigns the transport connection with a unique 16-bit connection identification (CID) [4]. Downlink connections are either unicast or multicast, while uplink connections are always unicast. Every service flow is mapped to a connection and the connection is associated with a QoS level. The MAC layer schedules the usage of airlink resources and provides QoS differentiation. At SS initialization, three pairs of management connections are established between the SS and the BS in both the uplink and downlink direction:

■ Basic connection is used to exchange short-time urgent MAC management messages (e.g., DL burst profile change request/DL burst profile change response).

■ Primary management connection is used to exchange longer, more delay-tolerant management messages (e.g., registration request and registration response message).

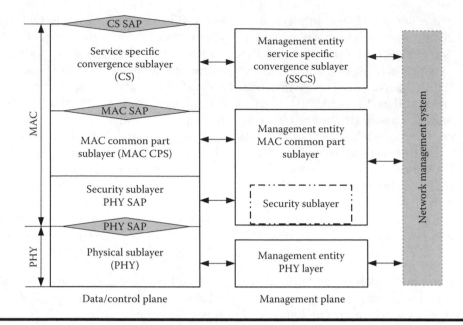

Figure 1.1   IEEE 802.16 MAC layer reference model.

- Secondary management connection is used for higher layer management messages and SS configuration messages (e.g., dynamic host configuration protocol [DHCP], trivial file transfer protocol [TFTP]).

1.2.1 MAC Sublayers

The 802.16 MAC layer is divided into three sublayers: the service specific convergent sublayer (CS), which interfaces to higher layers; the common part sublayer (CPS), which carries out the key MAC functions; and the privacy sublayer (PS), which provides authentication, secure key exchange, and encryption [1] (see Figure 1.1).

- **Service Specific Convergence Sublayer (CS):** The IEEE 802.16 standard defines multiple convergence sublayers. The WiMAX network architecture framework supports a variety of convergence sublayer types, including Ethernet, Internet Protocol version 4 (IPv4), and Internet Protocol version 6 (IPv6). The convergence sublayer transforms or maps external data received through the CP service access point (SAP) into the MAC service data unit (SDU) which is received by the CPS through the MAC SAP.
- **Common Part Sublayer (CPS):** The CPS represents the kernel of the MAC layer. It provides bandwidth allocation and establishes and maintains connections. It also provides a connection-oriented service to the subscriber stations. Moreover, the CPS preserves or enables transmission QoS and scheduling of data over the PHY layer. The CPS receives data from the various convergence sublayers through the MAC SAP classified to the proper MAC connection and associated with the CID.
- **Security Sublayer (SS):** This sublayer provides subscribers to the BS with privacy across the broadband wireless network. Privacy employs authentication of

the client/server key management protocol in which the BS (the server) controls the distribution of keying materials to SSs (the clients). The security sublayer supports encryption of data service flows by employing stronger cryptographic methods, such as the recently approved Advanced Encryption Standard (AES).

1.2.2 MAC Protocol Data Unit (PDU) Format

The MAC protocol data unit (PDU) is the data unit exchanged between the BS and the SS MAC layers. It consists of a fixed length generic MAC header which may be followed by a variable length payload. The payload, when it exists, must contain one or more subheaders. The payload information is either convergence sublayer data or MAC-specific management messages. In the first case, the payload may be one or more asynchronous transfer mode (ATM) cells, an 802 Ethernet frame, etc. In the second case, management messages include registration, ranging, bandwidth requests, and grant and dynamic service support. The flexible length of the MAC PDU enables it to deal with different types of higher traffic layers without needing to know the format or the bit pattern of these messages. Each MAC PDU begins with a fixed length generic MAC header which may be followed by a payload. The MAC PDU may also contain a cyclic redundancy check (CRC), as shown in Figure 1.2.

Six types of MAC subheader may exist in the MAC PDU following the generic MAC header: grant management, fragmentation, mesh, packing, fast feedback allocation, and extended.

1.2.2.1 Generic MAC Header

The generic MAC header begins each MAC PDU. In IEEE 802.16e [2], the downlink MAC PDU consists of only the generic MAC header with no payload data or CRC. The MAC PDU in the uplink direction defines two types of MAC headers: the generic MAC header with no payload and the generic MAC header with a payload or CRC data. The header type (HT) field is used to distinguish the two types of uplink MAC headers. The value of the HT is set to zero to indicate that the generic MAC header is followed by a payload subheader or CRC, and is set to one to indicate a signaling MAC header (a bandwidth request, for example) when no payload is required. In the downlink MAC PDU, the HT field is always set to zero, since only one MAC header format is used. The generic MAC header is illustrated in Figure 1.3 followed by a brief explanation of the next fields:

- **EC:** The encryption control (EC) field has two values—zero to indicate that the payload is not encrypted and one indicate that it is encrypted.
- **Type:** This field indicates the subheaders and special payload types present in the payload.
- **ESF:** The extended subheader field (ESF) is applicable both in the DL and the UL. The ESF indicates if the extended subheader is present or not. If the ESF is zero,

Generic MAC header	Payload (optional) *variable length*	CRC (optional)

Figure 1.2   IEEE 802.16 MAC PDU format.

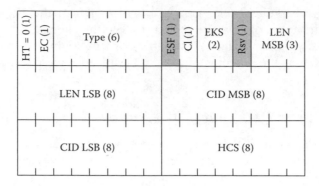

Figure 1.3   IEEE 802.16 generic MAC header format.

the extended subheader is absent. If the ESF is one, the extended subheader is present and will immediately follow the generic MAC header.

■ **EKS:** The encryption key sequence (EKS) represents the method of encryption used to encrypt the payload data. It is meaningful only if the EC field is set to one. It should be mentioned here that not all extended subheaders are encrypted.

■ **LEN:** The length (LEN) field indicates in bytes the total length of the MAC PDU, including the MAC header and CRC data.

■ **CID:** The CID is the unique connection identifier assigned by the BS to each connection.

■ **HCS:** The header check sequence (HCS) is an eight-bit field used to detect errors in the header.

1.2.2.2   MAC Signaling Header Format: Bandwidth MAC Header

The MAC header without payload is applicable to the UL only. As mentioned above, the HT field is set to one to indicate that the generic MAC header is not followed by a payload or CRC data. Payload is optional and is used to send subheader data in the network, and thus this header format is used for signaling messages that do not need to have additional subheaders, such as bandwidth requests and grants, ranging messages, etc.

The standard defines two types of MAC signaling header formats: signaling header type I and signaling header type II [2]. In the type I header, the type field consists of three

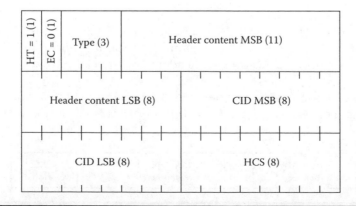

Figure 1.4   IEEE 802.16 MAC signaling header type I format.

Table 1.1   Type Field Encodings for MAC Signaling Header Type I

Type Field (three bits)	MAC Header Type (with HT/EC = 0b10)
000	BR incremental
001	BR aggregate
010	PHY channel report
011	BR with UL Tx power report
100	Bandwidth request and CINR report
101	BR with UL sleep control
110	SN report
111	CQICH allocation request

bits (see Figure 1.4) and is used to define different bandwidth requests: incremental or aggregate bandwidth requests, PHY channel reports, and other signaling messages. The bandwidth request header consists of a bandwidth request alone and does not contain a payload. The CID indicates the connection for which uplink bandwidth is requested.

Table 1.1 summarizes the type field encoding for MAC signaling header type I. From the table and for sending an aggregate bandwidth request, the field type should be equal to "001."

Type II has a one bit (type) following the EC field. The feedback header is an example of this type. The feedback header may come with or without a CID. When the UL resource used to send the feedback header is requested through bandwidth request ranging, the CID field appears in the header, otherwise the feedback header without CID field is used.

1.3   IEEE 802.16 PHY Layer

The physical layer of WiMAX can support the 10 to 66 GHz frequency range. Propagation of data in such systems needs a direct LOS. Because of this limitation, the IEEE opted for single-carrier modulation (WirelessMAN-SC). LOS transmission requires a fixed antenna on the roof of a home or business, with a strong and stable connection capable of transmitting large amounts of data. These antennas and their installation are costly. For WiMAX technology to be accessible by small and residential applications, an NLOS system is needed. The IEEE 802.16a physical layer is specifically designed to support frequencies below 11 GHz, which enables waves to bend over obstacles like houses and trees. The 2 to 11 GHz frequency band supports three specifications:

- WirelessMAN-SCa: using single-carrier modulation.
- WirelessMAN-OFDM: using orthogonal frequency division multiplexing with 256-point transform. Access is by time division multiple access (TDMA).
- WirelessMAN-OFDMA: using orthogonal frequency division multiple access with 2048-point transform.

Because of the PMP architecture, the BS transmits a time division multiplex (TDM) signal in the downlink channel, with individual SSs allocated time slots serially [5]. In the uplink direction, SSs use TDMA signals for accessing the channel. The WiMAX PHY layer supports two modes of duplexing: time division duplex (TDD) and frequency division duplex (FDD). The PHY layer specification operates in a framed format. Each frame consists of a downlink subframe and a uplink subframe. The downlink subframe begins with information necessary for synchronization and control. In the TDD mode, the downlink subframe comes before the uplink subframe. In the FDD mode, the uplink burst may occur concurrently with the downlink subframe.

1.3.1 Downlink Subframe

During the downlink subframe, only the BS transmits data packets to all SSs and each SS picks the packets destined to it in the MAC header. The BS broadcasts information elements (IEs) through the uplink mobile application part (UL-MAP) message at the beginning of each frame to all SSs. The IE defines the transmission opportunity for each SS (i.e., the number of time slots that each SS is allowed to transmit in an uplink subframe). Figure 1.5 shows the downlink subframe [5].

The DL subframe has two parts: the first is a preamble that starts the subframe and is used by the PHY layer for synchronization and equalization. This is followed by a frame control section that contains downlink mobile application part (DL-MAP) and UL-MAP messages. The DL-MAP message defines the access to the downlink information and is applicable to the current frame, while the UL-MAP message allocates access to the uplink channel and is applicable to the next frame.

The data burst comes after the control section and the downlink burst profile is determined by the BS according to the quality of the signal received from each SS. The downlink subframe typically contains the TDM portion, which contains information data for each SS multiplexed onto a single stream of data and is received by all SSs. In the FDD system, the TDM portion may be followed by the TDMA portion, which is used to transmit data to any half-duplex SSs. The TDMA is preceded by an additional preamble at the start of each burst profile.

1.3.2 Uplink Subframe

The uplink subframe contains contention-based allocations for initial system access or bandwidth requests. At the beginning of the uplink subframe, slots are reserved for initial ranging intervals and request intervals. Collisions may occur during these intervals. If this happens, then the SS selects a random number within its back-off window. This random number indicates the number of contention transmission opportunities to resolve this

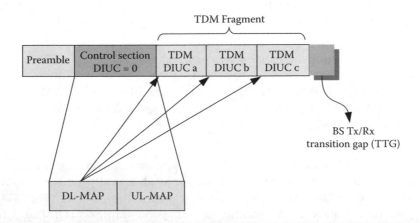

Figure 1.5   TDD downlink subframe structure.

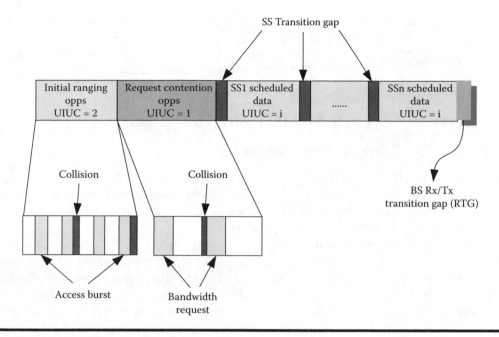

Figure 1.6   Uplink subframe structure.

problem. The BS controls bandwidth allocation on the uplink channel through a UL-MAP message and determines which minislots are involved in the collision. The remaining transmission slots are used by SSs to transmit the burst profile. Each SS is scheduled to transmit in a specific time slot predefined by the BS in the UL-MAP message.

An SS transmission gap (SSTG) followed by a preamble is used to separate the transmission of various SSs during the uplink subframe. The use of a preamble allows the BS to synchronize to the new SS (see Figure 1.6).

1.3.3   Duplexing Techniques

Regarding duplexing, a burst design was selected that allows both FDD and TDD. In the FDD method, the uplink and downlink operate on separate channels and sometimes simultaneously, while in the TDD method, the uplink and downlink transmissions share the same channel and do not transmit simultaneously.

- ■ **TDD:** In this mode, the downlink subframe comes before the uplink subframe. The uplink and downlink bursts are separated by a transition gap. This gap allows the time needed for the BS to switch from transmission mode to receive mode. Two types of gaps can be differentiated at the antenna port of the BS. The transition gap that separates the last sample of the UL burst and the first sample of the subsequent DL burst is called the BS receive/transmit transition gap (RTG), and the one that separates the last sample of the DL burst and the first sample of the subsequent UL burst is called the BS transmit/receive transition gap (TTG).
- ■ **FDD:** This duplexing method allows both the BS and the SS to transmit at the same time but at different frequencies. A fixed duration frame is used for both uplink and downlink transmissions, which results in the use of less sophisticated bandwidth allocation mechanisms. The FDD method adopted by WiMAX

is comparatively expensive because it requires two separate channels. Subscriber stations are permitted to use half-FDD (H-FDD), which is less expensive to implement than FDD and the subscriber stations don't transmit and receive simultaneously.

1.4 Services Supported by IEEE 802.16

1.4.1 Quality of Service (QoS) Support

It is much more difficult to guarantee QoS in wireless networks than wired networks. This is mainly due to the variable and unpredictable characteristics of wireless networks. Thus to achieve greater utilization of wireless resources for users, bandwidth adaptations and scheduling mechanisms should be adopted.

The WiMAX is designed to support QoS. The MAC, which is based on the concept of service flows, specifies a QoS signaling mechanism for both bandwidth request and bandwidth allocation in both uplink and downlink channels. However, the standard left the QoS scheduling algorithm that determines the uplink and downlink bandwidth allocation undefined. Transmission in the downlink channel is relatively simple, since only the BS broadcasts messages on this channel to all its associated SSs, and each SS picks up only those packets destined to itself. In the UL channel, multiple SSs send their packets using the TDMA mechanism explained in the previous section. Each SS transmits in the time slots predefined to it by the BS. The number of time slots specified for each SS is determined by the BS and broadcast through the UL-MAP at the beginning of each frame. The UL-MAP contains information elements (IEs), which include transmission opportunities (time slots) in which the SS can transmit during the uplink subframe. In their time slots, SSs can send their bandwidth requests or data packets in the uplink subframe. The SSs request their bandwidth connection requirements from the BS. The BS is responsible for allocating the required bandwidth for the new connection, if available, while maintaining the QoS for current connections.

There are two ways in which the BS grants bandwidth: grant per connection (GPC) and grant per subscriber station (GPSS) [5]. In the first method (GPC), the BS grants the requested bandwidth explicitly to the connection in the SS, which uses the grant only for this connection. In the second method (GPSS), the BS grants the whole bandwidth to the SS, and the SS must be more intelligent in order to manage the distributed resources between the different types of service flows and to maintain the QoS. This allows hierarchical and distributed scheduling to be used. The GPC method is less efficient than GPSS method, especially in the case of SSs with multiple entries. Moreover, the GPSS method reacts more quickly to changes in QoS requirements, and it is the only method allowed in the 10 to 66 GHz PHY layer.

The BS assigns a unique identifier for each connection or service flow. These connections can be created, changed, or deleted through the issue of dynamic service addition (DSA), dynamic service change (DSC), and dynamic service delete (DSD), respectively. These signaling messages may be initiated by either the SS or BS and carried out through a two- or three-way handshaking process. Activation of a service flow is done by an authorization module that resides in the BS. This process has two phases: admit and activate. Once the service flow is admitted, both the SS and BS can allocate resources for this connection. QoS parameter changes are requested through exchanging dynamic service flow messages (DSA, DSC, DCD) between the BS and the SS and approved by the authorization module [6].

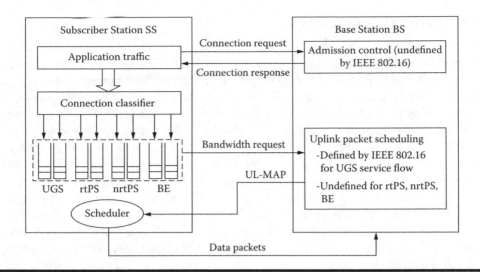

Figure 1.7   QoS architecture.

Figure 1.7 shows the existing QoS architecture defined by the IEEE 802.16 standard. As seen in Figure 1.7, admission control and the UL scheduler are both situated in the BS.

Admission control is responsible for accepting or rejecting the connection according to the available bandwidth that satisfies the connection and guarantees the required QoS without degrading the QoS for other existing connections. Admission control is not defined in the standard, although many propositions are made by different authors to establish admission control in the BS [4,7,8]. Since the IEEE 802.16 MAC protocol is connection oriented, the application first establishes the connection with the BS as well as the associated service flow (UGS, rtPS, nrtPS, or BE). Connection request/response messages are exchanged between the SS and the BS in order to establish the connection. The SS sends a connection request message asking for its bandwidth needs. If the required resources (bandwidth) are available, then the BS replies with a connection response message and the connection is established for further communication processes. Once the connection is established between the SS and the BS, the SS sends a bandwidth request message to the BS. The UL-MAP message is then generated by the uplink packet scheduler in the BS. This message carries IEs about transmission opportunities or the time slot in which the SS can transmit during the uplink subframe. On the SS side, the scheduler retrieves packets from the queue and sends them on the uplink channel according to the IE received in the UL-MAP message sent by the BS. Uplink packet scheduling, which is found on the BS side, controls all uplink packet transmissions. The IEEE 802.16 standard defines uplink packet scheduling in the BS for UGS service flows; rtPS, nrtPS, and BE service flows are undefined.

To provide priority to certain flows, they first must be classified. The classifier is a set of matching criteria applied to each packet entering the IEEE 802.16 network [1]. All data packets from the application layer in the SS are associated with a unique CID. The classifier then classifies these connections depending on their CIDs and forwards them to the appropriate queue. At the SS, the scheduler retrieves the packets from the queues and transmits them to the network in the appropriate time slots as defined by the UL-MAP message. Downlink classifiers are applied by the BS to packets it is transmitting, while uplink classifiers are applied at the SS.

1.4.1.1 Service Flow Concepts

The concept of service flows on a transport connection is central to the operation of the MAC protocol, and thus service flows provide a mechanism for both uplink and downlink QoS management [2]. In the 802.16 standard, all service flows have a 32-bit service flow identifier (SFID). Active service flows also have a 16-bit CID. Since multiple service flows may need to share a common set of QoS parameters, the 802.16 standard defines the concept of service classes or service class names. A service class is normally defined in the BS to have a particular QoS parameters set. A service flow with a certain QoS parameters set being referenced to certain service class may increase or even override the QoS parameters setting of that service class.

The IEEE 802.16 MAC layer enables classification of traffic flow and maps them to connections with specific scheduling services. Each connection is associated with a single scheduling data service and each data service is associated with a set of QoS parameters that quantify aspects of its behavior. Four types of scheduling services are defined by the 802.16 standard [1]:

- Unsolicited grant service: supports real-time uplink service flows that transport fixed-size data packets on a periodic basis, such as voice over internet protocol (VoIP). The service offers fixed-size grants on a real-time periodic basis.
- Real-time polling service: supports real-time uplink service flows that transport variable-size data packets on a periodic basis, such as Moving Picture Experts Group (MPEG) video. The service offers real-time, periodic, unicast request opportunities that meet the flows real-time needs and allow the SS to specify the size of the desired grant. This service requires more request overhead than UGS, but supports variable grant sizes for optimum data transport efficiency.
- Non-real-time polling service: supports non-real-time flows such as file transfer protocol (FTP). The BS polls the nrtPS CIDs periodically so that the uplink service flow requests opportunities even during network congestion.
- Best effort: provides efficient service for best effort traffic in the uplink, such as hypertext transfer protocol (HTTP). In order for this service to work correctly, SS is allowed to use contention request opportunities.

1.4.1.2 Uplink Scheduling

Scheduling services represent the data handling mechanisms supported by the MAC scheduler for data transport on a connection. Each connection is associated with a single scheduling service. A scheduling service is determined by a set of QoS parameters that quantify aspects of its behavior. These parameters are managed using the DSA and DSC message dialogs [2].

Uplink request/grant scheduling is performed by the BS with the intent of providing each subordinate SS with bandwidth for uplink transmissions or opportunities to request bandwidth. By specifying a scheduling type and its associated QoS parameters, the BS scheduler can anticipate the throughput and latency needs of the uplink traffic and provide polls or grants at appropriate times.

The access control, which is based on the concepts of service flows, specifies a QoS signaling mechanism for both bandwidth request and bandwidth allocation in both the UL and DL channels. IEEE 802.16 defines only uplink scheduling for UGS service flows and leaves the QoS scheduling algorithm for other service flows to be defined by the constructor.

IEEE 802.16 provides scheduling services for both uplink and downlink traffic. In the downlink direction, the flows are simply multiplexed, and therefore the standard scheduling algorithms can be used. Uplink scheduling is complex, as it needs to be in accordance with uplink QoS requirements provided by the standard. In fact, the standard defines the required QoS signaling mechanism, such as bandwidth request (BW-REQ) and UL-MAP, but it does not define the UL scheduler. The uplink scheduling mechanism, which is situated in the BS, is responsible for fair and efficient allocation of assigned time slots in the uplink direction. To guarantee a certain level of QoS, the scheduling mechanism is applied to different types of traffic in the network, including real-time and non-real-time applications. In recent years many scheduling mechanisms have been proposed for real-time and non-real-time traffic. It should be mentioned here that the scheduling algorithm can guarantee the QoS only if the number of connections is limited by the admission control. The sum of the minimum bit rates of all connected flows must be lower than the capacity allocated to the applications within the same QoS level [9].

1.4.1.3 Proposed Scheduling Methods

In computer networks, admission control and scheduling are very necessary to allocate sufficient resources for users while satisfying the QoS. In the IEEE 802.16 standard, admission control is not defined, while scheduling is only defined for UGS flows. MANs research have focused on scheduling service flows and defining mechanisms of admission control in order to enhance the QoS and to provide wireless end-users with a similar level of QoS as wired users.

WiMAX supports all types of service flows, including real-time and non-real-time data traffic. Real-time traffic issued at periodic intervals consists of either variable-size data packets, such as MPEG video, or fixed-size data packets, such as VoIP. Non-real-time data traffic requires a lesser degree of QoS such as FTP or HTTP data packets. In recent years WiMAX network QoS research has focused on defining the admission control mechanism and uplink scheduling algorithms. The rest of this section presents uplink scheduling mechanisms proposed by several authors.

In Hawa and Petr [10], a new mechanism is proposed for scheduling service flows. They use three types of queues for queuing service flows, with type 1 having the highest priority and type 3 having the lowest. Type 1 is used to store not only UGS flows, but also unicast rtPS and nrtPS flows. Semiprimitive priority is assigned to type 1 queues, which means that the service grant is not interrupted unless the type 1 grant arrives with a dangerously early deadline. In the work of Hawa and Petr, the proposed QoS architecture addresses only the UL channel. Scheduling of data traffic on the downlink channel is not considered. On the other hand, the presented solution is a hardware implementation that is difficult to apply and costly in terms of money, although it is much faster than a software solution.

In Wongthavarawat and Ganz [4], a hierarchical QoS architecture is defined. This architecture consists of two layers. A strict priority is used in the first layer, where the entire bandwidth is distributed between the different service flows. UGS has the highest priority, then rtPS, nrtPS, and finally BE. In the second level of this hierarchal architecture, different mechanisms are used to control the QoS for each class of service flow. The uplink packet scheduler allocates fixed bandwidth to the UGS connection based on their fixed bandwidth requirements. Earliest deadline first (EDF) is used to schedule rtPS service flows, in which packets with the earliest deadline are scheduled first. nrtPS service flows are scheduled using the weight fair queue (WFQ) based on the weight of the connection. The remaining bandwidth is equally allocated to each BE connection.

The disadvantage of this hierarchical structure is that high-level connections (like real-time data traffic) may starve low priority ones. In Chen et al. [7], a new hierarchial scheduling algorithm is proposed that also consists of two layers. To overcome the stated problem, the deficit fair priority queue (DFPQ) replaces the strict priority mechanism in the first layer. The DFPQ schedules bandwidth application services in an active list. If the queue is not empty, it stays in the active list, otherwise it is removed. This new algorithm serves different types of service flows in both the uplink and downlink and provides more fairness to the system. Priority in the second layer is distributed between different service flows using a similar method to the one presented in Wongthavarawat and Ganz [4]. Since UGS service flow is assigned a fixed bandwidth, it does not appear in the scheduling architecture. The round robin (RR) method is used to schedule BE service flows.

We noticed that most authors present a new QoS architecture by modifying the MAC layer in both the SS and BS. They add new modules that are not defined in the IEEE 802.16 standard. Because of the similarity of the proposed QoS architectures and the modules added, only the work done by Wongthavarawat and Ganz [4] will be explained here (see Figure 1.8).

First, the authors define the admission control that decides whether a new connection can be established based on the available resources while maintaining the same QoS of current connections. On the BS side, they define a new scheduling mechanism for rtPS, nrtPS, and BE by adding different modules responsible for collecting the queue size information from the bandwidth requests received during the previous time frame and then to generate the UL-MAP message. On the SS side, a traffic policing module is added. The role of this module is to enforce traffic based on the connection traffic contract and to control the delay violation for rtPS.

In the presented work, we noticed that some authors [4,7] define an admission control module in the BS which has to control the number of admitted service flows. While the scheduling mechanism presented by Hawa and Petr [10] did not take into consideration the downlink burst, most of proposed solutions that came after, pay more attention to

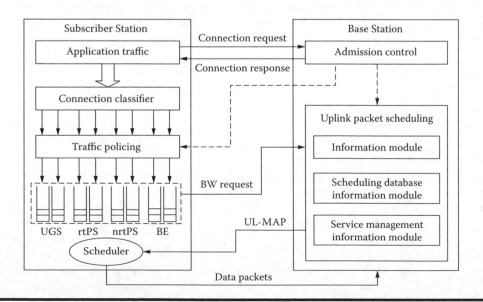

Figure 1.8   Proposed QoS architecture.

this point. Another important issue is that the implemented uplink scheduler is always found on the BS side, which means that the SS will only send its bandwidth requirements for each connection and the BS has to schedule the different service flows following the criteria of priority predefined by the mechanism.

1.4.2 Mobility Management Support

Ongoing development and industry support for the new standard led to the development of a mobile version of IEEE 802.16 (known as mobile WiMAX). This version is based on amendment 802.16e and provides support for handoff and roaming [11]. This new amendment aims to keep mobile clients connected to the MAN and supports portable devices from smart mobile phones and PDAs to notebook and laptop computers. IEEE 802.16e can also use the network to provide fixed network services. Mobile WiMAX profiles cover 5, 7, 8.75, and 10 MHz channel bandwidths for licensed worldwide spectrum allocations in the 2.3, 2.5, 3.3, and 3.5 GHz frequency bands [12].

Mobility management solutions can be classified into two categories [13]: macromobility and micromobility management solutions. The first category refers to the movement of mobile stations (MSs) between two network domains, while in the second case the MS moves between two subnets within the same network domain.

In a mobile WiMAX system, in which MSs are moving within the BS's sector, battery life and handoff are essential criteria for mobile applications. To extend battery life for mobile devices, the system is designed to support power saving. Sleep mode operation for power saving is one of the most important features to extend battery life for MSs. In addition to minimizing MS power usage, sleep mode decreases usage of BS air interface resources. In the sleep period, the MS is considered unavailable to serving BS [2]. Implementation of sleep mode is optional for the MS and mandatory for the BS.

The 802.16e amendment defines two operational modes for the MS when registering with the serving BS: sleep mode and awake mode [14]. In the awake mode, the MS is available to the serving BS and can send or receive data. The MS is considered unavailable to the serving BS during the sleep mode. Under sleep mode operation, the MS initially sleeps for a fixed interval called the sleep window. The initial maximum window size is negotiated between the MS and the BS. After waking up, if the MS finds that there is no buffered DL traffic destined to it from the BS, then it doubles the sleep window size up to the maximum sleep window size. On the other hand, if the MS has packets to send on the UL channel, it can wake up prematurely to prepare for the uplink transmission (e.g., bandwidth request) and then it can transmit its pending packets in the allocated time slots assigned by the BS. When the MS enters the sleep mode, the BS does not transmit data to the MS, which may power down or perform other operations that do not require connection with the BS.

To manage power usage in a more efficient way, the IEEE 802.16e standard also defines the idle mode [2]. In this mode, the MS becomes periodically available to receive DL broadcast traffic messaging without the need to register with a specific BS as the MS traverses an air-link environment populated by multiple BSs. This mode allows the MS to conserve power and resources by restricting its activity to scanning at discrete intervals and thus eliminates the active requirement for handover operation and other normal operations. On the BS and network side, idle mode provides a simple and timely method for alerting the MS to pending DL traffic directed to the MS and thus eliminates air interface and network handover traffic from essentially inactive MSs.

1.4.2.1 Handover Mechanism

Handover is the process by which a MS moves from the air interface provided by one BS to the air interface provided by another BS [2]. The MS needs to perform this process if it moves out of the serving BS's transmission range or if the serving cell is overloaded. A serving BS periodically broadcasts an advertisement message (MOB-NBR-ADV), which is decoded by the MS to obtain information about the characteristics of the neighboring BS. A BS may allocate a time interval to a MS called a scanning interval. A MS then sends a scan request message (MOB-SCN-REQ). In response to this messages, the serving BS indicates a group of neighbor BSs through a scanning response message (MOB-SCN-RSP). The MS then selects one suitable target BS for handover (recommended by the serving BS) from this group. Through ranging, the MS can acquire the timing, power, and frequency adjustment information of the neighboring BS. The target BS-MS association information is reported to the serving BS.

The MAC layer (L2) handover is divided into two phases: the handover preregistration phase and the real handover phase [15]. During handover preregistration, the target BS is selected and preregistered with the MS. However, the connection to the currently serving BS is maintained and packets may be exchanged during the preregistration phase. In the real handover, the MS releases the serving BS and reassociates with the target BS.

Either the MS or BS can start the handover process, which consists of several stages, starting with the cell reselection followed by a decision to make the handover. After that, the MS synchronizes to the target BS downlink and performs handover ranging followed by termination of MS contact as a final step in the handover process. Handover cancellation may be done by the MS at any time [2] (see Figure 1.9).

In the first step, cell reselection, the MS acquires information about the neighboring BSs from the advertisement message. The MS then selects a BS as a handover target. The serving BS may schedule scanning intervals or sleep intervals to conduct cell reselection activity. A handover begins when a MS decides to migrate from the serving BS to the target BS. This decision may originate at the MS, the serving BS, or the network backbone. The serving BS may notify potential target BSs or send them MS information over the backbone network to expedite handover. Once the target BS is selected, the MS must synchronize with the downlink transmission of the target BS. During this phase, the MS receives DL and UL transmission parameters. If the MS previously received information about this BS (through network topology acquisition), the length of this process may be shortened. After the synchronization step, the MS needs to perform initial ranging or handover ranging. Ranging is a procedure whereby the MS receives the correct transmission parameters, such as time offset and power level. The target BS may acquire MS information from either the serving BS's or the backbone network. Depending on the target BS's knowledge about the MS, some parts of the network reentry process may be omitted. The final step in handover is the termination of MS contact. The serving BS terminates all connections belonging to the MS; in other words, information in the queue, automatic repeat request (ARQ) state machine, counters, timers, etc., are all discarded.

The MS can cancel the handover process at any time and resume normal operation with the serving BS. The only condition is that the cancellation process must take place before expiration of the resource-retain-time interval after transmission of the handover indication message (MOB-HO-IND) [2].

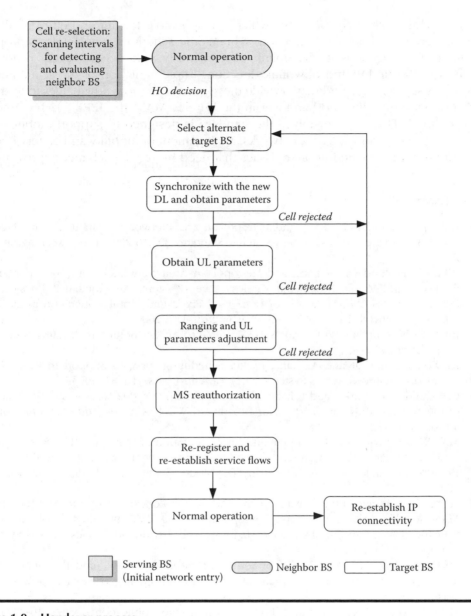

Figure 1.9   Handover process.

1.5   Conclusion

This chapter presented the IEEE 802.16 standard and the different versions developed since its appearance in 2001. The standard modifies the MAC and PHYS layers. The MAC layer PDU format defines two headers: the generic MAC header, which may be followed by payload or CRC data, and the signaling MAC header with no payload or CRC data. The MAC is connection oriented and provides the concept of service flows, with each one associated with a QoS level. Admission control and uplink scheduling mechanisms are not specified by the standard. Research has been aimed at enhancing the QoS and defining the admission control mechanism and the uplink and downlink scheduling architectures for the different types of connections. The scheduling mechanisms presented in this chapter are mainly situated in the BS, which is responsible for fair and efficient allocation of resources in the uplink direction. While some scheduling mechanisms propose scheduling of connections in the uplink direction only, other more recent proposals work in both the uplink and downlink direction. Implementation of the scheduler on the SS side is still possible and may provide the system with more advantages in terms of QoS. To provide mobility and handover in the WirelessMAN, the IEEE 802.16e amendment is defined. This new version of the standard is designed to support mobility and handover in broadband wide area networks. Management of mobility and battery energy consumption of the MS are important issues that need more research and improvement.

References

1. IEEE, IEEE Standard for Local and Metropolitan Area Networks—Part 16: Air Interface for Fixed Broadband Wireless Access Systems, Standard 802.16-2004, IEEE, Washington, DC, 2004.
2. IEEE, IEEE Standard for Local and Metropolitan Area Networks—Part 16: Air Interface for Fixed and Mobile Broadband Wireless Access Systems Amendment 2: Physical and Medium Access Control Layers for Combined Fixed and Mobile Operation in Licensed Bands, Standard 802.16e-2005, IEEE, Washington, DC, 2006.
3. Intel, Understanding Wi-Fi and WiMAX as Metro-Access Solutions, White paper, Intel Corporation, 2004.
4. K. Wongthavarawat and A. Ganz, Packet Scheduling for QoS Support in IEEE 802.16 Broadband Wireless Access Systems, *Int. J. Commun. Sys.*, 16, 81, 2003.
5. C. Eklund, R.B. Marks, and K.L. Stanwood, IEEE Standard 802.16: A technical Overview of the WirelessMAN Air Interface for Broadband Wireless Access, *IEEE Commun. Mag.*, 40(6), 98, 2002.
6. M.C. Wood, Analysis of the Design and Implementation of QoS Over IEEE 802.16, 2006.
7. J. Chen, W. Jiao, and H. Wang, A Service Flow Management Strategy for the IEEE 802.16 Broadband Wireless Access Systems in TDD Mode, White paper, IEEE, Washington, DC, 2005.
8. H. Wang, W. Li, and D.P. Agrawal, Dynamic Admission Control and QoS for 802.16 Wireless MAN, Wireless Telecommunications Symposium, p. 60, IEEE, Washington, DC, 2005.
9. J. Bostic and G. Kandus, MAC Scheduling for Fixed Broadband Wireless Access Systems, COST 263 TCM, 2002.
10. M. Hawa and D.W. Petr, Quality of Service Scheduling in Cable and Broadband Wireless Access Systems, Tenth International Workshop on Quality of Service, p. 247, IEEE, Washington, DC, 2002.
11. WiMAX Forum, Fixed, Nomadic, Portable and Mobile Applications for 802.16-2004 and 802.16e WiMAX Networks, WiMAX Forum, Beaverton, OR, 2005.
12. WiMAX Forum, Mobile WiMAX—Part 1: A Technical Overview and Performance Evaluation, WiMAX Forum, Beaverton, OR, 2006.
13. J.-Y. Hu and C.-C. Yang, On the Design of Mobility Management Scheme for 802.16-Based Network Environment, IEEE 62nd Vehicular Technology Conference (VTC-2005-Fall), vol. 2, pp. 720, IEEE, Washington, DC, 2005.
14. K. Han and S. Choi, Performance Analysis of Sleep Mode Operation in IEEE 802.16e Mobile Broadband Wireless Access Systems, IEEE 63rd Vehicular Technology Conference (VTC-2006-Spring), vol. 3, p. 1141, IEEE, Washington, DC, 2006.
15. K. Kim, C. Kim, and T. Kim, A Seamless Handover Mechanism for IEEE 802.16e Broadband Wireless Access, International Conference on Computational Science, vol. 2, p. 5, Springer, Berlin, 2005.

RF, SIGNAL PROCESSING, AND MIMO

I

Chapter 2

RF System and Circuit Challenges for WiMAX

Balvinder Bisla

Contents

Broadband wireless access has occupied a niche in the market for about a decade, but with the signing of the 802.16d standard it could finally explode onto the mass market. Intel's baseband transceiver chip is flexible enough to accommodate radio frequency integrated circuit (RFIC) architectures of today and the future. With the emergence of this standard, an environment is developing that will allow multiple vendors to produce components that adhere to a standard specification and hence allow large-scale deployment. One of the major challenges of the 802.16d standard is the plethora of options that exist: Worldwide Interoperability for Microwave Access (WiMAX) addresses this issue by limiting options and ensuring interoperability. The result will allow manufacturers of radio frequency (RF) components and test equipment to mass market their products.

This chapter focuses on the various RF challenges that exist on a RF system level and show how such challenges can translate into circuit designs. RF is made more complicated by the fact that WiMAX addresses wireless markets across the world in both the licensed and unlicensed bands. Thus solutions have to be flexible enough to allow for the many RF frequency bands and different regulations around the globe. Several major RF architectures are discussed and the implications for WiMAX specifications are explored, in particular intermediate frequency (IF)- and I/Q-based structures are investigated.

Part of this discussion provides insight into the cost and performance trade-offs between time division duplex (TDD) and frequency division duplex (FDD) systems both in licensed and unlicensed bands. It is generally accepted that TDD systems offer cost advantages over their FDD counterparts; however, most licensed bands intended for data applications operate with FDD systems in mind. Some of the RF subsystem blocks that have stringent WiMAX specifications are also elaborated upon; these include synthesizers, power amplifiers, and filtering. These fundamental subsystem blocks are where most of the transceiver costs reside; the same blocks are also responsible for most of the RF performance.

The industry is moving toward using orthogonal frequency division multiplexing access (OFDMA) and either spatial diversity or beam-forming techniques to enhance link margins. I touch on the RF challenges associated with these techniques. Finally, some of the important WiMAX specifications for RF and implications for design of RF circuits, including signal-to-noise plus distortion ratio (SNDR), channel bandwidths, RF bands, noise figures, output power levels, and gain setting, are discussed. Some important differences between WiMAX and the IEEE 802.11 RF specifications are also highlighted.

2.1   Introduction

As the RF challenges mount, so do the costs of the radio. For WiMAX to be successful, the cost versus performance equation has to be carefully balanced. Two extreme examples of this cost and performance equation are a single-input single-output (SISO) system from Hybrid Networks (now defunct) requiring line-of-sight (LOS) radios. LOS radios require experienced technicians to set up the equipment. However, the cost of the radio is low due to its simplicity. In general, SISO radios require expensive installation and reliability is poor; link margins are typically 145 dB. On the other hand, Iospan Wireless (now defunct) demonstrated a multiple-input multiple-output (MIMO) radio with a 3 × 2 system (i.e., three receive and two transmit chains). It was able to support link margins of 165 dB that could penetrate inside homes in multipath environments. With this ability, the issue of costly installation is eliminated; however, the cost of the multiple radio chains becomes a deterrent. Still, as RFIC integration improves, costs will decrease. WiMAX, through the use of integration and advanced techniques to increase link margins, should be able to achieve reliable wireless systems at a reasonable cost.

2.2   RF Architectures

This section describes the plethora of trade-offs and challenges for RF architectures in WiMAX-related radios. FDD and its cousin half FDD (HFDD) as well as TDD are discussed. IF, direct conversion or zero intermediate frequency (ZIF), as well as variants of these are presented. The interface between the baseband chip and the radio must be carefully designed, so these challenges are exposed. Methods to improve link margins, namely MIMO, and beam forming can be used in WiMAX. In addition, OFDMA, which allows for subchannelization, improves capacity efficiency. RF challenges inherent in the use of these methods are also discussed.

2.3   TDD, FDD, and HFDD Architectures

2.3.1   TDD

TDD systems utilize one frequency band for both transmit and receive. This concept requires only one local oscillator (LO) for the radio. In addition, only one RF filter is necessary and this filter is shared between the transmitter (TX) and the receiver (RX). The synthesizer and RF filters are major cost drivers in radios (Figure 2.1). Having one synthesizer saves on die area; a large part of the radio die size can be taken up by the LO, in particular the inductor, which is part of the resonant structure.

The RF filter in a TDD system is not required to attenuate its TX noise as severely as in FDD systems. The TDD mode prevents the TX noise from self-jamming the RX, since only one is on at any time. As well as relief of the RF filter specifications, having just one RF filter saves cost and space. It should be noted that to ensure transmitting radios do not interfere with nearby receiving radios, the specification for TX noise cannot be eased with abandon. The transmission noise from radio 1 will interfere with the received signal of radio 2. Thus, although self-jamming specifications are made easier, collocation specifications must be carefully considered. There is a notable savings in power from the TDD architecture, a direct result of turning the RX off while in transmit mode, and vice versa.

Several disadvantages exist, however. There is a reduction of data throughput, since there is no transmission of data while in the receiving mode, unlike FDD systems. The media access control (MAC) level software tends to have a more complicated scheduler than an FDD system since it must deal with synchronizing many users' time slots in both transmit and receive mode. It must be noted that while the RF filtering specifications are

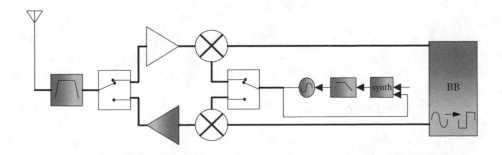

Figure 2.1   A TDD radio. The darkened blocks are the most costly in the radio.

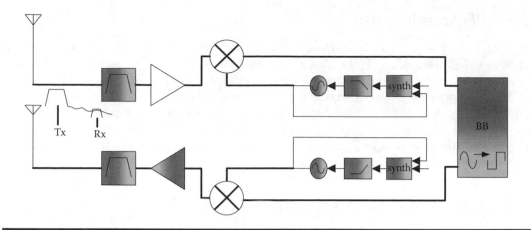

Figure 2.2   An FDD radio.

relaxed, this tends to imply that subscriber stations will have to be spaced further apart from each other to avoid interference. In essence, the system must handle fewer users in a given area than in FDD systems.

TDD systems are most prominent in unlicensed bands; in these bands the regulations for output noise are more relaxed than in licensed bands. Thus inexpensive RF filters can be specified. Since the unlicensed bands are free of cost, there is competition to drive for the lowest cost architecture, TDD.

2.3.2   FDD

A high-performance RF front end is required in FDD systems. Collocation issues from a TX noise perspective are solved since the worst-case scenario of self-jamming is not possible. FDD systems do not have to switch the RX or TX; this alleviates settling time specifications, which results in a simpler radio design (Figure 2.2). The MAC software is simpler because it does not have to deal with the time synchronization issues as in TDD systems.

The radio must be capable of data transmission while in receive mode without incurring any degradation in bit error rate (BER). To ease the burden on the filter, there is a gap between the TX frequency band and the RX band; however, carriers wish to minimize this space. Typically this is a separation of 50 to 100 MHz.

The TX noise is usually specified to be 10 dB below the RX input noise floor, in which case the TX noise will only degrade the RX by 0.5 dB. Unfortunately the specifications usually tie FDD systems to using cavity filters or four-pole ceramic filters. Cavity filters cost about $35 each, while ceramic filters are in the $8 range. Most licensed bands do not have one standard structure but are flexible (i.e., the TX and RX can be swapped in different geographical regions). This results in having to design several types of filters, something that does not lend itself to mass production of filters.

To get an idea of the filter requirements in FDD:

$$\text{Filter (dB)} = \text{Power output (dBm/Hz)} - \text{mask (dBc)}$$
$$- [-174 + \text{noise figure} - \text{cochannel}].$$

For example, if the power output is −33 dBm/Hz, in a 1 MHz signal bandwidth, output power is +27 dBm. The mask of the transmitter is 60 dBc (i.e., the thermal floor of the transmitter is 60 dB below the power output). The noise figure of the receiver is 5 dB. The cochannel is how far (in decibels) the undesired signal is below the desired signal, in this case, 10 dB below the desired signal. Thus the filter at the RX frequency is of 86 dB. If the RX is 100 MHz away from the TX, this filter is an expensive cavity filter.

The full duplex nature of the circuit requires a separate TX and RX synthesizer. The RFIC die area is significantly impacted by the inductor of a resonant circuit; this is part of a voltage controlled oscillator (VCO) which is used in the synthesizer. Thus these have a large impact on the cost of the RFIC.

A final note on FDD systems is that they are power hungry; this also increases the cost of the power system. Thus FDD is not an ideal platform for portable or mobile radios. FDD systems are typically deployed in licensed bands (e.g., 5.8 GHz, 3.5 GHz, 2.5 GHz) and this spectrum is expensive. The cost of the spectrum forces carriers to serve as many users as possible. Capacity must be optimized, which results in carriers favoring FDD architecture. Clearly it is very desirable to have the base station work in FDD, but to reduce costs the subscriber station could be an HFDD structure.

2.3.3 HFDD

The HFDD architecture combines the benefits of TDD systems while still allowing for frequency duplexing (Figure 2.3). The base station can operate in FDD and retain its capacity advantage over TDD systems. This can lower the cost of the radio significantly at the subscriber station, where the unit cost must be driven down. The cost reduction appears in the form of relief in the RF TX filter, and since there is one synthesizer, the die area of the RFIC shrinks. Power savings are also realized, as in TDD systems.

Once again, collocation issues have to be addressed carefully. Self-jamming is not a problem, as in TDD, but too much relief on the TX filter can result in interference between users. There is also a capacity loss at the subscriber station since the radio cannot simultaneously transmit and receive.

The HFDD structure can be used in both licensed and unlicensed bands. Transmit and receive can be at the same frequency, as in TDD systems, or separated by a frequency gap, as in FDD. This type of radio is very flexible. Its cost structure approaches that of a TDD radio.

In summarizing the duplexing schemes, Intel's baseband chip Rosedale can support both TDD and HFDD modes. This takes care of most of the subscriber stations. In a

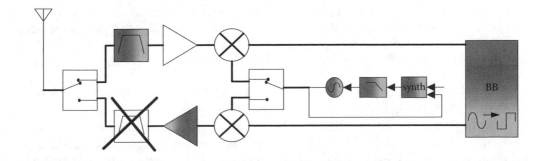

Figure 2.3 An HFDD radio.

typical deployment, the ratio between the base station and subscriber stations is 1:100 because of the low volume of the base station. The physical (PHY) and MAC layers need not be designed as a custom chip; a field programmable gate array (FPGA) can be cost effective. It is possible to connect two Rosedale baseband chips together to support an FDD scenario for the base station.

Various radio architectures are discussed in the following sections; these include IF- and I/Q-based architectures and some variants on these. Some of the interfaces between the radio and baseband chip are also discussed.

2.3.4   RF Interface

The baseband chip digitizes the analog signal and performs signal processing. This PHY layer chip contains the blocks for filtering, automatic gain control (AGC), demodulation of data, security, and framing of data. The algorithms that do power measurements, such as AGC and frequency selection can be taken care of by the lower level MAC. As can be seen, there are common parameters such as AGC that are shared across the PHY, MAC, and radio.

The major blocks within a radio that need control from the baseband integrated circuit (IC) are AGC, frequency selection, sequencing of the TX/RX chain, monitoring of TX power, and any calibration functions (e.g., I/Q imbalance). Each of these blocks are tightly coupled with the PHY and lower level MAC. A reasonable way to communicate with the radio is through a serial peripheral interface (SPI); it minimizes pins on the RFIC.

Usually the SPI is used to control the synthesizer. In order to make the interface more useful so that it can control the digital AGC of an RFIC and help perform measurements of power and temperature, the SPI needs to be a dedicated time-critical element. In this way the SPI can respond to AGC, measurements, and frequency commands in a timely and predictable manner. A note of caution, however: traffic on the SPI can cause interference with the incoming signal and put spurs on the TX signal. Therefore all SPI communication should only occur in the TX to RX time gaps. Other interface blocks are general purpose input/output (GPIO), pulse width modulators (PWMs), digital-to-analog converters (DACs), and analog-to-digital converters (ADCs).

The AGC is split into RX AGC and TX AGC. In the RX AGC, response times may have to be rapid to cope with the changing RF channel in a mobile environment (on the order of microseconds). However, in a fixed wireless application, the channel change is on the order of milliseconds. The TX AGC can be relatively slow in the steady state. However, in powering up the TX, the AGC may need to attain the correct power level in the microsecond time frame. Typically the AGC is controlled through a single-bit DAC (i.e., sigma delta converters). Either of these methods have clock noise that needs to be filtered out. The trade-off here is that for a large slope of the RF AGC, the clock noise must be filtered to avoid distorting the signal. However, the filtering introduces a delay that slows down the AGC response. To increase the time response of the AGC, multibit DACs can be used.

Selection of the RF is done through the SPI. For HFDD systems there is a settling time from the TX to RX frequency, and loading of the SPI is part of the timing budget.

Monitoring the temperature of the radio is a slow process; however, power measurements either from the TX or RX require synchronization with the TX/RX timing gaps. Interfacing to the radio must take into account the sequencing of the radio. For example, in the case of the TX we need to switch the antenna, enable the TX and load frequency,

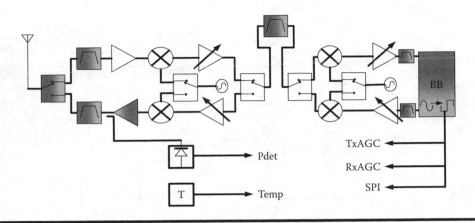

Figure 2.4   Details of the HFDD architecture.

change the TX gain, turn on the power amplifier, and finally ramp the modulation. Switching to the RX requires sequencing the TX down to avoid spurious emissions.

Two fundamental parameters drive radio design: noise and linearity. The goal is to attain as much dynamic range as possible in the presence of undesired signals. This requires a distribution of gain and filtering through the TX or RX chain. Many architecture designers struggle with the placement of this gain and filtering. We look at some of these radio architectures in the next sections.

2.3.5   HFDD Architecture

Figure 2.4 illustrates details of a HFDD architecture. There is a frequency separation between the TX and RX, so separate filtering is necessary in the RF front end. However, the IF is shared between the TX and RX. A surface acoustic wave (SAW) filter provides for excellent adjacent/alternate channel rejection. There is a final frequency conversion to a lower IF that can be handled by an ADC. Much of the AGC range is at the lower IF. An AGC range of 70 dB is required; the absolute gain is higher to overcome losses. For the TX AGC, a 50 dB range is required. The AGC can be controlled through PWMs for analog AGC or GPIO for step attenuators. Two synthesizers are necessary for the double conversion. The low-frequency synthesizer is fixed and does not have to be switched during the RX to TX change. The high-frequency synthesizer is the challenging block; it is required to settle within $100 \, \mu s$. The step size could be as low as 125 KHz in the 3.5 GHz band.

Several signals are also sent to the baseband IC: TX power level (sometimes RX power level), temperature, and synthesizer lock detect. The power level is most important since power output has to be as close as possible to the intended value and still within regulations.

2.3.6   TDD Architecture

TDD is a good example of direct conversion transceivers or ZIF. Figure 2.5 shows a block diagram of the ZIF architecture. The TX and RX frequencies are the same, so the RF filter can be shared. The down-conversion process is done with I/Q mixers; these consume a small area on the die. The issue with such mixers is they need to be matched, otherwise distortion is introduced. Also local oscillator feed-through effects tend to increase due

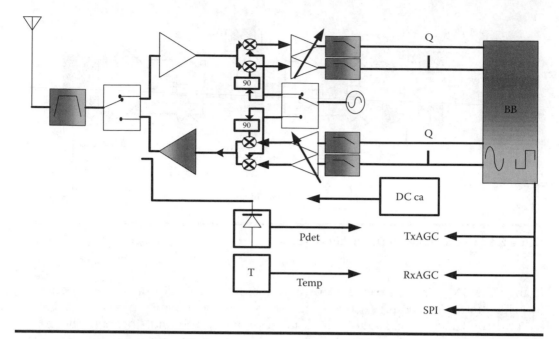

Figure 2.5   Block diagram of the ZIF architecture.

to direct current (DC) imbalances. These effects are significant since most of the gain is at the final conversion. The DC offset results in a reduction in the dynamic range of the ADC since extra bits are required for this offset. A DC calibration circuit can be implemented to reduce the effect. In addition, I/Q imbalance results in distortion. The problems are aggravated by temperature, gain changes, and frequency. By going to DC, low-pass filters can be used that are selective to channels. These can be implemented on-chip and can save on cost. It must be noted that on-chip low-pass filters consume a large die area. They can also introduce noise. WiMAX has variable bandwidths ranging from 1 to 14 MHz, but as the cutoff frequency is reduced there are significant challenges in the on-chip filter. For such ZIF schemes there must be an automatic frequency control (AFC) loop whereby the baseband IC controls the reference oscillator of the RFIC. This ensures that any DC leakage terms stay at DC and do not spill over into the desired tones of the orthogonal frequency division multiplexing (OFDM) waveform.

2.3.7   I/Q Baseband Architecture 1

A variant of the HFDD and TDD architectures mentioned above is shown in Figure 2.6. This structure has the advantage that some filtering is done at an IF, removing some of the strain on the DC filters. In addition, power can be saved by having the final stage operate at lower frequencies. The issues related to I/Q mismatch and DC leakage are lessened by having less gain at DC and operating the mixers at an IF instead of an RF. Savings can be realized at the TX: because the SAW can do most of the filtering, there is no need for TX low-pass filters. This has the added advantage that the I/Q mismatch from the low-pass filters is removed. One drawback is that two DACs and two ADCs are required.

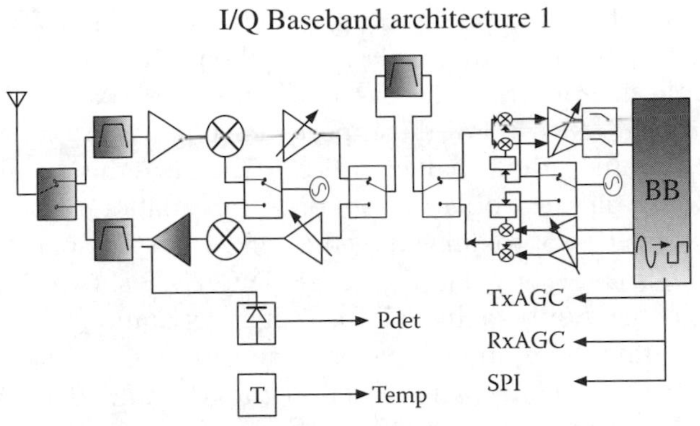

I/Q Baseband architecture 1

Figure 2.6   I/Q baseband architecture 1.

2.3.8   I/Q Baseband Architecture 2

To address the problems of I/Q baseband radios, another architecture is considered. Figure 2.7 shows a RX where the signal is mixed to DC then mixed up to a near-zero IF (NZIF). By going to DC, the IF filter is removed and filtering can be done on-chip. To avoid DC and I/Q problems, the signal is mixed to an IF. The choice of IF is greater than half the channel bandwidth. This structure allows the gain to be distributed between the DC and IF stages. Also, as an added benefit, only one ADC is required. For the TX stage, I/Q up-conversion is used.

2.3.9   RF Challenges for MIMO, AAS, and OFDMA

Antenna diversity is an important technique that can inexpensively enhance the performance of low-cost subscriber stations. It can help mitigate the effects of channel impairments like multipath, shadowing, and interference that severely degrade a system's performance, and in some cases make it inoperable. By using multiple antennas, a system's link budget can be significantly improved by reducing channel fading and in some implementations by providing array gain. There are several designs, that can be implemented, all of which yield excellent gains, ranging from low to high complexity.

I/Q Baseband architecture 2

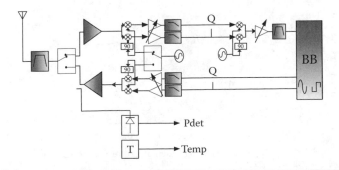

Figure 2.7   I/Q baseband architecture 2.

The basic designs are selection diversity combining (SDC), equal gain combining (EGC), and maximum ratio combining (MRC). SDC is a scheme for sampling the receive performance of multiple antenna branches and selecting the branch that maximizes the RX signal-to-noise ratio (SNR). To work properly, each antenna branch must have relatively independent channel fading characteristics. To achieve this, the antennas are either spatially separated, use different polarization, or a combination of both. The spatial correlation of antennas can be approximated by the zero-order Bessel function given by the equation $\rho = J_0^2(2\pi d/\lambda)$ and shown in Figure 2.8. Figure 2.8 shows that relatively uncorrelated antenna branches can be achieved for spatial separations greater than one-third a wavelength, supporting the requirement for small form-factor subscriber stations.

For optimal SDC performance, the selection process and data gathering must be completed within the coherence time. The coherence time is the period over which a propagating wave preserves a near-constant phase relationship both temporally and spatially. After the coherence time has elapsed, the antennas should be resampled to account for expected channel variations and to allow for reselection of the optimal antenna. For a TDD system, where reciprocal uplink (UL) and downlink (DL) channel characteristics are expected, the selected receive antenna can also be used as the transmit antenna. Although the SDC technique sounds rather simple, surprisingly large system gain improvements are possible if the algorithms can be designed effectively.

There are two figures of merit for judging the gain enhancement of an antenna diversity scheme. These are diversity gain and array gain. Under changing channel conditions, diversity gain is equivalent to the decrease in gain variance of local signal strength fluctuations of a multiantenna array system when compared to a single-antenna array system. The result of increased diversity gain is the reduction in fading depth. This is due to each antenna in a multiantenna system experiencing independent fading channels over frequency and time. The second figure of merit, array gain, is the accumulation of antenna gain associated with increased directivity via a multiantenna array system. In a typical system, as the number of antenna array elements grows, the gain increases $10 * \log(n)$, where n is the number of antenna array elements. This means a doubling of gain for every doubling of antenna elements.

The SDC scheme exhibits no array gain, as only one from n antennas is used at any instance. However, through spatial or polarization diversity, the SDC achieves stellar diversity gain, as shown in Table 2.1.

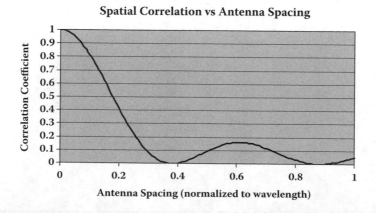

Figure 2.8   Bessel function approximation of the spatial correlation coefficient.

Another basic antenna diversity technique using multiple antennas is EGC. Instead of selecting one from n antennas, as in SDC, the algorithm combines the power of all antennas. The multiple independent signal branches are cophased, the gain of each branch set to unity (equal gain), and then all branches are combined. The EGC antenna diversity technique achieves diversity gain while also producing array gain. Thus EGC provides higher antenna diversity gain than SDC, as can be seen in Table 2.1. To achieve an antenna diversity benefit closer to optimal, MRC of the antenna elements can be used. This technique is similar to EGC, with the exception that the algorithm tries to optimally adjust both the phase and gain of each element prior to combining the power of all antennas. The summation of the signals may be done in either the analog or digital domain. When summation occurs in the digital domain, RF hardware for each independent antenna branch is required from RF to baseband. When MRC is realized in the analog domain, summation may occur directly at RF. Performance is better when processing is done in the digital domain, as frequency selective channel characteristics are compensated for in each branch. In an analog MRC, only the average channel distortion over the frequency is used to compensate for the amplitude and phase variation between array elements. In digital MRC, discrete frequency components across the signal bandwidth are cophased and individually weighted based on SNR at the receiver. MRC realizes the highest antenna diversity gain compared to the other techniques discussed (refer to Table 2.1). Although the complexity is high, MRC implementation costs are decreasing through better RF integration and reduced complementary metal oxide semiconductor (CMOS) geometries of the baseband processor IC.

The MIMO and adaptive antenna systems are used to improve link margins. Using MIMO requires multiple RF chains with multiple ADCs. With integration, the cost of these multiple chains should come down. Isolation between the receive chains needs to be in the order of 20 dB, which is easy to accomplish. There are no matching requirements for the gain and phase between the RX chains, which means that the radio design is simplified. MIMO works well in TDD or FDD, and its improvements to link margins are observed in multipath environments. In contrast, for adaptive antenna or beam-forming systems, the TX and RX chains need to be matched across frequency and over gain and phase. However, the subscriber station does not have multiple chains. Such systems work well in TDD mode since the TX frequency is the same as RX frequency. Adaptive antenna systems estimate the TX channel based on information they get from the RX channel, so having the same frequency improves these estimates.

OFDMA allows the RF channel to be split into subchannels. As a result, the power can be boosted since fewer tones are used. For users that do not transmit much data on the UL, a smaller bandwidth can be allocated. Thus more efficient use of the bandwidth

Table 2.1 Performance Enhancement of Antenna Diversity

Antenna Diversity Scheme (4 antenna branches)	*Antenna Gain (SUI3, SUI4 Model w/100 µs Rayleigh Delay Spread)*	*Implementation Complexity*
Selection diversity combining	8 dB	Low
Equal gain combining (analog)	9 dB	Mid
Maximum ratio combining (analog)	10 dB	High
Maximum ratio combining (digital)	14 dB	High

can be made on a per user basis. This technique does pose some challenges for the radio. Interference and noise between subchannels must be carefully considered over the whole transmit gain range. This problem is similar to the FDD case except there is no frequency separation. Therefore noise performance and linearity must be excellent since there is no help from filtering. Another issue with OFDMA is that the RF must be maintained to less than 1% accuracy; otherwise different users will collide with each other within the subchannels.

Various duplex schemes, including RF architectures, have been outlined and some methods to improve link margin considered. Next I discuss the particular circuit blocks within the RF system that are particular cost drivers.

2.4   RF System Blocks

There are three main areas of cost for a radio: the synthesizer, power amplifier, and filter.

2.4.1   Synthesizer

The synthesizer generates the local oscillator (LO) that mixes with the incoming RF to create a lower frequency signal that can be digitized and processed by the baseband IC. The WiMAX specifications call for a high-performance synthesizer. The synthesizer block takes up a large part of the RFIC die area and is therefore a costly component of the RFIC. The integrated phase noise is less than 1 deg rms with an integration frequency of 1/20 of the tone spacing (modulated carrier spacing) to 1/2 the channel bandwidth. Thus for the smaller bandwidths of 1.75 MHz, integration of the phase noise can start as low as 100 Hz. For HFDD architectures, the TX to RX frequency has to settle within $100\,\mu s$. The step size of the channel is 125 KHz in the 3.5 GHz band. In order to settle and maintain this step size, fractional synthesizers must be considered. It must be noted that as the RF increases, obtaining phase noise of less than 1 deg rms becomes a challenge. As well as all the radio LOs, the clock for the ADC must also be viewed as a LO that adds phase noise to the overall jitter specification.

2.4.2   Power Amplifier

Wideband digital modulation requires a high degree of linearity. Linearity implies higher power consumption. The trade-off between efficiency and linearity is a constant battle. For WiMAX, a power amplifier can work at 4% to 5% efficiency for about a 6 dB back-off from output P1dB. Such a back-off results in about a 2.5% error vector magnitude (EVM) or 32 dBc of SNDR. With a class AB power amplifier (PA), the efficiencies can run as high as 15% to 18% with similar EVM numbers. A much overlooked parameter in PA design is settling time. When a PA is switched on from cold, the power level will overshoot (or undershoot), then settle out. This settling time can be as poor as hundreds of milliseconds to get within 0.1 dB of the final value. For OFDM symbols, the RX has to estimate the power of a tone from the beginning of a frame to the end of a frame. If there is a drop in power from the beginning to the end of more than 0.1 dB across the frame, the BER for 64 quadrature amplitude modulation (QAM) will increase. The primary cause for this power drop is that the bias circuits and the output power field effect transistor

(FET) are at thermally different points. Since this phenomenon is thermal, the effect can last hundreds of milliseconds. To mitigate the power drop, the bias circuits have to be placed as close to the output FETs as possible so they see the same temperature. In some cases the PA may have to be turned on ahead of the TX cycle to allow the PA to stabilize and remove some of the drop. This implies having a trigger signal based on when data are to be transmitted. Having the MAC and PHY layers realize this trigger is not a simple matter. The budget of $100\,\mu s$ for HFDD is taken up by the synthesizer settling and any PA turn-on issues. A possible solution is to design the PA so that the PA settling is less than $5\,\mu s$.

2.4.3   Filtering

Filtering is required to eliminate undesired signals from adjacent or alternate channels. Any noise from these immediate signals can leak noise into the desired band. Filtering at the RX does not help; only a clean transmitted signal will prevent such degradation. Regulatory bodies control the transmitted mask.

For the adjacent channel problem, the challenge is between linearity and filtering complexity. If the undesired channels are filtered out, then less back-off in the radio is required and more of the analog-to-digital bits are available for the fading margin. SAW filters have decreased in cost and are now in the less than 2 range for high volume. SAWs provide the optimum filtering. A significant drawback is that the technology fixes the maximum channel bandwidth that can be supported. Another issue is that it is difficult to support a large array of RF bands with a fixed IF. For spurious analysis, the optimum IF depends on the RF. Filtering on-chip requires a large die area, and as the channel bandwidth is reduced, the die size increases. On-chip filters also produce more noise. A benefit is that the filter can be adjusted to accommodate the various bandwidths.

For I/Q-based designs, on-chip filters are necessary. The filters can be matched much more closely if they are on-chip. This minimizes the I/Q mismatch due to filtering. The final channel selectivity is performed in the baseband IC using digital filters.

Filtering, like gain, must be distributed between the RF and subsequent down-conversions. RF filtering is used to reduce the image and far blockers (i.e., out of the RF band). The RF front end must be linear enough to support the largest in-band blocker. In addition, reciprocal mixing of the LO with the undesired signal must be considered. RF filters are typically more than 50 MHz wide and are constructed from various technologies, each with different Qs. The larger the Q, the larger the size and the better the filter shape. In FDD systems, cavity filters may have to be used; these are large mechanical cavities and can cost more than 30 in high volume.

2.4.4   WiMAX Specifications

We highlight some of the WiMAX RF specifications and contrast them with 802.11 specifications where possible. The specifications are broken into RX and TX in Table 2.2 and Table 2.3, respectively. It should be noted that most designs aim to do better than the standards, hence these numbers should be viewed as the minimum requirements. In addition, I note the impact on the RFIC due to these specifications.

Table 2.2 RX Specifications

Parameter	802.11	WiMAX	Impact on RFIC
NF (dB)	10	7	The implication for the RFIC is that it may require an external LNA to meet a 5 dB NF.
SNDR-64QAM (dBc)	<29	29	The implication for the RFIC is excellent phase noise for tone spacing of 5 Khz and linearity. For 802.11 the tone spacing is larger; i.e., 300 Khz thus phase-noise requirement is less stringent.
Alternate channel rejection (dBc)	NA	30	The AD bits may be used for allowing the adjacent channel through and some of the alternate channel. The digital filter would perform the bulk of the close-in channel filtering. Results in increase in linearity for RFIC.
HFDD mode	No	Yes	More complicated synthesizer to support dual frequency.
Channel BW (MHz)	10, 20	1.25; 1.75; 3.5; 7; 14; 5; 10; 20	The implication for the RFIC is that the smaller bandwidths results in a complicated synthesizer due to the smaller step size. Filtering for an array of bandwidths introduces adjacent channel compromises.

Table 2.3 TX Specifications

Parameter	802.11	WiMAX	Impact on RFIC
Licensed band operation	No	Yes	The implication for RFIC is that the regulations are tighter and increase cost.
AGC range (dB)	NA	50	The implication for RFIC is that linearity must be maintained over the AGC range for 64-QAM.
SNDR (dBc)	<31	31	The implication for RFIC is NF of the TX chain, linearity, and phase noise.
OFDMA	No	Yes	Noise and linearity must be maintained over the AGC range for in-channel cases.
Smart antenna	No	Yes—Option	More RF chains for MIMO or matched RF chains for beam forming.
Power output (dBm)	Restricted in unlicensed bands	<24 dBm	The implication for RFIC is PAS require higher efficiency, or even smart PA technology.

2.5 Summary

WiMAX poses significant challenges to the RF subsystem. Several RF architectures were discussed both in FDD, HFDD, and TDD modes. The cost-performance trade-offs in the various architectures were discussed: these include IF- and baseband-type radios. Some of the more important RF system blocks, synthesizers, power amplifiers, and filtering that relate to cost and specifications were discussed. Finally, some of the WiMAX radio specifications were highlighted and contrasted with the 802.11 specifications and the impact to RFIC development was noted.

Glossary

802.11 A standard for communication of wireless devices over reduced distances and reliability than the 802.16d/e standards.

802.16d A standard for fixed wireless; it has progressed to 802.16e, a standard that allows for full mobility.

adapative antenna system (AAS) Wireless systems that include beam forming using multiple antennas.

analog-to-digital converter (ADC) Converts an analog signal to a digital signal.

automatic gain control (AGC) Control circuitry or software used to maintain a constant level at the analog and digital divide. AGC is used to maximize the dynamic range of the radio.

baseband The processor that converts the received analog signal from the radio into a digital signal and processes it and also converts the transmit digital stream into an analog domain for the radio.

bit error rate (BER) Allows wireless systems to be compared at the PHY level.

digital-to-analog converter (DAC) Converts a digital signal to an analog signal.

equal gain combining (EGC) A method of antenna diversity.

error vector magnitude (EVM) A measure of the fidelity of the signal.

frequency division duplex (FDD) Where transmit and receive information is allocated on a frequency domain basis. The same time slot is used for both transmit and receive information; however, the transmit and receive frequencies are different.

field programmable gate array (FPGA) An integrated circuit that allows digital circuits to be reprogrammed.

general purpose input/output (GPIO) Digital output for general usage when communicating with the radio (e.g., turning the power amplifier on and off).

half frequency division duplex (HFDD) Where the transmit and receive information is allocated on a frequency domain basis. The radio is in either transmit mode or receive mode. In addition, the transmit and receive frequencies are different.

I/Q Baseband frequency signals are signals that are centered at 0 Hz.

IF An intermediate frequency signal is one which is at a frequency higher than 0 Hz.

line of sight (LOS) Where the antenna view between the base station and the subscriber is unobstructed.

local oscillator (LO) A subcomponent of a radio that down-converts the RF to an IF or 0 Hz through at least one mixer, conversely it is used to up-convert an IF or 0 Hz signal to radio frequency.

maximim ratio combining (MRC) A method of antenna diversity.

medium access control (MAC) A layer of software that sits on top of the physical (PHY) layer. It serves to control transmit power, framing, synchronization of packets, network entry and quality of service.

multiple-in multiple-out (MIMO) Advanced communication system that improves throughput and reliability of a wireless system. It exploits multiple transmitters and recievers.

near zero intermediate frequency (NZIF) A very low IF.

orthogonal frequency division multiple access (OFDMA) Allows for bandwidth allocation across multiple users. In WiMax, scaleable OFDMA is used which allows carrier spacing to remain constant as channel bandwidths and FFT sizes are changed.

P1dB Compression of the gain of an amplifier.

power amplifier (PA) The final amplifier in the transmit chain that delivers power to the antenna.

pulse width modulators (PWM) Low-cost method for producing analog signals with a GPIO that can be used to affect radio circuits like amplifer gain.

quality factor (Q) The measure of inverse resistive loss in the material.

radio frequency integrated circuit (RFIC) A silicon chip or set of chips that form part of the radio. RFICs typically contain amplifiers, mixers, synthesizers, and calibration circuits.

receiver (RX) The radio chain that down-converts from RF to a low frequency. In addition, it amplifies the received signal so that the baseband can process it. The typical receive signal ranges from 1 μVrms to 22 mVrms.

selection diversity combining (SDC) A method of antenna diversity.

signal-to-noise plus distortion ratio (SNDR) A measure of the fidelty of a signal.

signal-to-noise ratio (SNR) A measure of the fidelity of a signal.

serial peripheral interface (SPI) A three-wire interface to control the radio.

surface acoustic wave (SAW) A filter that provides excellent characteristics in a small form factor that cannot be achieved in an RFIC.

single-in single-out (SISO) A wireless system with a single antenna input and a single antenna output.

time division duplex (TDD) Where transmit and receive information is allocated on a time domain basis. The same frequency is used for both transmit and receive information.

transmitter (TX) The radio chain that up-converts from low frequency to RF and delivers power to the antenna. Typical transmit chains have output of -30 dBm to 30 dBm.

voltage controlled oscillator (VCO) A component used for up-conversion or down-conversion in the synthesizer subsystem of a radio.

WiMax A standard that ensures interoperability between equipment vendors.

zero intermediate frequency (ZIF) Where the final frequency that is digitized by the baseband is at 0 Hz. Another name for an I/Q radio.

Bibliography

1. Abidi, A. "Direct-Conversion Radio Transceivers for Digital Communications." *IEEE J. Solid-State Circuits* 30(1995): 1399.
2. Crols, J., and M.S.J. Steyaert. "A 1.5 GHz Highly Linear CMOS Downconversion Mixture." *IEEE J. Solid-State Circuits* 30(1995): 736.

3. Gray, P., and R. Meyer. *Analysis and Design of Analog Integrated Circuits*. 3rd ed. New York: John Wiley & Sons, 1993.
4. IEEE, 802.16-Rev D/d5-2004.
5. Parissinen, A., J. Jussila, J. Ryynanen, L. Sumanen, and K.A.I. Halonen. "A 2-GHz Wide-Band Direct Conversion Receiver for WCDMA Applications." *IEEE J. Solid-State Circuits*, 34(2000): 1893.
6. Vizmuller, P. *RF Design Guide: Systems, Circuits and Equations*, Norwood, MA: Artech House, 1995.

Chapter 3

WirelessMAN Physical Layer Specifications: Signal Processing Perspective

---

Kun-Chien Hung, David W. Lin, Youn-Tai Lee,
and Kanchei Loa

Contents

This chapter considers the orthogonal frequency division multiple access (OFDMA) physical (PHY) layer specifications of the IEEE 802.16e standard, especially the subset adopted in the current mobile Worldwide Interoperability for Microwave Access (WiMAX) for time division duplex (TDD) operation.* We consider how the standard specifications impact the design of some key physical layer signal processing algorithms and how they may impact the design of backward-compatible mobile multihop relay (MMR) systems. Following an introduction, we give an overview of the Institute of Electrical and Electronics Engineers (IEEE) 802.16e OFDMA physical layer specifications from a signal processing perspective. We then turn our attention to the reception of such OFDMA signals and discuss how the main signal processing functions are designed. These functions include synchronization, channel estimation, and demoduation and decoding. We see that the required complexity of some functions is closely related to the frame structure, or more precisely, the property and placement of preamble and pilot subcarriers. We subsequently consider how the signal processing functions designed to the existing standard specifications might impact the design of backward-compatible relay systems. Again, the frame structure plays an important role. Aside from some general discussion, we present some simulation results on channel estimation to illustrate the fact that the physical layer specifications of relay systems can have a profound effect on the transmission performance of sophisticated devices.

3.1 Introduction

Since the publication of the first IEEE 802.16 standard for fixed broadband wireless access in 2001, a number of revisions and amendments have been added. Like other IEEE 802 standards, the 802.16 standards are primarily concerned with PHY layer and media access control (MAC) layer functionalities. Over time, four different kinds of modulation have been adopted for the PHY layer, of which two are of the single-carrier type (dubbed SC and SCa, respectively): one is orthogonal frequency division multiplexing (OFDM) and the other is OFDMA. Current WiMAX systems, both the fixed and the mobile versions, are built on a subset of the IEEE 802.16 specifications. The fixed version is based on the OFDM PHY in IEEE 802.16-2004 and the mobile version is based on the OFDMA PHY in IEEE 802.16e-2005 [1,2]. This chapter is concerned with OFDMA in a mobile communications environment.

As is well-known, the OFDM modulation technique, by its use of cyclic prefixing (CP), can easily combat multipath interference. OFDMA further divides the subcarriers into subchannels that can be allocated individually, thus providing flexible access control

* This work was supported by the National Science Council of R.O.C. under grant no. NSC 95-2219-E-009-003 and by the New Generation Broadband Wireless Communication Technologies and Applications Project of the Institute for Information Industry, sponsored by MOEA, R.O.C., under grant no. 96-EC-17-A-03-R7-0765.

Table 3.1   S-OFDMA Parameters Proposed by the WiMAX Forum

Parameter	Values			
Channel bandwidth (MHz)	1.25	5	10	20
FFT size (N_{FFT})	128	512	1024	2048
Subcarrier frequency spacing	10.94 kHz			
OFDMA symbol duration ($T_s = T_b + T_g$)	102.9 μs			
CP time ($T_g = T_b/8$)	11.4 μs			
Useful symbol time (T_b)	91.4 μs			

in a frequency-selective time-varying channel condition, but also introducing interesting signal processing problems at the PHY layer. One feature of the IEEE 802.16e OFDMA (and Mobile WiMAX) standard is the selectable fast Fourier transform (FFT) size, from 128 to 2048 in multiples of 2, excluding 256 to be used with OFDM. This has been termed scalable OFDMA (S-OFDMA). One use of S-OFDMA is that if the channel bandwidths are allocated based on integer powers of two times a base bandwidth, then one may consider making the FFT size proportional to the allocated bandwidth so that all systems are based on the same subcarrier spacing and the same OFDMA symbol duration, which may simplify system design. For example, Table 3.1 lists some S-OFDMA parameters proposed by the WiMAX Forum [3].

What is the spectrum efficiency of IEEE 802.16e OFDMA? Within the span of one OFDMA symbol, it depends on several parameters in addition to the modulation format and code rate: length of CP, percentage of subcarriers actually used to carry data, and sampling factor (ratio of sampling rate to the allocated bandwidth). Each parameter may assume one of several possible values depending on the design choice or the operating condition. For illustration, take a typical set of parameters for single-input single-output (SISO) transmission as follows: length of CP equal to 1/8 of the FFT size, data subcarriers being approximately 70% of total subcarriers (downlink [DL]) or approximately 55% (uplink [UL]), and sampling factor equal to 1.12. Taken together, the "raw efficiency" is approximately

$$\frac{8}{9} \times 1.12 \times \begin{Bmatrix} 70\% \\ 55\% \end{Bmatrix} \approx \begin{cases} 70\%, \text{ DL,} \\ 54\%, \text{ UL.} \end{cases} \tag{3.1}$$

Then, with 64-point quadrature amplitude modulation (64-QAM) (highest order of modulation specified), rate 3/4 coding (highest convolutional coding rate specified without using hybrid automatic repeat request [HARQ]), and a 10 MHz bandwidth, the peak information data rates are 31.5 and 24.5 Mbps, respectively, for DL and UL. Note that the above does not take into account the overhead due to the frame structure and higher layer protocols.

A recent development in the IEEE 802.16 Working Group is the formal establishment of the IEEE 802.16j Relay Task Group in March 2006, with the charge of developing a standard for MMRs that are backward compatible with mobile stations (MSs) designed to the specifications of IEEE 802.16e OFDMA. That is, such MSs should be able to work in a relay-enhanced system without needing to be modified. This requirement brings out interesting MMR design issues from the signal processing perspective; these will be elaborated later in the chapter.

Figure 3.1 illustrates two primary uses of relays, namely, coverage extension and throughput enhancement. Direct radio link between the base station (BS) and MS1 does

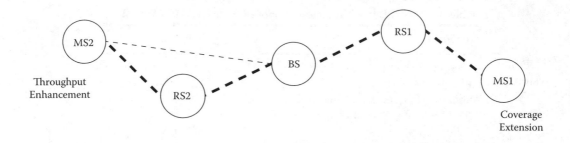

Figure 3.1 **Two primary uses of relays. Thickness of dashed line indicates radio link strength. No dashed line indicates no radio link.**

not exist (due to distance or shadowing, for example). With the aid of relay station 1 (RS1), MS1 can now gain access, hence coverage is extended. On the other hand, the direct radio link between the BS and MS2 exists but is weak. With the aid of RS2, MS2 can communicate with the BS at a higher rate. However, in this case a higher rate to or from MS2 does not necessarily mean higher throughput in the sense of better spectrum efficiency for the overall system. This is because both legs of the relayed link will consume spectrum. If the spectrum cannot be shared between the two legs (e.g., due to insufficient separation of antenna patterns in the two directions), and if neither leg can offer an efficiency at least twice that of the direct link between the BS and MS2, then the overall spectrum efficiency will degrade rather than be enhanced. In this situation there can be throughput enhancement when the system is lightly loaded with much spectrum to spare, but not when the system is heavily loaded. This is an interesting problem in relay system design and operation, but we shall not touch on it in this chapter.

In a broader sense, relayed communication can be viewed as a kind of cooperative communication that can improve wireless transmission performance [4–7]. At the least, relays can offer spatial diversity. They can work together in an ad hoc fashion without centralized control, or they can work under centralized control. The MMR systems currently under study in IEEE 802.16j are of the latter kind. In any case, the equipment cost of an RS should be much lower than that of a BS, otherwise there would not be much advantage deploying RSs rather than BSs.

The objective of this chapter is to look at the IEEE 802.16e OFDMA PHY specifications and the design of backward-compatible relay systems from the signal processing perspective. We examine how the standard specifications impact the design of signal processing algorithms at the PHY layer and how that may impact backward-compatible relay system design. We will concentrate on the TDD mode of operation, which is the mode of operation that the WiMAX Forum chose to focus on initially for the reasons of lower cost and more flexible capacity management between the DL and UL. In Section 3.2 we introduce the corresponding PHY specifications. In Section 3.3 we describe the main signal processing functions in an OFDMA receiver. In Sections 3.4 through 3.6 we elaborate on synchronization, channel estimation, and demodulation-decoding, respectively. In Section 3.7 we discuss how the desire for backward compatibility impacts relay system design due to signal processing issues in the PHY layer. We also illustrate the point with some simulation results on channel estimation. Finally, following a brief conclusion in Section 3.8, we list some open issues in Section 3.9.

The contributions of this chapter are as follows: (1) we provide an overview of the PHY layer specifications of the IEEE 802.16e OFDMA standard from the signal processing perspective; (2) we address the reception of IEEE 802.16e OFDMA signals and consider how the standard specifications affect the design of main receiver signal processing functions; and (3) we point out some impacts that the MS signal processing functions designed to the current standard may have on the design of backward-compatible relay systems.

3.2 PHY Layer Specifications of IEEE 802.16e OFDMA

3.2.1 Frame Structure and Types of Subchannel Organization

As mentioned, the OFDMA PHY defines four selectable FFT sizes: 2048, 1024, 512, and 128. The subcarriers are divided into three types: null (guard bands and DC), pilot, and data. The data subcarriers are organized into subchannels, which form the basic units of allocation for user data transmission. A data stream can be borne over one or more subchannels depending on its rate. Three basic types of subchannel organization are defined: partial usage of subchannels (PUSC), full usage of subchannels (FUSC), and adaptive modulation and coding (AMC); among them the PUSC is mandatory and the other two are optional. The DL and UL transmissions are divided into "zones" based on the type of subchannel organization used, as illustrated in Figure 3.2, where TUSC1 and TUSC2 are variants of PUSC used for multiple-input multiple-output (MIMO) transmission.

In PUSC DL, the entire channel bandwidth is divided into three segments to be used separately. Details of the PUSC DL and UL are given in the next subsection. The FUSC is employed only in the DL and it uses the full set of available subcarriers so as to maximize the throughput. In both PUSC and FUSC, the subcarriers that constitute a subchannel are pseudo-randomly distributed to attain frequency diversity. In AMC, in contrast, the subchannels are formed using adjacent subcarriers. A key idea is that if the transmitter does not know the channel response, then by distributing the transmitted data randomly over the subcarriers, it can maximize the frequency diversity. But if the relative quality of different subcarriers can be known, then there is the option of transmitting the data over the better subcarriers to maximize efficiency. Since the channel variation is typically smooth across subcarriers within the coherence bandwidth, it is natural to form subchannels using adjacent subcarriers.

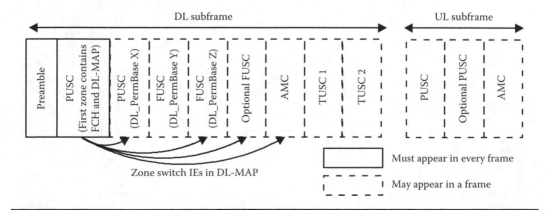

Figure 3.2 **Example OFDMA TDD frame with multiple zones, each employing a particular type of subchannel organization [2, Figure 219].**

The subcarrier quality can be obtained in several ways. In TDD, since the channels are reciprocal, the subcarrier quality measured at the local receiver can be fed to the local transmitter for its use. However, this requires that the subcarriers used to transmit data be a subset of that used to receive. Otherwise feedback of the subcarrier quality information from the remote end is needed. Consider a scenario where the MS informs the BS of the DL subcarrier quality information through a feedback channel. Since this information is available, at most, once per TDD frame, there arises the concern of possible channel variation in the frame duration. For example, with a 10 km/hr speed at a carrier frequency of 3.5 GHz, the coherence time T_c with omnidirectional antenna reception is approximately [8].

$$T_c \approx \frac{9}{16\pi f_m} \approx 5.52 \text{ ms}, \tag{3.2}$$

where f_m is the maximum Doppler spread. This indicates that AMC employing feedback subcarrier quality information may not perform well in this situation if the frame duration is much longer than 5 ms, unless the transmitter has some means to predict the channel variation. MIMO transmission with directional antenna beams can reduce the Doppler spread and result in better AMC performance.

In what follows, we concentrate on single-input single-output (SISO) transmission instead of MIMO transmission. And we concentrate on the mandatory PUSC subchannel organization.

3.2.2 Subchannel Organization in PUSC

Figure 3.3 illustrates the structure of a TDD frame containing only the mandatory PUSC zone. Excluding the preamble, the DL subframe consists of $2n$ OFDMA symbols and

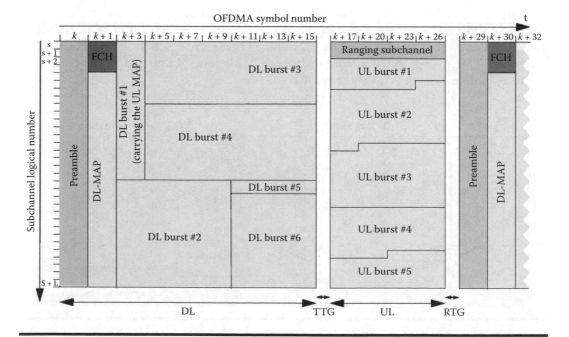

Figure 3.3   OFDMA TDD frame example, showing PUSC mandatory zone only [2, Figure 218].

the UL subframe $3m$, where n and m are some integers. The DL subframe starts with a preamble, which is an OFDM symbol that can use one of three disjoint subcarrier sets. In each set, the indexes of the used subcarriers are given by $3k + n$, where $n \in \{0, 1, 2\}$ indicates the subcarrier set and k runs over a range of values excluding the guard bands and DC. The preamble subcarrier set is binary phase-shift keying (BPSK) modulated with one of a set of selectable pseudo-noise (PN) sequences. The number of selectable sequences is 114 for each FFT size. The power level of each modulated subcarrier is 9 dB above the normal average data subcarrier power in subsequent OFDMA symbols. Among other things, the preamble sequence serves as a local identifier of the BS to the MSs. The preamble is also useful in synchronization and (to a lesser extent) in initial channel estimation for subsequent DL signal reception. These subjects will be discussed further later.

Following the preamble, the frame control header (FCH), downlink map (DL-MAP), and uplink map (UL-MAP) contain broadcast messages in PUSC mode that inform the MSs about how the subsequent DL and UL bursts are organized in time and in frequency, the associated coding and modulation schemes, and for which MS each burst is intended. Except for the preamble, the subcarriers are grouped into subchannels in a pseudo-random fashion. The subchannels are organized differently in the DL and UL.

In the DL, every two successive OFDMA symbols form one unit in division of subcarriers into subchannels. The subchannels are organized through several levels of clustering and numbering as shown in Figure 3.4. From right to left, the mappings are as follows:

1. *D*1: Divide the used subcarriers, excluding DC, into physical clusters of 14 adjacent subcarriers each.
2. *D*2: Renumber the physical clusters into logical clusters by a certain permutation formula.
3. *D*3: Divide the logical clusters into six major groups in a consecutive manner. The groups are numbered 0 to 5. The system may choose to use only a subset of the groups.
4. *D*4: Allocate the pilot subcarriers in each cluster of a used group according to Figure 3.5. Then organize the data subcarriers in a major group into subchannels in a pseudo-random fashion, with 24 data subcarriers per subchannel per OFDMA symbol.

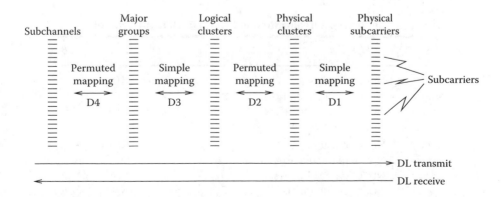

Figure 3.4   PUSC DL subchannel organization.

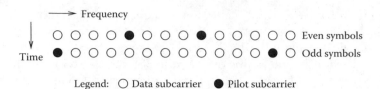

Figure 3.5   PUSC DL cluster structure.

(Thus "simple mapping" in Figure 3.4 means collection of consecutive elements into one group or its inverse operation.) Table 3.2 summarizes the DL subcarrier organization in PUSC. Note that the DL pilot subcarriers are associated with major groups, not with individual subchannels.

The major groups are organized into three segments numbered 0, 1, and 2. The default is that major groups 0, 2, and 4 are assigned to segments 0, 1, and 2, respectively. The organization of subchannels into three segments facilitates use of sectored antennas in a cell, with different sectors making use of different subchannel segments. In fact, each of the preamble subcarrier sets correspond to one such segment.

In the UL, every three successive OFDMA symbols form a unit in subchannel formation. Twelve subcarriers, four from each of the three successive OFDMA symbols, constitute a "tile," and six pseudo-randomly selected tiles form a subchannel. For each tile, the four subcarriers from each OFDMA symbol are contiguous in frequency and their frequency locations are the same over the three symbols. The 12 subcarriers in a tile are divided into 4 pilot subcarriers and 8 data subcarriers as shown in Figure 3.6. A subchannel thus contains 48 data subcarriers and 24 pilot subcarriers, distributed over three OFDMA symbols. Unlike the two-layer permutation in the DL, the pseudo-random tile selection is the only permutation in the UL. The mechanism of subchannel formation is illustrated in Figure 3.7, where the above-mentioned pseudo-random tile selection is realized through two mappings, namely, $U2$ and $U3$. Table 3.3 summarizes the UL subcarrier organization in PUSC.

In both the DL and UL, the pilot subcarriers are BPSK modulated with a pseudo-random binary sequence (PRBS). The DL pilot subcarrier power is 2.5 dB higher than the average data subcarrier power, whereas the UL pilot subcarrier power is the same as the average data subcarrier power.

Table 3.2   Subcarrier Organization in PUSC DL OFDMA Symbol

	FFT Size	*2048*	*1024*	*512*	*128*
Even Major Group	No. of clusters	24	12	10	2
	No. of pilot subcarriers	48	24	20	4
	No. of data subcarriers	288	144	120	24
	No. of subchannels	12	6	5	1
Odd Major Group	No. of clusters	16	8	N/A	N/A
	No. of pilot subcarriers	32	16	N/A	N/A
	No. of data subcarriers	192	96	N/A	N/A
	No. of subchannels	8	4	N/A	N/A

N/A = not applicable.

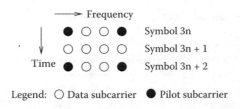

Figure 3.6   PUSC UL tile structure.

How pilot subcarriers are placed along the frequency and the time axes has much to do with the design of the channel estimation algorithm (which is needed for coherent OFDMA transmission). Within the coherence bandwidth (roughly equal to the inverse of the channel delay spread), the channel response usually shows smooth variation with frequency. Likewise, within the coherence time (inversely proportional to the maximum Doppler spread), the channel response usually shows smooth variation with time. Hence having pilots close to data subcarriers in frequency and in time is beneficial to channel estimation at the data subcarriers. For example, a delay spread of 10 μs corresponds to, very roughly, a coherence bandwidth of 100 kHz. (Depending on the definition, some may take the coherence bandwidth to be a smaller value. But we are only aiming at an order-of-magnitude argument here.) On the other hand, the coherence time at a 120 km/hr speed with a carrier frequency of 3.5 GHz is, according to Equation (3.2), approximately 460 μs. Then, for the 10.94 kHz subcarrier spacing and 102.9 μs OFDMA symbol duration considered in Mobile WiMAX, the coherence bandwidth spans a frequency range of approximately nine subcarriers and the coherence time spans a time span of approximately four to five symbols. The DL cluster structure illustrated in Figure 3.5 has pilots spaced in frequency by 12 subcarriers at most, which is about 1.3 times the coherence bandwidth and likely about the limit in which a low-order polynomial can be used to model satisfactorily the frequency domain variation of channel response in between. In contrast, the UL cluster structure has a three-subcarrier pilot spacing, and hence the frequency domain variation of channel response in between should be well modeled by a low-order polynomial. Along the time axis, the channel variation can be modeled as approximately linear over several symbols [9].

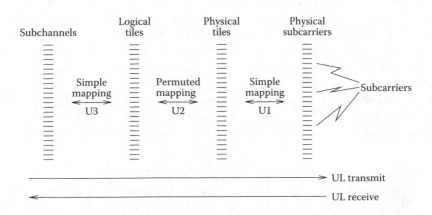

Figure 3.7   PUSC UL subchannel organization.

Table 3.3   Subcarrier Organization in PUSC UL OFDMA Symbol
($n = 0, 1, 2, \ldots$)

FFT Size	2048	1024	512	128
No. of tiles	420	210	102	24
No. of subchannels	70	35	17	4
No. of pilot subcarriers:				
Symbol $3n$	840	420	204	48
Symbol $3n + 1$	0	0	0	0
Symbol $3n + 2$	840	420	204	48
No. of data subcarriers:				
Symbol $3n$	840	420	204	48
Symbol $3n + 1$	1680	840	408	96
Symbol $3n + 2$	840	420	204	48

3.2.3   Coding and Modulation

Figure 3.8 illustrates the coding and modulation process. We briefly introduce the randomization, the forward error correction (FEC) coding, the bit interleaving, and the modulation methods below.

The randomizer does XORing of the input data with a PRBS whose generator is defined by $1 + x^{14} + x^{15}$. The randomizer is initialized every FEC coding block. The specified FEC schemes include tail-biting convolutional coding (CC), zero-tailed CC, block turbo coding (BTC), convolutional turbo coding (CTC), and low-density parity check (LDPC) coding. Only tail-biting CC is mandatory; the rest are optional. The generator vectors of the mandatory code are given by $(171, 133)_{OCT}$ and illustrated in Figure 3.9. The tail biting is effected by initializing the encoder memory with the last six bits of the encoder input data block. Hence the coding starts with the state that the encoder will end in. The mother code may be punctured to yield higher-rate coding, as shown in Table 3.4.

The bit interleaving is achieved in two steps. The first step is a block interleaver that maps adjacent coded bits to nonadjacently indexed subcarriers in the used subchannels. The second step maps adjacent coded bits to less or more reliable bits in the used QAM constellation in a cyclic manner, thus avoiding runs of lowly reliable bits. The above is a form of bit-interleaved coded modulation (BICM) [10] that increases the diversity order in fading rather than merely increasing the minimum Euclidean distance of the code: while a large Euclidean distance is good for handling additive white Gaussian noise (AWGN), a large diversity order is more effective in handling fading. The first step of bit interleaving spreads adjacent coded bits out in frequency, which should help enhance frequency diversity.

Three data modulation schemes are defined: quadrature phase-shift keying (QPSK), 16-QAM, and 64-QAM, where the 64-QAM is optional. The bit-to-constellation mapping is Gray. The mapping result is multiplied with a PRBS (with ± 1 values) before transmission.

Figure 3.8   Coding and modulation process [2, Figure 252].

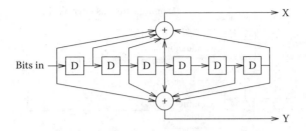

Figure 3.9   Rate-1/2 convolutional encoder.

3.3   Main Signal Processing Functions in OFDMA Receiver

Figure 3.10 depicts a typical OFDMA baseband receiver structure. Among its functions, the synchronizer, the channel estimator, and the demodulator-decoder block are more complex in the algorithm and play a major role in signal reception. The synchronizer estimates and compensates any offsets in carrier, sampling time, OFDMA symbol time, and frame time in the receiver in reference to the transmitter. The channel estimator acquires the channel response for use in data detection, and the demodulator-decoder block does the data detection.

Although it is possible to conceive decision-directed synchronization and channel estimation employing feedback from the decoder output, the decoding delay involved poses a stability and performance concern in time-varying mobile communication channels. Therefore, in practice, one would more likely consider a mix of blind and decision-directed methods for synchronization and channel estimation, where the decisions employed are not the decoder's final output, but some preliminary decisions.

Likewise it is also possible to conceive joint operation of the synchronizer and the channel estimator. However, due to the consequent complexity, a majority of the studies to date appear to have decoupled their operation. This we also do in the present chapter. The IEEE 802.16e OFDMA standard specifies that the MS needs to synchronize to the BS to within a relatively high accuracy. Specifically, the UL carrier frequency can deviate by at most $\pm 2\%$ of the subcarrier spacing and the UL symbol timing can only be off by $\pm 25\%$ of the minimum allowed CP duration (where the minimum allowed CP duration is 1/32 of the useful symbol time). Whether these are satisfied or not, the BS can instruct the MS to tune its frequency and timing to even greater accuracy through the ranging response (RNG-RSP) message.

In summary, the key tasks to be completed in the MS receiver in the DL subframe upon its initial entrance to the network are as follows:

Table 3.4   Punctured Convolutional Coding

Rate	1/2	2/3	3/4
Free distance (d_{free})	10	6	5
Puncture pattern for X[a]	1	10	101
Puncture pattern for Y[a]	1	11	110
Output sequence	$X_1 Y_1$	$X_1 Y_1 Y_2$	$X_1 Y_1 Y_2 X_3$

[a] "1" denotes retained bit and "0" denotes removed bit.

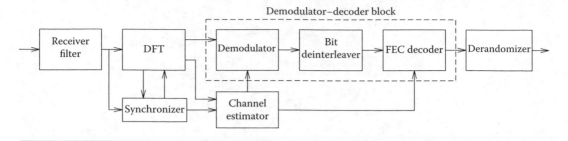

Figure 3.10 Typical OFDMA baseband receiver structure.

1. Synchronization of the MS carrier and timing (including estimation and compensation of offsets), and identification of the preamble index.
2. Optional initial channel estimation based on the received DL preamble.
3. Channel estimation (or channel tracking if the optional initial estimation above is conducted) in the broadcast message part of the signal.
4. Decoding of the broadcast message to identify the locations of the allocated data bursts.
5. Channel estimation or tracking for the allocated DL burst locations.
6. Decoding of data received in the allocated DL burst locations.

In subsequent frames, the synchronization task is easier, as the MS no longer has to reidentify the preamble index, nor does it have to do acquisition of the carrier and timing, but only needs to track them. But the other tasks are similar.

Disregarding the ranging operations, the key tasks in UL signal reception performed in the BS receiver during normal data transmission are

1. Synchronization of the carrier and timing for each MS.
2. Channel estimation for each MS for the allocated UL burst locations only.
3. Decoding of data received from each MS in its allocated DL burst locations.

In the next three sections we discuss synchronization, channel estimation, and demodulation-decoding, respectively. For DL synchronization, we focus on the initial synchronization rather than the later tracking.

3.4 Synchronization

In the OFDMA DL, initial synchronization (at the MS) involves carrier recovery, timing recovery, and preamble index identification, where carrier recovery involves estimation and compensation of the carrier frequency offset (CFO) and timing recovery, in principle, should include estimation and compensation of the sampling frequency offset (SFO) and the OFDMA symbol time offset. The carrier phase offset may be considered part of the channel response and taken care of in channel estimation. The sampling time offset can also be absorbed into the channel response if the CP is long enough. In the UL, synchronization (at the BS) involves carrier and timing recovery only. Our discussion concentrates on the estimation aspect of synchronization, omitting the compensation aspect.

Now, note that the IEEE 802.16e OFDMA requires not only that the MS carrier frequency be synchronized to the BS to within 2% of subcarrier spacing, but also that the transmitted center frequency and the sampling frequency of the MS be derived from one reference oscillator. When these are true, the sampling phase difference from the beginning of an OFDMA symbol to the end of it will only differ by at most $2\% \times (1 + 1/4)$ of the true sample period, where the factor $1/4$ accounts for the largest allowed CP time ratio. Therefore it appears unnecessary to perform separate SFO recovery in either the MS or the BS provided that the CFO can be accurately recovered. And this certainly would be the case after the startup period. What remains to be synchronized, besides preamble index identification in the DL, is the CFO and the OFDMA symbol timing.

Interestingly, even with a perfect carrier synchronizer, the MS may not be able to attain the specified 2% accuracy in high-speed motion without enlisting outside help (e.g., RNG-RSP feedback from the BS). An example should bear out the reason clearly. Consider a 3.5 GHz carrier frequency and that the MS is approaching the BS at a 120 km/hr speed. Let there be no other propagation paths but the line-of-sight (LOS) path. Then we have a Doppler shift of 389 Hz. For the WiMAX parameters of Table 3.1, this is about 3.6% of the subcarrier spacing. And a perfect carrier recovery circuit at the MS would yield exactly this amount of frequency deviation. Such a frequency "error" should be detected by the BS upon receiving transmission from the MS (which would be at a Doppler shift of $389 \times 2 = 778$ Hz). Then, through the RNG-RSP message, the BS can instruct the MS to adjust its frequency reference.

3.4.1 CFO and Symbol Timing Estimation

Nonzero CFO results in intercarrier interference (ICI) where nearby subcarriers interfere with one another. On the other hand, incorrect symbol timing estimation may lead to two different kinds of impairment. A negative timing error (or lead error, where the estimated timing is earler than the actual timing) amounts to adding a delay to the channel response. Hence a negative timing error causes no performance problem as long as the amount of error plus the original length of the channel response is still within the CP length. Nevertheless, a smaller negative timing error can better ensure proper system operation if the length of the channel response can vary from time to time. On the other hand, positive timing errors (i.e., lag errors, where the estimated timing is later than the actual timing) are more detrimental to system performance because some of next symbol's samples can be mistaken for part of the present symbol. Hence it is desirable to minimize the probability of positive timing errors.

Several approaches to CFO estimation have been proposed. One is the data-aided approach [11,12], applicable when the preamble consists of a known signal (or when a reliable decision on the preamble contents can be made). In the case of IEEE 802.16e OFDMA, it is not suitable because in the DL the preamble can be 1 of 114 choices, and in the UL the a priori known signal (i.e., the pilots) consists of one-third of the received signal. The second approach is based on subspace analysis (e.g., via the ESPRIT algorithm) [13,14]. While the resolution in CFO estimation of these methods can be high, the computational complexity can also be high. The third approach is completely blind estimation relying solely on the repetitive signal structure of OFDMA symbols (e.g., the presence of CP) [15,16]. This appears simplest and suitable for use in IEEE 802.16e OFDMA. Likewise, there are multiple approaches to symbol timing estimation. But again, one of the simplest, and appropriate for IEEE 802.16e OFDMA, is blind estimation based on the CP structure.

Indeed, the DL preamble in IEEE 802.16e OFDMA, by having its nonzero subcarriers spaced regularly in the frequency domain, gives rise to a quasiperiodic signal structure in the time domain. This quasiperiodicity may also be exploited for the benefit of CFO and symbol timing estimation. However, the following discussion considers the use of CP only.

Let N be the FFT size and L be the CP length in number of samples. Under the assumption that the received samples are jointly Gaussian, quasi-maximum-likelihood estimators in Rayleigh fading for symbol timing and CFO have been derived as [16]

$$\hat{\theta} = \arg\max\{|\lambda(\theta)|\} \tag{3.3}$$

and

$$\hat{\varepsilon} = -\frac{1}{2\pi}\tan^{-1}\left(\frac{\Im[\lambda(\hat{\theta})]}{\Re[\lambda(\hat{\theta})]}\right), \tag{3.4}$$

respectively, where

$$\lambda(\theta) = \sum_{k=\theta}^{\theta+L-1} r(k)r^*(k+N). \tag{3.5}$$

Figure 3.11 illustrates the method. A similar structure has also been derived in van de Beek et al. [15]. We can employ averaging over the symbols to obtain a more accurate estimate of the CFO and the symbol timing.

In applying the above method to multipath channels, note that Equation (3.3) tends to be determined by the strongest path rather than the first significant path. Therefore, starting at $\hat{\theta}$, one should search in the direction of smaller timing offsets to find the earliest significant path (e.g., by finding the smallest offset where $|\lambda(\theta)|$ is above a sizable fraction of $|\lambda(\hat{\theta})|$) in order to avoid lag error. The modified method can be employed directly in the DL. In the UL, however, since the signals from different MSs may be subject to different CFOs and different delays (as allowed by the standard), the situation becomes more complicated. But since the BS knows the subchannel allocation of each MS, it is possible to further modify the method based on the used subcarriers of each MS. The details are omitted. To minimize the required search range, however, it appears advisable for the BS to use the RNG-RSP message to adjust the CFOs and the timing offsets of the MSs to a greater accuracy than that allowed by the standard during the ranging process and as needed later.

Note that Equation (3.4) can only obtain the fractional part (normalized by subcarrier spacing) of the CFO. The integer part of the CFO needs to be estimated by another means. But this is needed only in the DL. Because of the 2% subcarrier spacing accuracy requirement on the carrier frequency of the MS, there is no integer CFO in the UL

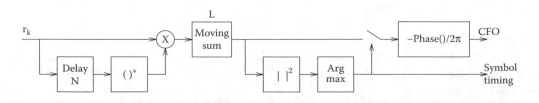

Figure 3.11   J.-C. Lin's simplified symbol timing and fractional CFO estimation method [16].

transmission. Now, for simplicity, consider estimating the DL CFO separately from the preamble index. Without knowing the preamble index, no data-aided methods can be used. The natural resort is blind estimation by finding which integer CFO value results in the highest total power in any of the three preamble subcarrier sets. Since the three preamble subcarrier sets are (nearly) indistinguishable in power content, there is a three-value ambiguity in the estimated integer CFO. One needs to wait until after finding the preamble sequence to determine the precise integer CFO. In fact, it may be better to consider a joint search of the integer CFO and the preamble index.

3.4.2 DL Preamble Index Search

There are a total of 114 possible preamble sequences. The most straightforward approach is apparently an exhaustive search for the best match. If the integer CFO is determined jointly, then the search range is 114 times the number of all possible integer CFO values. Complexity considerations aside, an important issue is how to measure the degree of match of each candidate solution. Correlation of the received preamble with a possible preamble sequence does not necessarily provide a good indicator of their degree of match because, for it to be a good indicator, the preamble cannot span a frequency range much beyond the coherence bandwith—but unfortunately this can be very far from the truth under the typical WiMAX system parameters discussed previously.

Toward a solution, note that since the coherence bandwidth may cover several subcarrier spacings, we may employ a technique resembling differential detection, but working in the frequency domain. Specifically, we can "differentially decode" the received preamble by taking the difference between each pair of neighboring subcarrier values in it. The differentially decoded sequence is compared to similarly "decoded" versions of all the possible preamble sequences at all possible integer CFO values. The one with the closest resemblance gives the search result. A technique in this vein has been proposed for WiBro [17]. A more general treatment of this approach can be found in Hung and Lin [18], which expounds the underlying theory to some depth and develops a number of variations to the fundamental technique.

After identifying the preamble index (and the integer CFO), we also obtain the subchannel segment used by the BS.

3.4.3 Summary

In summary, the initial DL synchronization presented above proceeds as follows. As stated before, we concentrate on the estimation aspect, omitting the compensation aspect.

1. In the time domain, perform OFDMA symbol timing and fractional CFO estimation employing blind correlation based on the repetitive signal structure in the preamble symbol, such as the CP structure.
2. Convert the signal into the frequency domain by FFT.
3. In the frequency domain, perform joint estimation of the integer CFO and the preamble index employing differentially decoded sequences.

Many alternative methods can be conceived, of which we have mentioned some in the foregoing discussion.

In normal UL signal reception after the ranging process, only OFDMA symbol timing synchronization and fractional CFO synchronization are needed. But these have to be done for each MS separately. To reduce the complexity, the BS may use the RNG-RSP message to control the timing and carrier frequency of each MS more tightly than permitted in the standard.

3.5 Channel Estimation

Channel estimation is needed for FEC decoding, as we will see later. For OFDMA systems subject to not very fast fading, the signal relationship in the frequency domain can be modeled simply as

$$r_k = H_k \cdot x_k + n_k, \tag{3.6}$$

where k is the subcarrier index, r_k is the received signal, H_k is the channel gain, x_k is the transmitted signal, and n_k is the additive noise (assumed white Gaussian). The task of the channel estimator is to estimate H_k. There exist a variety of methods for channel estimation, which differ greatly in complexity. We review a few representative ones below and elaborate on their interrelation with the PHY signal design. In an OFDMA system that employs preambles and pilots, a simple channel estimation method is the so-called least square (LS) method, which estimates the channel response at the subcarriers carrying known signal based on observation of only one OFDMA symbol [19]. The estimate can be interpolated in the frequency domain to obtain channel estimates at other subcarrier locations. It can also be interpolated or extrapolated, deterministically or adaptively, in the time domain to reduce the noise influence and obtain refined channel estimates. On the other hand, better estimates under various conditions can be effected by the more complex subspace-based methods.

3.5.1 LS Channel Estimation at Pilot Positions

The LS method of channel estimation, when operating on one received OFDM symbol alone, does the following optimization for each subcarrier carrying known signal:

$$\tilde{H}_k = \arg\min |r_k - H_k x_k|^2, \tag{3.7}$$

where x_k is the known signal. The solution is simply

$$\tilde{H}_k = r_k/x_k. \tag{3.8}$$

Assume that n_k is independent of x_k. Then the mean square error (MSE) is given by

$$\sigma_H^2 = E|\tilde{H}_k - H_k|^2 = \sigma_n^2/\sigma_x^2, \tag{3.9}$$

where σ_n^2 and σ_x^2 denote the variances of n_k and x_k, respectively, and we have assumed that the pilot signal power σ_x^2 is independent of k, a condition satisfied by the IEEE 802.16e OFDMA signal.

3.5.2 Frequency Domain Interpolation and Time Domain Averaging

Within the span of the channel coherence bandwidth, the channel frequency response can be considered largely constant. Therefore, if the pilot spacing is smaller than a small multiple of the coherence bandwidth, then simple low-order polynomial interpolation can be used to obtain channel estimates at nonpilot locations from an existing channel estimate at the pilot locations. Some related concepts were discussed in Section 3.2.2.

As also mentioned, time domain averaging (interpolation or extrapolation, deterministic or adaptive) over several OFDMA symbols can enhance the channel estimation performance, if the channel does not vary significantly over this time period. The simplest method of time averaging, based on the assumption that the channel response stays largely unchanged over a short period of time, is the exponential average:

$$\widehat{H}_k(t) = (1 - \beta)\widehat{H}_k(t - 1) + \beta\tilde{H}_k(t), \tag{3.10}$$

where β is the forgetting factor, which can be set to $1/T$, where T is the channel coherence time (i.e., the time span over which the channel response can be considered unchanged) in the number of OFDM symbols. That the channel may be modeled as linearly varying in a short time period [9] can be used to yield a predicted channel response at future OFDM symbol instants, for example,

$$\widehat{H}_k(t + 1) = \tilde{H}_k(t) + [\tilde{H}_k(t) - \tilde{H}_k(t - 1)]. \tag{3.11}$$

If the receiver latency is not a concern, time domain interpolation can be performed. This is obviously useful, for example, in UL signal reception according to the subcarrier arrangement illustrated in Figure 3.6, where a channel estimate for the second symbol can be obtained by averaging that of the first and the third symbols, as the first and the third symbols both contain pilots but the second symbol does not. But one can also conceive of ways of using this technique for DL signal reception as well as other ways of using it for UL signal reception. The simplest method of time domain interpolation is, of course, linear interpolation, such as

$$\widehat{H}_k(t) = \frac{1}{2}[\tilde{H}_k(t - 1) + \tilde{H}_k(t + 1)]. \tag{3.12}$$

Note that the time variation in pilot locations in the DL transmission of IEEE 802.16e OFDMA, as illustrated in Figure 3.5, leads to a little additional complexity in time domain averaging, but not significantly.

3.5.3 Subspace-Based Method

In the subspace-based method introduced in this subsection, we take the multipath structure of the channel response into consideration to eliminate the estimation noise outside the subspace defined by the multipath structure. To motivate the method, note that an L-path channel response can be described as

$$h(t) = \sum_{i=1}^{L} \alpha_i(t) \cdot g(t - \tau_i), \tag{3.13}$$

where $\alpha_i(t)$ is the coefficient of the ith path at time t, τ_i is the corresponding delay, and $g(t)$ is the combined response of the transmitter and the receiver filters. (That is, we treat the transmitter and the receiver filters as part of the continuous-time channel model.) In Rayleigh fading, $\alpha_i(t)$ has complex Gaussian distribution. In slow-enough fading, $\alpha_i(t)$ is constant over the time period of one OFDMA symbol and we may drop the time index over the period of any one OFDMA symbol. Let $\underline{g}_i$ be the vector of the samples of $g(t - \tau_i)$ over an OFDMA symbol. Then the sampled channel response over this period can be written as

$$\underline{h} = \sum_{i=1}^{L} \alpha_i \underline{g}_i = \left[\underline{g}_1 \ \underline{g}_2 \ \cdots \ \underline{g}_L \right] \underline{\alpha} \triangleq G\underline{\alpha}, \tag{3.14}$$

where G defines the "delay subspace" of the channel.

Let X_p denote the diagonal matrix in which the diagonal elements are the given values (such as pilot values) if they are known and zero otherwise. The vector of received signal, in frequency domain, with one element corresponding to each position where the transmitted signal is known (such as a pilot), is given by

$$\underline{r} = X_p \underline{H} + \underline{n} = X_p W G \underline{\alpha} + \underline{n}, \tag{3.15}$$

where W is the DFT matrix.

The subspace-based method contains two phases. First, we estimate the delay subspace (i.e., the G matrix). Then the channel response is estimated by estimating $\underline{\alpha}$. A variety of techniques can be used to estimate the delay subspace. If the pilots are regularly spaced, then we may employ, for example, MUSIC [20] and ESPRIT [21], or if they are not, then we may employ matching pursuit (MP) [22,23]. Although the computation cost of subspace estimation is great, fortunately in normal situations the path delays do not vary very quickly, and hence the delay subspace stays relatively invariant over a reasonably long period of time. Therefore it does not need to be estimated often. In TDD transmission, for example, if the frame period is short, then it may only need to be estimated once per frame or per several frames.

After acquiring the delay subspace, the channel response can be estimated via an LS approach [24] which results in

$$\tilde{\underline{H}} = WG \left(G^H W^H X_p^H X_p W G \right)^{-1} G^H W^H X_p^H \underline{r}. \tag{3.16}$$

If, in addition, the pilots have a constant envelope, then the equation can be simplified to

$$\tilde{\underline{H}} = \frac{1}{\sigma_x^2} WG \left(G^H W^H I_p W G \right)^{-1} G^H W^H X_p^H \underline{r}, \tag{3.17}$$

where $I_p = \frac{1}{\sigma_x^2} X_p^H X_p$ is the indexing matrix of the pilot locations. The resulting average MSE is given by

$$\sigma_H^2 \triangleq E \left\| \underline{H} - \tilde{\underline{H}} \right\|^2 = \frac{L}{P} \frac{\sigma_n^2}{\sigma_x^2}, \tag{3.18}$$

where P is the number of pilots. When $P > L$ (the typical situation in practice), the MSE by the subspace-based method is lower than that by simple LS estimation. Actually the MSE is the lower bound with known L and with the observation of one OFDMA symbol only.

3.5.4 Application to IEEE 802.16e OFDMA

We describe several channel estimation methods for the IEEE 802.16e OFDMA system based on the above-described techniques. Two methods are given for the DL and one for the UL. For the UL, since the data plus pilot subcarriers are organized in a tile structure that spans three successive OFDM symbols, it is appropriate to consider simple bilinear interpolation in both frequency and time within a tile after acquiring an estimate of the channel response at the pilot locations (the four corners). For the DL, the preamble is used to effect an initial channel estimation. Then we may "track" the channel estimate by one of the following two methods: simple interpolation-based tracking and subspace-based tracking. They have different computational complexity.

3.5.4.1 Interpolation-Based Channel Tracking

The procedure is as follows:

1. Initial channel estimation at the preamble: The initial channel estimation does an LS estimation followed by frequency domain interpolation, since only every third subcarrier is present in the preamble.
2. Channel tracking at subsequent OFDMA symbols: For each subsequent OFDMA symbol, repeat the process of LS channel estimation at pilot positions, frequency domain interpolations, and time domain averaging.

Aside from the computation needed for LS channel estimation, the computation complexity is only several real multiplications and additions per subcarrier.

3.5.4.2 Subspace-Based Channel Tracking

Interpolation-based channel tracking suffers from the problem of error floor as the signal: noise ratio (SNR) increases, especially when the MS moves at a fast speed. This is because of the channel estimation error by the interpolation method. The subspace-based method can mitigate this problem at the cost of a much higher complexity. The overall procedure, including initial channel estimation, is as follows:

1. Initial channel estimation at the preamble: First, LS channel estimation is performed on the subcarriers present in the preamble. Then the delay subspace is estimated. Finally, the subspace-based LS step is applied to eliminate the noise outside the delay subspace.
2. Linear prediction of current channel response: When a past response estimate is available, apply the linear prediction in Equation (3.11) to predict the current channel response. This only has to be done in the delay subspace, not the complete set of subcarriers.
3. Decision-directed refinement of the channel estimate: This step employs preliminary decisions on data subcarriers as well as the known pilots to improve the channel response estimate. This reduces the noise from prediction error. To make the preliminary decisions, simply do a hard-decision demodulation without regard to the fact that some constellation points may not exist at certain time instants due to channel coding. As long as the error probability is low, its average contribution to channel estimation accuracy is positive. This is especially the case with a high SNR, where the error floor phenomenon is particularly annoying. This step may use Equation (3.16) or Equation (3.14).

We mentioned that the delay subspace only needed to be updated once in a while. Hence in channel tracking, only steps 2 and 3 are executed. However, we note that if tracking of the delay subspace is desirable, then the method in Simeone et al. [25] may be considered. We reiterate that although the subspace-based method can yield a better channel estimate, its complexity is very high because of the need to perform matrix pseudo-inverses.

3.6 Demodulation and Decoding

Recall that the mandatory coding scheme in IEEE 802.16e OFDMA is tail-biting CC with puncturing, followed by bit interleaving and QAM. The punctured bits can be treated simply as erasures in the CC trellis which contribute nothing to any path metric during Viterbi decoding. The following considers how to handle tail-biting and bit-interleaved QAM.

3.6.1 Decoding of Tail-Biting CC

In principle, the optimal decoding of a tail-biting CC can be achieved by running as many parallel Viterbi decoders as there are states, one for each different starting and ending state [26]. The Viterbi decoder that yields the best performance among all gives the maximum likelihood (ML) solution. Since the code at hand has $2^6 = 64$ states, this optimal decoding method is apparently unpalatable.

A simple suboptimal decoding method with good performance exists [27,28]. It employs only one Viterbi decoder, but makes it work on a circularly extended input sequence. The idea is illustrated in Figure 3.12. The underlying philosophy is as follows: The encoding path through the code trellis can be considered one cycle of an infinitely long periodic sequence. Experience has shown that near-ML decoding performance can be attained with Viterbi decoding delay of about four to eight times the CC constraint length. Although we do not know the initial state of the received code sequence, if we execute the Viterbi decoding procedure long enough on the cyclically extended input sequence, it is likely that we will converge to the optimal solution after some time. Based on the above experience with Viterbi decoding, the leading segment and the trailing segment of the best performing path may not be reliable and should be discarded, where the lengths of these segments are expected to be about four to eight times the CC constraint length. The center segment likely gives the optimal solution. Hence, in the decoding process, we may do cyclic pre- and postfixing of the received sequence first, where the pre- and postfixes each have a length equal to about four to eight times the CC constraint length. After Viterbi decoding over the whole extended sequence, we drop the pre- and postfixed segments of the Viterbi output to obtain the final decoder output.

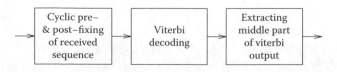

Figure 3.12   Principle of the suboptimal tail-biting CC decoder.

3.6.2   Computation of Branch Metrics under Bit-Interleaved QAM

For optimal soft-decision Viterbi decoding in additive white Gaussian noise (AWGN), the error metric is the Euclidean distance between the trellis path and the soft output of the demodulator. However, by having a bit interleaver between the CC encoder and the QAM modulator, the combined trellis of the CC and the modulator becomes greatly entangled to the point that truly ML metrics become very difficult to compute. A remedy, therefore, is to adopt suboptimal metric computation, which disregards the entangling of the code trellis due to bit interleaving, but considers the path metric as the sum of the bit metrics, one for each of the bits that constitute the path [29,30], although in reality the path metric is not a simple sum of the bit metrics due to the bit-interleaved modulation. A suitable bit metric for this is the log-likelihood ratio (LLR) [30].

Let $A = A_I + jA_Q$ denote a Gray-mapped QAM symbol for some subcarrier in IEEE 802.16e OFDMA and $\{b_{I1}, b_{I2}, \ldots\}$ and $\{b_{Q1}, b_{Q2}, \ldots\}$ be the corresponding bit sequences for A_I and A_Q, respectively. Then for QPSK, the LLRs of the two bits b_{I1} and b_{Q1} are

$$L_{I1} \approx -|H|y_I, \quad L_{Q1} \approx -|H|y_Q, \tag{3.19}$$

respectively, where

$$y_I = \Re\{r/H\}, \quad y_Q = \Im\{r/H\}, \tag{3.20}$$

with r being the received signal and H being the channel response at this subcarrier. For 16-QAM, the LLRs of the two in-phase bits are given by [30]

$$L_{I1} \approx -|H|y_I, \quad L_{I2} = |H|(|y_I| - 2), \tag{3.21}$$

respectively. The LLRs of the two quadrature-phase bits are similarly related to y_Q. For 64-QAM, the LLRs of the in-phase bits are given by, respectively,

$$L_{I1} \approx -|H|y_I, \quad L_{I2} \approx |H|(|y_I| - 4), \quad L_{I3} = |H|(||y_I| - 4| - 2). \tag{3.22}$$

The LLRs of the three quadrature-phase bits are similarly related to y_Q.

The above LLR computation actually constitutes the demodulator function block in Figure 3.10. After bit deinterleaving, the punctured tail-biting CC decoder of the last subsection fleshes out the "FEC decoder" in Figure 3.10.

3.7   Design of Backward-Compatible Relay Systems

As discussed previously, the IEEE 802.16j Relay Task Group is in charge of developing a standard for MMR systems that are compatible with MSs compliant to IEEE 802.16e OFDMA so that these MSs should be able to work in a relay-enhanced system without needing to be modified. Since, as of this writing, this standard is still in an early stage of development, it is too premature to say what the final specifications may look like. But it is of interest to examine from a signal processing perspective how the desired backward compatibility may impact relay system design. From the standpoint of standard development, the discussion below only provides a certain technical point of view, not necessarily what will be reflected in the standard.

In an MMR-enhanced system, one basic function of the RS is to serve as a repeater between the BS and the MS. It may assist the BS in the latter's network management functions, but does not manage the network itself. It is subordinate to the controlling BS. For this reason, various broadcast messages, including the preamble, the FCH, the DL-MAP, and the UL-MAP, remain to be managed by the BS. Concerning mobility, an RS can be fixed, nomadic, or mobile. A nomadic RS is similar to a nomadic BS in mobile phone systems which can be dynamically placed in areas showing high traffic load. A mobile RS can be placed in a public transit vehicle, such as a bus, to assist in mobile access.

We have seen in earlier sections that three key signal processing functions in the PHY layer are synchronization, channel estimation, and demodulation-decoding. Below we use synchronization and channel estimation to illustrate one important backward compatibility issue, namely, how can new RSs be deployed without degrading the transmission performance that can be attained by existing MSs in the absence of the RSs. It turns out that the solution lies to a large extent in the design of the frame structure, for the frame structure affects synchronization and channel estimation performance.

3.7.1 Design of Frame Structure and Its Impact on Synchronization and Channel Estimation

The frame structure governs how the radio resources (bandwidth and time) can be shared by multiple entities in the network, and it is one of the first issues faced in the development of the MMR protocol. This being so, an important factor to consider in its design is the potential impact on system latency and capacity, besides backward compatibility and MS performance. However, this is beyond the scope of the present chapter.

Consider a two-hop scenario, as illustrated in Figure 3.1. Besides BS-to-MS and MS-to-BS transmissions in a relayless system, we now also have BS-to-RS, RS-to-BS, RS-to-MS, and MS-to-RS transmissions. For backward compatibility, the RS should seem like a BS to the MS. The transmission mechanism to/from an RS from/to an MS should not be different from that to/from a BS. Assume that the BS, the RS, and the MS employ the same band for communication. Then both the DL subframe and the UL subframe will be shared by the BS and the RS. In addition, there should be periods of time reserved for BS-to-RS and RS-to-BS transmissions. One way to achieve this is to divide the frame into an MS subframe and an RS subframe, as shown in Figures 3.13 and 3.14.

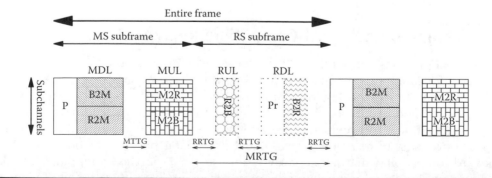

Figure 3.13 Example frame structure employing simple FDM between BS and RS.

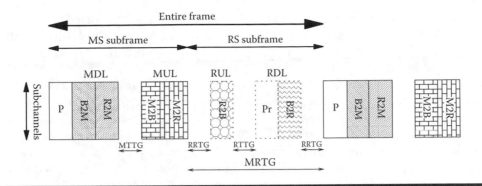

Figure 3.14   Example frame structure employing simple TDM between BS and RS.

Notations used in the figures are explained in Table 3.5. These figures also illustrate two ways of band sharing between the BS and the RS, namely, simple frequency division multiplexing (FDM) and simple time division multiplexing (TDM). The former is enabled by the subchannel structure of OFDMA. There, of course, are a plethora of ways to share the band in frequency and in time.

The mobile downlink subframe (MDL) starts with the same signal structure as the IEEE 802.16e OFDMA DL subframe so that the MS can lock into it, but the signal properties in the RS subframe should be such that the MS does not falsely lock into it. For example, the relay downlink subframe (RDL) may be designed to have a structure different from that of the broadcast portion (preamble, FCH, DL-MAP, and UL-MAP) of the IEEE 802.16e OFDMA DL subframe. The MAP fields in the MDL only contain allocation information for the MDL and mobile uplink (MUL); thus the RS subframe is in some sense invisible to the MS. The relay uplink (RUL) is placed immediately after the MUL to minimize the

Table 3.5   Notations in Frame Structures Shown in Figures 3.13 and 3.14

Notation	Explanation
P	Broadcast messages
Pr	Control messages to RS
B2M	BS-to-MS transmission
R2M	RS-to-MS transmission
M2B	MS-to-BS transmission
M2R	MS-to-RS transmission
R2B	RS-to-BS transmission
B2R	BS-to-RS transmission
MDL	DL subframe for MS
MUL	UL subframe for MS
RDL	DL subframe for RS
RUL	UL subframe for RS
MTTG	Transmit/receive transition gap of MS
MRTG	Receive/transmit transition gap of MS
RTTG	Transmit/receive transition gap of RS
RRTG	Receive/transmit transition gap of RS

UL transmission latency from the MS to the BS. Likewise, the RDL is placed immediately before the MDL to minimize the latency in DL transmission from the BS to the MS. In the case of coverage extension (see Figure 3.1), some of the broadcast messages should be transmitted by both the BS and the RS.

Now suppose an MS comes into the joint coverage of a BS and one of its RSs (thus in the throughput enhancement type of situation in Figure 3.1). Suppose the broadcast messages are transmitted by both the BS and the RS. Then the MS will hear these messages from both. Consider the TDM arrangement. The MS may face a problem in channel estimation because it hears the preamble from both the BS and the RS and may thus estimate the channel based on the received signal in this period of time. The estimated channel is the combined channel response of that from the BS to the MS and that from the RS to the MS. Halfway into the MDL, if the BS stops transmission, then the channel response suddenly becomes that from the RS alone. If the MS employs a channel estimation algorithm that takes some time to adapt to the changed channel response and if the DL burst to the MS is transmitted through the RS during the time when the MS is still adapting its channel estimate, then the signal reception performance will be affected adversely.

For an MS employing subspace-based channel estimation, there may be the added issue related to estimation of the delay subspace: We previously argued that, since the path delays often vary more slowly than the channel coefficients, it may be appropriate to estimate the delay subspace only once at the beginning of each DL subframe. But now, if the perceived channel response can switch from one to another in the middle of the subframe, one may need to track the variation in the delay subspace more frequently.

Now consider the issue of synchronization in the throughput enhancement scenario. In this situation, the power delay profile of the RS-MS channel will likely have a smaller initial delay and higher power average than that of the BS-MS channel, as illustrated in Figure 3.15. Suppose, in contrast to the last condition, only the BS transmits the broadcast messages. Then the MS will acquire its timing based on the BS-MS channel. When the RS joins in later and takes over the DL burst transmission to the MS, it effectively creates a lag-error condition which, as discussed previously, could be detrimental to the transmission performance.

From the above discussion, it appears that the simplest solution to avoid the synchronization and channel estimation problems is to let the RS duplicate the BS transmission in the whole MDL subframe. But there could be other solutions, depending on the BS-MS and RS-MS channel characteristics. In summary, when it comes to enhancing the system performance by addition of RSs, the DL subframe design needs to be examined closely from the perspective of signal processing functions in the PHY layer (such as synchronization and channel estimation). Otherwise there could be unexpected harm done to

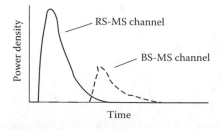

Figure 3.15   Likely relation between the power-delay profiles of the BS-MS channel and the RS-MS channel in the throughput enhancement scenario.

existing device transmission performance. In contrast, the UL transmission is not subject to similar concerns because there is neither a preamble nor a broadcast message part in the UL subframe structure. However, if rather than simply relaying all the uplink transmissions of the MS through the RS, the MS and the RS join hands in transmission of the signal to the BS at the same time, then similar problems may arise and similar solutions may also apply.

3.7.2 Numerical Illustration of the Impact of Frame Structure on Channel Estimation

To illustrate the possible impact of frame structure on transmission performance, we present some simulation results under some simplified conditions. For simplicity, we consider channel estimation only.

3.7.2.1 Simulated Conditions

Let the BS transmit the preamble, but the RS may or may not. Then the data signals (along with pilots) are transmitted to the MS by the RS alone. For simplicity, we omit the FCH and the MAP signals because the underlying PHY layer problem is the same with or without them. The BS stops transmission after the preamble.

Mathematically, in transmission of the preamble the received signal vector is given by

$$\underline{r}_p = (H_B + H_R)\underline{x}_p + \underline{n}, \tag{3.23}$$

where H_B and H_R are the diagonal matrices of the channel frequency responses of the BS-to-MS link and the RS-to-MS link, respectively, and $\underline{x}_p$ is the preamble. Subsequently only the RS transmits, and the received signal vector is given by

$$\underline{r}_d = H_R\underline{x}_d + \underline{n}. \tag{3.24}$$

Here the BS-to-MS link H_B disappears.

We consider the steady-state channel estimation MSE in mobile communication under both the IEEE 802.16e OFDMA and the relay-enhanced situation. Let the carrier frequency be 3.5 GHz, OFDMA system bandwidth be 10 MHz, and FFT size be 1024. Consider the mandatory PUSC mode of transmission employing the major group 0 of subcarriers [2]. And consider use of 64-QAM. The normalized power delay profiles of the BS-to-MS channel and the RS-to-MS channel are given by, respectively,

$$b_B = [0 \ \ 1 \ \ 0 \ \ 0 \ \ 0.5 \ \ 0 \ \ 0.2 \ \ 0 \ \ 0.1], \tag{3.25}$$

$$b_R = [1 \ \ 0 \ \ 0 \ \ 0.5 \ \ 0 \ \ 0.2 \ \ 0 \ \ 0.1 \ \ 0]. \tag{3.26}$$

Several power ratio (PR) values between these two channels are simulated, where the PR is defined as the power gain of the RS-to-MS channel relative to the BS-to-MS channel. We consider mobile speeds of 30, 120, and 240 km/h, whose corresponding normalized Doppler spread-symbol time products (F_dT_s) are equal to 0.01, 0.04, and 0.08, respectively. The PR values considered are 5, 10, and 20 dB. In terms of system capacity impacts, the first two more likely contribute to throughput enhancement and the third to coverage extension.

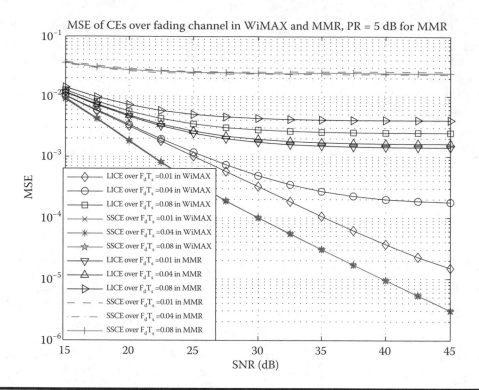

Figure 3.16   Average MSE of simulated channel estimation over the fading channel in IEEE 802.16e and RS-enhanced system (where RS also transmits the preamble) with PR = 5 dB.

A total of 200 experiment runs are conducted, each with 24 successive OFDMA symbols following a preamble. We simulate both the linear interpolation-based channel estimation (LICE) and the subspace-based channel estimation (SSCE).

3.7.2.2   Simulation Results

Consider first the case where the RS also transmits the preamble. Figure 3.16 shows the average MSE (over 24 symbols) of simulated channel estimation over a fading channel in IEEE 802.16e and in an RS-enhanced system with PR = 5 dB. The SSCE in IEEE 802.16e performs well even at $F_d T_s = 0.08$; at that point it still yields the same MSE as at $F_d T_s = 0.01$. But its performance is seriously limited in the RS-enhanced system. The reason is that its prediction mechanism highly depends on past estimation and it takes a long time to adapt to the new channel with a different delay subspace.

The LICE in IEEE 802.16e yields different amounts of error floor at different $F_d T_s$. Its performance consistently improves with SNR at $F_d T_s = 0.01$ and 0.04 up until about 30 dB. However, error floor gradually manifests at higher SNRs, which is particularly clear for $F_d T_s = 0.04$. The performance at $F_d T_s = 0.08$ is not good at all. The LICE working in an RS-enhanced system also exhibits error floors even if the speed is not high. However, our observation is that the performance floor is mainly due to the estimation error in the first OFDMA symbol after the preamble. The reason is that channel estimation for the first OFDMA symbol uses both the channel estimation results for the preamble and the pilots in the first OFDMA symbol, but the "effective channel" has changed be-

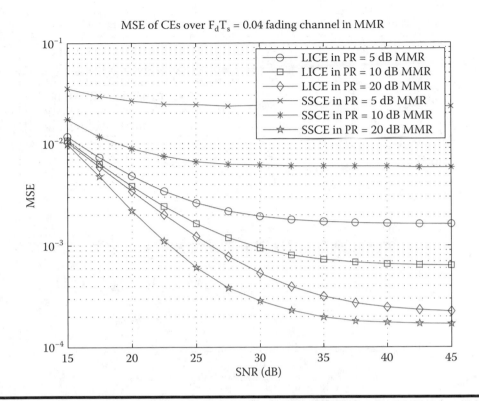

Figure 3.17 **Average MSE of simulated channel estimation over the fading channel in RS-enhanced system (where RS also transmits the preamble) with PR = 5, 10, and 20 dB.**

tween these symbols because the BS stops transmission after the preamble. After the first OFDMA symbol, the LICE adapts to the new channel situation quickly; its link adaptation capability is better than SSCE. A later figure (Figure 3.18) will bear this out clearly.

Figure 3.17 shows the average MSE of simulated channel estimation over a fading channel in an RS-enhanced system with PR values of 5, 10, and 20 dB. The results match our expectation in that larger PR values yields smaller performance losses because the smaller link power from the BS introduces less disturbance in channel estimation when the BS-to-MS channel suddenly disappears.

To see how the channel estimation MSE evolves with time, Figure 3.18 shows some results in an RS-enhanced system at PR values of 5, 10, and 20 dB. Note that the more complicated MS receiver needs longer recovery time when a channel suddenly "dies." Note also that the performance of LICE is similar in all PR values except for the first symbol; it is this difference that gives rise to the difference in the average performance curves of LICE shown in Figure 3.17.

Consider now the case where the RS does not transmit the preamble. Figure 3.19 shows the average channel estimation MSE and Figure 3.20 shows the time evolution of the MSE under conditions similar to those of the last two figures. In this case, LICE is able to maintain its performance, but SSCE fails to yield useful channel estimates due to the absence of preamble from the RS.

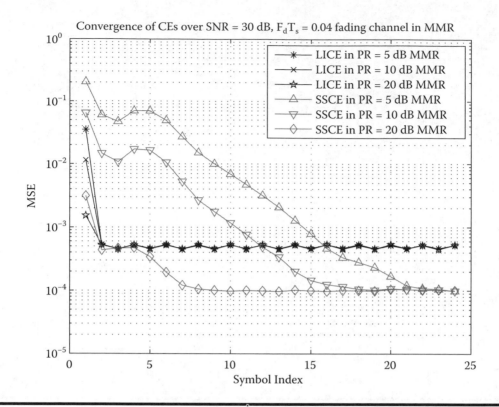

Figure 3.18 Evolution of channel estimation MSE in the fading channel in RS-enhanced system (where RS also transmits the preamble) at PR = 5, 10, and 20 dB.

3.8   Conclusion

We considered the OFDMA physical layer specifications of the IEEE 802.16e standard as well as the design of backward-compatible relay systems, both from the signal processing perspective. Specifically we examined how the standard specifications impacted the design of signal processing algorithms in the PHY layer. One conclusion is that there is a close relation between the required complexity of some such algorithms and the design and placement of preamble and pilot subcarriers in the overall frame structure. In addition, we examined how the signal processing functions designed to the existing standard specifications might impact the design of backward-compatible relay systems. For this we also presented some numerical results from simulated channel estimation under a simplified set of conditions to illustrate the point. An important conclusion is that, in the design of relay systems with backward compatibility, one needs to look at the physical-layer protocol (the frame structure in particular) from the signal processing point of view. Otherwise it is possible that a well-intended addition may turn out to degrade the transmission performance of sophisticated existing devices.

3.9   Open Issues

At several points in the foregoing discussion we have indicated interesting issues that can be further investigated. There are other open issues that can be listed regarding

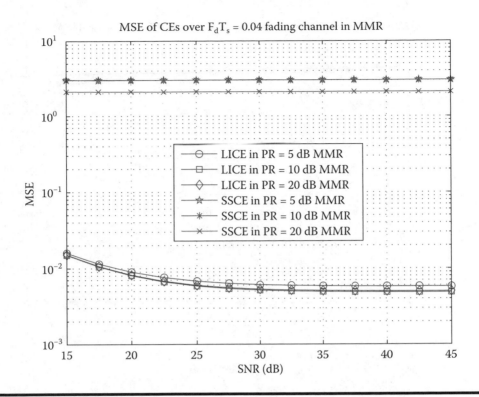

Figure 3.19 Average MSE of simulated channel estimation over the fading channel in RS-enhanced system (where RS does not transmit the preamble) with PR = 5, 10, and 20 dB.

signal processing for IEEE 802.16e OFDMA transmission. But here let us focus on some (largely) signal processing issues related to relay system design.

■ Frame structure: We have spent some time discussing frame structure and how it relates to synchronization and channel estimation. But we have barely scratched the surface. Many more frame structures can be conceived simply for two-hop systems than the ones illustrated, not to say multihop systems. How would they impact synchronization and channel estimation? What is a good design considering latency and overall capacity? Given a particular frame structure, can the RS adjust its transmission timing (and perhaps some other signal properties as well) to minimize the impact on the synchronization and the channel estimation functions in the MSs?

■ Frequency planning and reuse: Frequency planning and reuse (or equivalent issues) have long been studied in association with cellular systems employing frequency division multiple access (FDMA), time division multiple access (TDMA), and code division multiple access (CDMA). How can an OFDMA-based system, with relay stations, plan and reuse its spectrum to maximize capacity?

■ HARQ and quality of service (QoS) issues: The IEEE 802.16e supports HARQ and QoS. But it is easy to come up with some frame structures that will cause an increase in either the one-way or the roundtrip delay between the BS and the MS through multihop transmission via the RSs. How would HARQ perform in this case? How would it impact system behavior? Can it be improved? How about QoS-related issues?

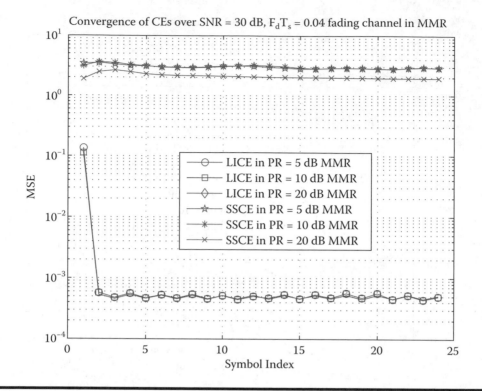

Figure 3.20 **Evolution of channel estimation MSE in the fading channel in RS-enhanced system (where RS does not transmit the preamble) at PR = 5, 10, and 20 dB.**

- AMC techniques: The IEEE 802.16e supports AMC. With the feedback of channel quality information, the throughput can be improved by employing the better subcarriers for data transmission. There are already open issues to resolve without RSs. With RSs, more issues emerge. Can AMC algorithms be designed to exploit spatial diversity when the coverage areas of multiple RSs overlap? Can AMC be employed in relayed transmission systems to any advantage?
- Distributed space-time coding and distributed MIMO techniques: Distributed space-time coding [31–33] can enhance spatial diversity by working with space-time coded signals from multiple relay stations. This can improve the throughput. In addition to further research into distributed space-time coding and distributed MIMO techniques, it would also be of interest to see how compatible these techniques are with IEEE 802.16e systems.

References

1. IEEE, IEEE Standard for Local and Metropolitan Area Networks—Part 16: Air Interface for Fixed and Mobile Broadband Wireless Access Systems, Standard 802.16-2004, IEEE, Washington, DC, October 2004.
2. IEEE, IEEE Standard for Local and Metropolitan Area Networks—Part 16: Air Interface for Fixed and Mobile Broadband Wireless Access Systems—Amendment 2: Physical and Medium Access Control Layers for Combined Fixed and Mobile Operation in Licensed

Bands and Corrigendum 1, Standard 802.16e-2005 and Standard 802.16-2004/Cor1-2005, IEEE, Washington, DC, February 2006.

3. WiMAX Forum, Mobile WiMAX—Part 1: A Technical Overview and Performance Evaluation, WiMAX Forum, Beaverton, August 2006.

4. A. Nosratinia, T.E. Hunter, and A. Hedayat, Cooperative communication in wireless networks, *IEEE Commun. Mag.*, 42(10), 74, 2004.

5. M. Yu and J. Li, Is amplify-and-forward practically better than decode-and-forward or vice versa? *IEEE International Conference on Acoustics, Speech, and Signal Processing*, vol. 3, p. 365, IEEE, Washington, DC, 2005.

6. M. Yuksel and E. Erkip, Broadcast strategies for the fading relay channel, *IEEE Military Communications Conference*, vol. 2, p 1060, IEEE, Washington, DC, 2004.

7. P.A. Anghel and M. Kaveh, On the diversity of cooperative systems, *IEEE International Conference on Acoustics, Speech, and Signal Processing*, vol. 4, p. 577, IEEE, Washington, DC, 2004.

8. T.S. Rappaport, *Wireless Communications*, 2nd ed., Prentice Hall, Upper Saddle River, N.J., 2002.

9. Y. Mostofi and D.C. Cox, ICI mitigation for pilot-aided OFDM mobile systems, *IEEE Trans. Wireless Commun.*, 4(2), 765, 2005.

10. G. Caire, G. Taricco, and E. Biglieri, Bit-interleaved coded modulation, *IEEE Trans. Inform. Theory*, 44(3), 927, 1998.

11. T.M. Schmidl and D.C. Cox, Robust frequency and timing synchronization for OFDM, *IEEE Trans. Commun.*, 45(12), 1613, 1997.

12. M. Morelli and U. Mengali, An improved frequency offset estimator for OFDM applications, *IEEE Commun. Lett.*, 3, 75, 1999.

13. U. Tureli, H. Liu, and M.D. Zoltowski, OFDM blind carrier offset estimation: ESPRIT, *IEEE Trans. Commun.*, 48, 1459, 2000.

14. X. Ma, C. Tepedelenlioğlu, G.B. Giannakis, and S. Barbarossa, Non-data-aided carrier offset estimators for OFDM with nullsubcarriers: identifiability, algorithms, and performance, *IEEE J. Select. Areas Commun.*, 19, 2504, 2001.

15. J.-J. van de Beek M. Sandell, and P.O. Borjesson, ML estimation of time and frequency offset in OFDM systems, *IEEE Trans. Signal Processing*, 45, 1800, 1997.

16. J.-C. Lin, Maximum-likelihood frame timing instant and frequency offset estimation for OFDM communication over a fast Rayleigh fading channel, *IEEE Trans. Vehic. Technol.*, 52(4), 1049, 2003.

17. H. Lim and D.S. Kwon, Initial synchronization for WiBro, *Asia-Pacific Conference on Communications*, p. 284, IEEE, Washington, DC, 2005.

18. K.-C. Hung and D.W. Lin, Joint detection of integral carrier frequency offset and preamble index in OFDMA WiMAX downlink synchronization, *IEEE Wireless Communications and Networking Conference*, IEEE, Washington, DC, 2007.

19. J.-J. van de Beek, O. Edfors, M. Sandell, S.K. Wilson, and P.O. Borjesson, On channel estimation in OFDM systems, *IEEE 45th Vehicular Technology Conference*, vol. 2, p. 815, IEEE, Washington, DC, 1995.

20. M. Oziewicz, On application of MUSIC algorithm to time delay estimation in OFDM channels, *IEEE Trans. Broadcast.*, 51(2), 249, 2005.

21. R. Roy and T. Kailath, ESPRIT-estimation of signal parameters via rotational invariance techniques, *IEEE Trans. Acoust. Speech Signal Process.*, 37(7), 984, 1989.

22. C.-J. Wu and D.W. Lin, Sparse channel estimation for OFDM transmission based on representative subspace fitting, *IEEE 61st Vehicular Technology Conference*, vol. 1, p. 495, IEEE, Washington, DC, 2005.

23. L. Zhiwei, A.B. Premkumar, and A.S. Madhukumar, Matching pursuit-based tap selection technique for UWB channel equalization, *IEEE Commun. Lett.*, 9(9), 835, 2005.

24. G.H. Golub, *Matrix Computations*, Johns Hopkins University Press, Baltimore, 1996.

25. O. Simeone, Y. Bar-Ness, and U. Spagnolini, Pilot-based channel estimation for OFDM systems by tracking the delay-subspace, *IEEE Trans. Wireless Commun.*, 3(1), 315, 2004.

26. H.H. Ma and J.K. Wolf, On tail biting convolutional codes, *IEEE Trans. Commun.*, 34(2), 104, 1986.

27. Y.-P.E. Wang and R. Ramésh, To bite or not to bite—a study of tail bits versus tail-biting, *IEEE International Symposium on Personal Indoor Mobile Radio Communications*, vol. 2, p. 317, IEEE, Washington, DC, 1996.

28. W. Sung and I.-K. Kim, Performance of a fixed delay decoding scheme for tail biting convolutional codes, *IEEE Asilomar Conference on Signals, Systems, and Computers*, vol. 1, p. 704, IEEE, Washington, DC, 1996.

29. E. Zehavi, 8-PSK trellis code for a Rayleigh channel, *IEEE Trans. Commun.*, 40, 837, 1992.

30. F. Tosato and P. Bisaglia, Simplified soft-output demapper for binary interleaved COFDM with application to HIPERLAN/2, *IEEE Global Telecommunications Conference*, vol. 2, p. 664, IEEE, Washington, DC, 2002.

31. J.N. Laneman and G.W. Wornell, Distributed space-time-coded protocols for exploiting cooperative diversity in wireless networks, *IEEE Trans. Inform. Theory*, 49(10), 2415, 2003.

32. A. Bletsas, A. Khisti, D.P. Reed, and A. Lippman, A simple cooperative diversity method based on network path selection, *IEEE J. Select. Areas Commun.*, 24(3), 659, 2006.

33. R.U. Nabar, H. Bolcskei, and F.W. Kneubuhler, Fading relay channels: performance limits and space-time signal design, *IEEE J. Select. Areas Commun.*, 22(6), 1099, 2004.

Chapter 4

MIMO for WirelessMAN

Xiaopeng Fan, Steven Y. Lai, Yuan Zheng,
and Jiannong Cao

Contents

With the development of the last mile access in wireless networks, multiple-input multiple-output (MIMO) has become one of the most important technologies for wireless metropolitan area networks (WirelessMANs) based on Worldwide Interoperability for Microwave Access (WiMAX). In this chapter we discuss the relationship between MIMO and WirelessMANs from three aspects: the capacity of MIMO, space-time signal processing, and diversity techniques. For the first problem we focus on the capacity of MIMO systems in WiMAX for both single-user and multi-user models. For the space-time coding, we first review the codes and then introduce the Alamouti code applied in WiMAX for IEEE 802.16-2004. We also describe various diversity schemes for enhancing the performance of wireless channels. Among them, space diversity is an effective scheme for combating multipath fading in WiMAX. Finally, we conclude this chapter with a brief summary.

4.1   Introduction

As the possible next development in wireless Internet protocol (IP) offers a possible solution to the last-mile access problem, WirelessMANs based on the 802.16 standards [1] have recently captured the interest of vendors and Internet service providers (ISPs). With a theoretical speed of up to 75 Mbps and a range of several miles, 802.16 broadband is expected to be an alternative to cable modems and digital subscriber line (DSL) in the near future. Promoters of 802.16 elected to form an organization called the WiMAX Forum (http://www.wimaxforum.org) to test and certify products for interoperability and standards compliance.

WiMAX specifies a technology devoted to making broadband wireless communication commercially available to the mass market. WiMAX is an IEEE 802.16 standards-based technology, which supports point-to-multipoint (PMP) broadband wireless access. The IEEE 802.16 specification includes IEEE 802.16-2004 [2] and 802.16e amendment [1] as the physical (PHY) layer specifications. The IEEE 802.16-2004 standard is primarily intended for stationary transmission, while the IEEE 802.16e amendment is primarily intended for both stationary and mobile deployments. Based on the IEEE 802.16-2004 Air Interface Standard, fixed WiMAX has proven to be a cost-effective fixed wireless alternative to cable and DSL services. Furthermore, the IEEE ratified the 802.16e amendment [1] to the 802.16 standard in December 2005 to add the features and attributes necessary for supporting mobility. The WiMAX Forum is now defining system performance and certification profiles based on the IEEE 802.16e Mobile Amendment.

WiMAX is a WirelessMAN technology that connects IEEE 802.11 (Wi-Fi) hotspots to the Internet and provides a wireless extension to cable and DSL for last-mile broadband access. However, this technology faces a lot of challenges in terms of coverage, data rate, and mobility. As for coverage, WiMAX is designed to provide up to 50 km of linear service area and allow users connectivity to a base station without a direct line of sight (LOS). For data rate, WiMAX should provide enough bandwidth to simultaneously support more than 60 businesses with T1-type connectivity and well over a thousand homes at the 1 Mbit/s DSL-level connectivity. For mobility, WiMAX needs to support a system of combined fixed and mobile broadband wireless access, with subscriber stations moving at vehicular speeds. To meet these new requirements, WiMAX adopts MIMO and other new technologies. In this chapter we will focus on MIMO and its related technologies, including orthogonal frequency division multiplexing (OFDM), space-time coding, and diversity.

The MIMO technology is a natural extension of developments in antenna array communication [3,4]. Sometimes referred to as "volume-to-volume" wireless links, MIMO systems are important because they have the potential to play a significant role in resolving traffic capacity bottlenecks in future wireless networks. Communication theory suggests that they can provide a potentially very high capacity that, in many cases, will grow approximately linearly with the number of antennas. Recently MIMO systems have attracted more and more attention because of their implementation in wireless communication systems, especially in WirelessMANs, and because of the proposal of a number of different MIMO system structures by industrial organizations in the Third-Generation Partnership Project (3GPP) standardizations.

MIMO systems can be defined as systems that contain multiple transmitter antennas and multiple receiver antennas. As an example, we might consider an arbitrary wireless communication system in which the links on both the transmitting and receiving ends are equipped with multiple antenna elements. The idea behind MIMO is that the signals on the transmit (TX) antennas at one end and the receive (RX) antennas at the other end are "combined" in such a way as to improve the quality (bit error rate [BER]) or the data rate (bits/sec) of the communication for each MIMO user. This has significant advantages in terms of both the network's quality of service and the operator's revenues.

When we analyze the capacity of MIMO, it is important to note that OFDM is particularly well suited to MIMO technology. OFDM is a multiplexing technique that subdivides bandwidth into multiple frequency subcarriers [5]. In an OFDM system, the input data stream is divided into several parallel substreams of reduced data rate (thus increased symbol duration) and each substream is modulated and transmitted on a separate orthogonal subcarrier. Because the narrowband subcarriers in the OFDM signals experience flat fading, MIMO reception does not require complex channel equalization schemes. When we discuss the capacity of the multiuser MIMO-OFDM system in Section 4.2, the basic ideas of OFDM will be introduced.

Space-time signal processing [6] is one of the core ideas in MIMO. This involves complementing the use of time, the natural dimension of digital communication, with the use of the spatial dimension inherent in the use of multiple spatially distributed antennas. Space-time codes (STCs) [7] are used for coding signals in both the temporal and spatial domains. MIMO systems can also be viewed as an extension of the so-called smart antennas, a popular technology using antenna arrays for improving wireless transmission dating back several decades.

Transmission diversity is one of the most important techniques to improve the transmission reliability in MIMO systems [8]. A signal can be transmitted redundantly over time, frequency, and space. Consequently the diversity techniques can take full advantage of the redundancy on the transmitting and receiving terminals, which can sample the transmitted signal independently and fuse the received samples to recover the original signal [9]. In MIMO systems, the most important diversity technique is space diversity.

This chapter is organized as follows. In Section 4.2 we analyze the capacity of MIMO systems to understand how much MIMO can improve the capacity when used in WiMAX. We begin with a single-user MIMO system model. We highlight the capacity analysis under two different physical specifications in WiMAX. In Section 4.3 we discuss various kinds of space-time block codes (STCs) used in MIMO and how they are applied in WiMAX. In Section 4.4 we discuss the diversity technique to show how MIMO can improve the reliability in WiMAX. In the last section we conclude the chapter with a brief summary.

4.2 Multiple-Input Multiple-Output (MIMO) Wireless Communication

In this section we begin with a single-user MIMO system model. Based on this model, we generalize the discussion on capacity to cases that encompass transmitters having some prior knowledge of a channel [10]. Then we consider a single MIMO user communicating over a fading channel with additive white Gaussian noise (AWGN). Finally we discuss the capacity of MIMO systems in WiMAX by using a multiuser model.

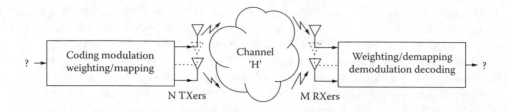

Figure 4.1   Diagram of a wireless MIMO communication system.

4.2.1   MIMO System Model

A single-user MIMO system consists of n_T transmitter antennas (Tx) and n_R receiver antennas (Rx). During each symbol time slot (STS), the transmitted signals are presented as an $n_T \times 1$ column vector x, whose entry x_i, $i = 1, \ldots, n_T$, is the transmitted signal at the ith TX antenna during the considered STS. Figure 4.1 is a diagram of a wireless transmission system. The transmitter and receiver are equipped with multiple antenna elements. Coding, modulation, and mapping of the signals onto the antennas may be realized jointly or separately.

We consider here an additive Gaussian MIMO channel for which the optimal distribution of the transmitted signals in x is also Gaussian, that is, the transmitted signals x_i, for $i = 1, \ldots, n_T$, are zero-mean, identically independently distributed (i.i.d.) complex random variables. The covariance matrix of x is $R_{xx} = E\{xx^H\}$, where $E\{.\}$ denotes the expectation, and $(.)^H$ denotes the Hermitian transposition operation (i.e., the transpose-conjugate operation). The total power of transmitted signals (during each STS) is constrained to P, regardless of the number of transmitter antennas n_T. This implies that $P = tr(R_{xx})$, where $tr(.)$ denotes the trace operation on the argument matrix.

In the following sections we assume that channel coefficients (or transmission coefficients) are perfectly known at the receiver, but may or may not be known at the transmitter.

When channel coefficients are unknown at the transmitter (but known at the receiver), we assume that the transmitted power at each TX antenna is the same and equal to $P_{tj} = \frac{P}{n_T}$, for $j = 1, \ldots, n_T$. When the channel coefficients are known at the transmitter, the transmitted power is unequally assigned to the TX antennas following the water-filling rule [11]. The scenario where channel coefficients are unknown at both the transmitter and receiver is mentioned in Marzetta and Hochwald [12]. Most of the results in this section can be found in Tran et al. [13].

The channel is represented by an $n_R \times n_T$ complex matrix H, whose elements h_{ij} are the channel coefficients between the jth TX antenna ($j = 1, \ldots, n_T$) and the ith RX antenna $i = 1, \ldots, n_R$. Channel coefficients h_{ij} are assumed to be zero-mean i.i.d. complex Gaussian random variables with a distribution $CN(0, 1)$. Noise at the receiver is represented by an $n_R \times 1$ column vector n whose elements are zero-mean, i.i.d. complex Gaussian random variables with identical variances σ^2.

Let r denote the column vector of signals received at RX antennas during each STS. The transmission model is represented as follows:

$$r = Hx + n.$$

If we assume that the average total power P_r received by each RX antenna (regardless of noise) is equal to the average total transmitted power P from n_T TX antennas, the signal-to-noise ratio (SNR) at each RX antenna is given by the following equation:

$$\rho = \frac{P_r}{\sigma^2} = \frac{P}{\sigma^2}.$$

To guarantee the assumption of a channel with fixed channel coefficients and the equal transmitted power per TX antenna P/n_T (i.e., channel coefficients are known at the receiver, but unknown at the transmitter), we have the following constraint:

$$\sum_{j=1}^{n_T} |h_{ij}|^2 = n_T \qquad (4.1)$$

for $i = 1, \ldots, n_R$.

The system capacity, $Cbits/s$ is defined as the maximum possible transmission rate such that the error probability is arbitrarily small. In this chapter we also consider the normalized capacity, $C/Wbits/s/Hz$, which is the system capacity C normalized to the channel bandwidth W.

4.2.2 Capacity of MIMO

In this section we consider the additive Gaussian noise channels with fixed channel coefficients. We derive the most general formula to calculate the channel capacity for both cases where channel coefficients are known and unknown at the transmitters.

The general formula for calculating the channel capacity in the case where channel coefficients are either known or unknown at the transmitter is given by the Shannon capacity:

$$C = W \sum_{i=1}^{r} \log_2 \left(1 + \frac{P_{ri}}{\sigma^2}\right), \qquad (4.2)$$

where W is the bandwidth of each subchannel, r is the rank of the channel coefficient matrix H (r is equal to the number of nonzero eigenvalues of $H^H H$), and P_{ri} is the received power at each RX antenna from the ith subchannel, for $i = 1, \ldots, r$, during the symbol time slot under consideration. The rank r is less than or equal to $m = \min(n_T, n_R)$.

Then we calculate the channel capacity where there are unknown channel coefficients at the transmitter. Let Q be the Wishart matrix, defined as

$$f(n) = \begin{cases} HH^H & \text{if } n_R < n_T \\ H^H H & \text{if } n_R \geq n_T. \end{cases}$$

From Equation (4.2), it has been proved in Vucetic and Yuan [11] that the channel capacity for such a scenario is

$$C = W \log_2 \left[\det \left(I_r + \frac{\rho}{n_T} Q\right)\right], \qquad (4.3)$$

where det(.) denotes the determinant of the argument matrix.

The channel capacity can be increased if the channel coefficients are known at the transmitter. In this case the transmitted power is assigned unequally to the TX antennas, according to the "water-filling" rule (i.e., a larger power is assigned to a better subchannel and vice versa) (see Appendix 1.1 in Vucetic and Yuan [11]). The power assigned to the ith subchannel is

$$P_{ti} = \left(\mu - \frac{\sigma^2}{\lambda_i} \right)^+ \quad i = 1, \ldots, r,$$

where $(a)^+ = \max(a, 0)$, λ_i is the nonzero eigenvalues of the matrix $H^H H$ (also HH^H), and μ is determined to satisfy the power constraint

$$\sum_{i=1}^{r} P_{ti} = P. \tag{4.4}$$

For the ith subchannel, the received power P_{ri} at the receiver antenna is calculated as (see Equation (1.20) in Vucetic and Yuan [11]):

$$P_{ri} = \lambda_i P_{ti} = (\lambda_i \mu - \sigma^2)^+.$$

Thus, the channel capacity is

$$C = W \sum_{i=1}^{r} \log_2 \left[1 + \frac{(\lambda_i \mu - \sigma^2)^+}{\sigma^2} \right]. \tag{4.5}$$

4.2.3 Capacity of MIMO in WiMAX

As mentioned before, the WiMAX technology is based on the IEEE 802.16 specification, which includes IEEE 802.16-2004 [2] and 802.16e amendment as the physical (PHY) layer specifications. The IEEE 802.16-2004 standard is primarily intended for stationary transmission, while the 802.16e amendment is primarily intended for both stationary and mobile deployments. In this section we will examine the PHY layer in WiMAX in detail. We discuss two different PHY layer specifications [1].

The 10 to 66 GHz bands provide a physical environment where, due to the short wavelength, LOS is required and multipath is negligible. In the 10 to 66 GHz band, channel bandwidths of 25 or 28 MHz are typical. With raw data rates in excess of 120 Mb/s, this environment is well suited to PMP access serving applications from small office/home (SOHO) through medium to large office applications.

With the condition that LOS propagation is needed, single-carrier modulation is easily selected; the air interface is designated as "WirelessMAN-SC." However, many fundamental design challenges remain. Because of the PMP architecture, the base station basically transmits a time division multiplex (TDM) signal, with individual subscriber stations allocated time slots serially. Access in the uplink direction is by time division multiple access (TDMA). Following extensive discussions regarding duplexing, a burst design was selected that allows simultaneously both time division duplex (TDD), where the uplink and downlink share a channel but do not transmit simultaneously, and frequency division duplex (FDD), in which the uplink and downlink operate on separate channels, sometimes simultaneously. This burst design allows both TDD and FDD to be handled in a similar fashion. Support for half-duplex FDD subscriber stations, which is less expensive because it does not require simultaneous transmission and receiving, was added at the expense of some slight complexity. Both TDD and FDD alternatives support adaptive burst profiles in which modulation and coding options may be dynamically assigned on a burst-by-burst basis.

In the above scenario, the channel model can be modeled as the flat Rayleigh fading channel. We assume that the channel coefficients are zero-mean i.i.d. complex Gaussian random variables with variance of 1/2 per dimension (real and imaginary). Hence each channel coefficient has a Rayleigh distributed magnitude and uniformly distributed phase. The expected value of the squared magnitude equals one (i.e., $E\{|h_{i,j}^2| = 1\}$). In all the following sections, channel coefficients are assumed to be known at the receiver, but unknown at the transmitter. Thus the transmitted power per TX antenna is assumed to be identical and equals $P_{tj} = \frac{P}{n_T}$, for $j = 1, \ldots, n_T$.

If the channel coefficient matrix H is random and its entries change randomly during every STS, then the channel is referred to as the fast flat Rayleigh fading channel. The capacity of MIMO systems in fast and block Rayleigh fading channels is calculated as follows (see Equation (1.56) in Vucetic and Yuan [11] or Theorem 1 in Ganesan and Stoica [14]):

$$C = E\left\{ W\log_2\left[\det\left(I_r + \frac{P}{n_T\sigma^2}\right)Q\right]\right\}, \tag{4.6}$$

where r is the rank of matrix H and matrix Q is the Wishart matrix defined above.

If H is random and its entries change randomly with each block containing a fixed number of STSs, then the channel is referred to as the block flat Rayleigh fading channel. If H is random, but is selected at the beginning of transmission, and its entries keep constant during the whole transmission, the channel is referred to as the slow flat Rayleigh fading channel. These results were originally derived by Foschini and Gans [10]. Consider a MIMO system where the channel coefficient matrix H is chosen randomly at the start of transmission and stays constant during the whole transmission. The entries of H follow the Rayleigh distribution. Examples of this scenario include wireless local area networks (WLANs) with high data rates and low fading and IEEE 802.16 for fix broadband access systems in WirelessMANs.

Now we consider the transmit and receive diversity. We first assume that $n = n_T = n_R$ and n is large. As shown by Equation (20) in Foschini and Gaus [10] or by Equation (1.82) in Vucetic and Yuan [11], the lower bound on the capacity is given by

$$\frac{C}{W_n} > \left(1 + \frac{\sigma^2}{P}\right)\log_2\left(1 + \frac{P}{\sigma^2}\right) - \log_2 e + \varepsilon_n, \tag{4.7}$$

where ε_n is a Gaussian random variable with the mean and variance as given below:

$$E\{\varepsilon_n\} = \frac{1}{n}\log_2\left(1 + \frac{P}{\sigma_2}\right)^{-1/2}$$

$$\text{var}\{\varepsilon_n\} = \left(\frac{1}{n\ln 2}\right)^2\left[\ln\left(1 + \frac{P}{\sigma^2}\right) - \frac{\frac{P}{\sigma^2}}{1 + \frac{P}{\sigma^2}}\right].$$

The other bands with frequencies below 11 GHz provide a physical environment where, due to the longer wavelength, LOS is not necessary and multipath may be significant. The ability to support near-LOS and non-LOS (NLOS) scenarios requires additional

PHY functionalities, such as the support of advance power management techniques, interference mitigation/coexistence, and multiple antennas.

The original WiMAX standard (IEEE 802.16) specified WiMAX in the 10 to 66 GHz range. IEEE 802.16a, updated in 2004 to 802.16-2004 (also known as 802.16d), added support for the 2 to 11 GHz range. IEEE 802.16d was updated to 802.16e in 2005. Revision 802.16e uses scalable OFDM as opposed to the nonscalable version used in revision 802.16d. This brings potential benefits in terms of coverage, self-installation, power consumption, frequency reuse, and bandwidth efficiency. Revision 802.16e also adds a capability for full mobility support. The design of the 2 to 11 GHz physical layer is driven by the need for NLOS operation. Because residential applications are expected, rooftops may be too low for a clear sight line to a base station antenna, possibly due to obstruction by trees. Therefore, significant multipath propagation must be expected.

IEEE 802.16-2005 (formerly named, but still best known as 802.16e or Mobile WiMAX) provides an improvement on the modulation schemes stipulated in the original (fixed) WiMAX standard. It allows for fixed wireless and mobile NLOS applications primarily by enhancing the orthogonal frequency division multiple access (OFDMA). Furthermore, outdoor-mounted antennas are expensive due to both hardware and installation costs. The three 2 to 11 GHz air-interface specifications [15] are

- WirelessMAN-SCa: uses a single-carrier modulation format.
- WirelessMAN-OFDM: uses OFDM with a 256-point transform by TDMA. This air interface is mandatory for license-exempt bands.
- WirelessMAN-OFDMA: uses OFDMA with a 2048-point transform. In this system, multiple access is provided by addressing a subset of the multiple carriers to individual receivers.

OFDM has become a popular technique for transmission of signals over wireless channels. It converts a frequency-selective channel into a parallel collection of frequency-flat subchannels, which makes the receiver simpler. The time domain waveforms of the subcarriers are orthogonal, yet the signal spectra corresponding to different subcarriers overlap in frequency. Hence the available bandwidth is used very efficiently. Using adaptive bit loading techniques based on the estimated dynamic properties of the channel, the OFDM transmitter can adapt its signaling to match channel conditions and approach the ideal water-pouring capacity of a frequency-selective channel. The increased symbol duration improves the robustness of OFDM to delay spread. Furthermore, introduction of the cyclic prefix (CP) can completely eliminate intersymbol interference (ISI) as long as the CP duration is longer than the channel delay spread. The CP is typically a repetition of the last samples of data of the block that is appended to the beginning of the data payload.

MIMO is known to boost channel capacity [28]. For high data-rate transmission, the multipath characteristic of the environment causes the MIMO channel to be frequency selective. OFDM can transform such a frequency-selective MIMO channel into a set of parallel frequency-flat MIMO channels and therefore decrease receiver complexity. A combination of the two powerful techniques, MIMO and OFDM, is very attractive and has become one of the most promising broadband wireless access schemes. In the following we will consider the capacity of MIMO-OFDM.

In the scheme proposed in Uthansakul and Bialkowski [16], at the transmitter, the bit streams for each of n_T antennas are coded separately and then mapped to their corresponding symbols. These symbols are then grouped into N_F symbols with a serial-to-parallel (S/P) converter and spread with an N_C size Walsh spreading code, where $N_F > N_C$. Next, the N_F-point inverse fast fourier transform (IFFT) is performed and time domain symbols are parallel-to-serial (P/S) converted and transmitted. At the receiver, n_R antennas receive signals from user k and pass them through a complex channel matrix whose characteristic is described by the i.i.d. complex Gaussian random matrix H_k. In all the following cases we assume that the realization of H_k is known to the receiver perfectly but unknown to the transmitter. L is defined as the maximum number of interfered symbols corresponding with maximum delay spread. Thus the channel matrix of user k, $\tilde{H}_k$, is written as

$$
\tilde{H}_k = \begin{bmatrix}
H_k^1 & 0 & \cdots & 0 \\
\vdots & H_k^1 & \cdots & 0 \\
H_k^L & \vdots & \ddots & 0 \\
0 & H_k^L & \ddots & H_k^1 \\
0 & 0 & \ddots & \vdots \\
0 & 0 & \cdots & H_k^L
\end{bmatrix}.
$$

For a single-user channel in MIMO-OFDM, we can calculate the capacity of the considered system as

$$
C = E\left[\log_2 \det\left(I_m + \frac{\rho}{n_T}\tilde{H}_k''\tilde{H}_k''^+\right)\right],
\tag{4.8}
$$

where $E[.]$ denotes expectation, m is $\min(n_R, n_T)$, and ρ is an average SNR at each RX antenna. H^T denotes the transpose of matrix H, H^+ denotes the conjugate and transpose of matrix H.

For a multiuser channel in MIMO-OFDM, we can calculate the capacity of the considered system as

$$
C = \bigcup\left\{(R_1, \ldots, R_K) : \sum_{i\in S} R_i \le E\left[\log_2 \det\left(I_m + \sum_{i\in S}\frac{\rho}{n_T}\tilde{H}_i''\tilde{H}''^+\right)\right]\right\},
\tag{4.9}
$$

where K is the number of users, $\forall S \subseteq \{1, 2, \ldots, K\}$. We assume that the receiver knows the realization of every user matrix channel. Also, we assume that all the transmitting devices generate equal power and use the same number of antennas.

4.3 Space–Time Block Codes

An STC is a channel coding method used in multiple-antenna wireless communications. The objective of STC is to improve the reliability of high data-rate transmission in wireless communication systems by transmitting multiple and redundant copies of a data stream in the hope that some of them may arrive at the receiver in a better state than others. The data streams are sent on multiple antennas with multiple consecutive time slots. Therefore STCs allow information to be transmitted in space and time.

Most of the earlier works on wireless communications focused on having an antenna array at only one end of the wireless link—usually at the receiver. Some works extended the scope of wireless communication possibilities by showing that substantial capacity gains are enabled when antenna arrays are used at both ends of a link. Examples are D-BLAST [15] and V-BLAST [29]. Later the STC [7] was proposed as an alternative approach that relies on having multiple transmit antennas and only optionally multiple RX antennas. It has been shown that STC achieves significant error rate improvements over single-antenna systems. Its original scheme was based on trellis codes. However, the cost for this scheme is additional processing, which increases exponentially as a function of bandwidth efficiency (bits/s/Hz) and the required diversity order. A simpler approach called block codes was proposed by Alamouti [9], and was later extended to develop STCs [17].

An STC is usually represented by a matrix. Each row represents a time slot and each column represents one antenna's transmissions over time:

$$
\text{time slots} \Bigg\downarrow \begin{bmatrix} s_{11} & s_{12} & \cdots & s_{1n_T} \\ s_{21} & s_{22} & \cdots & s_{2n_T} \\ \vdots & \vdots & \ddots & \vdots \\ s_{T1} & s_{T2} & \cdots & s_{Tn_T} \end{bmatrix}
$$

(transmit antennas →)

Here, s_{ij} is the modulated symbol to be transmitted in time slot i from antenna j. There are T time slots and n_T transmit antennas as well as n_R receive antennas.

The code rate of an STC measures how many symbols per time slot it transmits on average over the course of one block [17]. If a block encodes k symbols, the code rate is

$$
r = \frac{k}{T}.
$$

4.3.1 Different Kinds of STCs

There are different kinds of STCs in terms of different numbers of transmit antennas. In WiMAX, the simplest transmission scheme—the Alamouti scheme—is used on the downlink to provide space transmit diversity.

4.3.1.1 Alamouti's Code

Alamouti invented the simplest of all the STCs [9]. It was designed for a two-transmit-antenna system and has the coding matrix:

$$
C_2 = \begin{bmatrix} s_1 & s_2 \\ -s_2^* & -s_1^* \end{bmatrix},
$$

where ∗ denotes complex conjugate.

It is readily apparent that this is a full-rate code. It takes two time slots to transmit two symbols and the BER of this STC is equivalent to $2nR$-branch maximal ratio combining (MRC). This is a result of the perfect orthogonality between the symbols after receive processing—there are two copies of each symbol transmitted and n_R copies received.

4.3.1.2   Higher Order STCs

Tarokh et al. discovered a set of higher order STCs [17,18]. They also proved that no code for more than two transmit antennas could achieve fullrate. Their codes have been improved upon (both by the original authors and by many others). Nevertheless, they serve as clear examples of why the rate cannot reach one and what other problems must be solved to produce "good" STCs. They also demonstrated the simple, linear decoding scheme that goes with their codes under the perfect channel state information assumption.

Two STCs for three transmit antennas are

$$
C_{3,1/2} = \begin{bmatrix}
s_1 & s_2 & s_3 \\
-s_2 & s_1 & s_4 \\
-s_3 & s_4 & s_1 \\
-s_4 & -s_3 & s_2 \\
s_1^* & s_2^* & s_3^* \\
-s_2^* & s_1^* & s_4^* \\
-s_3^* & s_4^* & s_1^* \\
-s_4^* & -s_3^* & s_2^*
\end{bmatrix}
\quad \text{and} \quad
C_{3,3/4} = \begin{bmatrix}
s_1 & s_2 & \frac{s_3}{\sqrt{2}} \\
-s_2^* & s_1^* & \frac{s_3}{\sqrt{2}} \\
\frac{s_3^*}{\sqrt{2}} & \frac{s_3^*}{\sqrt{2}} & \frac{(-s_1-s_1^*+s_2-s_2^*)}{2} \\
\frac{s_3^*}{\sqrt{2}} & -\frac{s_3^*}{\sqrt{2}} & \frac{(s_2+s_2^*+s_1-s_1^*)}{2}
\end{bmatrix}.
$$

These codes achieve the rates of 1/2 and 3/4, respectively. The two matrices give examples of why codes for more than two antennas must sacrifice rate—it is the only way to achieve orthogonality. One particular problem with $C_{3,3/4}$ is that it has uneven power among the symbols it transmits. This means that the signal does not have a constant envelope and the power that each antenna must transmit has to vary, both of which are undesirable. Modified versions of this code that overcome this problem have since been designed.

Two STCs for four transmit antennas are

$$
C_{4,1/2} = \begin{bmatrix}
s_1 & s_2 & s_3 & s_4 \\
-s_2 & s_1 & s_4 & s_3 \\
-s_3 & s_4 & s_1 & -s_2 \\
-s_4 & -s_3 & s_2 & s_1 \\
s_1^* & s_2^* & s_3^* & s_4^* \\
-s_2^* & s_1^* & s_4^* & s_3^* \\
-s_3^* & s_4^* & s_1^* & -s_2^* \\
-s_4^* & -s_3^* & s_2^* & s_1^*
\end{bmatrix}
\quad \text{and} \quad
C_{4,3/4} = \begin{bmatrix}
s_1 & s_2 & \frac{s_3}{\sqrt{2}} & \frac{s_3}{\sqrt{2}} \\
-s_2^* & s_1^* & \frac{s_3}{\sqrt{2}} & -\frac{s_3}{\sqrt{2}} \\
\frac{s_3^*}{\sqrt{2}} & \frac{s_3^*}{\sqrt{2}} & \frac{(-s_1-s_1^*+s_2-s_2^*)}{2} & \frac{(-s_2-s_2^*+s_1-s_1^*)}{2} \\
\frac{s_3^*}{\sqrt{2}} & -\frac{s_3^*}{\sqrt{2}} & \frac{(s_2+s_2^*+s_1-s_1^*)}{2} & \frac{(s_1+s_1^*+s_2-s_2^*)}{2}
\end{bmatrix}.
$$

These codes achieve the rates of 1/2 and 3/4, respectively, as for their three-antenna counterparts. $C_{4,3/4}$ exhibits the same uneven power problems as $C_{3,3/4}$. An improved version [14] of $C_{4,3/4}$ is

$$
C_{4,3/4} = \begin{bmatrix}
s_1 & s_2 & s_3 & 0 \\
-s_2^* & s_1^* & 0 & s_3 \\
-s_3^* & 0 & s_1^* & -s_2 \\
0 & -s_3^* & s_2^* & s_1
\end{bmatrix}.
$$

which has equal power from all antennas in all time slots.

4.3.2 Orthogonal STC

STCs, as originally introduced, are orthogonal, meaning that the STC is designed in such a way that the vectors representing any pair of columns taken from the coding matrix are orthogonal. The result of this is simple, linear, optimal decoding at the receiver. Its most serious disadvantage is that all but one of the codes that satisfy this criterion must sacrifice some proportion of their data rates.

There are also "quasi-orthogonal STCs" that allow some intersymbol interference but can achieve a higher data rate, and even a better error rate performance, in harsh conditions. Quasi-orthogonal STCs exhibit partial orthogonality and provide only part of the diversity gain. An example given by Jafarkhani [19] is

$$C_{4,1} = \begin{bmatrix} s_1 & s_2 & s_3 & s_4 \\ -s_2^* & s_1^* & -s_4^* & s_3^* \\ -s_3^* & -s_4^* & s_1^* & s_2^* \\ s_4 & -s_3 & -s_2 & s_1 \end{bmatrix}.$$

The orthogonality criterion only holds for columns 1 and 2, 1 and 3, 2 and 4, and 3 and 4. Crucially, however, the code is full rate and still only requires linear processing at the receiver, although decoding is slightly more complex than for orthogonal STCs. Results show that this quasi-STC outperforms (in a BER sense) the fully orthogonal four-antenna STC over a good range of SNRs. At high SNRs, though (above about 22 dB in this particular case), the increased diversity offered by orthogonal STCs yields a better BER. Beyond this point, the relative merits of the schemes have to be considered in terms of useful data throughput. More quasi-STCs have also been developed from the basic example shown above.

4.3.3 Space-Time Codes for WiMAX

In IEEE 802.16-2004 OFDM-256, the Alamouti code is applied to a specific subcarrier index k. For instance, suppose that in the uncoded system $S_1[k]$ and $S_2[k]$ are sent in the first and second OFDM symbol transmissions. The Alamouti encoded symbols send $S_1[k]$ and $S_2[k]$ off the first and second antennas in the first transmission and $-S_2^*[k]$ and $S_1^*[k]$ off the first and second antennas in the next transmission.

There are a number of features of IEEE 802.16-2004 OFDM-256 Alamouti transmission that are of interest. The first is that the preamble for Alamouti transmission is transmitted from both antennas with the even subcarriers used for antenna 1 and the odd subcarriers used for antenna 2. This means that each set of data needs to be appropriately smoothed. The second feature is that the pilots have certain degenerate situations: for the first Alamouti transmitted symbol, the pilots destructively add, and for the second Alamouti transmitted symbol, the pilots constructively add. Hence the pilots are not always useful. The pilot symbols must be processed properly.

Figure 4.2 shows the detailed flow of an Alamouti implementation [30]. This implementation has two parts. The first calculates the parameters that are necessary for data demodulation, such as channel estimates. The second part is the actual data demodulation and tracking. It has been shown that under various conditions, the error rates can be greatly reduced when the Alamouti code is used.

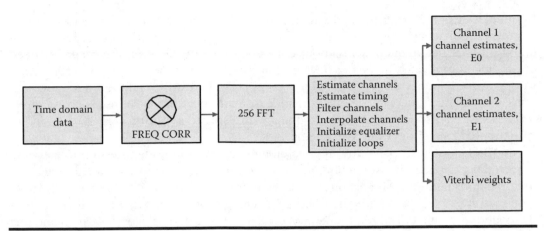

Figure 4.2 Alamouti implementation in WiMAX.

4.4 Transmission Diversity Techniques

Multipath fading is a significant problem in wireless communications [8]. In a fading channel, signals suffer fading (i.e., the signal strength is significantly decreased), which leads to the failure of receiving signals. One of the important solutions to combat fading is diversity [9], which takes advantage of the signal redundancy over time, frequency, or space on the transmitting and receiving terminals. In general, the effectiveness of any diversity technique depends on the precondition that the receiver can provide independent samplings of the transmitted signal. Under this precondition we can assume that two or more relevant parts of a signal will not deeply fade at the same time. The diversity techniques need to optimally fuse the received diversified samples in order to greatly improve the quality of the obtained signals.

4.4.1 Classification of Transmission Diversity Techniques

According to the dimensions of the signal domain providing the redundancy, diversity techniques can be divided into time diversity, frequency diversity, and space diversity. Time diversity is a diversity technique where identical signals are transmitted during different time slots. Because the channel must provide sufficient variations in time, the time slots can be uncorrelated (i.e., the temporal separation between those slots is greater than the coherence time of the wireless channel) [20]. Thus the interleaving symbol duration is independent of the previous symbol and then the completely new replica of the original signal can be obtained. However, this technique will incur considerable redundancy in the time domain and cause a negative consequence of loss in bandwidth efficiency. The loss in bandwidth is due to guaranteeing that the time duration is larger than the coherence time between the time slots. In practice, interleavers and error control coding, such as forward error correction (FEC) codes, are applied to provide time diversity for the receiver. In addition, a rake receiver in code division multiple access (CDMA) systems is an example of a modern implementation of time diversity [8].

Frequency diversity is a technique that uses several carriers with different frequencies to transmit the same signals. The frequency separation between these carrier frequencies is on an order of several times the coherence bandwidth of the channel to achieve uncorrelated carriers, which do not experience the same fading. Similar to time diversity, frequency diversity using the redundancy in the frequency domain will lead to a loss in spectral efficiency. The loss is due to guaranteeing that enough spectral bands exist among the carrier frequencies without coherence [21]. In addition, the receivers need to employ complicated receiving devices that can work with a number of frequencies. In practice, frequency diversity is often used in LOS microwave channels. Example systems employing frequency diversity include spread spectrum systems, such as direct sequence spread spectrum (DS-SS), frequency hop spread spectrum (FH-SS), and multi-carrier spread spectrum (MC-SS) systems [8].

In MIMO systems, space diversity techniques are most important to improve system performance. Space diversity techniques, also called antenna diversity techniques, use multiple transmitting and receiving antennas to transmit and/or receive signals [22]. These antennas must be separated by half of the wavelength of the signal, which is defined as the coherent distance. The requirement of spatial distance allows every involved antenna to be regarded as an independent sampling channel [23].

Different from time diversity and frequency diversity, space diversity techniques will not cause a loss in spectral or bandwidth efficiency. However, space diversity needs a larger space to install multiple interference-free antennas at the transmitters and receivers compared with the time and frequency techniques.

In practical wireless communication systems, space diversity techniques are often combined with other diversity techniques to achieve multidimensional diversity. For instance, in MIMO systems [22], a combination of multiple antennas at the base station (space diversity) and time coding (time diversity) is utilized to provide two-dimensional diversity for receivers (mobile users) [8].

Space diversity techniques can be further classified according to different criteria. In terms of how to combine the redundant signals at the receivers, space diversity techniques are classified as MRC [9], selection combining [24], scanning combining [25], and equal-gain combining techniques [25]. Depending on whether a technique is applied to the transmitter or the receiver, it can be classified as using either receiver diversity or transmitter diversity [6]. In addition, there are two other techniques—polarization diversity [26] and angle diversity [27]. In the next subsection we will describe these techniques.

4.4.2 Combination Techniques for Space Diversity

There are four important space diversity combination techniques: MRC, selection combining, scanning combining, and equal-gain combining. All of these are designed to achieve a high diversity gain, which means that the combination replicas of signals can effectively improve the signal transmission in channels.

In the WiMAX standards [2] and the relevant practical systems, MRC is recommended and expected to be widely used. MRC techniques can use samples of all incoming signals by assigning different weights to different incoming signal branches and then summing them together to obtain the final incoming signal. Generally the weighting factor of a signal branch is proportional to the strength ratio of the incoming signal (including both signal and noise) to the noise. In addition, before being summed the signals must carry out phase alignment to provide coherence voltage addition. The average SNR of the output signal is simply the sum of individual SNRs of all branches. This technique can provide a satisfying output signal with the expected SNR, even when there is no acceptable incoming signal branch. However, the cost of MRC devices is higher than any other combining technique because there is an independent radio frequency channel for every signal branch. MRC has been widely accepted by MIMO systems to improve transmission bandwidth.

Selection combining techniques are the simplest spatial diversity combining methods. They requires only an SNR monitoring action and an antenna switch at the receiver. The receiver needs M antennas and demodulators to provide M branches of signal samplings. The receiver selects the incoming signal sampling with the highest SNR to demodulate at every sampling instant. In practice, because the instantaneous SNR is difficult to measure, as a substitution the signals with the highest strength (including the strength of both the signal and the noise are measured. Obviously the SNR of the selected signals is larger than the average SNR of all signals provided by all pairs of antennas and demodulators, since the signal with the highest SNR will be selected in every sampling instant. However, this technique does not use all the diversified signals simultaneously to provide the best received signal. Thus, it is not the optimal method for combining signals.

Scanning combining techniques employ multiple receiving antennas, but only one antenna switch and one demodulator. A receiver scans all antennas following a certain order to obtain the SNR of every branch and selects a specific branch with a SNR above a predetermined SNR threshold. The receiver then uses the signal of this branch as the output signal. Once the SNR of the selected signal becomes lower than the predetermined threshold, the receiver starts the searching process again and selects a new branch as the output signal. The receiver using this technique need not continuously monitor the SNRs of all branches at every sampling instant. In addition, the receiver needs just one set of devices for demodulation and switching. Thus the cost will be less. However, this method does not always select the signal with the highest SNR. Similar to selection combining, scanning combining does not use all the branches to obtain the output signal. Therefore the SNR of the output signal is less than with MRC.

Equal gain combining techniques assign the same weight of one to every input signal branch. However, since the technique considers all the input branches, its performance is a bit lower than the MRC method.

4.4.3 *Receiver Diversity and Transmitter Diversity*

Depending on the terminals where the diversity techniques are employed, transmission diversity techniques can be categorized into receiver diversity and transmitter diversity. When using receiver diversity, it is assumed that the receivers have complete knowledge of the channels. According to the fading properties, a receiver can benefit from the combining gains of the SNR from channel coding and diversity. The most popular method used in receiver diversity is MRC. In general, as for terminals in cellular networks, the receiver diversity is thought to be highly costly and impractical because it is impossible to install several independent antennas in a cell phone. However, with the development of personal communication systems, the dual antennas are expected to be widely used in personal wireless communication devices (e.g., personal digital assistants, [PDAs], phones, and laptop computers).

When using transmitter diversity, transmitters are assumed to have complete knowledge of channels. According to the fading properties of a channel, transmitters can be elaborately controlled to provide signal redundancy, which can then be exploited by receivers to improve the efficiency of signal collection. With the advent of MIMO systems using time-space codings like Alamouti's scheme, the transmitters can combine channel coding techniques with diversity techniques and effectively use the diversity techniques even without any knowledge of channels.

4.4.4 Polarization Diversity and Angle Diversity

Space diversity includes two other types of diversity techniques: polarization diversity and angle diversity. When using polarization diversity, a transmitter employs horizontal and vertical antennas to transmit horizontal and vertical signals and the receiver also needs horizontal and vertical antennas to receive the polarized signals. There is no correlation between different polarized signals and thus there is no need to consider the coherent distance for separating the same signals.

Angle diversity is based on the widely accepted fact that signals can be highly spatially scattered when the frequencies of their carriers are higher than 10 GHz. Using such a carrier, a transmitter can transmit signals via two highly directional antennas that are facing in totally different directions. The receiver can use two antennas facing in the corresponding directions and then collect samples of the same signal, which will not be coherent with each other.

4.5 Summary

MIMO is an important technology for WiMAX-based WirelessMAN. In this chapter we briefly introduced the WiMAX and IEEE 802.16 standard. For MIMO, we focused on system capacity, which is approximately proportional to the number of antennas. We generalized the discussion on capacity to cases that encompass transmitters having some prior knowledge of a channel. Our discussion covered a fading channel with additive white Gaussian noise and a low flat Rayleigh fading channel. When discussing MIMO in WiMAX, we analyzed the capacity of a multiuser system model. Space-time coding is one of the schemes for the transmission of signals via MIMO systems. We introduced Alamouti transmission because the Alamouti code is applied in WiMAX for IEEE 802.16-2004 OFDM-256. We also described various diversity schemes for enhancing the performance of wireless channels. Space diversity is an effective scheme for combating multipath fading.

References

1. IEEE, IEEE Standard for Local and Metropolitan Area Networks—Part 16: Air Interface for Fixed and Mobile Broadband Wireless Access Systems Amendment for Physical and Medium Access Control Layers for Combined Fixed and Mobile Operation in Licensed Bands, Standard 802.16e-2005, IEEE, Washington, DC, 2005.
2. IEEE, IEEE Standard for Local and Metropolitan Area Networks—Part 16: Air Interface for Fixed Broadband Wireless Access Systems, Standard 802.16-2004, IEEE, New York, 2004.
3. W.C. Jakes, *Microwave Mobile Communications*, Wiley, New York, 1974.
4. V. Weerackody, Diversity for Direct-Sequence Spread Spectrum Using Multiple Transmit Antennas, *IEEE International Communications Conference* 3, p. 1775, IEEE, Washington, DC, 1993.
5. R. Van Nee and R. Prasad, *OFDM for Wireless Multimedia Communications*, Artech House, Norwood, MA, 2000.
6. M. Jankiranman, *Space-Time Codes and MIMO Systems*, Artech House, Norwood, MA, 2007.
7. V. Tarokh, N. Seshadri, and A.R. Calderbank, Space-time codes for high data rate wireless communication: performance analysis and code construction, *IEEE Trans. Inform. Theory*, 44(2), 744, 1998.

8. T.S. Rappaport, *Wireless Communications: Principles and Practice*, Prentice Hall, Upper Saddle River, NJ, 2002.

9. S.M. Alamouti, A simple transmit diversity technique for wireless communications, *IEEE J. Select. Areas Commun.*, 16(8), 1451, 1998.

10. G.J. Foschini and M.J. Gans, On limits of wireless communications in a fading environment when using multiple antennas, *Wireless Pers. Commun.*, 6, 311, 1998.

11. B. Vucetic and J. Yuan, *Space-Time Coding*, Wiley, Hoboken, NJ, 2003.

12. T.L. Marzetta and B.M. Hochwald, Capacity of a mobile multiple-antenna communication link in Rayleigh flat fading, *IEEE Trans. Inform. Theory*, 45(1), 139, 1999.

13. L. Tran, T.A. Wysochi, A. Metrins, and J. Seberry, *Complex Orthogonal Space-Time Processing in Wireless Communications*, Springer, New York, 2006.

14. G. Ganesan and P. Stoica, Space-time block codes: a maximum SNR approach, *IEEE Trans. Inform. Theory*, 47(4), 1650, 2001.

15. G.J. Foschini, Layered space-time architecture for wireless communications in a fading environment when using multi-element antennas, *Bell Labs Tech. J.*, 1, 41, 1996.

16. P. Uthansakul and M.E. Bialkowksi, Multipath signal effect on the capacity of MIMO, MIMO-OFDM and spread MIMO-OFDM, *15th International Conference on Microwaves, Radar, and Wireless Communications*, vol. 3, p. 989, IEEE, Washington, DC, 2004.

17. V. Tarokh, H. Jafarkhani, and A.R. Calderbank, Space-time block codes from orthogonal designs, *IEEE Trans. Inform. Theory*, 45(5), 1456, 1999.

18. V. Tarokh, H. Jafarkhani, and A. Robert Calderbank, Space-time block coding for wireless communications: performance results, *IEEE J. Select. Areas Communi.* 17(3), 451, 1999.

19. H. Jafarkhani, A quasi-orthogonal space-time block code, *IEEE Trans. Commun.*, 49(1), 1, 2001.

20. A.M.D. Turkmani and A.F. de Toledo, Time diversity for digital mobile radio, *IEEE 40th Vehicular Technology Conference*, p. 576, IEEE, Washington, DC, 1990.

21. J.G. Proakis, *Digital Communications*, 4th ed., McGraw-Hill, Boston, 2000.

22. D. Gesbert, M. Shafi, D.-S. Shiu, P. J. Smith, and A. Naguib, From theory to practice: an overview of MIMO space-time coded wireless systems, *IEEE J. Select. Areas Commun.*, 21(3), 281, 2003.

23. J. Salz and J.H. Winters, Effect of fading correlation on adaptive arrays in digital mobile radio, *IEEE Trans. Vehic. Technol.*, 43(4), 1049, 1994.

24. L. Yue, Analysis of generalized selection combining techniques, *IEEE 51st Vehicular Technology Conference*, vol. 2, p. 1191, IEEE, Washington, DC, 2000.

25. D.G. Brennan, Linear diversity combining techniques, *Proc. IEEE*, 91(2), 331, 2003.

26. B. Lindmark and M. Nilsson, Polarization diversity gain and base station antenna characteristics, vol. 1, p. 590, *IEEE 49th Vehicular Technology Conference*, IEEE, Washington, DC, 1999.

27. J.B. Carruther and J.M. Kahn, Angle diversity for nondirected wireless infrared communication, *IEEE Trans. Commun.*, 48(6), 960, 2000.

28. D. Bliss, K. Forsythe, and A. Chan, MIMO Wireless Communication, Lincoln Laboratory Journal, vol. 1 and Wireless Access Systems.

29. G.D. Golden, G.J. Foschini, R.A. Valenzuela, and P.W. Wolniansky, V-BLAST: an Architecture for Realizing very High Data Rates over the Rich-scattering Wireless Channel. 1998 URSI International Symposium on signals, Systems, and Electronics, 1998. ISSSE 98. (V-BLAST).

30. A. Salvekar, S. Sandhu, Q. Li, M. Vuong, and X. Qian, Multiple-Antenna Technology in WiMAX Systems, Intel Technology Journal, 2004.

Chapter 5

MIMO Spectral Efficiency of the Mobile WiMAX Downlink

*Alexei Davydov, Roshni Srinivasan, Shailender Timiri,
Yang-seok Choi, Hujun Yin, Atul Salvekar,
and Apostolos Papathanassiou*

Contents

The growing demand for ubiquitous and mobile access to Internet services with quality comparable to that of the fixed Internet poses stringent requirements on next-generation mobile communications systems. Consequently, evolving mobile wireless networks require a number of radical enhancements to cost-effectively meet these requirements. Foremost among these enhancements are the use of orthogonal frequency division multiple access (OFDMA) as the air interface, support for large channel bandwidths, and the application of advanced antenna technologies. Mobile WiMAX, the mobile broadband air interface based on the IEEE 802.16 standards and promoted by the WiMAX Forum, is based on OFDMA, supports scalable channel bandwidths of 10 MHz and higher, and incorporates multiple-input multiple-output (MIMO) as one of its basic features. In this chapter we analyze and evaluate by extensive simulations the performance of Mobile WiMAX in the MIMO downlink. Based on the performance evaluation of Mobile WiMAX at the link and system level, it is shown that the use of OFDMA, large channel bandwidths, and MIMO techniques leads to a cellular communications system suitable for mobile broadband services.

5.1   Introduction

The promise of mobile Internet is the provision of user experience similar to that offered by the fixed Internet with ubiquitous access and in the presence of mobility. Access to fixed Internet services through cable and digital subscriber lines (DSL) [1] is also increasing. This translates into a number of technology requirements for the wide area wireless communications system designed to offer ubiquitous broadband services to mobile users [2].

Assuming a typical cellular wide area mobile network [3], the aforementioned technology requirements are usually associated with a multitude of interrelated performance metrics dealing with coverage, capacity, and service quality. These include metrics such as peak user data rates, end-to-end packet delay, and delay jitter, especially for real-time applications such as gaming and voice over Internet protocol (VoIP), to name a few. However, the most essential characteristic of a mobile broadband wireless access system is the provision of high aggregate sector throughput. The aggregate sector throughput is defined as the total number of successfully transmitted information bits per second that a sector of the cellular network can serve for a given system deployment scenario in a fully loaded network characterized by the number of simultaneously active users, site-to-site distance, outage, and fairness criterion. Closely related to the aggregate sector throughput is the sector/site spectral efficiency measured in bits per second per hertz per sector/site, and defined as the sector throughput offered to all users per spectrum block assignment bandwidth for a given system deployment scenario.

The full-buffer data traffic assumption, accompanied by outage and fairness criteria for all active users in a fully loaded network, provides a relatively straightforward manner to evaluate the aggregate throughput and spectral efficiency per sector/site for a wireless network. This model predicts the performance for best-effort data traffic and is widely used to study system performance. It also provides a means for fair comparison of the system-level performance of different air interfaces.

Third-generation (3G) mobile radio systems, based predominantly on code division multiple access (CDMA) [4], are being enhanced with features to support higher data rates, such as high-speed downlink packet access (HSDPA)/high-speed uplink packet access (HSUPA) for wideband code division multiple access (WCDMA) in Third-Generation Partnership Project (3GPP) and 1x evolution-data only (1xEVDO) for CDMA2000 in Third-Generation Partnership Project 2 (3GPP2) [5]. However, to meet the requirements for true mobile broadband services, work is under way to further evolve the 3G systems to alternative technologies that can support larger bandwidths and multiple antenna technologies in a cost-effective manner.

Mobile WiMAX defines the air interface for mobile broadband wireless systems based on IEEE 802.16-2004 [6], the air interface for fixed broadband wireless access systems, and IEEE 802.16e-2005 [7], its amendment for physical (PHY) and medium access control (MAC) layers for combined fixed and mobile operation in licensed bands. Mobile WiMAX [8] specifies system profiles based on mandatory and optional features of IEEE 802.16-2004 and IEEE 802.16e-2005 that are subject to certification by the WiMAX Forum. The WiMAX Forum is a nonprofit trade organization comprising hundreds of member companies that include service providers, equipment vendors, and device manufacturers. Its aim is to promote the worldwide adoption of broadband wireless solutions based on the IEEE 802.16 air interface standards with guaranteed interoperability. The Mobile WiMAX system profiles enable the deployment of mobile broadband wireless systems based on a common basic feature set and are designed to ensure interoperability of a complying mobile station (MS) and base station (BS). While all elements of the MS profiles are mandatory in order to enable unrestricted roaming in Mobile WiMAX networks, some elements of the BS profiles are optional so as to provide deployment flexibility for different capacity or coverage optimization configurations.

Mobile WiMAX broadband systems are based on OFDMA [9,10], a multiple access scheme that enables the multiplexing of data streams from multiple users onto the downlink (DL) and the uplink (UL). OFDMA is regarded as the extension of orthogonal frequency division multiplexing (OFDM) to the multiuser case and thus inherits the benefits of OFDM. These benefits include high spectral efficiency, resilience in high time dispersion environments due to multipath propagation such as non-line-of-sight (NLOS) urban environments, and cost-effective implementation in semiconductor devices for both single and multiple antenna system configurations. The Mobile WiMAX system profiles rely on scalable OFDMA (S-OFDMA), the concept introduced in IEEE 802.16e-2005 to support scalable channel bandwidths. The first release covers channel bandwidths of 5, 7, 8.75, and 10 MHz for licensed worldwide spectrum allocations in the 2.3, 2.5, and 3.5 GHz frequency bands. Based on its main system characteristics (i.e., high spectral efficiency, multipath robustness, support of large channel bandwidths, ease of implementation of multiple antenna technologies at both ends of the wireless communications link, and an all Internet protocol [IP] architecture), Mobile WiMAX is designed to fulfill the requirements for a true mobile broadband access system.

Since the wireless Internet imposes more demands on the downlink due to the asymmetric nature of its services, this chapter focuses on and evaluates the link- and system-level performance of the Mobile WiMAX DL. The performance results are based on the MIMO system profiles specified by the WiMAX Forum. This chapter addresses the issues of MIMO PHY abstraction required for simulation at the system level, as well as adaptive MIMO switching (AMS). AMS refers to the adaptive choice between the two supported MIMO modes of Mobile WiMAX, space-time block coding (STC) and spatial multiplexing (SM), depending on the instantaneous conditions of the wireless link. It is one of the key features of Mobile WiMAX for achieving high aggregate sector throughput and spectral efficiency. A number of other critical issues for system-level performance, such as MIMO channel and interference modeling, BS scheduling achieving fairness among the supported users, and hybrid automatic repeat request (HARQ), are also considered in the analysis and simulation.

The chapter is structured as follows. In Section 5.2, an overview of the Mobile WiMAX air interface is given according to the documentation available from the WiMAX Forum [8] and the IEEE 802.16-2004 and IEEE 802.16e-2005 standards [6,7]. The focus is on the system profile parameters relevant to the downlink MIMO simulation results presented in this chapter. Section 5.3 deals with the adaptive MIMO switching concept of Mobile WiMAX. In Section 5.4, after a description of the MIMO channel model employed in the simulations, link-level simulation (LLS) results for both modes of downlink MIMO—STC and SM—are presented and discussed. Further, the PHY abstraction approach adopted for enabling a realistic, yet computationally attractive representation of the PHY performance in the system-level simulations (SLS) is presented. Section 5.5 contains extensive results for the system performance of the MIMO-based Mobile WiMAX downlink. After describing the framework and parameters of the SLS, the behaviors of numerous system performance metrics are investigated, such as user throughput statistics, MCS selection probabilities, and HARQ overhead, under different deployment scenarios. Finally, the achieved downlink aggregate sector/site throughput and spectral efficiency of the MIMO-based Mobile WiMAX downlink are presented and compared with other system alternatives such as single-input single-output (SISO) and single-input multiple-output (SIMO) deployments. Section 5.6 gives a summary of the most important findings and concludes the chapter.

5.2   Overview of Mobile WiMAX

As mentioned in Section 5.1, Mobile WiMAX is the air interface for mobile broadband wireless systems based on the IEEE 802.16-2004 and IEEE 802.16e-2005 standards, which specifies system profiles enabling the deployment of mobile broadband wireless systems based on a common feature set with guaranteed interoperability between the network devices. Mobile WiMAX offers scalability in both radio access technology and network architecture, which provides a great deal of flexibility in network deployment options and service offerings. Some of the salient features supported by the Mobile WiMAX PHY and MAC layers are

■ **Scalability:** Despite the efforts targeting global spectrum harmonization, spectrum resources for wireless broadband worldwide are still quite disparate in their allocations. Therefore Mobile WiMAX is designed to be able to scale to operate in different channel bandwidths to comply with varied requirements worldwide. This also allows diverse economies to realize the multifaceted benefits of the Mobile WiMAX technology for their specific geographic needs, such as providing affordable Internet access in rural settings versus enhancing the capacity of mobile broadband access in urban and suburban areas.

- **High data rates:** MIMO antenna techniques along with flexible subchannelization schemes, larger MAC frames, and advanced coding and modulation schemes enable Mobile WiMAX to support DL peak data rates up to 70 Mbps and UL peak data rates up to 32.67 Mbps in a 10 MHz channel.

- **Quality of service (QoS):** The fundamental premise of the IEEE 802.16 MAC architecture is the provision of QoS. Appropriate schemes for resource utilization and control signaling enable a flexible mechanism for optimal scheduling based on space, frequency, and time resource blocks that can be allocated in a different manner from frame to frame. This flexibility ensures that robustness is maintained along with adaptive modulation and coding schemes (MCS). Further, Mobile WiMAX defines service flows that can map to DiffServ code points or multi-protocol label switching (MPLS) flow labels that allow for end-to-end IP-based QoS.

- **Security:** The features provided for Mobile WiMAX security are the best in their class, with extensible authentication protocol (EAP)-based authentication, advanced encryption standard counter with cipher-block chaining message authentication code (AES-CCM)–based authenticated encryption, and block cipher-based message authentication code (CMAC) and keyed hash message authentication code (HMAC)–based control message protection schemes. Support for a diverse set of user credentials exists including subscriber identity module/universal subscriber identity module (SIM/USIM) cards, smart cards, digital certificates, and username/password schemes based on the relevant EAP methods for the credential type.

- **Mobility:** Mobile WiMAX supports optimized handover schemes with latencies less than 50 ms to ensure real-time applications such as VoIP without service degradation. Flexible key management schemes ensure that security is also maintained during handover.

In the following sections, a brief overview of the physical and MAC layers of Mobile WiMAX is given, where the focus is on those features and techniques used in the simulations for evaluating the link-and system-level performance of the downlink MIMO to be presented in Sections 5.4 and 5.5. For a detailed description of the system profiles of Mobile WiMAX, the reader is directed to the documentation provided by the WiMAX Forum [8] as well as the related IEEE 802.16-2004 and IEEE 802.16e-2005 standards [6,7].

5.2.1 PHY Layer Description

5.2.1.1 Brief Description of OFDMA

OFDMA is a multiple access scheme that enables the coexistence of multiple users whose data streams are modulated according to the OFDM technique [9]. OFDMA subdivides the available system bandwidth into multiple orthogonal frequency subcarriers in the frequency domain and into symbol periods in the time domain. In each symbol period, the serial input data stream of a specific user is divided into parallel data streams and each stream modulates a specific subcarrier. Each user occupies a subset of the available frequency subcarriers in the band, which enables the coexistence of multiple users in the same frequency band. In this way, user-specific data substreams are created that are orthogonal in frequency and time. Each of the parallel streams are associated with a lower data rate than the original user's serial input data stream.

The reduced data rate of the orthogonal user data substreams facilitates an increase in the OFDMA symbol duration time, leading to improved robustness of OFDMA to propagation environments with high multipath channel delay spreads. It also enables the choice of increased duration for the cyclic prefix (CP) (e.g., on the order of 10 μs), which eliminates intersymbol interference (ISI) between successively transmitted OFDMA symbols.

OFDM exploits the frequency diversity of the multipath mobile broadband channel by coding and interleaving the information across the subcarriers prior to transmission. After organizing the time and frequency resources in an OFDMA system into resource blocks for allocation to the individual MSs, the coded and interleaved information bits of a specific MS are modulated onto the subcarriers of its resource blocks. Then OFDM modulation is cost effectively realized by the inverse fast fourier transform (IFFT) that enables the use of a large number of subcarriers—up to 1024 according to the Mobile WiMAX system profiles—to be accommodated within each OFDMA symbol. Prior to transmission, each OFDMA symbol is extended by its CP followed by digital-to-analog (D/A) conversion at the transmitter. At the receiver end, after analog-to-digital (A/D) conversion, the CP is discarded and OFDM demodulation is applied through the fast fourier transform (FFT). Due to the orthogonality of the data substreams modulated onto different subcarriers, channel estimation and data equalization takes place on a per-subcarrier basis, which reduces the digital signal processing complexity of the OFDMA receiver significantly compared to time division multiple access (TDMA) and CDMA-based multiple access schemes. In the DL, the MS then decodes the information bits of its allocation that may span multiple subcarriers over multiple OFDMA symbols. In the UL, the BS decodes the information bits of all users simultaneously active within the allocated OFDMA symbols for UL transmission. In this way, OFDMA offers highly cost effective and flexible DL and UL operation for broadband wireless communications.

5.2.1.2 OFDMA Symbol Structure and Parameters

An OFDMA symbol of the Mobile WiMAX air interface consists of three types of sub-carriers, as shown in Figure 5.1. Data subcarriers are used for data transmission. Pilot subcarriers are used for estimation and synchronization purposes and null subcarriers are used as guard bands for spectrum mask requirements. As is typical of OFDM systems, the direct current (DC) subcarrier is not modulated. Data and pilot subcarriers are grouped into sets of subcarriers called subchannels. The WiMAX OFDMA PHY supports subchannelization in both the DL and UL.

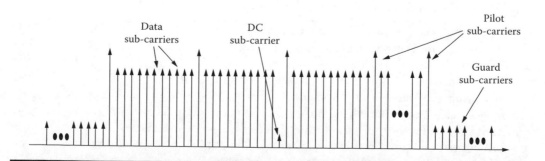

Figure 5.1   OFDMA symbol structure (frequency-domain).

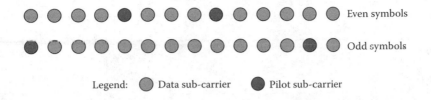

Figure 5.2   DL PUSC subchannel (frequency domain on *x*-axis, time domain on *y*-axis).

In Mobile WiMAX there are two types of subcarrier permutations applied at the BS or MS transmitter: diversity and contiguous permutations. Diversity permutations draw subcarriers in a pseudo-random manner when forming a subchannel, and provide frequency diversity as well as intercell interference averaging. They are expected to perform well in mobile applications where there is a rapid change of both signal and interference powers. In DL partially used subchannelization (PUSC) permutations, which are of interest in this chapter, the total available subcarriers used for transmission (i.e., data and pilot subcarriers) are grouped into clusters, each containing 14 contiguous subcarriers per OFDMA symbol, with pilot and data allocations as shown in Figure 5.2 for even and odd symbols, respectively.

Contiguous permutations, termed adaptive modulation and coding (AMC) in Mobile WiMAX terminology, group a block of contiguous subcarriers to form a subchannel. AMC permutations enable multiuser diversity when choosing those subchannels for transmission to or from a specific user exhibiting a favorable signal-to-interference plus noise ration (SINR). They are well suited for fixed, portable, or low-mobility environments, where the relatively constant behavior of the SINR can enable the choice of appropriate allocations for each user.

The IEEE 802.16e-2005 standard defines the concept of S-OFDMA. S-OFDMA supports a wide range of bandwidths to flexibly address the need for various spectrum allocations and usage model requirements. The scalability is supported by adjusting the FFT size while fixing the subcarrier frequency spacing at 10.9375 kHz. The parameters for the 10 MHz Mobile WiMAX system profile, used for the simulation results presented in Sections 5.4 and 5.5, are shown in Table 5.1.

5.2.1.3   TDD Frame Structure

Although IEEE 802.16e-2005 supports time division duplex (TDD), frequency division duplex (FDD), and half-duplex FDD (HFDD) operation, the initial certification profiles

Table 5.1   S-OFDMA Parameters

Parameters	Values
System channel bandwidth (MHz)	10
Sampling frequency (F_s in MHz)	11.2
FFT size (N_{FFT})	1024
Subcarrier frequency spacing (Δ_f)	$10.9375\ kHz$
Useful symbol time ($T_b = 1/\Delta_f$)	$91.43\ \mu s$
Guard time ($T_g = T_b/8$)	$11.43\ \mu s$
OFDMA symbol duration ($T_s = T_b + T_g$)	$102.86\ \mu s$
Number of OFDMA symbols (in a 5 ms frame)	48

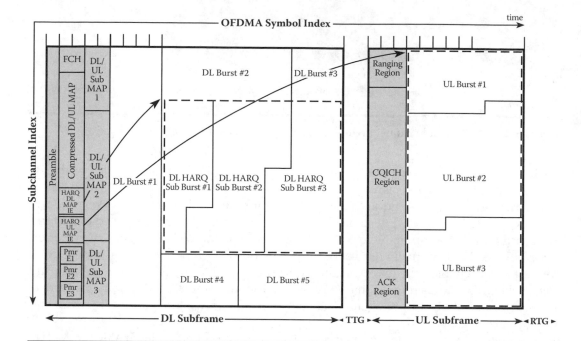

Figure 5.3   Mobile WiMAX OFDMA frame structure.

of Mobile WiMAX only include TDD. TDD is the preferred duplexing mode of Mobile WiMAX for the following reasons:

■ Unlike FDD, which requires a pair of channels, TDD only requires a single channel for both downlink and uplink, providing greater flexibility for adaptation to varied global spectrum allocations.

■ TDD enables adjustment of the downlink/uplink ratio to efficiently support asymmetric downlink/uplink traffic, while with FDD, the downlink and uplink always have fixed and generally equal channel bandwidths.

■ TDD ensures channel reciprocity, leading to better support for link adaptation, smart antennas, and other closed-loop advanced antenna technologies.

■ Transceiver designs for TDD, especially at the radio frequency side, since there is no need for simultaneous UL and DL operation, are less complex and therefore more cost effective.

Figure 5.3 illustrates the OFDM frame structure for a TDD implementation. Each frame is divided into DL and UL subframes separated by transmit/receive and receive/transmit transition gaps (TTG and RTG, respectively) to prevent DL and UL transmission collisions. In a frame, the following control information is used to ensure optimal system operation:

■ **Preamble:** The preamble, which can be used for timing and frequency synchronization, as well as for SINR and channel estimation purposes, is the first OFDM symbol of the frame.

■ **Frame control header (FCH):** The FCH follows the preamble and provides frame configuration information such as the mobile application part (MAP) message length and coding scheme and usable subchannels.

Table 5.2   Supported Coding and Modulation Schemes in the DL

Modulation		QPSK, 16-QAM, 64-QAM
Code rate	CTC	1/2, 2/3, 3/4, 5/6
	Repetition	2×, 4×, 6×

- **DL-MAP and UL-MAP:** The DL-MAP and UL-MAP provide subchannel allocation and other control information for the DL and UL subframes, respectively.
- **UL ranging:** The UL ranging subchannel is allocated to MSs for the purpose of closed-loop time, frequency, and power adjustments, as well as bandwidth requests.
- **UL channel quality information channel (CQICH):** The UL CQICH is allocated so that an MS can feed back channel-state information.
- **UL acknowledge (ACK):** The UL ACK channel is allocated so that the MS feeds back DL HARQ acknowledgments.

5.2.1.4   Other Advanced PHY Layer Features

AMC, HARQ, and fast channel feedback (CQI channel) were introduced with Mobile WiMAX to enhance the coverage and capacity for WiMAX in mobile applications. Support for quadrature phase-shift keying (QPSK), 16-quadrature amplitude modulation (QAM), and 64-QAM are mandatory in the DL in Mobile WiMAX. In the UL, 64-QAM is optional. Both convolutional coding (CC) and convolutional turbo coding (CTC) with variable code rates and repetition coding are also supported. The repetition coding is concatenated with either CC or CTC. The effective code rate is the corresponding CC or CTC rate divided by the repetition factor. Table 5.2 summarizes the coding and modulation schemes supported in the Mobile WiMAX system profiles in the DL.

The combinations of various modulation and coding rates according to Table 5.2 provide a fine resolution of data rates. Table 5.3 shows the achievable DL data rates for

Table 5.3   Mobile WiMAX PHY DL Data Rates with PUSC Permutations

Modulation	Code Rate	Data Rate (Mbps) Non-SM/SM	MCS # Non-SM/SM
QPSK	1/2 CTC, 6x	1.13/—	2/—
	1/2 CTC, 4x	1.69/—	3/—
	1/2 CTC, 2x	3.38/—	4/—
	1/2 CTC, 1x	6.77/13.54	5/13
	3/4 CTC	10.15/20.30	6/14
16-QAM	1/2 CTC	13.54/27.08	7/15
	3/4 CTC	20.30/40.60	8/16
64-QAM	1/2 CTC	20.30/40.60	9/17
	2/3 CTC	27.07/54.14	10/18
	3/4 CTC	30.46/60.92	11/19
	5/6 CTC	33.84/67.68	12/20

a 10 MHz channel bandwidth when partial usage of subchannels (PUSC) permutations are employed. The PHY data rate is defined here as the ratio of the number of used data subcarriers multiplied by number of information bits per symbol achieved by the considered MCS to the OFDMA symbol period. In the case of the 2 × 2 spatial multiplexing (SM) MIMO mode, the achievable data rates are doubled. The BS scheduler determines the appropriate MCS for each burst allocation based on the channel propagation conditions. A CQI is utilized to provide channel state information (CSI) from the MS to the BS. CSI can be fed back by the MS in the CQI channel and can appear in the following forms: physical carrier to interference plus noise ratio (CINR), effective CINR, MIMO mode selection, and frequency selective subchannel selection. The effective CINR reveals information about the MCS that can be supported at the MS under the current channel conditions.

HARQ is also supported by the Mobile WiMAX system profiles. HARQ is enabled using the N channel "stop-and-wait" protocol which provides fast response to packet errors and improves cell edge coverage. Chase combining HARQ is supported to further improve the reliability of a retransmission. A dedicated ACK channel is also provided in the UL for HARQ ACK/NACK signaling. Multichannel HARQ operation is also supported. Multichannel stop-and-wait automatic repeat request (ARQ) with a small number of channels is an efficient, simple protocol that minimizes the memory required for HARQ and stalling. Mobile WiMAX provides signaling to allow fully asynchronous HARQ operation. HARQ combined with the CQI channel and adaptive modulation and coding offers a powerful mechanism for robust link adaptation in mobile environments at vehicular speeds up to 120 km/hr.

5.2.2   MAC Layer Description

The Mobile WiMAX MAC layer provides a standard medium-independent interface to the PHY layer. The MAC protocol is designed to support PMP and mesh network deployments. Upon entering the network, each MS creates one or more connections over which data are transmitted to and from the BS. The MAC layer schedules the usage of the air-link resources and provides QoS differentiation. It performs link adaptation and ARQ and HARQ functions to maintain the target PER (packet error rate) while maximizing link utilization. The MAC also handles network entry for MSs that enter and leave the network, encryption, and standard tasks associated with protocol data unit (PDU) creation. The MAC layer provides a convergence sublayer that supports asynchronous transfer mode (ATM) cell- and packet-based network layers.

5.2.2.1   802.16 MAC Connections

The Mobile WiMAX MAC provides a connection-oriented service to upper layers of the protocol stack. Unidirectional connections are established between the BS and the MS to control transmission ordering and scheduling on the air interface. Each connection may be either a management or a transport connection. Management connections carry only management messages, while transport connections carry data traffic. Each connection is identified by a unique connection identification (CID) number. Every MS, when joining a network, sets up a basic connection, a primary management connection, and a secondary management connection. Once all the management connections are established, transport connections are set up.

5.2.2.2 MAC PDU

The generic MAC header (GMH) contains details of the contents of MAC packet data units (MPDUs) that are transported over the PHY layer. A 32-bit Comité Consultatif International Telephonique et Telegraphique (CCITT) (now the International Telecommunication Union [ITU]) standard cyclic redundancy check (CRC) of the entire MPDU may be appended to the frame if required. The payload can contain either a management message or transport data. A payload in a transport connection can contain a MAC service data unit (MSDU), fragments of MSDUs, aggregates of MSDUs, aggregates of fragments of MSDUs, bandwidth requests, or retransmission requests according to the MAC rules on bandwidth requesting, fragmentation, packing, and ARQ.

5.2.2.3 Scheduling and QoS Support

Centralized scheduling at each BS enables efficient resource allocation in every OFDMA frame in response to traffic dynamics and time-varying channel conditions. The CQI channel provides fast channel information feedback to enable link adaptation. AMC combined with HARQ provides robust transmission over the time-varying and frequency-selective channel. Resource allocation on the DL and UL in every OFDMA frame is communicated in MAP messages at the beginning of each frame. This allows the BS scheduler to dynamically adapt changing traffic and channel conditions on a frame-by-frame basis. Since both the DL and the UL are scheduled, the UL must feed back accurate and timely information as to the traffic conditions and QoS requirements. UL bandwidth requests are supported through ranging, piggybacking, and polling.

The scheduler also supports resource allocation in multiple subchannelization schemes. For frequency diversity permutations such as PUSC, where subcarriers in the subchannels are pseudo-randomly distributed across the bandwidth (Section 5.2.1), subchannels are of similar quality. Frequency diversity scheduling can support a QoS with fine granularity and flexible time-frequency resource scheduling. With contiguous permutations such as AMC (Section 5.2.1), the subchannels may experience different quality.

Mobile WiMAX supports QoS requirements for a wide range of data services and applications by mapping those requirements to unidirectional service flows that are carried over UL or DL connections. Table 5.4 describes the five QoS classes used to provide service differentiation by the MAC scheduler.

5.2.2.4 Mobility Management, Encryption, and Security

The IEEE 802.16e-2005 standard supports hard handoff (HHO), fast base station switching (FBSS), and macrodiversity handover to support user mobility within the network. In Mobile WiMAX, HHO is mandatory while FBSS and macrodiversity handover are optional. In order to conserve MS power during the inactive state and to minimize usage of the serving BS air-interface resources, IEEE 802.16e-2005 supports two power-saving modes: sleep mode and idle mode. Sleep mode is a state in which the MS conducts prenegotiated periods of absence from the serving BS air interface. These periods are characterized by the unavailability of the MS, as observed from the serving BS, to DL or UL traffic. While in the sleep mode, the MS has the flexibility to scan other BSs to collect information to assist hand-off. Idle mode provides a mechanism for the MS to become periodically available for DL broadcast traffic messaging without registration at a specific BS as the MS traverses between areas of coverage of multiple BSs. Mobile WiMAX supports mutual device/user authentication, flexible key management protocol, traffic

Table 5.4 QoS Classes Supported by Mobile WiMAX

Service	Description	QoS Parameters
Unsolicited grant service (UGS)	Support for real-time service flows that generate fixed-size data packets on a periodic basis, such as VoIP without silence suppression	Maximum sustained rate Maximum latency tolerance Jitter tolerance
Real-time packet service (rtPS)	Support for real-time service flows that generate transport variable size data packets on a periodic basis, such as streaming video or audio	Minimum reserved rate Maximum sustained rate Maximum latency tolerance Traffic priority
Extension of real-time polling service (ertPS)	Extension of rtPS to support traffic flows such as variable rate VoIP with voice activation detection (VAD)	Minimum reserved rate Maximum sustained rate Maximum latency tolerance Jitter tolerance Traffic priority
Non-real-time packet service (nrtPS)	Support for non-real-time services that require variable size data grants on a regular basis	Minimum reserved rate Maximum sustained rate Traffic priority
Best effort (BE)	Support for best-effort traffic	Maximum sustained rate Traffic priority

encryption, control and management plane message protection, and security protocol optimizations for fast handovers.

5.3 MIMO for Mobile WiMAX: The Concept of Adaptive MIMO Switching

Multiple antenna techniques constitute one of the most effective means to increase the system spectral efficiency exploiting the spatial diversity of the wireless communications channel [11]. As mentioned in Section 5.2.1, the application of multiple antenna techniques is substantially facilitated in TDD systems due to the reciprocity of the wireless channel in the UL and DL. Multiple antenna approaches are usually classified in two categories: smart antennas, also known as beam-forming [12,13], and MIMO techniques [3,14,15]. In smart antenna techniques, the BS is usually equipped with two or more antennas and determines so-called antenna weights for both uplink reception and downlink transmission targeting at improving the postprocessing SINR and thus enhancing link and system performance. On the other hand, MIMO techniques require that both communication ends be equipped with more than two antennas, which offers the possibility of transmitting either space-time codes for increasing the link performance or more than one spatially multiplexed data stream, referred to as spatial streams in what follows, which leads to an increase of the link throughput compared to traditional smart antenna systems.

Although smart antenna techniques provide a substantial performance increase in wireless networks in a relatively cost-effective manner, since the computational burden at both the UL and DL is undertaken by the BS, they are associated with two serious deficiencies: First, since they optimize the link for a single spatial stream, they cannot provide a throughput increase of the link. Second, being closed-loop techniques, they rely on CSI (Section 5.2.1), obtained from the UL phase of the network operation, which, even in TDD systems, inevitably leads to downlink performance degradation in the presence of mobility.

Those shortcomings associated with the application of smart antennas are overcome by the 2×2 (two transmit, two receive) open-loop MIMO techniques adopted by the Mobile WiMAX system profiles, which include two modes: STC, also known as the Alamouti code [3,16], and SM [17,18]. STC is a powerful technique for effectively implementing open-loop transmit diversity, while its performance is further increased in Mobile WiMAX since a second antenna is also present at the MS receiver. STC offers favorable performance in all propagation environments; that is, it is not constrained by the MIMO channel quality usually represented by the spread of the MIMO channel eigenvalues. However, STC cannot lead to link throughput increase because it transmits two spatial streams over two OFDMA symbols.

In contrast to STC, 2×2 SM doubles the offered data rate because it transmits two spatial streams over a single OFDMA symbol. However, the performance of SM depends on the eigenvalue spread of the MIMO channel. This is especially true when the zero-force (ZF) or minimum mean square error (MMSE) equalizers [19] are employed since they are associated with self-interference that can render their application prohibitive, especially in MIMO channels exhibiting high correlation factors. Fortunately this deficiency of ZF/MMSE receivers is to a large extent alleviated by the application of maximum-likelihood (ML) types of SM equalizers, called ML decoders (MLDs). Section 5.4 contains performance comparisons of the MMSE equalizer and MLD in different realistic scenarios.

To enable the use of the favorable characteristics from each of the two MIMO modes, Mobile WiMAX allows for adaptive switching between STC and SM depending on the instantaneous link condition. This operation is termed AMS and, as is shown in Figure 5.4, should be considered as part of the link adaptation typically applied in a wireless communications system at the BS transceiver. By relying on the CSI provided by the MS in the CQI channel (Section 5.2.1), the BS determines both the MIMO mode (STC or SM) and the MCS (from QPSK 1/2 to 64-QAM 5/6 according to Table 5.2) to be transmitted

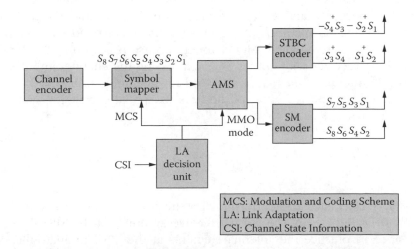

Figure 5.4   Illustration of the AMS concept of Mobile WiMAX.

to be selected in the next transmission opportunity to the MS in the next transmission opportunity provided by the BS scheduler.

The BS needs to utilize channel quality metrics for reliably determining the appropriate MIMO mode and MCS for each supported MS. The availability of multiple MCSs and two MIMO modes provides a wide range of operating points. While this flexibility allows the data transmission to be efficiently adapted to the instantaneous channel conditions, a robust link adaptation mechanism is required for taking advantage of this system capability. To this end, it is stressed that the CSI type can substantially assist the BS in making the decision regarding the supported MIMO mode and MCS. In particular, the performance of the MIMO SM mode depends critically on the type of receiver employed at the MS receiver. However, information about the MS receiver type may not be available at the BS. The BS then relies on a combination of the effective CINR and MIMO mode types of the CSI for the next transmission to the considered MS. In this way, Mobile WiMAX specifies all appropriate mechanisms for MIMO operation achieving link robustness, due to the STC mode, high throughput, due to the SM mode, and high spectral efficiency, due to the possibility of accurate link adaptation offered by the AMS approach.

5.4 Performance Evaluation at the Link Level

This section presents the LLS results for the Mobile WiMAX DL MIMO modes. The channel model employed in the simulations is described first, followed by the PER performance of STC and SM. In order to model the interference from each interfering BS of the network accurately, the MIMO interference model in the SLS is the same as the MIMO channel model described in this section. In other words, both information-carrying and interference signals are filtered at the system level according to the same channel model.

5.4.1 MIMO Channel Model

The used MIMO channel model is a stochastic channel model for MIMO systems that extends the SISO channel model to the MIMO case by utilizing the transmit and receive spatial correlation matrices. Let M and N be the number of transmit (TX) and receive (RX) antennas, respectively, and let R_{TX}, of dimensions $M \times M$, and R_{RX}, of dimensions $N \times N$, be the correlation matrices at the transmit and receive side, respectively. If **H** denotes the $N \times M$ discrete-time MIMO channel impulse response matrix between the transmitter and the receiver with entries $H_{n,m}$ expressing the channel impulse response between transmit antenna $m, m = 1, \ldots, M$, and receive antenna $n, n = 1, \ldots, N$, then the elements $[R_{TX}]_{i,j}, i, j = 1, \ldots, M$, of the $M \times M$ spatial correlation matrix R_{TX} are defined according to

$$[R_{TX}]_{i,j} = \langle H_{i,l}, H_{j,l} \rangle, \tag{5.1}$$

where $\langle H_{i,l}, H_{j,l} \rangle$ calculates the correlation coefficient between $H_{i,l}$ and $H_{j,l}$ and is independent of $l, l = 1, \ldots, N$ (i.e., of the receive antennas at the MS). This is a valid assumption since in practice the antenna elements at the MS illuminate the same surrounding scatterers and thus create the same power azimuth spectrum (PAS) at the BS. The elements of the $N \times N$ correlation matrix R_{RX} are defined similarly.

Since the correlation coefficients between any two channel impulse responses connecting two different sets of antennas can be expressed as the product of the correlation coefficients at the transmit and the receive antennas (see Kermoal et al. [19] for a detailed proof), the spatial correlation matrix of the MIMO channel matrix **H** can be expressed as the Kronecker product [20] of the spatial correlation matrices at the transmit and receive side:

$$\mathbf{R}_{MIMO} = \mathbf{R}_{TX} \otimes \mathbf{R}_{RX} = \mathbf{C}_{TX}^{*T} \mathbf{C}_{TX} \otimes \mathbf{C}_{RX}^{*T} \mathbf{C}_{RX}, \qquad (5.2)$$

where $\mathbf{C}_{TX}$ and $\mathbf{C}_{RX}$ represent the Cholesky decomposition [20] of R_{TX} and R_{RX}, respectively. This property of the MIMO channel matrix **H** means that the effects of multipath propagation and mobility can be modeled by generating $M \times N$ uncorrelated channel impulse responses, each according to the SISO power delay profile (PDP) and the desired mobility model. For each multipath component $w, w = 1, \ldots, W$, the MIMO channel matrix is determined as

$$\mathbf{H}_w = \mathbf{C}_{RX}^{*T} \mathbf{H}_{un,w} \mathbf{C}_{TX}^*, \qquad (5.3)$$

where $\mathbf{H}_{un,w}, w = 1, \ldots, W$ denotes the $N \times M$ MIMO channel matrix created by the NM uncorrelated channel impulse responses at delay $w, w = 1, \ldots, W$.

Although the correlation matrix of each multipath component may be selected independently depending on the propagation environment, a more straightforward approach is used here that selects all correlation matrices to be the same. In the case of uniformly spaced antenna elements, the correlation matrices can be selected as Hermitian Toeplitz and determined by a single correlation factor α, with their first row being equal to $[\alpha, \alpha^2, \ldots, \alpha^{K-1}]$ and K being equal to M or N. Then, in the practical 2×2 MIMO case of the Mobile WiMAX system profiles considered in this chapter, R_{TX} and R_{RX} can be chosen as follows (see Equation (5.1) and Equation (5.2)):

$$\mathbf{R}_{TX} = \begin{bmatrix} 1 & \rho \\ \rho^* & 1 \end{bmatrix}, \quad \mathbf{R}_{RX} = \begin{bmatrix} 1 & \mu \\ \mu^* & 1 \end{bmatrix}, \qquad (5.4)$$

where ρ and μ are the complex correlation factors at the transmit (BS) and receive (MS) side, respectively. Letting ρ and μ in Equation (5.4) take values in [0, 1] for the amplitude and $[0, 2\pi]$ for the phase enables investigation of the influence of different propagation environments on MIMO performance (e.g., in LOS links, ρ and μ in Equation (5.4) can be chosen to have amplitudes close to one). As a final remark regarding the practical parameters of the described MIMO channel model, the International Telecommunications Union (ITU) power delay profiles and the Doppler spectrum according to ITU-R M.1225 [21] are employed, while ρ and μ in Equation (5.4) take values in [0, 1] for the amplitude and $[0, 2\pi]$ for the phase.

5.4.2 Performance of the STC MIMO Mode

This section presents link-level results from Monte Carlo simulations for the STC mode of the Mobile WiMAX MIMO system profiles. Figure 5.5 plots the PER versus the signal-to-noise ratio (SNR) for all CTC-coded MCSs for the 2×2 STC case. To provide a comparison with the two-branch transmit diversity scheme of the Alamouti code without any receive diversity (i.e., by using only a single antenna at the MS), Figure 5.6 presents the corresponding curves to Figure 5.5 for the 2×1 STC case. The simulations assume DL

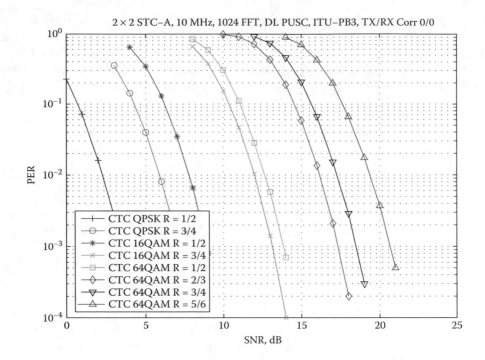

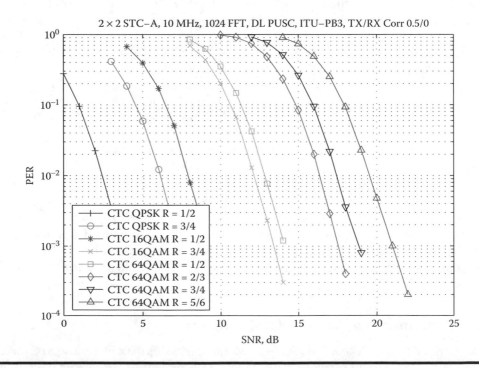

Figure 5.5 Average PER versus average receive SNR; Mobile WiMAX DL 2 × 2 STC; 10 MHz channel bandwidth; PUSC permutations; ITU pedestrian B, 3 km/hr; CTC; top graph: TX/RX correlation factor = 0/0; bottom graph: TX/RX correlation factor = 0.5/0; legend entries top to bottom correspond to curves left to right.

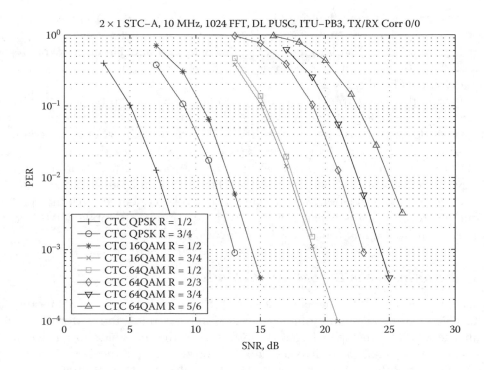

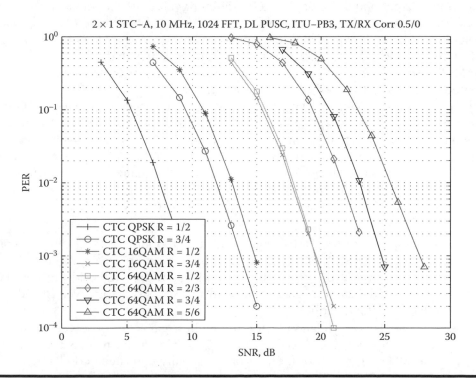

Figure 5.6   Average PER versus average receive SNR; DL 2 × 1 STC illustrated for comparison purposes with the 2 × 2 STC case of Figure 5.6; 10 MHz channel bandwidth; PUSC permutations; ITU pedestrian B, 3 km/hr; CTC; top graph: TX/RX correlation factor = 0/0; bottom graph: TX/RX correlation factor = 0.5/0; legend entries top to bottom correspond to curves left to right.

PUSC permutation, channel bandwidth of 10 MHz, and the MIMO channel model of Section 5.4.1. Spatial correlation factors [19] are $\rho = \mu = 0$ for the top graphs and $\rho = 0.5$, $\mu = 0$ in the bottom graphs in both Figure 5.5 and Figure 5.6. Interference is modeled as additive white Gaussian noise (AWGN) and is uncorrelated between the antenna elements for the 2×2 STC case.

Comparing the two graphs in Figure 5.5 and Figure 5.6, it can be seen that STC offers a remarkably stable performance independent of the antenna correlation, since the average SNR degradation is less than 0.5 dB for all MCSs. On the other hand, comparing Figure 5.5 (the 2×2 STC case) with Figure 5.6 (the 2×1 STC case), the use of the second antenna at the MS offers an average SNR gain on the order of 5 dB for all supported MCSs. As the observations from the results of Figure 5.5 and Figure 5.6 also hold for higher values of the correlation factors ρ and μ, it can be concluded that the combined transmit and receive diversity gains of the STC MIMO mode of Mobile WiMAX lead to high performance and link robustness in all environments.

5.4.3 Performance of the SM MIMO Mode

This section presents the link level simulation (LLS) results for the 2×2 SM mode of the Mobile WiMAX MIMO system profiles. As argued in Section 5.3, the performance of SM critically depends on the spatial correlation factors at the transmit and receive side of the communications link, as well as the equalizer used at the MS receiver. Therefore this section investigates the performance of the MMSE equalizer and a suboptimum version of the maximum-likelihood detector (MLD) for different values of the transmit and receive correlation factors ρ and μ, respectively [19]. The used suboptimum MLD approach is associated with an SNR degradation of approximately 0.5 dB for MCSs based on QPSK and 16-QAM (Table 5.2 of Section 5.2.1), and approximately 1 dB for MCSs based on 64-QAM. The rest of the simulation assumptions are the same as for the simulation results presented for the STC mode of Mobile WiMAX in Section 5.4.2.

Figure 5.7 through Figure 5.10 presents the PER versus SNR performance of both the MMSE equalizer and the suboptimum MLD for transmit and receive correlation factors ρ/μ equal to 0/0, 0.5/0, 0.5/0.5, and 0.8/0.8, respectively. To assist in the evaluation of the presented results from Figure 5.7 to Figure 5.10, Table 5.5 summarizes the SNR gains of the suboptimum MLD compared to the MMSE equalizer for all MCSs and all investigated combinations of the correlation factors ρ and μ. The following conclusions can be drawn from the results of Figure 5.7 to Figure 5.10:

1. First, compared to the MMSE equalizer, the suboptimum MLD offers a consistently increasing SNR gain with increasing correlation factors, where the SNR gain ranges from 2 dB to 8.5 dB, depending on the MCS and the correlation factor, and exhibits an average SNR gain in excess of 4 dB over all MCSs and correlation factors. Further,
   - For the same correlation factor, the highest gains are observed for the high code rates of the CTC (i.e., 3/4 and 5/6). This result reveals the stability of the MLD regardless of the rate of the serially concatenated channel coding scheme.
   - The impact of the transmit correlation factor ρ is similar to that of the receive correlation factor μ. Comparing Figure 5.7 and Figure 5.8 to Figure 5.9 and Figure 5.10 over all supported MCSs, the average increase of the SNR gain of the suboptimum MLD compared with the MMSE equalizer is on the order of

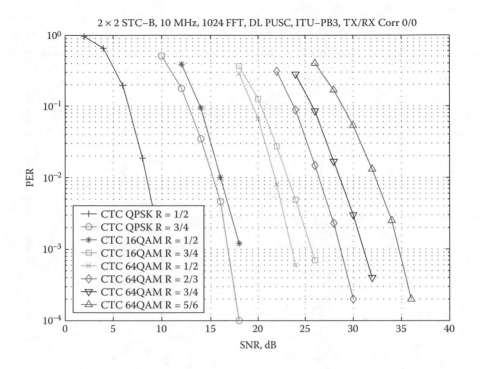

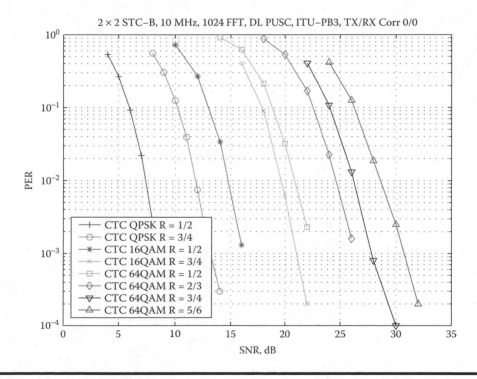

Figure 5.7 Average PER versus average receive SNR; Mobile WiMAX DL 2 × 2 SM; 10 MHz channel bandwidth; PUSC permutations; ITU pedestrian B, 3 km/hr; TX/RX correlation factor = 0/0; CTC; top graph: MMSE receiver; bottom graph: suboptimum MLD; legend entries top to bottom correspond to curves left to right.

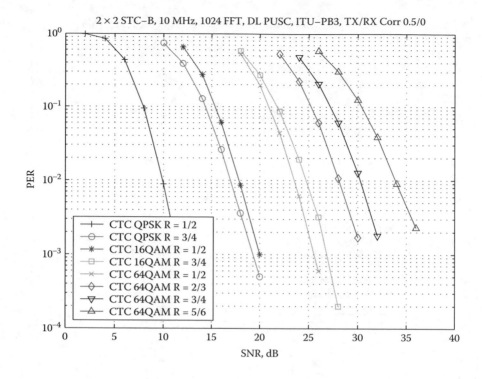

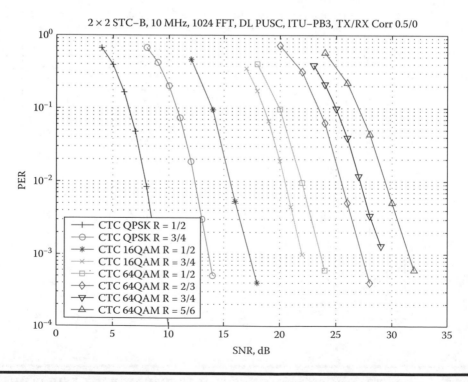

Figure 5.8 Average PER versus average receive SNR; Mobile WiMAX DL 2 × 2 SM; 10 MHz channel bandwidth; PUSC permutations; ITU pedestrian B, 3 km/hr; TX/RX correlation factor = 0.5/0.0; CTC; top graph: MMSE receiver; bottom graph: suboptimum MLD; legend entries top to bottom correspond to curves left to right.

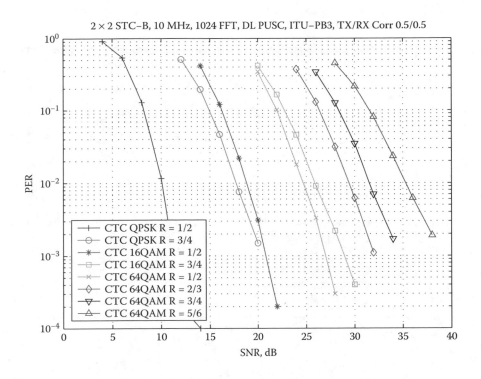

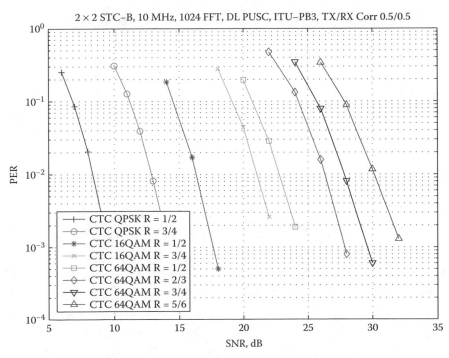

Figure 5.9 **Average PER versus average receive SNR; Mobile WiMAX DL 2 × 2 SM; 10 MHz channel bandwidth; PUSC permutations; ITU pedestrian B, 3 km/hr; TX/RX correlation factor = 0.5/0.5; CTC; top graph: MMSE receiver; bottom graph: suboptimum MLD; legend entries top to bottom correspond to curves left to right.**

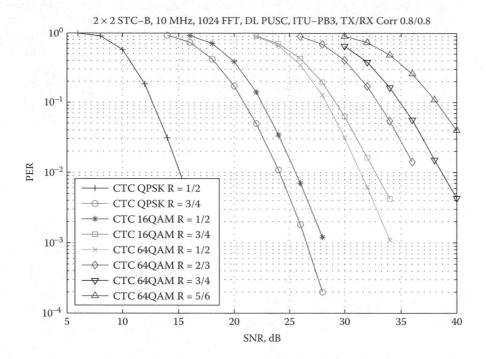

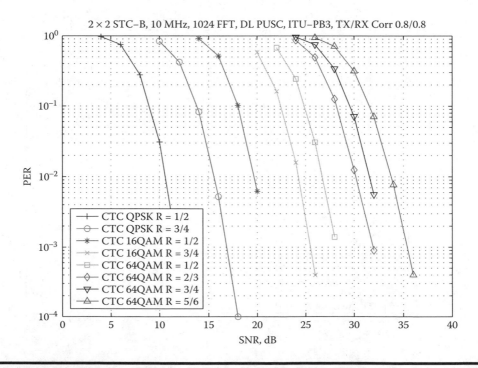

Figure 5.10   Average PER versus average receive SNR; Mobile WiMAX DL 2 × 2 SM; 10 MHz channel bandwidth; PUSC permutations; ITU Pedestrian B, 3 km/hR; TX/RX correlation factor = 0.8/0.8; CTC; top graph: MMSE receiver; bottom graph: suboptimum MLD; legend entries top to bottom correspond to curves left to right.

Table 5.5   Gain of Suboptimal MLD Versus MMSE Per MCS and TX/RX Correlation Factor for the Mobile WiMAX DL 2 × 2 SM Model

MCS	SNR_{MMSE} $-SNR_{subMLD}$ (*TX/RX Corr.* = 0.0/0.0) at *PER* = 0.01	SNR_{MMSE} $-SNR_{subMLD}$ (*TX/RX Corr.* = 0.5/0.0) at *PER* = 0.01	SNR_{MMSE} $-SNR_{subMLD}$ (*TX/RX Corr.* = 0.5/0.5) at *PER* = 0.01	SNR_{MMSE} $-SNR_{subMLD}$ (*TX/RX Corr.* = 0.8/0.8) at *PER* = 0.01
CTC QPSK R = 1/2	2 dB	2 dB	2.5 dB	4.5 dB
CTC QPSK R = 3/4	4 dB	4.5 dB	5 dB	8 dB
CTC 16-QAM R = 1/2	2 dB	2 dB	2.5 dB	5.5 dB
CTC 16-QAM R = 3/4	4.5 dB	4.5 dB	4.5 dB	8.5 dB
CTC 64-QAM R = 1/2	1 dB	2.5 dB	2.5 dB	5 dB
CTC 64-QAM R = 2/3	2 dB	2.5 dB	3 dB	6 dB
CTC 64-QAM R = 3/4	3 dB	3 dB	3.5 dB	7 dB
CTC 64-QAM R = 5/6	3.5 dB	4.5 dB	5 dB	8.5 dB

0.5 dB in both cases. This result shows that in SM the transmit and receive antenna correlations have a more or less similar impact on the spatial stream decoding performance. Therefore, although practical space limitations may not allow for increased distance between antennas at the MS receiver, the robustness of the spatially multiplexed link in mobile broadband applications can still be improved by using large antenna interelement distances at the BS (e.g., in excess of 10 wavelengths), which leads to small transmit correlation factors. The use of cross-polarized antennas at both communication ends is associated with substantial gains independently of the used equalization technique, without diminishing the value of using MLD-type equalizers for decoding of the 2 × 2 MIMO SM mode.

2. Second, performance of the suboptimum MLD for the decoding of spatially multiplexed streams is robust up to correlation factors 0.5/0.5. Since most mobile broadband applications are usually connected with NLOS operation and low correlation factors at the BS transmitter, the suboptimum MLD fulfills the requirements for high performance and robustness of the Mobile WiMAX MIMO SM mode. In the extreme cases of high correlation factors or low SINRs, the link adaptation mechanism of Mobile WiMAX (Section 5.3) ensures that the more robust STC mode is selected for link reliability.

5.4.4   MIMO PHY Layer Abstraction

One of the biggest challenges of system-level simulation is the use of a reliable PHY layer abstraction methodology. The PHY layer abstraction enables running simulations at the system level without the need to simultaneously run the complete PHY layer transceiver for all connections in the network, which demands a prohibitively high computational effort. Therefore the objective of the PHY layer abstraction methodology is to accurately predict the link layer performance utilizing a computationally simple model. In practice, the PHY layer abstraction model is implemented as a look-up table (LUT) containing a probabilistic description of the PHY layer performance. The LUT is obtained from PHY layer simulation results such as those in Section 5.4.2 and Section 5.4.3, with the difference being that the interference is now modeled according to the same way the information-carrying received signal is generated (i.e., by using the same channel model as the one described in Section 5.4.2). The PHY layer abstraction represents the mobile broadband communications link as a set of the performance curves, which describe the dependence of a system quality indicator (SQI) versus a channel quality indicator (CQI).

The PER is usually employed as the SQI, which is justified because the overall quality of the system operation can be characterized by whether a packet is correctly or erroneously decoded at the receivers of the mobile network. The CQI is chosen depending on the transmission and reception method (i.e., it is determined according to the number of transmit and receive antennas), as well as the transmit and receive PHY layer signal processing techniques.

In the case of the Mobile WiMAX DL MIMO considered in this chapter, the CQI for PHY layer abstraction can be based on the mean instantaneous postprocessing capacity (MIPPC) for the STC mode. Considering a specific resource allocation in an OFDMA frame, the MIPPC is the average of the instantaneous capacity per subcarrier averaged over all subcarriers used for the considered resource allocation. The instantaneous capacity per subcarrier is calculated according to the well-known capacity formula of Shannon, where the SNR parameter required by Shannon's expression is replaced by the postprocessing SINR. The postprocessing SINR is defined as the SINR at the output of the mobile broadband OFDMA receiver (see Section 5.2.1 for a description of the main OFDMA receiver architecture), which is defined in a straightforward manner for linear MIMO receivers. In the case of the STC MIMO mode of Mobile WiMAX, where the signal processing at the receiver end takes place in a linear manner, the MIPPC can be used according to the description given above for generating PHY layer abstraction curves to be used in the system-level simulation (SLS).

In the case of the SM MIMO mode of Mobile WiMAX, MIPCC can be used as the CQI for the MMSE-based MIMO receiver (Section 5.4.3), since the required signal processing for its implementation is linear. However, if the MLD (Section 5.4.3) is employed at the MS receiver, the postprocessing SINR cannot be defined and thus alternative methods have to be found for abstracting the performance of ML-based receivers for decoding spatially multiplexed data streams. To this end, the information-theoretic capacity formula for MIMO channels can be used as the basis for generating PHY abstraction curves for ML-based SM MIMO receivers. This approach can be considered as intuitively appropriate since ML-based receivers are expected to theoretically attain the capacity offered by spatially multiplexed streams.

Figure 5.11 to Figure 5.14 present representative PHY abstraction curves for the MIMO modes of the Mobile WiMAX used in SLS. Figure 5.11 and Figure 5.12 are valid for the STC mode, while Figure 5.13 and Figure 5.14 correspond to the SM mode, where the use of suboptimum MLD at the MS receiver (Section 5.4.3) is assumed. Table 5.3 contains the mapping of the MCS numbers shown in Figure 5.11 to Figure 5.14 to the STC/SM modulation and coding schemes.

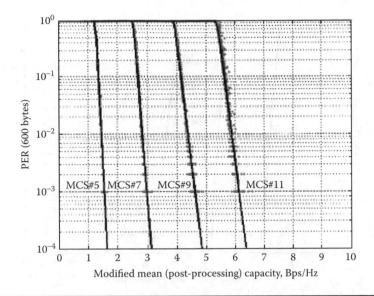

Figure 5.11   PHY abstraction curves (scatter plot and approximation curves); Mobile WiMAX DL 2 × 2 STC; 10 MHz channel bandwidth; PUSC permutations; TX/RX correlation factor = 0.5/0.5; CTC; MCSs 5, 7, 9, and 11 (see Table 5.3).

For the STC mode, PHY layer abstraction curves in Figure 5.11 and Figure 5.12, the dots represent the PER scattering plots per MCS that illustrate the instantaneous PER versus the CQI for a given channel and interference realization. The solid curves illustrate the approximated curves to be used for the generation of the MIMO STC PHY layer abstraction LUT (see above in this section). It can be observed that the selected

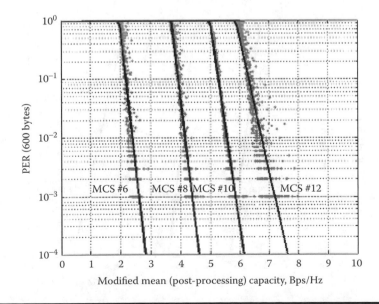

Figure 5.12   PHY layer abstraction curves (scatter plot and approximation curves); Mobile WiMAX DL 2 × 2 STC; 10 MHz channel bandwidth; PUSC permutations; TX/RX correlation factor = 0.5/0.5; CTC; MCSs 6, 8, 10, and 12 (see Table 5.3).

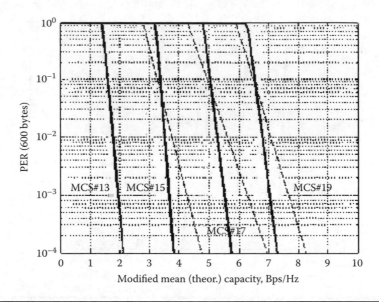

Figure 5.13   PHY abstraction curves (scatter plot and approximation curves); Mobile WiMAX DL 2 × 2 SM; 10 MHz channel bandwidth; PUSC permutations; TX/RX correlation factor = 0.5/0.5; CTC; MCSs 13, 15, 17, and 19 (see Table 5.3).

CQI, which is based on the mean postprocessing capacity, constitutes a favorable metric for abstracting the link-level performance of STC in SLS since there is a clear separation between the scattering plots of different MCSs. This is especially true at high values of the PER (e.g., 10%) which, as will be shown in the system-level simulation results of

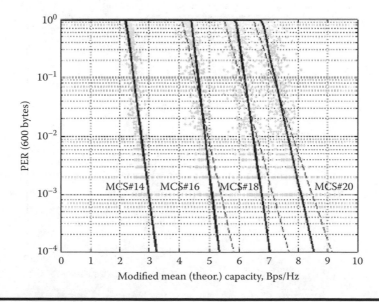

Figure 5.14   PHY layer abstraction curves (scatter plot and approximation curves); Mobile WiMAX DL 2 × 2 SM; 10 MHz channel bandwidth; PUSC permutations; TX/RX correlation factor = 0.5/0.5; CTC; MCSs 14, 16, 18, and 20 (see Table 5.3).

Section 5.5, is of practical interest for the real-world deployment of a Mobile WiMAX network.

The corresponding PHY layer abstraction curves for the SM mode are presented in Figure 5.13 and Figure 5.14 when MLD is employed at the MS MIMO receiver: The dots show the PER scattering plots for each MCS, while the solid and dashed curves represent the approximated PER curves following two different approximation methods. As for the STC case (Figure 5.11 and Figure 5.12), the selected CQI for SM, which is based on the mean theoretical capacity, provides a favorable metric for abstracting the link-level performance of MLD-based SM in SLS, especially at high values of the PER.

As a final remark to this section, it is stressed that the used CQIs in Figure 5.11 through Figure 5.14 have been modified to include the impact of PHY layer nonidealities on the PHY abstraction curves. These PHY nonidealities are associated with the use of practical timing and frequency synchronization, channel estimation, and suboptimum equalization techniques at the MS receiver, and fortunately only lead to modifications of the theoretical CQIs for use as PHY layer abstraction metrics in SLS.

5.5 Performance Evaluation at the System Level

This section describes the adopted system-level simulation framework and the system-level simulation results of the Mobile WiMAX downlink employing the 2×2 MIMO system profiles.

5.5.1 Network Topology and Frequency Reuse

The Mobile WiMAX system is modeled as a cellular network with two tiers around the central cell, leading to a total of 19 cells in the network. A frequency reuse of 3 (reuse-3) refers to three mutually exclusive frequency allocations, one per sector of a cell. Similarly a frequency reuse of 1 (reuse-1) refers to a single frequency allocation reused between sectors of a cell. In both cases, the frequency reuse pattern repeats across cells. Table 5.6 lists the main parameters for the network topology used in the Mobile WiMAX SLS for the MIMO DL. The BS-to-BS or site-to-site distance is chosen depending on the pathloss model assumed in the simulations (see Section 5.5.3 for a classification of the considered models).

5.5.2 Deployment Scenario and Link Budget

In the simulations, the network deployment models specified by IMT-2000 [21] and 3GPP2 [22] are considered for the urban and suburban macrocell scenario. This scenario

Table 5.6 Network Model for SLS

Parameters	Value
Number of cells	19
Number of sectors per cell	3
Total number of sectors	57
BS-BS distance	0.9/1.4 km (urban)
Center frequency	2.5 GHz
Frequency reuse	3×3 (reuse-3), 3×1 (reuse-1)

Table 5.7   DL Link Budget Parameters

Parameters	Value
Transmission power/sector	43 dBm
BS height	40 m
Number of TX antennas	2
TX antenna pattern	70° (-3 dB) with 20 dB front-to-back ratio
TX antenna gain	15 dBi
MS height	1.5 m
Number of RX antennas	2
RX antenna pattern	Omnidirectional
RX antenna gain	0 dBi
MS noise figure	8 dB
Hardware losses	2 dB

is characterized by large cells, BS antennas above rooftop height (e.g., 20 m or more), and high transmit power. It is assumed that all BS and MS equipment is configured identically across the network. The BS and MS equipment models are characterized by their transmit power, height, number of antennas, and antenna pattern. Table 5.7 contains the values of those parameters as used in the SLS of the Mobile WiMAX DL MIMO.

5.5.3   Propagation Model

The Mobile WiMAX system-level simulator models the important characteristics of the wireless propagation in a cellular deployment scenario (Section 5.5.1). The slow fading characteristics are captured through the adopted pathloss model and the variance of the log-normal shadowing. Shadowing is modeled as a log-normal random variable with zero mean and standard deviation that depends on the propagation environment. The fast fading effects are modeled by the frequency-selective fading resulting from multipath and the time-selective fading resulting from the mobility model. The spatial MIMO channel model presented in Section 5.4.1 provides the means for a complete characterization of the MIMO channel present between the BS transmitter and the MS receiver with respect to frequency, time, and spatial selectivity of the mobile broadband communications channel.

■ **Modified COST231 Hata Urban Pathloss Model:** This model is used by 3GPP-3GPP2 [23] and 1xEV-DV [24] to compute the pathloss in urban/suburban macro-cell environments for frequencies around 2 GHz. According to this model, the pathloss is given by

$$PL_{modHata}(d) = 45.5 + 0.7H_{MS} - 13.82\log_{10}(H_{BS}) + (44.9 - 6.55\log_{10}(H_{BS}))$$
$$\times \log_{10}(d/1000) + (35.46 - 1.1H_{MS})\log_{10}(f) + C, \qquad (5.5)$$

where H_{BS} and H_{MS} are the BS and MS heights in meters, d is the distance between the BS and the MS in meters, f is the carrier frequency in megahertz, and C is a constant factor set to 0 dB for suburban scenarios and 3 dB for urban scenarios. In the system-level simulator, the modified COST231 Hata model is used according to Equation (5.5), with C chosen to be equal to 3 dB. In the following, this model is referred to as the urban pathloss model.

Table 5.8   Propagation Model

	Slow Fading
Path loss model	Suburban (ITU-Vehicular)/Urban (modified COST231 Hata)
Log-normal shadowing	$\mu = 0$ dB, $\sigma_{SF} = 8.9$ dB
Shadowing correlation	100% intersector, 50% inter-BS
	Fast fading
Channel model	Channel mix based on ITU PedB and VehA
Time correlation	Jakes spectrum
	Spatial model (MIMO)
Spatial correlation	Use of spatial correlation factor depending on antenna spacing and BS-MS distance

■ **ITU Vehicular Test Environment Pathloss Model:** According to the ITU recommendation for evaluation of radio technologies for IMT-2000 [21], a pathloss model is proposed for test scenarios in urban and suburban areas outside the high-rise core where the buildings are of nearly uniform height. According to this model, the pathloss is calculated by

$$PL_{ITUv}(d) = 40(1 - 4.10^{-3}H_{BS})\log_{10}(d) - 18\log_{10}H_{BS} + 21\log_{10}(f) + 80, \quad (5.6)$$

where d is the BS-to-MS separation in kilometers, and H_{BS} is the BS antenna height above average rooftop level in meters. In the following, this model is referred to as the suburban pathloss model. Table 5.8 summarizes the main parameters of the propagation model used in the SLS.

5.5.4   Interference Model and Loading

The interference in the system-level simulator is generated using the same propagation models (Section 5.5.3) and MIMO channel model (Section 5.4.1) as that used for the information-carrying signals (i.e., it includes the effects of pathloss, shadowing, and the fast fading frequency, time, and spatially selective components). Although it is recognized that interfering signals arriving from BSs of different sectors may exhibit slightly different characteristics due to different propagation conditions among the interfering sectors and the MSs in the considered sector (e.g., less frequency and spatial selectivity), the assumption here is that the interference model is the same as that used for the information-carrying signals of the considered sector.

Frequency loading is defined as the fraction of subchannels allocated to an interferer from the set of available subchannels. It is assumed that all interferers are active for the entire duration of the simulation drop. Frequency loading, which is used in the SLS, has been shown to be the best model for OFDMA interference because it realistically captures interference characteristics of diversity permutations such as the PUSC permutations modeled in the system-level simulator, and it can take advantage of the transmit power gain achievable when only a fraction of the totally available subchannels is utilized (subchannelization gain).

5.5.5 OFDMA Air-Interface Modeling

The system simulation models the complete DL subframe, which is assumed to have 24 data-carrying OFDMA symbols. It is assumed that all BSs are synchronized to maintain common frame start times and frame lengths, as well as using the same type of permutations. PUSC permutations are explicitly modeled to explore the inherent frequency diversity of the PUSC scheme. In the case of the considered DL PUSC permutations, the scheduler assigns subchannels to sectors based on the load requirements of the sector and the system. Scenarios with frequency reuse of 1 and 3 are simulated. For the parameter values of the air interface used in the SLS, the reader is referred to Table 5.1 and Reference [8].

5.5.6 Simulation Methodology

The methodology for system-level evaluation of Mobile WiMAX is based on the methodology developed by the 3GPP2 study groups [22,24] with the difference being that the impact of timing and frequency synchronization as well as channel estimation is considered in the used PHY layer abstraction curves (Section 5.4.4). System-level performance statistics are collected for MSs served by the sectors of the center cell, while users in the outer two tiers generate interference to the center cell. As already mentioned in Section 5.4.1 and Section 5.5.1, the evolution of the MIMO channel in time is modeled for both the information-carrying signals and the interference.

At the start of the simulation, subscribers are randomly dropped in sectors served by the center cell. A simulation drop is defined as a simulation run for a given set of MSs over a specified number of time frames. At the beginning of each drop the MSs determine the specific sector of the center cell serving them. In each drop, the MIMO channel model of each MS served by the center cell is evolved in time for the duration of the drop, while bulk parameters such as MS location, pathloss, and shadowing remain fixed. Similarly, in each drop the MIMO channel model of each interfering BS is evolved in time for the duration of the drop. For each BS-to-MS link, the system-level simulator computes the channel and interference power on the loaded data subcarriers. The received signal power level at a specific subcarrier of an MS is determined by the BS transmit power, the pathloss and shadowing between the serving BS and the considered MS, the transmit and receive antenna gains, and the total number of loaded subcarriers in an OFDMA symbol.

The received interference power is similarly calculated for all interfering BSs and summed up on a per-subcarrier basis. Then, by assuming the presence of an AWGN-related power level per subcarrier, the SINR at each loaded subcarrier of each active MS in the OFDMA DL subframe of each frame of the simulation drop is calculated. In this way the effects of frequency, time, and spatial selectivity are captured in a most realistic way for both user and interfering signals in the cellular mobile broadband network. Finally, based on the PHY layer abstraction LUTs of both Mobile WiMAX MIMO modes (Section 5.4.4), the MIMO mode and the MCS for each MS served by the center cell are selected, where the target of the selection process is the achievement of the highest spectral efficiency while simultaneously satisfying the target PER.

5.5.7 OFDMA Scheduler and HARQ

An OFDMA scheduler at the BS partitions the two-dimensional frequency subchannel/OFDMA symbol grid between active users. The scheduler makes decisions for the entire

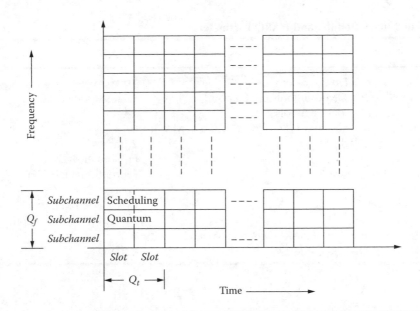

Figure 5.15 **Selection of a scheduling quantum in an OFDMA frame; depending on the requirements regarding the supported number of MSs and the target MS data rate, a scheduling quantum can be chosen to span multiple frequency subchannels over multiple OFDMA symbols.**

TDD frame at frame boundaries using the available CQI from all MSs. The objective is to allocate resources in a fashion that maximizes link utilization while managing QoS requirements and controlling signaling overhead. It is assumed that the resource available for allocation is the TDD frame, and scheduling decisions are made at frame boundaries. A single MS is assigned a quantum consisting of one or more slots. The size of the quantum determines the number of users that can be accommodated within the frame as well as the control signaling overhead. Figure 5.15 illustrates an OFDMA frame where the *x*-axis is in units of OFDMA symbols and the *y*-axis is in units of frequency subchannels. The scheduler at the BS uses the PHY layer abstraction to make a decision on which MSs will be active within each frame based on the proportional fair metric. Based on the specifics of Mobile WiMAX, it is assumed that the CQI is known at frame boundaries with a two frame delay.

In the system-level simulator, synchronous nonadaptive HARQ is modeled. The stop-and-wait protocol is implemented with multiple HARQ streams in each frame for every served MS. The scheduler considers the status of the HARQ streams when computing the scheduling metric and deciding the priority for scheduling streams. Chase combining is used for successive HARQ retransmissions. To track HARQ performance and effectiveness, the residual PER after successive HARQ retransmissions and the HARQ retransmission overhead are used. Table 5.9 lists the main parameters for scheduling and HARQ.

The distribution of MCSs from all served MSs over both multiple frames within a simulation drop and multiple drops is used to calculate the throughput statistics and the system spectral efficiency. Multiple time frames enable capturing the dynamics of the mobile broadband channel and interference, while multiple simulation drops enable capturing the multitude of scenarios with respect to the placement of MSs within the Mobile WiMAX network.

Table 5.9   Scheduler and HARQ Parameters

Parameters	Value
Number of active users per sector	10
Number of OFDMA symbols per frame	47 (excluding TTG/RTG)
Number of DL data symbols	24
Number of symbols for control overhead (DL-MAP)	6
CQI feedback delay	2 frames
Traffic type	Full buffer data only
Scheduling algorithm	Proportional fair (PF)
Target PER	10%
HARQ type	Chase combining
HARQ mode	Synchronous nonadaptive
Maximum number of HARQ retransmissions	4
HARQ retransmission delay	4 frames

5.5.8   System-Level Simulation Results for the MIMO DL of Mobile WiMAX

A wide range of deployment scenarios were studied that took into consideration different frequency reuse factors, spectrum allocations, channel bandwidths, antenna configurations, cell sizes, propagation environments, and a host of other parameters. The results presented in this section are a representative subset of these results, with the goal of establishing system performance benchmarks. The system-level performance of the Mobile WiMAX MIMO DL is evaluated according to the framework and parameters presented in Section 5.5.1 through Section 5.5.7.

In order to gain greater insight into the propagation environment characteristics determining the system-level performance of Mobile WiMAX, the SINR statistics are first presented for the suburban and urban propagation model (Section 5.5.3). In the top graph of Figure 5.16, the cumulative distribution functions (CDFs) of the SNR, signal-to-interference ratio (SIR), and SINR are presented for the suburban propagation environment when the reuse pattern is 3 × 1 (reuse-1 scheme). The CDFs in Figure 5.16 show that the Mobile WiMAX system, as a cellular network, is interference limited, which is also evident in the reuse-3 case (the bottom graph of Figure 5.16). Since similar conclusions regarding the interference limitation of Mobile WiMAX are drawn in the urban propagation environment, Figure 5.17 illustrates only the CDF of the SINR for the reuse-1 and reuse-3 deployment patterns.

Comparing the SINR statistics between the suburban and urban propagation environment, it can be observed that the urban propagation environment exhibits lower SINR values across the whole geographical area. In particular, at the high SINR region of the CDF of the SINR, the gap between the observed SINRs increases. This implies that in addition to selecting lower order MCSs for MSs that are not in the vicinity of the serving BS, the probability of selecting higher order MCSs in the urban environment, which require higher SINR values to achieve a specific PER, is smaller than in the suburban environment. Because of this, it is expected that the aggregate sector throughput and spectral efficiency will attain smaller values in the urban propagation environment.

Figure 5.18 [25] illustrates the CDF of user throughput when 10 users/sector with fully loaded queues compete for air-link resources in the urban (0.9 km BS-to-BS separation)

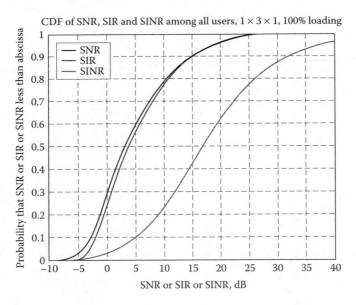

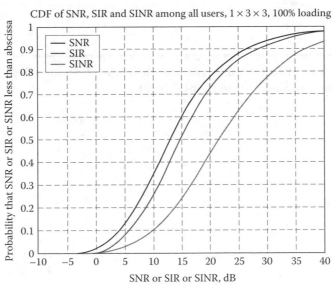

Figure 5.16 CDF of the SINR, SIR, and SNR (plots from left to right) in the suburban propagation environment for a single carrier frequency; the top graph refers to reuse-1 (reuse pattern 3 × 1) and the bottom graph to reuse-3 (reuse pattern 3 × 3).

propagation environment. As described in Table 5.9, a six-symbol overhead is applied to every DL subframe to account for PHY and MAC overheads. User throughput is then defined as the average goodput delivered by the MAC layer to the higher layers over the entire frame period. The channel-aware OFDMA scheduler exploits multiuser diversity by scheduling multiple users in a frame when their channel conditions are most favorable. Proportional fairness is maintained over a fairly short time scale to ensure that the air-link resource is shared equitably among active users in time and frequency. From Figure 5.18, it can be seen that the median user throughput for a 10 MHz system with 10 users/sector

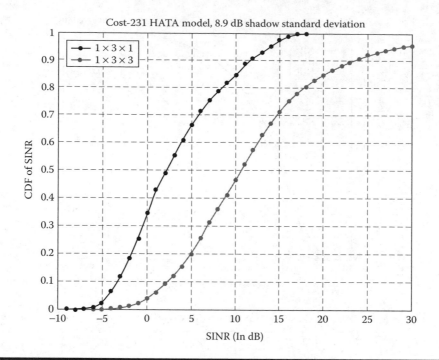

Figure 5.17   CDF of the SINR in the urban propagation environment for a single carrier frequency, 0.9 km BS-BS distance; reuse-1 (left) and reuse-3 (right).

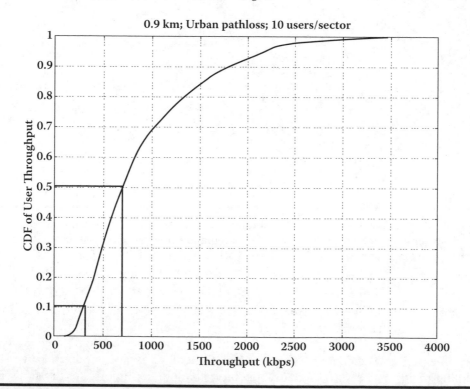

Figure 5.18   CDF of the user throughput in the Mobile WiMAX DL MIMO; 10 users/sector; urban propagation environment; reuse-1, BS-to-BS separation of 0.9 km.

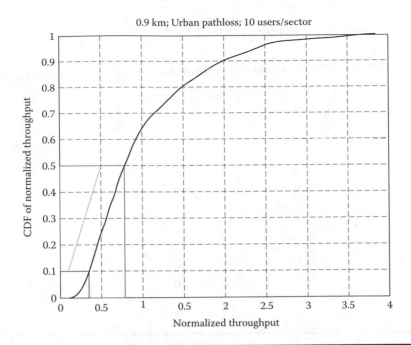

Figure 5.19 CDF of the normalized user throughput in the Mobile WiMAX DL MIMO; 10 users/sector; urban propagation environment; reuse-1, BS-to-BS separation of 0.9 km.

is 720 kbps, while users who are not in outage (10th percentile) see throughputs above 320 kbps.

In addition to offering a favorable user throughput distribution, a mobile broadband cellular system such as Mobile WiMAX should achieve fairness with respect to the allocation of system resources. The fairness of the scheduler can be evaluated through the CDF of normalized throughput. It can be observed from Figure 5.19 that the CDF of the normalized user throughput, which is defined as the ratio of the instantaneous user throughput to the average user throughput, lies to the right of the line that defines the fairness criterion as established in 3GPP2.

Figure 5.20 presents the MCS selection probability for the supported MCSs in Mobile WiMAX as given by Table 5.3. Based on the reported CQI, the BS scheduler selects the appropriate MIMO mode for transmission. In the cases where both STC and SM can be selected, the scheduler picks STC for greater reliability. Note that MCSs with the same spectral efficiency (e.g., STC MIMO mode with 16-QAM 1/2 and SM MIMO mode with QPSK 1/2 having spectral efficiency of 2 bits/s/Hz) are considered in an additive manner in the bar graph of Figure 5.20.

It must also be observed that the effect of HARQ retransmissions is illustrated in the results of Figure 5.20. This implies that the probability of selecting an MCS with a specific spectral efficiency also accounts for the number of HARQ retransmissions required for successfully transmitting this MCS in the SLS. Figure 5.20 reveals that due to the SINR values present in the urban propagation environment, the probability of selecting MCSs with high spectral efficiency (i.e., greater than or equal to 4 bits/s/Hz attained by STC 64-QAM 2/3 and SM 16-QAM 1/2) is low compared to the probability of selecting MCSs with lower spectral efficiency (up to 3 bits/s/Hz attained by STC 16-QAM 3/4, STC 64-QAM 1/2, and SM QPSK 3/4). Although the selection probability of high spectral efficiency

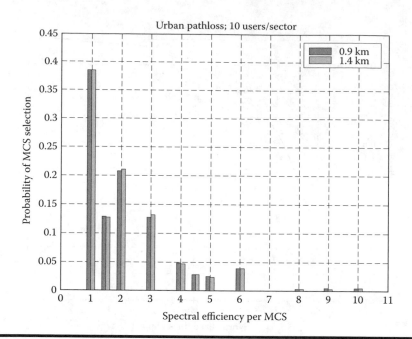

Figure 5.20 Probability of MCS selection in the urban propagation environment; comparison of 0.9 km (left bars) with 1.4 km (right bars) BS-to-BS separation; reuse-1, 10 users/sector.

MCSs does not exceed 20% in most deployment scenarios, selection of the SM modes enables a dramatic improvement of the offered user data rate and the overall spectral efficiency.

Figure 5.21 highlights the impact of the selection of high spectral efficiency MCSs by plotting the average spectral efficiency per MCS, which is defined as the product of the MCS selection probability with the spectral efficiency of each MCS. As shown in Figure 5.21, the use of MCSs with high spectral efficiencies has a substantial impact on the overall spectral efficiency. There is a remarkable similarity between the spectral efficiencies of the deployments with BS-to-BS separation equal to 0.9 km and 1.4 km (left and right bars, respectively, in Figure 5.20 and Figure 5.21), suggesting that the interference-limited urban propagation environment does not result in appreciable changes of the SINR for typical ranges of BS-to-BS separation.

One of the most important performance evaluation metrics for a cellular system is its spectral efficiency. Table 5.10 compares the performance of the DL MIMO with SIMO, which refers to the use of dual antenna diversity at the MS receiver and is considered the baseline system in Mobile WiMAX. As is the conventional practice for cellular systems, the aggregate sector throughput and the spectral efficiency per cell are used as performance metrics. Note that the results for both reuse-1 and reuse-3 are presented in Table 5.10. The reuse-3 deployment scenarios of Table 5.10 use three frequency allocations, which can be realized either by allocating one-third of the total available subcarriers of a single channel (i.e., of 10 MHz bandwidth) to each sector of a cell or by allocating one of three available channels (i.e., each of 10 MHz bandwidth) to each sector of a cell. The latter deployment, which is referred to as the nonsegmented reuse-3 scenario, results in utilization of three times more bandwidth than the former, which is referred to as the segmented reuse-3 scenario. According to Table 5.10, the superiority of DL MIMO

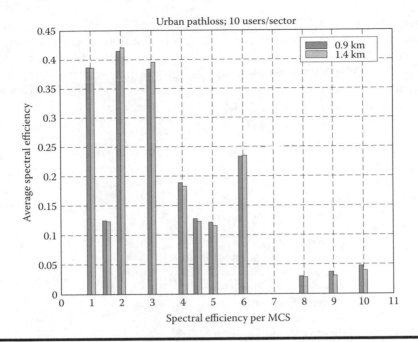

Figure 5.21   Average spectral efficiency per MCS in the urban propagation environment; comparison of 0.9 km (left bars) with 1.4 km (right bars) BS-to-BS separation; reuse-1, 10 users/sector.

with respect to both the aggregate sector throughput and the spectral efficiency per cell is evident when comparing the various deployment scenarios, with reuse-1 MIMO offering the highest spectral efficiency of 4.95 bits/s/Hz cell for the urban propagation environment.

5.6   Conclusion

In this chapter, the link- and system-level performance of the Mobile WiMAX MIMO downlink has been investigated through extensive simulations based on a framework that captures the most important characteristics of a WiMAX deployment. Based on the performance achieved by the MIMO profiles of the Mobile WiMAX DL for both the aggregate sector throughput and the spectral efficiency, it can be concluded that the application of MIMO techniques in Mobile WiMAX constitute the means for enabling the

Table 5.10   Comparison of the Aggregate Sector Throughput and the Spectral Efficiency of the Mobile WiMAX Downlink for SIMO and MIMO

	Aggregate Sector Throughput (in Mbps)	*Spectral Efficiency (in bits/s/Hz/cell)*
Reuse-1 SIMO	7.37	3.42
Reuse-1 MIMO	10.65	4.95
Reuse-3 MIMO segmented/nonsegmented	4.97/15.70	2.31/2.43

first generation of true mobile broadband services, while simultaneously fulfilling the requirements for a flexible and cost-effective next-generation wireless communications system. The results presented in this chapter highlight the benefits of advanced PHY layer techniques for the DL using full-buffer traffic and proportional-fair multiuser scheduling. Further studies are under way on system performance for the uplink, as well as the application of traffic profiles such as VoIP and web browsing.

Acronyms

3GPP	3G Partnership Project
3GPP2	3G Partnership Project 2
A/D	Analog-to-Digital
AES	Advanced Encryption Standard
AMC	Adaptive Modulation and Coding
AMS	Adaptive MIMO Switching
ARQ	Automatic Repeat reQuest
ATM	Asynchronous Transfer Mode
AWGN	Additive White Gaussian Noise
BE	Best Effort
BS	Base Station
CC	Chase Combining (also Convolutional Code)
CCITT	Now renamed ITU (International Telecommunications Union)
CCM	Counter with Cipher-block chaining Message authentication code
CDF	Cumulative Distribution Function
CDMA	Code Division Multiple Access
CID	Connection IDentifier
CINR	Carrier to Interference + Noise Ratio
CMAC	block Cipher-based Message Authentication Code
CP	Cyclic Prefix
CQI	Channel Quality Indicator
CQICH	Channel Quality Indicator CHannel
CRC	Cyclic Redundancy Check
CSI	Channel State Information
CTC	Convolutional Turbo Code
D/A	Digital-to-Analog
DL	Downlink
DSL	Digital Subscriber Line
EAP	Extensible Authentication Protocol
EVDO	Evolution Data Optimized or Evolution Data Only
FCH	Frame Control Header
FDD	Frequency Division Duplex
FFT	Fast Fourier Transform
FTP	File Transfer Protocol
FUSC	Fully Used Sub-Channelization
GMH	Generic MAC Header

HARQ	Hybrid Automatic Repeat reQuest
HFDD	Half-Duplex FDD
HHO	Hard Hand-Off
HMAC	keyed Hash Message Authentication Code
HO	Hand-Off or HandOver
HSDPA	High Speed Downlink Packet Access
HSUPA	High Speed Uplink Packet Access
HTTP	Hyper Text Transfer Protocol
IEEE	Institute of Electrical and Electronics Engineers
IETF	Internet Engineering Task Force
IFFT	Inverse Fast Fourier Transform
IMT-2000	International Mobile Telecommunications-2000
ISI	Inter-Symbol Interference
ITU	International Telecommunications Union
LOS	Line of Sight
LUT	Look-Up Table
MAC	Medium Access Control
MCS	Modulation and Coding Scheme
MD5	Message-Digest algorithm 5
MDHO	Macro Diversity Hand Over
MIMO	Multiple-Input Multiple-Output (Antenna)
MIPPC	Mean Instantaneous Post-Processing Capacity
MLD	Maximum Likelihood Detector or Maximum Likelihood Decoder
MMSE	Minimum Mean Square Error
MPLS	Multi-Protocol Label Switching
MPDU	MAC Packet Data Unit
MSDU	MAC Service Data Unit
MS	Mobile Station
NLOS	Non-Line-of-Sight
nrtPS	Non-Real-Time Packet Service
OFDM	Orthogonal Frequency Division Multiplexing
OFDMA	Orthogonal Frequency Division Multiple Access
PAS	Power Azimuth Spectrum
PedB	(ITU) Pedestrian B channel
PDP	Power Delay Profile
PDU	Packet Data Unit
PER	Packet Error Rate
PF	Proportional Fair (Scheduler)
PHY	PHYsical layer
PKMv2	Privacy and Key Management protocol Version 2
PMP	Point to MultiPoint
PUSC	Partially Used Sub-Channelization
QAM	Quadrature Amplitude Modulation
QPSK	Quadrature Phase Shift Keying
RB	Resource Block
RTG	Receive/transmit Transition Gap
rtPS	Real-Time Packet Service
Rx	Receive

SE	Spectral Efficiency
SF	Shadow Fading
SIM	Subscriber Identity Module
SIMO	Single-Input Multiple-Output (Antenna)
SINR	Signal to Interference + Noise Ratio
SISO	Single-Input Single-Output (Antenna)
SM	Spatial Multiplexing
SMS	Short Message Service
SNR	Signal to Noise Ratio
S-OFDMA	Scalable Orthogonal Frequency Division Multiple Access
SQI	System Quality Indicator
SS	Subscriber Station
STC	Space Time Block Coding
TDD	Time Division Duplex
TDMA	Time Division Multiple Access
TTG	Transmit/receive Transition Gap
Tx	Transmit
UGS	Unsolicited Grant Service
UL	Uplink
USIM	Universal Subscriber Identity Module
VoIP	Voice over Internet Protocol
WCDMA	Wideband Code Division Multiple Access
WiMAX	Worldwide Interoperability for Microwave Access
ZF	Zero Forcing

References

1. S. Tekinay, ed., *Next Generation Wireless Networks*, Springer, New York, 2001.
2. *Mobile and Wireless Internet: Protocols, Algorithms, and Systems*, N. Pissinou, K. Makki, and E.K. Park, eds., Springer, New York, 2004.
3. S.G. Glisic, *Advanced Wireless Communications: 4G Technologies*, John Wiley, New York, 2004.
4. R. Prasad, *CDMA for Wireless Personal Communications*, Artech House, Norwood, MA, 1996.
5. B.S. Valluri, *Spectral Efficient Technologies in 3G for Packet Access*, TechOnLine, 2005; available at http://www.techonline.com/community/ed_resource/feature_article/37982.
6. IEEE, *IEEE 802.16-2004, Part 16: Air Interface for Fixed Broadband Wireless Access Systems*, IEEE, Washington, DC, 2004.
7. IEEE, *IEEE 802.16e-2005, Part 16: Air Interface for Fixed Broadband Wireless Access Systems, Amendment for Physical and Medium Access Control Layers for Combined Fixed and Mobile Operation in Licensed Bands*, IEEE, Washington, DC, 2005.
8. WiMAX Forum, *Mobile WiMAX—Part I: A Technical Overview and Performance Evaluation*, White Paper, WiMAX Forum, 2006; available at http://www.wimaxforum.org/news/downloads/Mobile_WiMAX_ Part1_Overview_and_Performance.pdf.
9. S. Pietrzyk, *OFDMA for Broadband Wireless Access*, Artech House, Norwood, MA, 2006.
10. J. Chuang and N. Sollenberger, Beyond 3G: wideband wireless data access based on OFDM and dynamic packet assignment, *IEEE Commun. Mag.*, 38(7), 78, 2000.
11. J.J. Blanz, A. Papathanassiou, M. Haardt, I. Furi, and P.W. Baier, Smart antennas for combined DOA and joint channel estimation in time-slotted CDMA mobile radio systems with joint detection, *IEEE Trans. Vehic. Technol.*, 49(2), 293, 2000.

12. A.J. Paulraj and B.C. Ng, Space-time modems for wireless communications, *IEEE Personal Commun.*, 5(1), 36, 1998.
13. C. Wen, Y. Wang, and J. Chen, Adaptive spatio-temporal coding scheme for indoor wireless communication, *IEEE J. Select. Areas Commun.*, 21, 161, 2003.
14. V. Tarokh, H. Jafarkhani, and A.R. Calderbank, Space-time block codes from orthogonal designs, *IEEE Trans. Inform. Theory*, 45, 1456, 1999.
15. J. Foschini and M. Gans, On the limits of wireless communication in a fading environment when using multiple antennas, *Wireless Personal Commun.*, 6(3), 311, 1998.
16. S.M. Alamouti, A simple transmitter diversity scheme for wireless communications, *IEEE J. Select. Areas Commun.*, 16, 1451, 1998.
17. D. Tse and P. Viswanath, *Fundamentals of Wireless Communication*, Cambridge University Press, Cambridge, 2005.
18. E.A. Jorswieck and H. Boche, Performance analysis of capacity of MIMO systems under multiuser interference based on worst-case noise behavior, *EURASIP J. Wireless Commun. Networking*, 2004(2), 273, 2004.
19. J.P. Kermoal, L. Schumacher, K.I. Pedersen, P.E. Mogensen, and F. Frederiksen, A Stochastic MIMO radio spatial channel model with experimental validation, *IEEE Trans. Select. Areas Commun.*, 20(6), 1211, 2002.
20. G.H. Golub and C.F. van Loan, *Matrix Computations*, Johns Hopkins University Press, Baltimore, 1996.
21. ITU, *Guidelines for Evaluation of Radio Transmission Technologies for IMT-2000*, Recommendation ITU-R M.1225, ITU, Geneva, 1997.
22. 3GPP2, *CDMA2000 Evaluation Methodology*, C.R1002-0, 3GPP2, 2004.
23. 3GPP-3GPP2 Spatial Channel Model Ad Hoc Group, *Spatial Channel Model Text Description*, v7.0 (SCM-135), 3GPP2, 2003.
24. Working Group 5, Evaluation Ad Hoc Group, *1xEV-DV Evaluation Methodology—Addendum (V6)*, 3GPP2, 2001.
25. R. Srinivasan, S. Timiri, A. Davydov, and A. Papathanassiou, *Downlink Spectral Efficiency of Mobile WiMAX*, IEEE VTC, Dublin, Ireland, 2007.

PROTOCOL ISSUES

Chapter 6

Medium Access Control in WirelessMAN

Mehmet S. Kuran, Fatih Alagoz, and Tuna Tugcu

Contents

Wireless metropolitian area networks (WirelessMANs) have gained importance in recent years. While the Institute of Electrical and Electronics Engineers (IEEE) 802.16 standard is the most important and promising WirelessMAN technology, European Telecommunications Standards Institute (ETSI) and Telecommunications Technology Association (TTA) of South Korea have also developed their own WirelessMAN standards. In this chapter the media access control (MAC) layer of these standards is explained. Since IEEE 802.16 is the most important WirelessMAN technology and HiperMAN and WiBro were developed based on IEEE 802.16, this chapter focuses on IEEE 802.16. Both operating modes (point-to-multipoint [PMP] and mesh) of these standards are described. In addition to the mechanisms included in these standards, newly proposed MAC layer mechanisms in the literature are included. Also, the current problems and open issues of the WirelessMAN technologies are addressed in this chapter.

6.1   Introduction

The MAC layer is responsible for accessing the physical (PHY) layer and one hop transmission in the same WirelessMAN. All of the current WirelessMAN technologies have connection-oriented MAC layers. The link state is one of the most important aspects of wireless networks. MAC layer protocols of WirelessMAN systems are designed to be responsive to variations in the link state. Thus MAC layer protocols are able to change the modulation and coding of connections as the link state changes.

There are two types of devices in a WirelessMAN network, the base station (BS) and subscriber stations (SSs). The central station that connects the network to external networks is called the BS. The main responsibility of the BS is the management of traffic in the network. All other devices belong to the users in the network. These user devices are called SSs. There can be only one BS in the network, whereas there can be many SSs in the same WirelessMAN.

The main topology in all four WirelessMAN standards described in this chapter is the PMP topology. In this topology, all the SSs directly connect to the BS. Thus all transmissions are between SSs and the BS. With the exception of HiperACCESS, all three standards also support mesh topology. In mesh topology, SSs are either directly or indirectly connected to the BS. SSs that are not directly connected to the BS send their transmissions to the BS through other SSs. In this topology SSs can also communicate directly between themselves. Similar to the PMP topology, there is only one BS in the network in this mode.

6.2   MAC Layer of IEEE 802.16

The IEEE 802.16 standard was initially developed in 2001 [1]. The MAC layer of the initial standard only supports the PMP operating mode (Figure 6.1). The MAC layer of the current standard, the IEEE 802.16-2005 [2], also supports the mesh operating mode. There are several differences between the two operating modes, such as frame structure and bandwidth allocation.

Transmissions are described by service flows (SFs) and connections in IEEE 802.16. A transport service that provides the transmission of MAC protocol data units (PDUs) between two nodes is called a SF. Each SF is associated with a connection in the IEEE 802.16 standard.

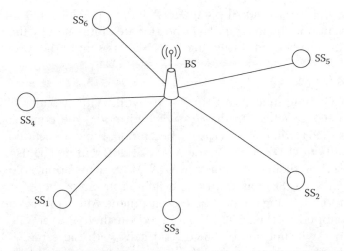

Figure 6.1   PMP topology of IEEE 802.16 [1].

6.2.1   Sublayers of the MAC Layer of IEEE 802.16

There are three sublayers in the MAC layer stack of IEEE 802.16 (Figure 6.2). Each sublayer has its own mechanisms and is responsible for different tasks.

6.2.1.1   Convergence Sublayer

The convergence sublayer (CS) is the highest sublayer of the IEEE 802.16 MAC layer. Unlike the other two sublayers, there are different CSs in the MAC layer for different network layer protocols used in the network. Currently the IEEE 802.16 standard defines two CSs. The asynchronous transfer mode (ATM) CS is used for handling ATM cells and the packet CS is used for Ethernet, point-to-point protocol (PPP), and transmission control protocol/internet protocol (TCP/IP) packets. On the transmitter side, the CS is mainly responsible for converting network layer packets into MAC segment data units (SDUs) and vice versa on the receiver side. First, the CS maps the transmission parameters

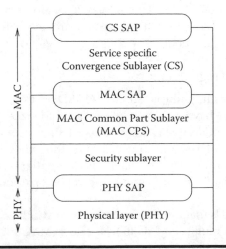

Figure 6.2   Layers of IEEE 802.16 [1].

(including quality of service [QoS] parameters) into applicable IEEE 802.16 SFs. Then repetitive information in the network layer header are eliminated by the payload header suppression (PHS) mechanism. Lastly, the MAC SDU is sent to the lower sublayer.

6.2.1.1.1 ATM CS

There are two switching modes in the ATM network: virtual path (VP) switched and virtual channel (VC) switched. In the VP-switched mode, the connection is identified by VP identifiers (VPI), whereas in VC-switched mode, the connection is identified by a VPI and VC identifier (VCI) couple. In the ATM CS, different mechanisms are applied to ATM packets based on their switching modes. ATM CS maps applicable identifiers (e.g., ATM QoS constraints) to SFs based on the switching mode of the ATM network. These mappings are handled during connection establishment. After connection establishment, ATM cells are mapped to IEEE 802.16 SFs based on their VPI or VPI and VCI values depending on the switching mode. Since ATM cells with the same VPI values or VPI and VCI values have mostly the same information in their cell headers, these repetitive fields are suppressed when the PHS mechanism is used. Only header information that is independent of VPI and VCI values is not suppressed. Therefore, if the service provider wants to keep the intralayer transparency, the PHS mechanism should not be used.

6.2.1.1.2 Packet CS

The packet CS defines a set of classifiers (e.g., source and destination addresses, protocol version) for network layer packets. These classifiers are used for mapping higher layer packets into IEEE 802.16 SFs. Each classifier is also assigned with priorities. These priorities indicate in which order the classifiers are applied to packets. An SF can receive packets from more than one classifier. If a network layer packet does not match any of the classifiers in the packet CS, that packet is discarded.

6.2.1.2 *Common Part Sublayer*

The second part of the MAC layer is the common part sublayer (CPS). This sublayer is the main sublayer of the MAC layer of IEEE 802.16. It handles media access, bandwidth allocations, connection establishment, connection management, and QoS management. MAC SDUs are gathered from the CS sublayer via the MAC service access point (SAP) and are converted into MAC PDUs. This sublayer is also responsible for packing and fragmentation of MAC SDUs. These mechanisms are explained below. Media access, connections, and QoS management are explained in Section 6.2.2 and Section 6.2.3.

6.2.1.2.1 Packing and Fragmentation

With the packing mechanism, multiple small MAC SDUs are packed into one MAC PDU. MAC SDUs and PDUs can have either fixed or variable sizes. These size selections are based on the SF used. If the MAC SDUs have fixed sizes, the receiving node can easily calculate where the MAC SDUs start and end. On the other hand, if MAC SDUs have variable sizes, a special subheader—the packing subheader—is used in each MAC PDU. This subheader is responsible for informing the receiver node how to access the MAC SDUs inside the MAC PDU.

In the case of MAC SDUs that are larger than one MAC PDU, the transmitter node fragments the MAC SDU into multiple MAC PDUs. These MAC PDUs are sent in an orderly fashion. Similar to the packing subheader, a special subheader, the fragmentation subheader, is inserted into each MAC PDU that is part of the same MAC SDU. This subheader informs the CPS of the receiving node of the order in which these MAC PDUs will be integrated.

Packing and fragmentation can be used in conjunction with the automatic response request (ARQ) mechanism. With the help of PHS and packing mechanisms, the standard eliminates bandwidth waste due to repeating information.

6.2.1.2.2 ARQ

In the PMP mode, during SF establishment, the use of ARQ is negotiated between the SSs and the BS on an SF basis, whereas in the mesh mode, ARQ is enabled on a MAC PDU basis. While some MAC PDUs that are being transmitted on a link use the ARQ mechanism, other MAC PDUs do not. SSs working in the PMP mode are not required to implement the ARQ mechanism. However, SSs in the mesh mode must implement ARQ mechanism whether they will use it or not.

6.2.1.3 Security Sublayer

The third sublayer, the security sublayer, is responsible for maintaining security in the network. Security is maintained with encryption of data packets, secure key distribution via privacy key management (PKM), authorization of PKM, and identification of nodes via X.509 profiles. Usage of these mechanisms are described with *security associations* (SAs). These SAs are identified by 16-bit SA identifiers (SAIDs). Each connection can be assigned a different SAID, or a single SAID can be associated with a number of connections. Two types of SAs are defined: data SAs and authorization SAs. With the standardization of IEEE 802.16e [2], the security mechanisms are improved in order to cope with the threats arising from mobile profiles. Johnston and Walker [3] studied the security mechanisms in IEEE 802.16. They state that the security mechanisms defined in the IEEE 802.16 standard have many flaws, especially regarding the authorization process, since there is no explicit definition for authorization SAs in the standard. While the new security mechanisms introduced in IEEE 802.16e provide better protection against attacks, the authorization problem still exists and must be addressed.

6.2.2 Service Flows and Connections

As described in Section 6.2, a transport service that provides the transmission of MAC PDUs between two nodes is called an SF. Each SF has a 32-bit identifier called an SF identifier (SFID). An SS has a number of SFs at the same time, each with different service parameters. An SF defines various characteristics regarding the traffic supported by itself, such as QoS parameters, the SA used for the traffic, the choice of MAC SDU and MAC PDU sizes, and the choice of ARQ.

Three QoS parameter sets are associated with SFs: ProvisionedQoSParamSet, AdmittedQoSParamSet, and ActiveQoSParamSet. These QoS parameter sets include the following QoS parameters: the maximum sustained traffic rate, the minimum reserved traffic rate, traffic priority, tolerated jitter, and the maximum latency parameters. The first parameter set is defined by higher level protocols (e.g., multiprotocol label switching [MPLS]) and cannot be changed by the MAC layer of IEEE 802.16. If an SF has only ProvisionedQoSParamSet, it is called provisioned SF. The second parameter set is the AdmittedQoSParamSet. This parameter set is used by the BS to allocate resources for SFs. If a provisioned SF has its resources allocated, it becomes an admitted SF. Lastly, if an SF is associated with an active connection (a connection that sends packets), it sends

Table 6.1 QoS Parameters of Scheduling Services

	UGS	rtPS	nrtPS	BE	ertPS
Maximum sustained traffic rate	+	+	+	+	+
Maximum latency	+	+	−	−	+
Tolerated jitter	+	−	−	−	−
Request/transmission policy	+	+	+	+	+
Minimum reserved traffic rate	+/−	+	+	−	+
Traffic priority	−	−	+	+	−

them based on the ActiveQoSParamSet of its SF. This kind of SF is called an active SF. Hence both admitted and active SFs are associated with a connection.

6.2.2.1 PMP Mode

In the PMP operating mode, each SF is associated with a scheduling service. These scheduling services define the data handling mechanism associated with the SF and also the resource allocation mechanisms available to the SF. There are four scheduling service types defined in IEEE 802.16 standard. A fifth scheduling service type is introduced in the IEEE 802.16e standard. Each scheduling service type uses different QoS parameters (Table 6.1). Traffic supported by each scheduling service and their bandwidth request mechanisms are as follows.

- **Unsolicited grant service (UGS):** Constant bit rate (CBR) traffic and T1/E1 services use this scheduling service. SSs declare their bandwidth requirements for their UGS SFs to the BS in SF establishment. In each frame the BS allocates exactly the same amount of bandwidth to each UGS SF, based on the initial bandwidth provisioning. This bandwidth is always allocated to the SS regardless of the actual traffic of lower priority SFs. There is a poll me bit (PMB) in the grant subheader of each MAC PDU belonging to a UGS SF. This field is used by the SS to send a polling request to the BS. If the PMB in a MAC PDU is set, the BS allocates a dedicated request slot to the SS in the next frame. This dedicated request is used by the SS to send its bandwidth request of non-UGS SFs. The bandwidth of the UGS SF is fixed and cannot be changed unless the SF is terminated and established once again.
- **Real-time polling service (rtPS):** This scheduling service is used for real-time variable bit rate (VBR) traffic. The BS periodically allocates dedicated slots in the uplink subframe as request opportunities for each rtPS SF in the network. The SF primarily requests bandwidth using these periodic request opportunities. Unlike UGS SFs, the size of the requested bandwidth varies from time to time up to a limit set during the SF establishment. Because of this request/grant mechanism, there is some MAC overhead introduced to the network for rtPS SFs.
- **Non-real-time polling service (nrtPS):** Non-real-time traffic uses nrtPS SFs. These scheduling services are similar to rtPS SFs regarding its bandwidth request mechanism. However, the period of the dedicated request opportunities is higher than for rtPS SFs. Unlike rtPS SFs, an nrtPS SF may also use contention periods to send its bandwidth request.

- **Best effort (BE):** Non-time-critical BE traffic uses this scheduling service. BE SFs can send bandwidth requests only in the contention periods. A BS never allocates dedicated slots for request opportunities to the BE SFs.
- **Extended real-time polling service (ertPS):** Lee et al. have shown that the four scheduling services in IEEE 802.16 are not appropriate for services like voice over Internet protocol (VoIP) [4]. Addressing this issue, the latest standard of IEEE 802.16 (i.e., IEEE 802.16e) introduced the ertPS scheduling service. The ertPS is similar to UGS regarding bandwidth allocation. In every frame the BS allocates the same amount of bandwidth to the SF. However, the bandwidth allocated to the SF can change in time. An ertPS SF can decrease or increase its allocated bandwidth up to the maximum bandwidth set in SF establishment based on its queue status. ertPS SFs usually have a higher priority than rtPS SFs.

6.2.2.1.1 Connections

A unidirectional transmission between the BS and a SS is defined by a connection in the PMP mode of the IEEE 802.16 standard. All admitted and active SFs are associated with a connection and every connection is associated with an SF. A connection can be unicast, multicast, or broadcast. Connections are identified by 16-bit connection identifiers (CIDs). Upon initialization, an SS establishes its connections including three pairs of management connections. These management connections are used for management messages (Table 6.2). Basic management connections are used for the most important and time-critical management messages. Primary management connections are used for more delay-tolerant management messages. Secondary management connections are only used in managed SSs. During the operation of the network, connections can be altered or terminated, as well as new connections can be established. Xhafa et al. [5] studied the effects of the number of connections. It is shown that if the number of connections increases, the MAC layer efficiency decreases considerably.

6.2.2.2 Mesh Mode

Transmission between nodes is handled differently in the mesh mode of the IEEE 802.16 standard. Mesh SSs (MSSs) do not need to be connected directly to the mesh BS (MBS) in the mesh mode. Also, they can communicate directly with each other. Every node in the network has its neighbor nodes. A node with one-hop distance to the central node is called a neighbor node of the central node. A node and all of its neighbor nodes form the neighborhood of the central node (Figure 6.3). In addition to neighborhoods, there is an extended neighborhood concept in the mesh mode. An extended neighborhood of a node consists of nodes with one- and two-hop distances to the central node.

Also, a parent node is selected among the nodes in the neighborhood. A node with a lower hop count to the MBS than the central node is called the parent node of the central node. There may be multiple parent node candidates in a neighborhood, but one of these candidate nodes is selected to be the parent node. The IEEE 802.16 standard describes a method utilizing signal-to-noise ratio (SNR) values for selecting parent nodes. This method selects the node with the highest SNR value among candidate nodes as the parent node. From the parent node's view, the central node is called a child node (Figure 6.3).

In addition to SFs and connections, there are links in the mesh mode. Upon initialization, an MSS establishes one link with each of its neighbors. These unidirectional links, identified by eight-bit link identifiers (link ID), are used for all data transmissions

Table 6.2   Management Messages in PMP Mode

Message Type	Management Connection Used
Ranging messages	Basic management
Burst profile change messages	Basic management
Reset command	Basic management
SS basic capability messages	Basic management
De/re-register command	Basic management
ARQ messages	Basic management
Channel measurement reports	Basic management
Adaptive antenna system messages	Basic management
Uplink channel descriptor (UCD)	Broadcast
Downlink channel descriptor (DCD)	Broadcast
Downlink access definition (DL-MAP)	Broadcast
Uplink access definition (UL-MAP)	Broadcast
SS network clock comparison	Broadcast
Fast power control	Broadcast
Privacy key management (PKM) messages	Primary management
Dynamic service addition/alteration/termination messages	Primary management
Multicast assignment messages	Primary management
Simple network management protocol (SNMP) messages	Secondary management
Dynamic host configuration protocol (DHCP) messages	Secondary management
Trivial file transfer protocol (TFTP) messages	Secondary management

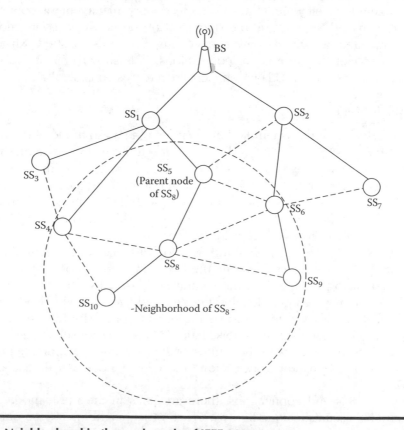

Figure 6.3   Neighborhood in the mesh mode of IEEE 802.16.

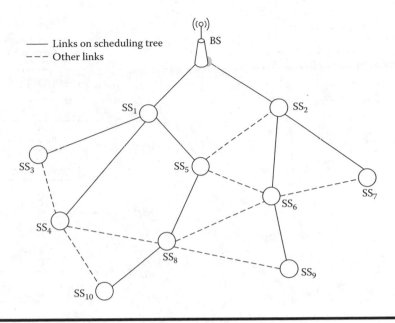

— Links on scheduling tree
--- Other links

Figure 6.4   Scheduling tree in the mesh mode of IEEE 802.16.

between the nodes in each end. Among all the links in the network, links between nodes and their parent nodes form a scheduling tree (Figure 6.4). However, the parent node selection method described in the standard does not guarantee that it finds the optimal scheduling tree. Shetiya and Sharma [6] developed an analytical solution to find the optimal routing for the mesh mode. In Wei et al. [7], an interference-aware routing algorithm is introduced. This mechanism utilizes a space division multiple access (SDMA) approach in parent node selection. These link IDs are also used for CID construction in the mesh mode. While there are not any QoS or service parameters associated with links in the mesh mode, there are three QoS-related fields used in the CID construction. They are used for handling QoS in mesh mode. Thus whenever a link is established, 64 connections are set up for that link. However, handling these QoS-related fields are left unstandardized in IEEE 802.16.

Since all transmissions between a node couple are sent using a single link, the five scheduling services developed for the PMP mode cannot be used in the mesh mode. Therefore different request/grant mechanisms are developed for the mesh mode. There are three different ways for handling data transmission scheduling in the mesh mode of IEEE 802.16: centralized scheduling, coordinated distributed scheduling, and uncoordinated distributed scheduling. These methods are used in conjunction with each other. While centralized scheduling is generally used for MSS-MBS traffic (i.e., Internet traffic), distributed scheduling is generally used for intranetwork transmissions (i.e., transmissions between two MSSs). In each method, the resulting data transmissions are sent in a collision-free manner.

6.2.2.2.1   Centralized Scheduling

This method is similar to the PMP operating mode. Transmission requests of each link on the scheduling tree are sent to the MBS. The MBS decides how much of each request is granted and in which part of the frame these transmissions take place. Then the MBS

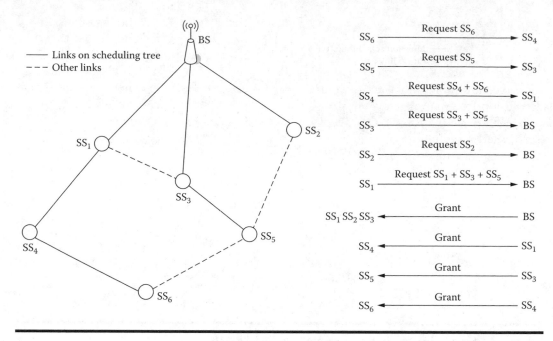

Figure 6.5 Steps of centralized scheduling.

broadcasts this grant information using grant messages. In this method, MSSs send their requests in a collision-free manner.

MSSs not directly connected to the MBS send their request messages to their parent nodes. Upon receiving request messages from their child nodes, MSSs send their requests to their parent nodes along with the requests of their child nodes. Thus request messages of all MSSs reach the MBS in several frames depending on the topology and number of MSSs. MSSs receiving grant messages forward these messages to their child nodes (Figure 6.5). Grant messages do not contain the actual grants for each link; instead, they include flow assignments. Nodes calculate the actual grants from these flow assignments in the grant message.

The request and grant messages are sent in a fixed order depending on the topology in the mesh mode. The grant messages follow the order of the scheduling tree. First, the MBS sends its message, followed by the MSSs that have a one-hop count to the MBS. After MSSs that are directly connected to the MBS send their messages, MSSs send their requests in groups that have the same hop count to the MBS, starting with MSSs with a hop count of two and ending with the leaf nodes. Request messages are sent in the reverse order of the scheduling tree, starting from the leaf nodes to the nodes directly connected to the MBS.

6.2.2.2.2 Distributed Scheduling

Distributed scheduling is used for handling traffic on the links not in the scheduling tree. Nodes in distributed scheduling coordinate their schedules with all nodes in their extended neighborhood. Scheduling is carried on a three-way handshaking mechanism. Each MSS broadcasts its requests, grants for earlier requests, and their available data slots to their neighbors. If the destination neighbor grants the request, it compares its available data slots with the data slots sent in the request message and allocates a subset of the intersection between these two slot sets to the source node. Then the destination node sends a grant message to the source node. Lastly, the source node sends a grant message to the destination node to finalize the handshake procedure. The difference between coordinated and uncoordinated scheduling is the manner in which the request/grant messages are sent. Similar to centralized scheduling, request/grant messages are sent in a collision-free manner in coordinated distributed scheduling, whereas these messages may collide with each other in uncoordinated distributed scheduling. Uncoordinated distributed scheduling is generally used for fast adhoc transmissions.

Nodes send their distributed scheduling messages and data packets of these schedules in the dedicated distributed scheduling part of the frame. Thus if there is any transmission using distributed scheduling in the network, the frame is divided into two parts, one for centralized scheduling and one for distributed scheduling. Cheng et al. [8] show that this partitioning results in unused data slots in the frame and develop a combined scheme that allows either method to send their data packets in both parts of the frame.

6.2.3   Frame Structure

Transmissions in an IEEE 802.16 cell is divided into fixed-size frames. The size of these frames varies between 0.5 ms and 25 ms depending on the PHY layer used and the choice of the service provider. The structure of frames is different in the two operating modes. While uplink and downlink transmissions are clearly differentiated inside a frame in the PMP mode, there is no such distinct differentiation in the mesh mode.

6.2.3.1   PMP Mode

Two different frame structures exist in PMP mode: frequency division duplex (FDD) and time division duplex (TDD). In FDD frame structure, downlink and uplink subframes are transmitted in different frequencies (Figure 6.6). These subframes occupy the same frequency in the TDD frame structure, but separated in time (Figure 6.7). Downlink and uplink subframes are used for transmissions initiated by the BS and SSs, respectively. Both subframes are formed of physical slots (PSs). PSs allocated to the same connection are located adjacent to each other and are called bursts.

Each downlink subframe includes a frame control header (FCH) that is composed of the downlink mobile application part (DL-MAP), uplink MAP (UL-MAP), and the periodic downlink channel descriptor (DCD) and uplink channel descriptor (UCD). DL-MAP and UL-MAP describe the mapping of the downlink and uplink PSs, respectively, to connections. The DCD and UCD fields do not reside in every frame; instead they are transmitted periodically. They describe the use of burst profiles to SSs. After the FCH, downlink data transmission is sent in bursts in a time division multiplex (TDM) manner. In the FDD mode, after the TDM downlink bursts, there is an additional time division multiple access (TDMA) part that is used for half-duplex SSs.

The data transmission is conducted in a similar manner in the uplink subframe as in the downlink subframe. In addition to allocations to unicast connections, there are allocations to multicast and broadcast connections. PS allocations for broadcast connections form a contention-based part in the uplink subframe. The first contention-based part is used for the initial ranging and initialization process of new SSs. Others are used by SSs for their nrtPS and BE connections. The sizes of these contention periods affect the overall performance of the uplink subframe. Unnecessary large contention periods decrease the size of the uplink subframe usable by data transmissions. On the other hand, if the sizes of these contention periods are smaller than necessary, nrtPS and BE connections

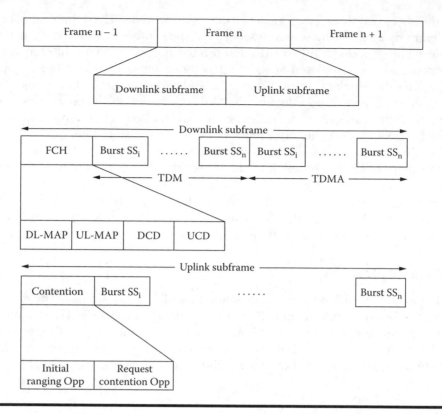

Figure 6.6   FDD frame structure in PMP mode.

suffer unacceptable delay values. In addition to these parts, there are periodic unicast PS allocations for each ertPS, rtPS, and nrtPS connection in the network. These allocations are small in size and are used to send only bandwidth requests to the BS. The effects of the contention window size are studied in Iyengar et al. [9] and Doha et al. [10].

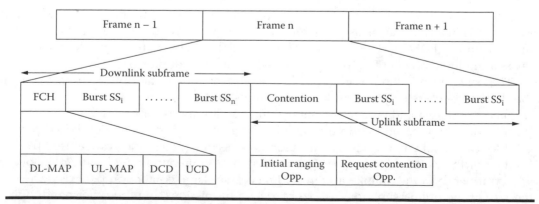

Figure 6.7   TDD frame structure in PMP mode.

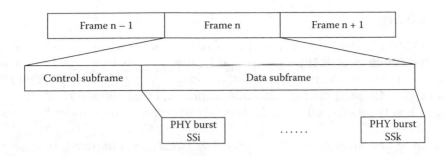

Figure 6.8   Frame structure of the IEEE 802.16 mesh mode.

6.2.3.2   Mesh Mode

The frame structure is quite different in the mesh mode of IEEE 802.16 than the PMP mode (Figure 6.8). Only a TDD frame structure exists in the mesh mode. The frame is divided into two parts: control subframe and data subframe. The first part is used to send network configuration and scheduling messages. The second part consists of data bursts of MSSs and the MBS.

The control subframe can be either a scheduling control subframe or a network control subframe. Request, grant, and scheduling configuration messages are sent using the scheduling control subframe (Figure 6.9). The order in which these messages are sent is described in Section 6.2.2. The scheduling configuration messages are sent similar to grant messages. Centralized scheduling and coordinated scheduling messages use this subframe. On the other hand, scheduling messages of the uncoordinated distributed scheduling method uses the data subframe. The second subframe is used periodically, similar to the UCD and DCD messages in the PMP mode (e.g., 1 network control subframe in 15 frames). The MBS informs MSSs about data subframe channel usage by sending the burst profiles in the network configuration messages. MSSs use entry messages to join the network (Figure 6.9).

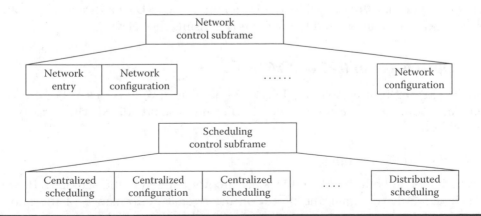

Figure 6.9   Control subframe of the IEEE 802.16 mesh mode.

6.2.4 Mobility

The IEEE 802.16 standard is developed for fixed users, therefore it does not support mobile users. Leung et al. [11] propose a mechanism that enables mobile user support to IEEE 802.16 networks. A handover mechanism is developed for mobile stations (MSs) in Leung et al. [11] based on the initialization procedure in the standard. IEEE 802.16 Task Group E was established to standardize similar efforts; the standard is finished in December 2005.

Unlike an SS and an MSS, an MS can switch between BSs during a transmission with a handover mechanism. This handover procedure can be initiated by both the MS and its current BS. The handover procedure is similar to the initialization of an SS to the network. The MS may listen to the target BS before the handover and learn about its system parameters, thus speeding up the handover procedure. Similarly the MS may notify its target BS through the backhaul network. In this case, the target BS allocates a dedicated ranging opportunity to the MS that will be used by the MS to enter the network instead of the contention-based ranging part. Lee et al. [12] studied the handover process of the standard and proposed a faster handover mechanism based on eliminating redundant work in the process.

Another major problem regarding the MSs is energy consumption. A mechanism called sleep mode was introduced in IEEE 802.16e to reduce the energy consumption of MSs. In this mechanism, the transmission between the BS and MS is divided into two parts: interval of unavailability and interval of availability. During the first interval, the transmitter of the MS is switched off and it does not receive any transmission from the BS. The BS buffers any packets that arrive in this interval destined to the MS. During the second interval, the BS sends these buffered packets as well as the packets that arrive in this interval to the MS. If there are no packets destined to the MS during an interval couple, the MS increases its sleep time and informs the BS about its new waking time. In Seo et al. [13], Xiao [14], and Zhang and Fujise [15], it is shown that this power-saving mechanism is effective.

The IEEE recently formed a new task group under Working Group 16 [16]. This task group—Mobile Multihop Relay (MMR), also known as IEEE 802.16j—is developing a new standard that works in the PMP mode and allows SSs not directly connected to the BS to connect to the network. Unlike the mesh mode of IEEE 802.16, this standard tries to achieve this goal with relay stations (RSs) in the network. These RSs are directly connected to the BS, and SSs connect to the BS through these stations. RSs are similar to signal repeaters and are only able to relay a transmission. Data allocations in both the downlink and uplink are altered to enable this transmission relaying.

6.2.5 Open Issues in IEEE 802.16

Although IEEE 802.16 was developed in 2002, there are several issues to be resolved in the standard. Some mechanisms are intentionally left unstandardized. These open issues are as follows.

6.2.5.1 BS and SS Scheduler

Data schedulers on both the BS and SS sides are not defined in the standard. However, these schedulers have significant effects on the overall performance of the network. The BS scheduler is responsible for allocating bandwidth to SSs based on the request messages. A good BS scheduler should consider various QoS requirements and fairness issues. The BS aggregates requests from the same SS and generates a single grant value for each SS. Thus an additional scheduler is necessary on the SS side to distribute this aggregate grant between the SS's connections. The distributed structure of this scheduling mechanism allows better responding to queue variations, especially in burst traffic.

In the literature, there are several proposals for SS and BS schedulers. Chu et al. [16] introduce an SS scheduler. In these schedulers, connections with the same scheduling services are integrated and different queuing policies are applied to the queue of each scheduling service. They propose using wireless packet scheduling for rtPS connections, weighted round robin for nrtPS connections, and a first in first out (FIFO) scheduler for BE connections. Wongthavarawat and Ganz [17] propose a BS scheduler that takes the deadline parameters of rtPS PDUs into consideration. Arrival times of rtPS PDUs are sent to the BS through the UGS connection of the same SS. The BS scheduler applies different queuing policies to different scheduling services. Earliest deadline first queuing is used for rtPS connections and weighted round robin for nrtPS connections. Bandwidth is allocated to each BE connection equally. Jiang and Tsai [18] develop another BS scheduler using token buckets to characterize traffic flows. Cicconetti et al. [19] study the QoS mechanisms of the PMP mode. In this work, a weighted round robin scheduler is used for uplink bandwidth allocation in the BS and a deficit round robin scheduler is used in the SS scheduler as well as the downlink bandwidth scheduler in the BS. Niyato and Hossain [20] propose a queue state aware SS scheduler for polling service connections. This scheduler informs the packet source of its queue status and tries to control the packet arrival rate. Kim and Ganz [21] introduced a BS scheduler for the mesh mode. This scheduler introduces a node-ordering mechanism among the nodes with the same hop count from the MBS. Moreover, an SDMA mechanism is used to further increase the throughput in the network. Another SDMA-based BS scheduler for centralized scheduling of the mesh mode is introduced in Wei et al. [7]. This scheduler considers the interference of transmissions in links and makes scheduling decisions based on this information. Shetiya and Sharma [22] develop a BS scheduler that is based on a dynamic programming framework. This framework maximizes the total reward of the scheduler. Various definitions regarding the meaning of the reward metric are introduced and their performances are evaluated.

6.2.5.2 MAC PDU Size

The MAC overhead resulting from packing and segmentation subheaders can be decreased by selecting an ideal MAC PDU size. Also the size of the MAC PDU affects the performance of the ARQ mechanism. The size of a MAC PDU can be fixed or dynamic and recent works show that adaptive rather than fixed MAC PDU sizes result in better link utilization. Sengupta et al. [23] develop a method in the PMP mode with ARQ mechanism that changes the MAC PDU size according to the wireless channel state to optimize the MAC PDU size for fewer PDU retransmissions. Hoymann [24] calculates optimal PDU sizes for given bit error rates. This calculation also considers the MAC overhead due to retransmissions and packet headers. The MAC PDU size is calculated for PMP mode only. In mesh mode, different methods must be used to find optimal MAC PDU sizes to optimize overall link utilization.

6.2.5.3 Effects of Contention Periods

The SFs with nrtPS or BE scheduling services contend with each other for bandwidth allocation. The number of collisions can be decreased by selecting a longer contention window, but this in turn may generate unnecessarily long contention periods. Thus the system throughput decreases. Iyengar et al. [9] and Doha et al. [10], respectively, analyzed the effects of contention window size. Both works assume that each SS sends one bandwidth request message in each frame instead of sending one bandwidth request for each active connection. According to these studies, contention window size should be selected close to the number of SSs in the network.

When bandwidth request messages collide, their SSs wait for several slots before retransmitting the request messages. This back-off mechanism is analyzed in Oh and Kim [25] and Vinel et al. [26]. These studies show that there are different optimal back-off values for different numbers of active SSs in the network.

6.2.5.4   QoS in Mesh Mode

Unlike the PMP mode, MAC PDUs are responsible for their own QoS constraints in the mesh mode. There are no QoS parameters set with connections in the mesh mode. Every MAC PDU carries its own QoS constraint. The standard does not introduce any mechanism to handle these QoS constraints. Also, these QoS parameters in the MAC PDU header increase the MAC overhead in the mesh mode. A QoS mechanism is developed for the centralized scheduling in mesh mode of IEEE 802.16 in Kuran et al. [27]. Upon initialization, the MBS allocates five node identifiers (IDs) to each MSS. Each of these five virtual nodes establish one link with the actual node's parent node, as in the default mesh mode. Each of these virtual nodes sends one request message to the MBS. These five virtual nodes represent the five scheduling services in the PMP mode and their request/grant mechanisms are similar to these five scheduling services.

6.2.5.5   Security

The security sublayer of IEEE 802.16 left the authorization SAs undefined. Without a substantial authorization module, the mechanisms in the rest of the security sublayer cannot protect the network against malicious users. Johnston and Walker [3] propose several changes that must be introduced to the standard in order to increase the security of IEEE 802.16.

6.3   MAC Layer of ETSI HiperACCESS

High-performance radio access (HiperACCESS) was standardized by ETSI in 2002 to provide broadband wireless access for line-of-site (LOS) SSs [28]. The MAC layer of HiperACCESS is similar to the MAC layer of IEEE 802.16 in many aspects. However, HiperACCESS does not have a mesh mode like IEEE 802.16. The MAC layer of HiperAC-CESS consists of two layers, the convergence layer (CL) and the data link control (DLC) layer. The DLC layer is also divided into two planes, the user plane and the control plane (Figure 6.10).

6.3.1   Convergence Layer

The CL of HiperACCESS is very similar to the CS layer of IEEE 802.16. This layer is responsible for converting higher layer data units into HiperACCESS data units. Also similar to the CS layer of IEEE 802.16, there are two CLs in HiperACCESS, a cell-based CL, used for ATM, and a packet-based CL that is used for TCP/IP.

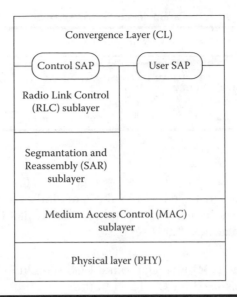

Figure 6.10   Layers of ETSI HiperACCESS.

6.3.2 DLC Layer

The DLC layer of HiperACCESS is connection-oriented, similar to IEEE 802.16's CPS. However, unlike IEEE 802.16, the sizes of the MAC PDUs are fixed. The MAC PDUs are 54 bytes long in the downlink and 55 bytes long in the uplink in HiperACCESS. There is also a 12-byte control MAC PDU used in ranging, bandwidth request messages, and queue status messages. These short messages can only be used in the uplink. The control plane DLC layer is composed of three sublayers—radio link control (RLC) sublayer, segmentation and reassembly (SAR) sublayer, and the MAC sublayer—while the user plane only includes the MAC sublayer.

The first sublayer—the RLC sublayer—deals with connection-related issues such as connection establishment and connection termination. Connections are associated with connection aggregates (CAs) and service categories in this sublayer. The initialization procedure used by new SSs joining the network is also handled in this sublayer. Finally, the security mechanisms, authorization, and encryption are also handled in this sublayer. The security methods used in HiperACCESS are similar to the ones used in IEEE 802.16.

Both data and management packets are segmented into fixed smaller PDUs in Hiper-ACCESS. The segmentation and reassembly (SAR) sublayer segments large data units and management messages into MAC PDUs. These segmented data units are reassembled at the receiving node by the reassembly mechanism in the SAR sublayer.

The request/grant and ARQ mechanisms are handled in this sublayer. Also, ARQ mechanisms can be used in HiperACCESS. However, only uplink connections can enable ARQ support and only on a connection basis.

6.3.3 Connections

The definition of a connection in HiperACCESS is the same as in the PMP mode of IEEE 802.16. Each connection is unidirectional, and there can be unicast, multicast, and broadcast connections in the network. Each SS has three management connection pairs

Table 6.3   QoS Parameters of Service Categories

	PRT	rRT	NRT	BE
Maximum bit rate	+	+	+	+/−
Guaranteed bit rate	+	+	+	−
Transfer delay	+	+	−	−
Delay variation	+	+	−	−

with the BS. Each connection has a service category in HiperACCESS. There are four types of service categories in HiperACCESS that are similar to IEEE 802.16 scheduling services. A connection has four QoS parameters. The availability of these parameters vary between service categories (Table 6.3):

- **Periodic real time (PRT):** Similar to the UGS scheduling service in IEEE 802.16, the PRT category is used for CBR data traffic.
- **Real time (RT):** This is the HiperACCESS QoS class counterpart of the rtPS scheduling service. It is used for VBR real-time traffic with bandwidth, delay, and jitter constraints.
- **Non-real time (NRT):** This is the HiperACCESS QoS class counterpart of nrtPS scheduling service. It is used for non-real-time traffic and has a minimum bandwidth constraint.
- **Best effort (BE):** Similar to the BE scheduling service in IEEE 802.16, there are no QoS constraints for this class.

Connections belonging to the same service category and SS are grouped in CAs. These CAs are identified by 16-bit CA Identifiers (CAIDs). The establishment of CAs is managed by the BS in connection establishment. The CAs are used for bandwidth requests. An SS sends its request messages on a CA basis. Otherwise the bandwidth request/grant mechanisms in HiperACCESS are similar to IEEE 802.16. The PRT CAs do not make any bandwidth requests because the BS makes continuous grants to these connections periodically. CAs with other service categories send their requests to the BS by either piggybacking their requests to the header of current data PDUs or sending request messages in dedicated transmission opportunities. These transmission opportunities are allocated to SSs periodically. Also, SSs can request additional dedicated transmission opportunities in the PMB. NRT and BE CAs may also send their requests in contention-based transmission opportunities.

6.3.4   Frame Structure

The frame structure of HiperACCESS is similar to the frame structure of the PMP mode of IEEE 802.16. The frame durations are fixed as 1 ms in HiperACCESS. The frame is divided into two parts, the downlink subframe and the uplink subframe. The downlink subframe includes DL-MAP, UL-MAP, the periodic general broadcast information (GBI) message, ARQ-MAP, TDM-based downlink data, and TDMA-based downlink data. The uplink subframe includes contention-based bandwidth requests, ranging bursts, and uplink data bursts.

The GBI message replaces the DCD and UCD messages in IEEE 802.16. Changes in PHY layer modes are transmitted to SSs with this message. The ARQ-MAP is used when there is at least one ARQ-enabled uplink connection. In case of an erroneous MAC PDU in the uplink subframe, the BS requests a retransmission of the corrupted MAC PDU by indicating it in the ARQ-MAP.

6.4   MAC Layer of ETSI HiperMAN

Initially published in 2003, the high-performance radio metropolitan access network (HiperMAN) standard is another ETSI standard in the broadband radio access networks (BRAN) project. The MAC layer of ETSI HiperMAN is very similar to the MAC layer of IEEE 802.16 [29]. The standard supports a PMP mode as well as a Mesh mode. While the initial standard targeted only fixed SSs, the current standard allows MSs as well. Support for MSs is based on the mechanisms introduced in IEEE 802.16e [30].

6.5   MAC Layer of TTA WiBro

Another recently developed WirelessMAN solution from the Telecommunications Technology Association (TTA) of Korea—WiBro—is designed to provide broadband wireless access to stationary and mobile personal SSs (PSSs) (low-to medium-mobility users up to 60 kmhr). Also known as the Portable Internet Service, development of WiBro started in 2003 in South Korea and finished its first phase in 2004. The second phase that enables WiBro's collaboration with IEEE 802.16e was finished in 2005. Based on IEEE 802.16 and IEEE 802.16e Draft 12, WiBro includes all mandatory mechanisms of IEEE 802.16. Since WiBro is based on IEEE 802.16, the MAC layer of WiBro is very similar to the MAC layer of IEEE 802.16. ARQ and PHS mechanisms, packing, fragmentation, and the five scheduling services of IEEE 802.16 are also available for WiBro [31].

References

1. IEEE, IEEE Standard for Local and Metropolitan Area Networks—Part 16: Air Interface for Fixed Broadband Wireless Access Systems, Standard 802.16-2004, IEEE, Washington, DC, 2004.
2. IEEE, IEEE Standard for Local and Metropolitan Area Networks—Part 16: Air Interface for Fixed Broadband Wireless Access Systems for Mobile Users, Standard 802.16-2005, IEEE, Washington, DC, 2005.
3. D. Johnston and J. Walker, Overview of IEEE 802.16 security, *IEEE Security Privacy*, 2(3), 40, 2004.
4. H. Lee, T. Kwon, and D. Cho, An efficient uplink scheduling algorithm for VoIP services in IEEE 802.16 BWA systems, *IEEE 60th Vehicular Technology Conference*, vol. 5, p. 3070, IEEE, Washington, DC, 2004.
5. A.E. Xhafa, S. Kangude, and X. Lu, MAC performance of IEEE 802.16e, *IEEE 62nd Vehicular Technology Conference*, vol. 1, p. 685, IEEE, Washington, DC, 2005.
6. H. Shetiya and V. Sharma, Algorithms for routing and centralized scheduling to provide QoS in IEEE 802.16 Mesh networks, *1st ACM Workshop on Wireless Multimedia Networking and Performance Modeling*, p. 140, IEEE, Washington, DC, 2005.
7. H.Y. Wei, S. Ganguly, R. Izmailov, and Z.J. Haas, Interference-aware IEEE 802.16 WiMax Mesh networks, *IEEE 61st Vehicular Technology Conference*, vol. 5, p. 3102, IEEE, Washington, DC, 2005.

8. S.M. Cheng, P. Lin, D.W. Huang, and S.-R. Yang, A study on distributed/centralized scheduling for wireless Mesh network, *International Wireless Communications and Mobile Computing Conference*, p. 599, IEEE, Washington, DC, 2006.

9. R. Iyengar, V. Sharma, K. Kar, and B. Sikdar, Analysis of contention-based multi-channel wireless MAC for point-to-multipoint networks, *IEEE International Symposium on a World of Wireless, Mobile and Multimedia Networks*, p. 453, IEEE, Washington, DC, 2006.

10. A. Doha, H. Hassanein, and G. Takahara, Performance evaluation of reservation medium access control in IEEE 802.16 networks, *IEEE International Conference on Computer Systems and Applications* p. 369, IEEE, Washington, DC, 2006.

11. K.K. Leung, S. Mukherjee, and G.E. Rittenhouse, Mobility support for IEEE 802.16d wireless networks, *IEEE Wireless Communications and Networking Conference*, vol. 3, p. 1446, IEEE, Washington, DC, 2005.

12. D.H. Lee, K. Kyamakya, and J.P. Umodi, Fast handover algorithm for IEEE 802.16e broadband wireless access system, *1st International Symposium on Wireless Pervasive Computing*, 2006 (ISWPC '06), p. 1,6, Phuket, Thailand, 2006.

13. J.B. Seo, S.Q. Lee, N.H. Park, H.W. Lee, and C.H. Cho, Performance analysis of sleep mode operation in IEEE 802.16e, *IEEE 60th Vehicular Technology Conference*, vol. 2, p. 1169, IEEE, Washington, DC, 2004.

14. Y. Xiao, Energy saving mechanism in the IEEE 802.16e WirelessMAN, *IEEE Commun. Lett.*, 9(7), 595, 2005.

15. Y. Zhang and M. Fujise, Energy management in the IEEE 802.16e MAC, *IEEE Commun. Lett.*, 10(4), 311, 2006.

16. M.J. Lee, J. Zheng, Y.-B. Ko, and D.M. Shrestha, Emerging Standards for Wireless Mesh Technology, *IEEE Wireless Communications*, 13(2), 56–63, 2006.

17. G. Chu, D. Wang, and S. Mei, A QoS architecture for the MAC protocol of IEEE 802.16 BWA system, *IEEE International Conference on Communications, Circuits, and Systems*, vol. 1, p. 435, IEEE, Washington, DC, 2002.

18. K. Wongthavarawat and A. Ganz, IEEE 802.16 based last mile broadband wireless military networks with quality of service support, *IEEE Military Communications Conference*, vol. 2, p. 779, IEEE, Washington, DC, 2003.

19. C.-H. Jiang and T.-C. Tsai, Token bucket based CAC and packet scheduler for IEEE 802.16 broadband wireless access networks, *IEEE 3rd Consumer Communications and Networking Conference*, vol. 1, p. 183, IEEE, Washington, DC, 2006.

20. C. Cicconetti, L. Lenzini, E. Mingozzi, and C. Eklund, Quality of service Support in IEEE 802.16 networks, *IEEE Network*, 20(2), 50, 2006.

21. D. Niyato and E. Hossain, Queue-aware uplink bandwidth allocation and rate control for polling service in IEEE 802.16 broadband wireless networks, *IEEE Trans. Mobile Comput.*, 5(6), 668, 2006.

22. D. Kim and A. Ganz, Fair and efficient multihop scheduling algorithm for IEEE 802.16 BWA systems, *2nd International Conference on Broadband Networks*, vol. 2, p. 833, IEEE, Washington, DC, 2005.

23. H. Shetiya and V. Sharma, Algorithms for routing and centralized scheduling in IEEE 802.16 Mesh networks, *IEEE Wireless Communications and Networking Conference*, IEEE, Washington, DC, 2006.

24. S. Sengupta, M. Chatterjee, S. Ganguly, and R. Izmailov, Exploiting MAC flexibility in WiMAX for media streaming, *IEEE 6th International Symposium on World of Wireless Mobile and Multimedia Networks*, p. 338, IEEE, Washington, DC, 2005.

25. A. Vinel, Y. Zhang, M. Lott, and A. Tiurlikov, Performance analysis of the random access in IEEE 802.16, *16th International Symposium on Personal, Indoor and Mobile Radio Communications*, vol. 3, p. 1596, IEEE, Washington, DC, 2005.

26. C. Hoymann, Analysis and performance evaluation of the OFDM-based metropolitan area network IEEE 802.16, *Elsevier Computer Networks*, 49(3), 341, 2005.

27. S.M. Oh and J.H. Kim, The optimization of the collision resolution algorithm for broadband wireless access network, *8th International Conference on Advanced Communication Technology*, vol. 3, p. 1944, IEEE, Washington, DC, 2006.

28. M.S. Kuran, B. Yilmaz, F. Alagoz, and T. Tugcu, Quality of service in Mesh mode IEEE 802.16 networks, *14th International Conference on Software, Telecommunications and Computer Networks*, 2006 (SoftCOM '06), Split, Dubrovnik, Croatia, 2006.

29. ETSI HiperACCESS, Broadband radio access networks (BRAN): HiperACCESS DLC protocol specification, ETSI, Sophia-Antipolis, France, 2004.

30. ETSI HiperMAN, Broadband radio access networks (BRAN): functional requirements for fixed wireless access systems below 11 GHz: HiperMAN data link control (DLC) Layer, ETSI, Sophia-Antipolis, France, 2006.

31. TTA WiBro, Specifications for 2.3GHz band portable internet service—physical and medium access control layer, TTA, Seoul, Korea, 2005.

Chapter 7

MAC and QoS in WiMAX Mesh Networks

Maode Ma and Yan Zhang

Contents

7.1 Introduction

A Worldwide Interoperability for Microwave Access (WiMAX) [1] network utilizes a shared medium to provide efficient transmission services. The point-to-multipoint (PMP) and mesh topology wireless networks are examples for sharing wireless media. The medium is radio waves in space.

In the PMP mode, the downlink from the base station (BS) to the subscriber stations (SSs) operates on a PMP basis. Within a given frequency channel and coverage of the BS, all SSs receive the same transmission, or parts of it. The BS is the only transmitter operating in this direction. So it transmits without having to coordinate with other stations. The downlink is used for information broadcasting. In cases where the message downlink mobile application part (DL-MAP) does not explicitly indicate that a portion of the downlink subframe is for a specific SS, all SSs are able to listen to that portion. The SSs check the connection identifiers (CIDs) in the received protocol data units (PDUs) and retain only those PDUs addressed to them. SSs share the uplink to the BS on a demand basis. Depending on the class of service at the SSs, the SSs may be issued continuing rights to transmit or the transmission rights may be granted by the BS after receipt of requests from the SSs. In addition to individually addressed messages, messages may also be sent by multicast to groups of selected SSs and broadcast to all SSs. In each sector, SSs are controlled by the transmission protocol at the media access control (MAC) layer. And they are enabled to receive services tailored to the delay and bandwidth requirements of each application. This is accomplished by four types of uplink sharing schemes: unsolicited bandwidth grants, polling, and bandwidth request contention.

The transmission scheme of the MAC layer is connection oriented. All data communications are defined in the context of a connection. Service flows can be provisioned at an SS and connections are associated with these service flows, each of which is to provide transmission service with requested bandwidth to a connection. The service flow defines the quality of service (QoS) parameters for the PDUs that are exchanged on the connection. The concept of a service flow on a connection is a key issue in the operation of the MAC protocol. Service flows provide a mechanism for uplink and downlink QoS management as bandwidth allocation processes. An SS requests uplink bandwidth on a per connection basis. Bandwidth is granted by the BS to an SS as an aggregate of grants in response to per connection requests from the SS. Connections may require active maintenance, and three connection management functions are supported: static configuration and dynamic addition, modification, and deletion of connections. The termination of a connection is stimulated by the BS or SS.

7.2   Services Provisioning

7.2.1   Services and Parameters

Scheduling services represent the data handling mechanisms supported by the MAC scheduler for data transport on a connection. Each connection is associated with a single data service. Each data service is associated with a set of QoS parameters that quantify aspects of its behavior. These parameters are managed using the dynamic service addition (DSA) and dynamic service change (DSC) message dialogs. Four services are supported: unsolicited grant service (UGS), real-time polling service (rtPS), non-real-time polling service (nrtPS), and best effort (BE).

The UGS is designed to support real-time data streams consisting of fixed-size data packets issued at periodic intervals, such as the voice over Internet protocol (VoIP) without silence suppression. The mandatory QoS service flow parameters for this scheduling service are maximum sustained traffic rate, maximum latency, tolerated jitter, and request/transmission policy. If present, the minimum reserved traffic rate parameter will have the same value as the maximum sustained traffic rate parameter.

The rtPS is designed to support real-time data streams consisting of variable-size data packets that are issued at periodic intervals, such as Moving Picture Expert Group (MPEG) video. The mandatory QoS flow parameters for this scheduling service are minimum reserved traffic rate, maximum sustained traffic rate, maximum latency, and request/transmission policy.

The nrtPS is designed to support delay-tolerant data streams consisting of variable-size data packets for which a minimum data rate is required, such as the file transfer protocol (FTP). The mandatory QoS flow parameters for this scheduling service are minimum reserved traffic rate, maximum sustained traffic rate, traffic priority, and request/transmission policy.

The BE service is designed to support data streams for which no minimum service level is required and therefore may be handled on a space-available basis. The mandatory QoS flow parameters for this scheduling service are maximum sustained traffic rate, traffic priority, and request/transmission policy.

7.2.2 Service Implementation Schemes

7.2.2.1 Uplink Scheduling Schemes

Uplink request/grant scheduling is performed by the BS with the intention of providing each SS with bandwidth for uplink transmissions or opportunities to request bandwidth. By specifying a scheduling service and its associated QoS parameters, the BS scheduler can anticipate the throughput and latency needs of the uplink traffic and provide polls or grants at the appropriate times.

7.2.2.1.1 UGS Service

The UGS is designed to support real-time service flows that generate fixed-size data packets on a periodic basis, such as T1/E1 and VoIP without silence suppression. The service offers fixed-size grants on a real-time periodic basis, which eliminate the overhead and latency of SS requests and ensure that grants are available to meet the flow's real-time needs. The BS provides data grant burst information elements (IEs) to the SS at periodic intervals based on the maximum sustained traffic rate of the service flow. The size of these grants should be sufficient to hold the fixed-length data associated with the service flow, but may be larger at the discretion of the BS scheduler.

In order for this service to work correctly, the request/transmission policy setting should be such that the SS is prohibited from using any contention request opportunities for this connection. The key service IEs are the maximum sustained traffic, maximum latency, tolerated jitter, and request/transmission policy. If present, the minimum reserved traffic rate parameter should have the same value as the maximum sustained traffic rate parameter.

The grant management subheader is used to pass status information from the SS to the BS regarding the state of the UGS service flow. The most significant bit of the grant management field is the slip indicator (SI) bit. The SS sets this flag once it detects that this service flow has exceeded its transmit queue depth. Once the SS detects that the service flow's transmission queue is back within limits, it clears the SI flag. The flag allows the BS to provide for long-term compensation for conditions, such as lost maps or clock rate mismatches, by issuing additional grants. The poll-me (PM) bit may be used to request polling for a different, non-UGS connection.

The BS should not allocate more bandwidth than the maximum sustained traffic rate parameter of the active QoS parameter set, excluding the case when the SI bit of the grant management field is set. In this case, the BS may grant up to 1% additional bandwidth for clock rate mismatch compensation.

7.2.2.1.2 rtPS Service

The rtPS is designed to support real-time service flows that generate variable-size data packets on a periodic basis, such as MPEG video. The service offers real-time, periodic, unicast request opportunities which meet the flow's real-time needs and allow the SS to specify the size of the desired grant. This service requires more request overhead than UGS, but supports variable grant sizes for optimum data transport efficiency.

The BS provides periodic unicast request opportunities. In order for this service to work correctly, the SS is prohibited from using any contention request opportunities for that connection. The BS may issue unicast request opportunities as prescribed by this service, even if prior requests are currently unfulfilled. This results in the SS using only unicast request opportunities in order to obtain uplink transmission opportunities (the SS could still use unsolicited data grant burst types for uplink transmission as well). All other bits of the request/transmission policy are irrelevant to the fundamental operation of this scheduling service and should be set according to network policy. The key service IEs are the maximum sustained traffic rate, minimum reserved traffic rate, maximum latency, and request/transmission policy.

7.2.2.1.3 nrtPS Service

The nrtPS offers unicast polls on a regular basis, which ensures that the service flow receives request opportunities, even during network congestion. The BS typically polls nrtPS CIDs on an interval on the order of 1 s or less. The BS provides timely unicast request opportunities. In order for this service to work correctly, the request/transmission policy setting should be set such that the SS is allowed to use contention request opportunities. This results in the SS using contention request opportunities as well as unicast request opportunities and unsolicited data grant burst types. All other bits of the request/ transmission policy are irrelevant to the fundamental operation of this scheduling service and should be set according to network policy.

7.2.2.1.4 BE Service

The intent of the BE service is to provide efficient service for BE traffic. In order for this service to work correctly, the request/transmission policy setting should be set such that the SS is allowed to use contention request opportunities. This results in the SS using contention request opportunities as well as unicast request opportunities and unsolicited data grant burst types. All other bits of the request/transmission policy are irrelevant to the fundamental operation of this scheduling service and should be set according to network policy.

7.2.2.2 Bandwidth Allocation Schemes

During network entry and initialization, every SS is assigned up to three dedicated CIDs for the purposes of sending and receiving control messages. These connection pairs are used to allow differentiated levels of QoS to be applied to the different connections carrying MAC management traffic. Changing bandwidth requirements is necessary for all services except constant bit rate UGS connections. Demand assigned multiple access (DAMA) services provide resources on a demand assignment basis. When an SS needs to ask for bandwidth on a connection with BE scheduling service, it sends a message to the BS containing the immediate requirements of the DAMA connection. The QoS for the connection was established at connection establishment and is looked up by the BS. There are numerous methods by which the SS can get the bandwidth request message to the BS.

7.2.2.2.1   Requests

Requests are for SSs to indicate to the BS that they need uplink bandwidth allocation. A request may come as a stand-alone bandwidth request header or it may come as a piggyback request. Because the uplink burst profile can change dynamically, all requests for bandwidth should be made in terms of the number of bytes needed to carry the MAC header and payload, but not the physical (PHY) layer overhead. The bandwidth request message may be transmitted during any uplink allocation, except during any initial ranging interval. Bandwidth requests may be incremental or aggregate. When the BS receives an incremental bandwidth request, it adds the quantity of bandwidth requested to its current perception of the bandwidth needs of the connection. When the BS receives an aggregate bandwidth request, it replaces its perception of the bandwidth needs of the connection with the quantity of bandwidth requested. The piggybacked bandwidth requests are always incremental. The self-correcting nature of the request/grant protocol requires that SSs periodically use aggregate bandwidth requests. The period may be a function of the QoS and of the link quality. Due to the possibility of collisions, bandwidth requests transmitted in broadcast or multicast request IEs should be aggregate requests.

7.2.2.2.2   Grants

For an SS, bandwidth requests are not to individual connections; each bandwidth grant is addressed to the SS's basic CID. In all cases, based on the latest information received from the BS and the status of the request, the SS may decide to perform back-off and request again or to discard the MAC service data unit (SDU). An SS may use request IEs that are broadcast, directed at a multicast polling group it is a member of or directed at its basic CID. In all cases, the request IE burst profile is used, even if the BS is capable of receiving the SS with a more efficient burst profile. To take advantage of a more efficient burst profile, the SS should transmit in an interval defined by a data grant IE directed at its basic CID. Because of this, unicast polling of an SS is normally done by allocating a data grant IE directed at its basic CID. Also note that in a data grant IE directed at its basic CID, the SS may make bandwidth requests for any of its connections.

7.2.2.3   *Request Transmission Schemes*

There are two ways to issue the bandwidth requests. In rtPS and nrtPS, the requests are issued by control of the polling scheme or contention. In the BE service, requests are issued mainly by contention.

7.2.2.3.1   Polling

Polling is the process by which the BS allocates to the SSs bandwidth specifically for the purpose of making bandwidth requests. These allocations may be to individual SSs or to groups of SSs. Allocations to groups of connections or SSs actually define bandwidth request contention IEs. The allocations are not in the form of an explicit message, but are contained as a series of IEs within the uplink mobile application part (UL-MAP). Polling is done on an SS basis. Bandwidth is always requested on a connection basis and bandwidth is allocated on an SS basis.

When an SS is polled individually, a unicast polling scheme without explicit message is transmitted to poll the SS. Rather, in the UL-MAP, the SS is allocated sufficient bandwidth to respond with a bandwidth request. If the SS does not need bandwidth, the allocation is padded. SSs that have an active UGS connection of sufficient bandwidth are not polled individually unless they set the PM bit in the header of a packet on the UGS connection. This saves bandwidth over polling all SSs individually. Note that unicast polling is normally done on a per SS basis by allocating a data grant IE directed at its basic CID.

If insufficient bandwidth is available to individually poll many inactive SSs, some SSs may be polled in multicast groups or a broadcast poll may be issued. As with individual polling, the poll is not an explicit message, but bandwidth allocated in the UL-MAP. The difference is that rather than associating allocated bandwidth with an SS's basic CID, the allocation is to a multicast or broadcast CID.

When the poll is directed at a multicast or broadcast CID, an SS belonging to the polled group may request bandwidth during any request interval allocated to that CID in the UL-MAP by a request IE. In order to reduce the likelihood of collision with multicast and broadcast polling, only SSs needing bandwidth reply. They take the contention resolution algorithm to select the time slot in which to transmit the initial bandwidth request. The SS assumes that the transmission has been unsuccessful if no grant is received in the number of subsequent UL-MAP messages specified by the parameter contention-based reservation timeout. Note that with a frame-based PHY layer with UL-MAPs occurring at predetermined instants, erroneous UL-MAPs may be counted toward this number. If the request is made in a multicast or broadcast opportunity, the SS continues to run the contention resolution algorithm.

7.2.2.3.2 Contention Resolution

The mandatory contention resolution method is the truncated binary exponential back-off, with the initial back-off window and the maximum back-off window controlled by the BS. When an SS has information to send and wants to enter the contention resolution process, it sets its internal back-off window equal to the request back-off start defined in the uplink channel descriptor (UCD) message referenced by the (UCD) count in the UL-MAP message currently in effect. The SS randomly selects a number within its back-off window. This random value indicates the number of contention transmission opportunities that the SS defers before transmitting. An SS considers only contention transmission opportunities for which this transmission would have been eligible. These are defined by request IEs in the UL-MAP messages.

The SS now increases its back-off window by a factor of two, as long as it is less than the maximum back-off window. The SS randomly selects a number within its new back-off window and repeats the deferring process described above. This retry process continues until the maximum number of retries (i.e., request retries for bandwidth requests and contention ranging retries for initial ranging) has been reached. At this time, for bandwidth requests, the PDU is discarded.

For bandwidth requests, if the SS receives a unicast request IE or data grant burst type IE at any time while deferring for this CID, it stops the contention resolution process and uses the explicit transmission opportunity. The BS has much flexibility in controlling the contention resolution. At one extreme, the BS may choose to set up the request (or ranging) back-off start and request (or ranging) back-off end to emulate an Ethernet-style back-off, with its associated simplicity and distributed nature as well as its fairness and efficiency issues.

A transmission opportunity is defined as an allocation provided in a UL-MAP or part thereof intended for a group of SSs authorized to transmit bandwidth requests or initial ranging requests. This group may include either all SSs having an intention to join the cell or all registered SSs or a multicast polling group. The number of transmission opportunities associated with a particular IE in a map is dependent on the total size of the allocation as well as the size of an individual transmission. The size of an individual transmission opportunity for each type of contention IE is published in each transmitted UCD message. The BS always allocates bandwidth for contention IEs in integer multiples of these published values.

7.3 A QoS Framework

The MAC layer of the IEEE 802.16 standard [2] defines service provisioning mechanisms and functions that can control the communication between the BS and SSs in the PMP topology. On the downlink, the transmission is relatively simple because the BS is the only one to broadcast the control information and payload data in the downlink sub-frame. The information is broadcast to all SSs. Each SS takes the information destined to it from the broadcast. For the uplink transmission, the BS determines the number of time slots in which each SS is allowed to transmit its PDU in an uplink subframe. The scheduling result is broadcast by the BS through the UL-MAP message at the beginning of each downlink subframe. The UL-MAP contains information on the transmission opportunities for all SSs. After receiving the UL-MAP, each SS transmits its data in predefined time slots. The uplink bandwidth allocation scheduling algorithm resides at the BS to schedule and manage the uplink PDU transmissions for each SS. The IEEE 802.16 MAC protocol is connection oriented. Each traffic flow for any application needs to establish the connection with the BS as well as the different associated service flows, including UGS, rtPS, nrtPS, and BE. Any communication between the BS and an SS over the connection has to provide a bandwidth request for its PDU transmission.

Different services at each SS have different ways of issuing their requests. The requests will finally be approved by the BS in uplink scheduling through the UL-MAP. At each SS, the local scheduler assigns the bandwidth granted by the BS to different connections and manages the transmission of PDUs from the queue of each connection in the time slots specified by the UL-MAP. The standard specifies the services, such as UGS, rtPS, nrtPS, and BE, provided by the MAC layer protocol for information exchange between the BS and SS. The standard also specifies the schemes to provide those services, such as connection establishment, the behaviors of each service, and bandwidth request and allocation. However, the standard does not define detailed uplink scheduling schemes for rtPS, nrtPS, and BE service flows at the BS and local SSs. Also, the standard does not define the different necessary schemes to ensure that the specified QoS can be obtained.

To fill the gap between the specification of the standard and the schemes needed to provide QoS guarantees, a solution on the framework of the QoS architecture has been proposed [3], as shown in the Figure 7.1. At the BS, a few components have been proposed for ensuring the QoS guarantee. They are the admission control scheme, uplink bandwidth allocation scheduling algorithm, packet analysis module, traffic management table, and packet allocation module. When a connection is established with the proposed QoS architecture, an application that originates at an SS establishes a connection with a BS through connection negotiation. The admission control scheme makes a decision to accept or reject the new connection. Once the new connection can be accepted, it notifies the uplink bandwidth allocation scheduling algorithm to perform scheduling, and it controls the assistant modules to reach the QoS guaranteed bandwidth allocation. Traffic management enforces traffic based on the traffic specification of the connection. The data packet analysis module collects the bandwidth request information from the

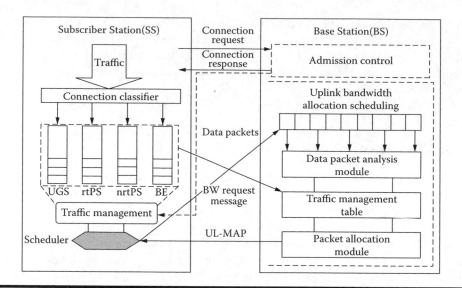

Figure 7.1 Structure of the QoS framework.

bandwidth requests from SSs in the previous uplink subframe to the uplink scheduler at the beginning of each uplink subframe. It updates the traffic management table to prepare for uplink scheduling. The packet allocation module retrieves the information from the traffic management table and generates the UL-MAP under the control of the bandwidth allocation scheduling algorithm. At each SS there is a traffic connection classifier, which differentiates the arrival of the PDUs into different traffic streams and transmission queues. The traffic management module at each SS schedules the uplink data transmission among different types of service and different connections for each type of traffic according to the local scheduling algorithm and the available granted bandwidth.

7.4   QoS Scheduling

Although the solution in Cho et al. [3] presents a framework of QoS, it does not discuss the scheduling algorithms at both the BS and SSs for uplink bandwidth allocation. Detailed scheduling is left unspecified in the standard. The uplink bandwidth scheduler at the BS should be in charge of the total available uplink bandwidth allocation among SSs in the same sector, while the local uplink bandwidth scheduler at each SS is in charge of allocating uplink bandwidth to the SS among different service types and different connections. Scheduling at the BS of uplink sharing for data transmission from multiple SSs is a crucial issue. And scheduling of the granted uplink bandwidth to the SS among different service types and different connections is also very important for the QoS guarantee.

To support all types of service flows (UGS, rtPS, nrtPS, and BE), the proposed uplink packet scheduling [4] uses a combination of strict priority service discipline, earliest deadline first (EDF), and weight fair queue (WFQ). The structure of the bandwidth allocation is shown in Figure 7.2.

The scheduling principles are as follows:

1. Overall bandwidth allocation: Bandwidth allocation on the different types of service follows a strict priority from highest to lowest: UGS, rtPS, nrtPS, and BE.

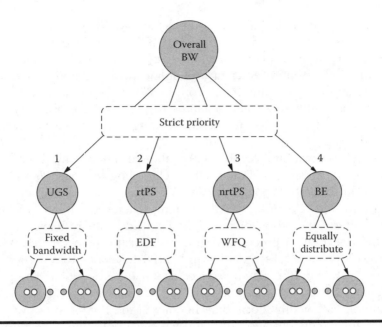

Figure 7.2   The structure of bandwidth allocation.

One disadvantage of the strict priority service discipline is that higher priority connections can starve the bandwidth of lower priority connections. To overcome this problem, a traffic policing module in each SS is included which forces the connection's bandwidth demand to stay within its traffic contract. This prevents higher priority connections from using more bandwidth than their allocation. And the overall bandwidth allocation is performed at the BS.

2. Bandwidth allocation within UGS connections: The BS allocates fixed bandwidth (fixed time duration) to UGS connections at each SS based on fixed bandwidth requirements. This policy is determined by the IEEE 802.16 standard without further scheduling. The bandwidth for the UGS connections allocated to each SS is further divided into pieces to assign to the individual connections based on negotiation at the connection establishment.

3. Bandwidth allocation within rtPS connections: The EDF service discipline schedules transmission among the rtPS connections at each SS. The connection with the shortest deadline at each SS is scheduled for transmission first. The scheduling is performed at each SS.

4. Bandwidth allocation within nrtPS connections: The WFQ service discipline is applied to the scheduling of transmissions among nrtPS connections at each SS. The nrtPS connections are scheduled for transmission based on the weight of the connection, which is the ratio between the connection's nrtPS average data rate and the total nrtPS average data rate. The scheduling is performed at each SS.

5. Bandwidth allocation within BE connections: The remaining bandwidth is equally allocated to the individual BE connections at each SS.

The policy proposed in Wongthavarawat and Gauz [4] presents a general consideration on the scheduling issue in WiMAX systems. However, transmission scheduling over only the traffic flows or connections within different types of services at each SS ignores the burst nature of real-time and non-real-time traffic because each burst in one traffic flow can have its individual time constraint, which could be different from others in the same flow, and the burst has its individual burst size for either real-time or non-real-time traffic. Further transmission scheduling over the burst blocks in one traffic flow is necessary to produce more efficient scheduling results. Another shortcoming from the observation on the scheduling framework in WiMAX systems presented in the IEEE 802.16d standard is that there are two polling schemes (rtPS and nrtPS) employed which can cause a great deal of control overhead, resulting in a complicated scheduler at the BS as well as the SSs.

To target efficient scheduling for the QoS guarantee, a hierarchical scheduling scheme to provide real-time and non-real-time traffic scheduling for the rtPS and nrtPS at both the BS and SSs in the PMP topology of WiMAX systems has been proposed [5]. Since, in the standard for the rtPS and nrtPS, the polling scheme has been specified to handle transmission requests, the proposed scheduling considers the rtPS and nrtPS at the same time with only one polling process in order to allocate them for uplink sharing. The beauty of the proposal is that the rtPS and nrtPS have been combined as a unique polling process, while differentiated services can also be obtained. The proposal has a special design at both the BS scheduler and SS local scheduler.

For the uplink bandwidth allocation scheduler at the BS, the proposal combines the rtPS and nrtPS for bandwidth allocation and allocates only one collection of time slots for the polling-based service, which includes the rtPS and nrtPS among different SSs. It results in only one polling-based service. However, the nrtPS still exists, which is included in the one service. The service differentiation is performed at each SS local scheduler.

At the local scheduler for the granted bandwidth allocation among different types of connections, in order to meet the time constraints of real-time messages as much as possible and avoid scheduling starvation of the non-real-time messages, the scheduling of data transmission is further divided into two levels—interclass scheduling and intraclass scheduling—when multiple classes of traffic exist in the network.

The interclass scheduler multiplexes the messages from different classes for transmission, and the corresponding scheduling algorithm is supposed to provide service differentiation according to different QoS requirements from different classes of traffic. The intraclass scheduler decides the service priorities of messages within the same class. The interclass scheduler selects the messages for service from class i ($1 \leq i \leq N$) and the intraclass scheduler for class i decides which waiting message in the queue i will be transmitted first when a channel is available. There are a total of $N+1$ schedulers if there exist N different classes in the network. The scheduling schemes of different intraclass schedulers i and j $j(1 \leq i, j \leq N)$ can be different or the same, and are dependent on the particular requirements of classes i and $j(1 \leq i, j \leq N)$. In this study, only two different types of traffic are considered in the network, where $N = 2$. One traffic class is real-time traffic with time constraints. The other is non-real-time traffic.

7.4.1 Interclass Scheduling

With the objective of providing service differentiation between the real-time and non-real-time classes of traffic, the proportional delay differentiation (PDD) model is suggested to provide scalable, predictable, and controllable service architecture. The PDD model aims to control the ratios of the average class queuing delays based on generic delay differentiation parameters (DDPs) $\{\delta_i, i = 1, \ldots, N\}$. Specifically, let $\bar{d}_i$ be the average queuing delay or simply the average delay of class i traffic. By the PDD model, the ratio of average delays between two classes i and j is fixed to the ratio of the corresponding DDPs:

$$\frac{\bar{d}_i}{\bar{d}_j} = \frac{\delta_i}{\delta_j} \qquad i \leq i, \quad j \leq N. \tag{7.1}$$

The DDP can represent the priority of one class. Since higher priority classes provide better service with lower queuing delays, we will have $\delta_1 > \delta_2 > \cdots > \delta_N > 0$. If class 1 is the reference class and $\delta_1 = 1$, then the average delay of class i is a certain fraction δ_i of the average delay of class 1:

$$\bar{d}_i = \delta_i \bar{d}_1. \tag{7.2}$$

The beauty of the PDD model is that it makes the average delay of class i traffic independent of the aggregate traffic load or the class load distribution. We propose employing the PDD model for interclass scheduling and call it proportional average delay (PAD) scheduling. The PDD model provides normalized average delays, defined as $\tilde{d}_i = \bar{d}_i / \delta_i$, which must be equal in all classes, that is,

$$\tilde{d}_i = \frac{\bar{d}_i}{\delta_i} = \frac{\bar{d}_j}{\delta_j} = \tilde{d}_j \quad 1 \leq i, \quad j \leq N. \tag{7.3}$$

Let $B(t)$ be the set of backlogged classes at time t, $D_i(t)$ be the sequence of class i messages that departed before time t, in order of their departure, and d_i^m be the delay of the mth message in $D_i(t)$. Assuming that there is at least one departure from class i before t, the normalized average delay of class i at t is

$$\tilde{d}_i(t) = \frac{1}{\delta_i} \frac{\sum_{m=1}^{|D_i(t)|} d_i^m}{|D_i(t)|} = \frac{1}{\delta_i} \frac{S_i}{P_i}, \tag{7.4}$$

where S_i is the sum of queuing delays of all messages in $D_i(t)$, and $P_i = |D_i(t)|$ is the number of messages in $D_i(t)$.

Suppose that a message has to be selected for transmission at time t. The PAD algorithm works according to the PDD model to select the backlogged class j with the maximum normalized average delay,

$$j = \arg \max_{i \in B(t)} \tilde{d}_i(t). \tag{7.5}$$

The message at the head of queue j is transmitted, after a queuing delay $d_j^{P_j+1}$. The variables S_j and P_j are then updated as $S_j + d_j^{P_j+1}$ and $P_j = P_j + 1$. The new normalized average $\tilde{d}_j(t)$ is then calculated from Equation (7.4). The basic idea in the PAD algorithm is that if a message from class j with the maximum normalized average delay is served, the delay of that message will not increase any more. So serving a message from class j tends to reduce the difference of $\tilde{d}_j(t)$ from the normalized average delays of the other classes.

7.4.2 Intraclass Scheduling

The interclass scheduler by the PAD algorithm supports a delay-based relative differentiated service that relies on the ratio of average delays among the different classes.

In order to achieve the absolute performance of certain traffic streams, we have to choose different queuing disciplines for the intraclass schedulers. The intraclass schedulers work at different levels from the interclass schedulers. Therefore it is possible for them to shorten the absolute delays of the individual messages of different classes of traffic while the ratios of delay differences among different classes are unchanged.

The tardy rate is the major performance measurement for real-time messages in this study. It is the ratio of the number of the messages later than their deadlines to the total number of transmitted messages. Let's assume that the kth real-time message has its transmission time of l^k and delay bound d^k, which is the time tolerance of the message. The message deadline is defined as the delay bound after the arrival of the message as $D^k = a^k + d^k$, where D^k is the deadline of message k and a^k is the arrival time of message k. In order to achieve the lower tardy rate for the network, the scheduler should reduce the number of late messages as much as possible. To achieve this, both the message deadline and message transmission time should be considered for queue scheduling. We set the real-time traffic queue as a dynamic priority queue with a combined priority, P_{RT}^k, which is assigned as

$$P_{RT}^k = a^k + d^k + l^k = D^k + l^k. \tag{7.6}$$

Since non-real-time traffic is time insensitive, there is no time constraint with each message. However, the network should serve non-real-time traffic as soon as possible to prevent the starvation of non-real-time traffic. Let's assume that w^i is the waiting time of the ith message delayed in the non-real-time traffic queue, l^i is the message's transmission time, and T^i denotes the total time spent in the network by the message: $T^i = w^i + l^i$. Let c^i be the ratio of transmission time over the total time in the network, thus

$$c^i = \frac{l^i}{T^i} = \frac{l^i}{w^i + l^i}. \tag{7.7}$$

The idea is to have the scheduler of the non-real-time traffic queue allocate the highest priority to the message with the smallest c^i value. Then the priority of message i is determined by the following:

$$P_{NRT}^i = c^i = \frac{l^i}{w^i + l^i} = \frac{1}{\frac{w^i}{l^i} + 1}. \tag{7.8}$$

By this priority assignment scheme, a message with less transmission time will have a higher priority to be served. However, at the same time, a message which has been delayed a longer time in the queue will also have higher priority to be transmitted. With this scheme, the total time for messages to stay in the network may be reduced and the starvation of the messages with longer transmission time may be avoided.

A series of extensive simulation experiments has been conducted to evaluate the performance of the proposed hierarchical scheduling algorithm for rtPS and nrtPS in WiMAX systems. The entire network has been designed to have $N = 100$ SSs. The messages are generated as a Poisson process. The length of the messages follows exponential distribution, with an average message length of 10 time slots. The delay bound of real-time messages also follows exponential distribution, with an average value of five times the transmission time of the average message length. The time slot for data transmission is the transmission time of 640 bits. The average delay ratio between the non-real-time and

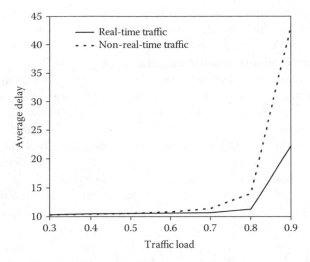

Figure 7.3   Average delay versus traffic load.

real-time traffic has been set to be two. In the simulation experiments, there are only two traffic flows considered. One is a real-time traffic flow and the other is a non-real-time traffic flow. Both of these share the entire traffic load equally.

Figure 7.3 presents the relationship between the average message delay and total traffic load in order to show the performance of interclass scheduling and the effectiveness of PAD scheduling. Intraclass scheduling in both real-time and non-real-time traffic transmission queues is first in first out (FIFO). From Figure 7.3, it is clear that the PAD interclass scheduling scheme functions by making the delay to the non-real-time traffic two times that of the real-time traffic under heavy traffic load.

Figure 7.4 shows the relationship between the average message delay and the traffic load. Intraclass scheduling for real-time traffic takes the proposed dynamic real-time

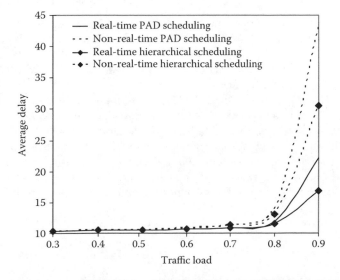

Figure 7.4   Average delay versus traffic load for hierarchical scheduling.

priority assignment scheme by which the priority of a message is determined by both the deadline of the message and its transmission time with the goal of reducing the average message delay and average tardy rate. Intraclass scheduling for non-real-time traffic takes the proposed dynamic priority assignment scheme by which the priority of a message is determined by both the transmission time of the message and the time the message has been delayed with the goal of reducing the average message delay. From Figure 7.4, it is very clear that hierarchical scheduling can not only differentiate the integrated traffic and provide corresponding services, but can also improve the performance of real-time and non-real-time traffic. Using hierarchical scheduling, the average delay to real-time traffic is half that of non-real-time traffic due to the delay ratio of two. At the same time, the delay in real-time and non-real-time traffic is reduced due to the effective intraclass scheduling schemes employed in the different traffic streams. Thus, the performance of hierarchical scheduling is much better than that of interclass scheduling alone.

7.5   Conclusion

In this chapter, the services and mechanisms to provide the QoS specified by the MAC layer protocol in the PMP topology of WiMAX systems have been reviewed. In order to achieve efficient scheduling and reduce the scheduling overhead introduced by two polling schemes, a novel proposal for transmission scheduling has been proposed. For the uplink bandwidth allocation scheduler at the BS and local scheduler at SSs, a hierarchical scheduling algorithm has been designed to provide differentiated service for rtPS and nrtPS together under a unique polling service framework in WiMAX networks. The virtue of the proposal is that the rtPS and nrtPS can be combined as one polling service without impairing the rtPS to real-time traffic. This service is controllable according to performance requirements. The proposal is significant in that it paves the way for further QoS guaranteed service.

References

1. IEEE, IEEE Standard for Local and Metropolitan Area Networks—Part 16: Air Interface for Fixed Broadband Wireless Access Systems, Standard 802.16-2004, IEEE, Washington, DC, 2004.
2. T. Cooklev, *Wireless Communication Standards: A Study of IEEE 802.11, 802.15, and 802.16*, IEEE Press, Washington, DC, 2004.
3. D.H Cho, J.H. Song, M.S. Kim, and K.J. Han, Performance analysis of the IEEE 802.16 wireless metropolitan area network, *First International Conference on Distributed Frameworks for Multimedia Applications* p. 130, IEEE, Washington, DC, 2005.
4. K. Wongthavarawat and A. Ganz, Packet scheduling for QoS support in IEEE 802.16 broadband wireless access systems, *Int. J. Commun. Syst.*, 16, 81, 2003.
5. M. Ma and B.C. Ng, Supporting differentiated services in wireless access networks, *10th IEEE International Conference on Communication Systems*, p. 1, IEEE, Washington, DC, 2006.

Chapter 8

Radio Resource Management in IEEE 802.16 WiMAX Mesh Networks

Hung-Yu Wei

Contents

The IEEE 802.16 Worldwide Interoperability for Microwave Access (WiMAX) mesh network is a promising wireless networking architecture for future broadband wireless access. The multihop relay technique in broadband wireless networks could enhance network capacity and extend service coverage. Radio resource management (RRM) is essential for efficient wireless spectral resource utilization and networking capacity enhancement in these emerging wireless networks. In this chapter, two RRM frameworks for IEEE 802.16 WiMAX mesh networks are introduced. In the first RRM framework, interference-aware RRM techniques are applied to tree topology multihop relay 802.16 networks for efficient resource utilization. The cross-layer approach considers physical (PHY) layer interference conditions for efficient multihop routing and scheduling design. The second RRM framework can be applied to mesh topology IEEE 802.16 WiMAX mesh backhaul networks. Multipath routes are constructed for pipelined data dispatching in mesh topology 802.16 networks. Adaptive mesh base station (BS) scheduling, mesh route selection, and mesh access point selection are proposed for integrated RRM. The described RRM frameworks are not only applied to 802.16 mesh mode, but also readily extended to the IEEE 802.16j mobile multihop relay (MMR) standard.

8.1 Introduction

The IEEE 802.16 Working Group and WiMAX Forum are the driving forces for the next-generation wireless broadband access standard. IEEE 802.16 technology is best known for its single-hop point-to-multipoint (PMP) mode operation. Nevertheless, IEEE 802.16 can also be operated in multihop relay scenarios. The IEEE 802.16a standard [1] first specified the basic signaling flows, message formats, mesh network formation, and operating mechanisms. Subsequently the mesh mode specifications were integrated into the IEEE 802.16-2004 revision [2]. Recently the IEEE 802.16 Working Group has formed the IEEE 802.16 Relay Task Group to study the feasibility of simplified multihop relay extension that is compatible with PMP mode 802.16-2004 [2] and 802.16e-2005 [3] PHY and MAC layer specifications. IEEE 802.16 multihop relay networks have the advantages of adaptive configuration and low-cost deployment. IEEE 802.16 multihop relay networks can be used to extend network coverage and enhance system capacity.

RRM is an important issue for wireless broadband network operation. In 802.16 multihop relay networks, RRM can be conducted through multihop route selection and time frame scheduling. To achieve better system performance and high spectral efficiency, route construction, route selection, and data frame dispatching should be coordinated under the RRM framework. In this chapter I describe two RRM frameworks that adopt a cross-layer design approach to optimize overall system performance across multiple protocol layers. The first RRM framework is designed for tree topology 802.16 multihop relay networks. The second RRM framework is designed for mesh topology multihop relay networks, where multipath routes can be exploited for better transmission efficiency. These two RRM schemes are designed under the 802.16 mesh mode context; however, these RRM techniques can be readily extended to the 802.16j MMR standard that is currently being developed by the IEEE 802.16 Relay Task Group.

8.2 IEEE 802.16 Mesh Mode Operations

In this section I will first introduce the basic operations in the IEEE 802.16 mesh mode. In the 802.16 wireless network, a wireless user terminal is termed a subscriber station (SS). In the PMP mode, an SS directly connects to a BS through a one-hop wireless connection. In mesh mode, an SS can route traffic to the BS via multiple intermediate SSs or transmit to another SS directly, as shown in Figure 8.1. In general, an IEEE 802.16

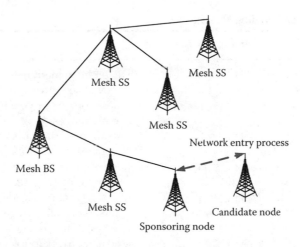

Figure 8.1 An IEEE 802.16 mesh network: a new node joins the network.

multihop relay network consists of a node with backhaul connection toward the outside network and several other nodes. The node with backhaul connection is termed the mesh BS, while other nodes are termed mesh SSs. When a new node would like to join in an existing active mesh, the new node, which is termed the candidate node (CN), initiates the network entry process. In the existing active mesh network, a node termed the sponsoring node (SN) sponsors and manages the network entry process. After the network entry process completes, the CN is attached to the SN. Table 8.1 summarizes some of the terminology that is used in this chapter.

Several control signaling message flows are described in this section. Some signaling messages have several options and can be used in multiple contexts. These signaling messages include information elements (IE) for applications in various scenarios. Table 8.2 and Table 8.3 list the signaling messages that are used in the 802.16 mesh

Table 8.1 Summary of Terminology

BS	Base station
SS	Subscriber station
MSH	Mesh
SN	Sponsoring node
CN	Candidate node
PMP	Point-to-Multipoint
IE	Information elements
CINR	Carrier-to-interference-and-noise ratio
CID	Connection ID
PKM	Privacy key management
AK	Authorization key
TEK	Traffic encryption key
KEK	Key encryption key
HMAC	Hashed message authentication code
SA	Security association
SAID	Security association identifier
TFTP	Trivial file transfer protocol

Table 8.2   Overview of Signaling Messages

Message Type	Acronym	Mesh Only	Use
Mesh network configuration	MSH-NCFG	Y	Disseminate mesh network configuration. Advertise network and open/reject sponsorship for new node in network entry process.
Mesh network entry	MSH-NENT	Y	Used in mesh network entry process (request, acknowledge, and close).
Mesh distributed schedule	MSH-DSCH	Y	Request/grant bandwidth in distributed scheduling scheme.
Mesh centralized schedule	MSH-CSCH	Y	Request/grant bandwidth and update link information in centralized scheduling scheme.
Mesh centralized schedule configuration	MSH-CSCF	Y	Update network topology and parameters in centralized scheduling scheme.
Registration request	REG-REQ	N	New node request for registration.
Registration response	REG-RSP	N	Reply for registration request.
SS basic capability request	SBC-REQ	N	Request to establish SS basic capabilities.
SS basic capability response	SBC-RSP	N	Reply for establishing SS basic capabilities.
Privacy key management request	PKM-REQ	N	Request for authentication and auhorization.
Privacy key management response	PKM-RSP	N	Reply for authorization request.
Configuration file TFTP complete message	TFTP-CPLT	N	SS notifies BS that TFTP configuration file download is compete.
Configuration TFTP complete response	TFTP-RSP	N	Reply for configuration file TFTP complete.
Reset command	RES-CMD	N	Reset command that forces SS to reset.

mode. The following paragraphs discuss the scheduling schemes in the 802.16 mesh network and then describe the network entry process when a new node joins an existing mesh. Later, tunneling control messages over multiple hops and security mechanisms are addressed.

Table 8.3   Signaling Messages with Multiple Use

Message	Type
MSH-NCFG	MSH-NCFG network descriptor MSH-NCFG NetEntryOpen MSH-NCFG NetEntryAck MSH-NCFG NetEntryReject
MSH-NENT	MSH-NENT NetentryRequest MSH-NENT NetEntryClose MSH-NENT NetEntryAck
PKM-REQ	PKM-REQ authorization request PKM-REQ authorization information
PKM-RSP	PKM-RSP authorization reply PKM-RSP authorization reject

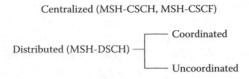

Figure 8.2   Scheduling schemes in the IEEE 802.16 mesh mode.

8.2.1   Scheduling

In the IEEE 802.16 mesh mode, radio resources are allocated in time slots. As illustrated in Figure 8.2, time slot scheduling and allocation can be conducted in a centralized fashion or in a distributed fashion. For distributed scheduling, scheduling schemes can be further categorized as coordinated and uncoordinated.

8.2.1.1   Centralized Scheduling

In the centralized scheduling scheme, all of the nodes periodically update the network topology and the traffic request profile. The mesh BS determines the schedules of all transmissions and disseminates the allocated schedules to all nodes in the mesh network. Two controlling messages—the mesh centralized schedule configuration (MSH-CSCF) and the mesh centralized schedule (MSH-CSCH)—are used in the centralized scheduling scheme. MSH-CSCF messages are used to update mesh network topology and transmission parameters. MSH-CSCH messages are mainly used for requesting and granting radio resources, but can also be used for link information updates.

In IEEE 802.16, burst profiles are used to describe transmission parameters such as modulation, coding, and preamble structure. Both uplink and downlink burst profiles are contained in MSH-CSCF messages to specify transmission parameters. MSH-CSCF messages are distributed via broadcasting. The mesh BS will first broadcast to neighboring nodes. The nodes that receive MSH-CSCFs will rebroadcast MSH-CSCFs to their neighbors.

An MSH-CSCF message indicates the number of available channels, which can be used for centralized scheduling, and the logical index of these physical channels. Logical indexes of the physical channels are described in the network descriptor (MSH-NCFG) message. An MSH-CSCF can be used to update network topology information. An MSH-CSCF message describes the total number of nodes in the mesh network, indexes of nodes, and each node's children nodes. Uplink burst profiles and downlink burst profiles between the parent node and the child node are also included.

In the centralized scheme, every mesh SS estimates and sends its resource request to the mesh BS, and the mesh BS determines the amount of granted resources for each link and communicates. The request and grant process uses the MSH-CSCH message type. An SS sends its capacity requests using the MSH-CSCH request message to the SS's parent node and then forwards them to the mesh BS. After the mesh BS determines the resource allocation, an MSH-CSCH grant message is propagated along the route from the mesh BS toward the SS.

An MSH-CSCH message can be transmitted either downlink or uplink. The grant/request flag in the MSH-CSCH message determines if the message is intended for bandwidth grants or requests. When the grant/request flag is set to zero, the mesh

BS announces the granted bandwidth to nodes within the mesh network; thus the MSH-CSCH request message is generated and broadcast from the mesh BS in the downlink direction. The MSH-CSCH also indicates the maximum number of hops the message will be rebroadcast. When the grant/request flag is set to one, the MSH-CSCH request is sent uplink to the mesh BS from an SS for requesting bandwidth allocation.

An MSH-CSCH message can also be utilized for link information updating. The mesh BS can add information regarding the parent node index, child node index, and corresponding uplink/downlink burst profiles in a MSH-CSCH. In an MSH-CSCH message, the configuration flag indicates whether the next centralized scheduling control message will be an MSH-CSCH or MSH-CSCF. Nodes can verify the validity of scheduling by comparing the received MSH-CSCH and MSH-CSCF messages and the previous configuration flag value.

8.2.1.2   Distributed Scheduling

In distributed scheduling, mesh distributed schedule (MSH-DSCH) messages are sent to request and grant bandwidth between neighboring nodes. Distributed scheduling can be either coordinated or uncoordinated. The coordination flag in an MSH-DSCH message indicates whether a coordinated or uncoordinated scheme will be used. In both schemes, a valid schedule is established based on a three-way handshake that includes request, grant, and grant confirmation.

Here a simple two-node example where node A requests bandwidth allocation from node B is considered. First, node A sends an MSH-DSCH request message with availability IE to indicate the empty slots for replying and actual transmission. Then node B replies with an MSH-DSCH grant to describe the usable time slots. Neighbors of node B notice that these slots are busy.

Finally, node A acknowledges with an MSH-DSCH grant that contains the allocated slots. Neighbors of node A notice that these granted slots are busy. In coordinated distributed scheduling, MSH-DSCH messages are scheduled in a collision-free control subframe. On the other hand, an MSH-DSCH is transmitted in a data subframe and may collide with other transmissions in the uncoordinated scheme. In the uncoordinated case, before sending a grant message, node B should wait several minislots so that neighboring nodes have time to respond for any time schedule conflict. Node A should send the grant confirmation message right after receiving the grant message.

In the coordinated distributed scheduling scheme, all nodes will transmit an MSH-DSCH periodically to determine transmission schedules. In uncoordinated distributed scheduling, an MSH-DSCH will only be transmitted in unused minislots. Both an MSH-DSCH and a mesh network entry (MSH-NENT) are transmitted in control subframes. The scheduling algorithm to determine the periodic transmission of an MSH-DSCH is the same as the transmission of an MSH-NENT.

The IEEE 802.16 standard defines four types of IEs for MSH-DSCH messages used in different signaling purposes. Depending on the context, an MSH-DSCH might contain different combinations of IEs. In a coordinated scheme, every node distributes its own and all neighboring nodes' next transmission time information in an MSH-DSCH scheduling IE. An MSH-DSCH request IE is used to request bandwidth allocation of a link. An MSH-DSCH availability IE describes the available minislots that neighboring nodes can use. An MSH-DSCH grant IE is used for granting and confirming bandwidth allocation.

8.2.2 Network Entry Process

The process of a new node, the CN, joining an active mesh network is termed the network entry process. A new node needs a sponsor, the SN, to join an existing mesh. Nodes that are available to sponsor new nodes for the network entry process periodically advertise the network information. A CN will first obtain timing information and transmission parameters from the network advertisement. Then the CN and the SN will establish a sponsor channel to exchange network entry information. The CN should be authorized and then registered with the network. After registration, the CN will obtain an Internet protocol (IP) address through dynamic host configuration protocol (DHCP) and time of day information through the time protocol. Finally, service configuration parameters are transferred.

Two mesh control messages—mesh network configuration (MSH-NCFG) and MSH-NENT—play important roles in the network entry process. An MSH-NCFG message can be used in various contexts such as in network advertisement and distributing network configuration information. There are nine IE types that can be used in conjunction with MSH-NCFG messages. Neighboring node information can be described by an Nbr Physical IE and an Nbr Logical IE. A network descriptor IE is used to advertise the network and provide the necessary information for new joining nodes. Two types of channel IEs are used to describe the operating spectrum information in the license-exempt band and licensed band. A network entry open IE and a network entry reject IE are used in the network entry process for accepting and rejecting a CN joining the mesh. A neighbor link establishment IE can be applied to establish connections to neighboring nodes.

MSH-NENT messages are mainly used in the network entry process. The MSH-NENT message format can be used in the following three scenarios: network entry request (NetEntryRequest), network entry acknowledge (NetEntryAck), and network entry close (NetEntryClose). The type field of an MSH-NENT indicates if the MSH-NENT message is used as a NetEntryRequest, NetEntryAck, or NetEntryClose. All MSH-NENT messages include the sponsor node ID, transmission power, and logical index of the transmission antenna, which specifies the antenna to be used if a node has multiple antennas. An MSH-NENT NetEntryRequest message contains a request IE, which indicates the MAC address of the CN, network operator configuration, and operator authentication information. The network entry signaling flow is illustrated in Figure 8.3.

Step 1: Periodic network advertisement from the sponsor node
Active nodes in a mesh network advertise the network by periodically sending MSH-NCFG network descriptor messages. The timestamp field in an MSH-NCFG message provides network time information. To facilitate the network entry process, the network descriptor IE includes crucial information such as frame structure of the mesh control subframe, logical channel information, and burst profiles that detail the modulation scheme, forward error correction (FEC) coding, and carrier-to-interference plus noise ratio (CINR) threshold information.

Step 2: CN acquires coarse synchronization and network parameters
The new node listens to MSH-NCFG network descriptor messages before joining an active 802.16 mesh. Based on the received MSH-NCFG message, the new node tries to synchronize to the starting time of frames. A CN will first synchronize at the PHY layer and then obtain network parameters at the MAC

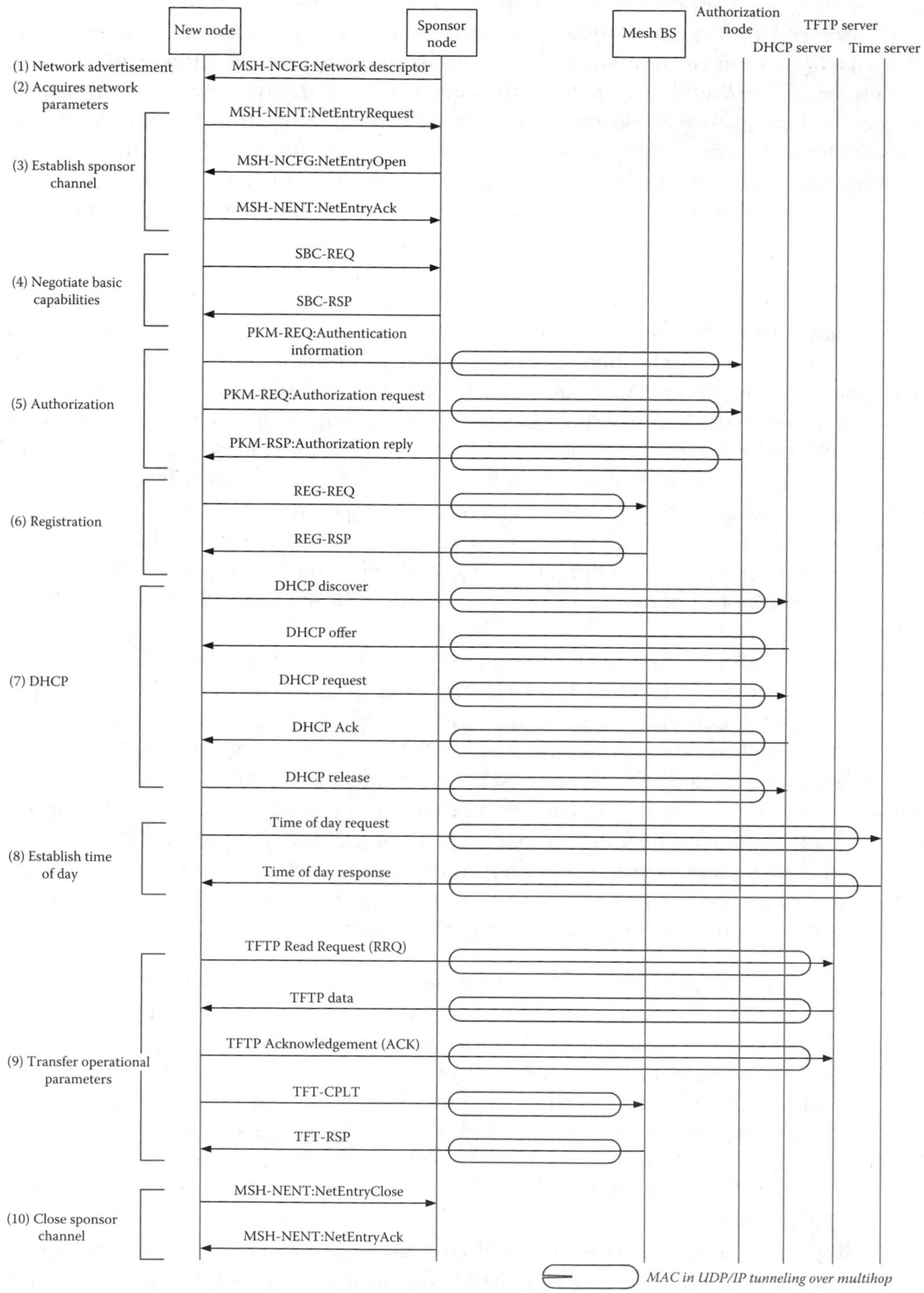

Figure 8.3 Signaling flows of the network entry process.

layer. The CN should build a physical neighbor list by scanning MSH-NCFG messages. Then the CN should pick a suitable SN with proper operator ID, which is described in a network descriptor IE in an MSH-NCFG.

Step 3: Establish a sponsor channel

The CN tries to negotiate a sponsor channel with the SN by sending an MSH-NENT NetEntryRequest message, which includes the CN's 48-bit universal MAC address and the obtained network operator information from the received network advertisement. The CN will set its temporary node ID to zero before obtaining a unique node ID. Upon receiving an MSH-NENT NetEntryRequest, the SN should decide whether to accept the request. If the SN accepts the request, it will prepare to open a sponsor channel and send an MSH-NCFG NetEntryOpen to the CN. Then the CN confirms the acceptance by replying an MSH-NENT NetEntryAck. If the SN rejects the request, the SN sends an MSH-NCFG NetEntryReject to the CN. The CN then finds another candidate SN when the network entry request is rejected.

After successfully establishing a sponsor channel, the CN and the SN use the sponsor channel for the following network entry process (i.e., Steps 4 through 9). The sponsor channel scheduling is described in the MSH-NCFG NetEntryOpen message. At the end of the network entry process (i.e., after Step 9), the CN should send an MSH-NENT NetEntryClose to the SN to close the sponsor channel. The SN acknowledges the closure of the sponsor channel by replying with an MSH-NENT NetEntryAck.

Step 4: Negotiate basic capabilities

In the basic capabilities exchange process, the SN and the CN exchange the basic transmission capabilities information and agree on the PHY layer transmission parameters that both nodes support. Similar to the negotiation of basic capabilities between the SS and BS in the PMP mode, SS basic capability request (SBC-REQ) messages and SS basic capability response (SBC-RSP) messages are used to negotiate basic capabilities in mesh mode. The CN sends an SBC-REQ that describes its connection identification (CID) and the supported basic capabilities (e.g., half-duplex/full-duplex, orthogonal frequency division multiplex (OFDM) mode, transmitted power, modulation, forward error correction (FEC), power control capability, interleaving, framing, space-time coding, etc.). Based on the capabilities that the SN supports, the SN selects a subset of capabilities described in an (ss basic capability request SBC-REQ) and replies with an SBC-RSP indicating the capabilities that are supported by both the SN and the CN. The future communications between the SN and the CN will follow the agreed capabilities given in the SBC-RSP message.

Step 5: Authorization

The privacy key management (PKM) protocol is used during the node authorization process. In this process, a CN is authenticated and authorized by an authorization node in the mesh network. Tunneling is used to forward signaling messages over a multihop mesh route during the node authorization process. Based on the request for comments (RFC) 3280 standard, X.509 digital certificates, which might be issued by a hardware manufacturer or third-party security authority, are used for client identity authentication. The authorization node first authenticates the CN's identity based on the CN's certificate. Then the authorization node negotiates the cryptographic suite to be used and exchanges a shared secret, which is termed the authorization key (AK), with the authenticated CN. The key encryption key (KEK), traffic encryption key (TEK), and hashed message authentication code (HMAC) message authentication keys are calculated from the AK. Then the authorization node provides a security association identifier (SAID), primary security association (SA), and static security association to the authenticated node.

The authorization process begins with the CN sending a PKM request (PKM-REQ) authentication information message to the authorization node. The authentication information message informs the AN about the identity of a user client. Right after sending the PKM-REQ authentication information message, the CN sends an authorization request message to the authorization node. The PKM-REQ authorization request message includes an X.509 certificate for authentication, information about the supported cryptographic suites, and the CID of the client.

The authorization node replies to the PKM-REQ authorization request message with an authorization reply message after authenticate the client's X.509 certificate. If the authentication fails, a PKM response (PKM-RSP) authorization reject message is sent by the AN. After successful authentication, a PKM-RSP authorization reply message is sent. The message contains the agreed cryptographic suite capabilities information, generated AK, key sequence number, key lifetime, and SAID. After the authorization process is complete, the CN generates a TEK for every neighbor for each SAID given in the PKM-RSP authorization reply message.

Step 6: Registration

In the registration process, the new node registers with the mesh network. The network entity that is responsible for registration is termed the registration node. The new node first sends a registration request (REG-REQ) message to the SN. Like the node authorization process, the SN tunnels an REG-REQ message toward the registration node. The REG-REQ message should include the new node's media access control (MAC) address and HMAC message digest to ensure network security. The registration node replies to a registration response (REG-RSP) message upon receiving an REG-REQ message. If the HMAC information is valid, the registration node assigns a node ID to the new node in the REG-RSP message. If the HMAC information is invalid, the registration node replies with a REG-RSP indicating message authentication failure.

Step 7: Establish IP connectivity

After successful registration, the new node establishes IP connectivity. The DHCP is used to acquire an IP address. As described in RFC 2132, DHCP Discovery, DHCP Offer, DHCP Request, DHCP Ack, and DHCP Release are used during the IP address acquisition process. Between the CN and the SN, the sponsor channel is used for the DHCP signaling procedure.

Step 8: Establish time of day

The new node establishes the time of day using the time protocol (RFC 868), which uses the sponsor channel for signaling. A time of day request message is sent to the time server and the server replies with a time of day response message.

Step 9: Transfer operational parameters

After establishing IP connectivity and the time of day, the new node downloads the operational trivial file transfer protocol (TFTP; RFC 1350). TFTP is a lightweight file transfer protocol, which uses the user datagram protocol

(UDP) for transmission. The IP address of the TFTP server is given by the DHCP during IP connectivity establishment. The new node should use the TFTP read request to initiate downloading the configuration file and will send an acknowledgement message to the TFTP server after successfully download-ing the file. The new node notifies the mesh BS of the completion of configura-tion file downloading by sending a configuration file TFTP complete message (TFTP-CPLT). The mesh BS then replies with a configuration file TFTP com-plete response (TFTP-RSP) message. The authorized quality of service (QoS) parameters, which are included in the configuration file, are essential for QoS provisioning. In the IEEE 802.16 mesh mode, packet-by-packet QoS can be provisioned based on the mesh CID.

Step 10: Close the sponsorship
The new node initiates closing of the sponsor channel at the end of the network entry process. The new node sends an MSH-NENT NetEntryClose message to the SN. The SN then closes the sponsor channel and replies with an MSH-NENT NetEntryAck to acknowledge closure of the sponsor channel. Now the network entry process is complete and the new node becomes a regular mem-ber of the mesh network. The SN is available to sponsor another new node and sponsorship availability is indicated within MSH-NCFG messages.

8.2.3 Tunneling

The 802.16 mesh mode uses a MAC over UDP/IP tunneling mechanism to forward a MAC message over multiple hops. The tunneled packet is a UDP/IP packet with tunnel subheader and the encapsulated MAC message including its MAC header. Tunneling is mainly used to transport a MAC management message in scenarios where a mesh net-work node functions similar to a BS in PMP mode. For example, when a new node joins a mesh network, the new node has to be authenticated and authorized. The autho-rization node, which is located within the mesh network, performs the authentication and authorization function similar to the BS node in PMP mode. The MAC management message for authentication and authorization is transported by multihop via tunneling between the new SS node and the authorization node.

The following three types of messages are transmitted from the CN to the SN and are tunneled in the multihop mesh (e.g., toward the authorization node): PKM-REQ authorization request, PKM-REQ authentication information, and REG-REQ. The follow-ing three types of messages are tunneled through multiple hops and are forwarded from the SN to the CN: PKM-RSP authorization reply, PKM-RSP authorization reject, and REG-RSP.

8.2.4 Security

Public key cryptography is applied in the IEEE 802.16 standard. As previously described, a new node has to be authenticated and authorized when joining an existing mesh. In 802.16 networks, X.509 certificates issued by manufacturers are used in authentication. The authorization node is responsible for the authentication process in the 802.16 mesh. After verifying the identity of a client, the authorization node provides the authenticated user with an AK, which is the shared secret to generate KEK and HMAC. To verify the authenticity of the MAC management message (e.g., key request and key reply message), HMAC-Digest, a one-way hashing value calculated based on the HMAC (RFC

2104) algorithm with the SHA-1 algorithm, is included in these messages. Currently, to ensure data privacy, three encryption algorithms are specified in the standard, including 128-bit 3-DES, 1024-bit RSA, and 128-bit AES.

To guarantee system security in 802.16 systems, periodic key refreshing is needed. In a PSM-RSP authorization reply message, the AK lifetime information denotes the requirements of the reauthorization process. The AK reauthorization process is similar to the authorization process except that no PSM-REQ authentication information message is sent. The client first sends a PSM-REQ authorization request message to the authorization node. Then authorization node replies with a PSM-RSP authorization reply message with an AK encrypted with the public key of the client.

To refresh traffic encryption keys between neighboring nodes, the exchange of key request and key reply messages is applied. Traffic encryption keys (TEKs) are refreshed more frequently than AKs. Typical AK lifetimes range from 1 day to 70 days, with 7 days as the default value. On the other hand, typical TEK lifetimes range from 30 minutes to 7 days, with 12 hours as the default value. The client node periodically sends a key request message to its neighbor, and the neighbor replies with a key reply message that includes a TEK for a given SAID. Like the AK in the PSM-RSP authorization reply message, the TEK in the key reply message is encrypted by the client's public key, which is given by the digital certificate. A TEK is established between a pair of neighboring nodes for SAID. To maintain uninterrupted communications, two sets of TEKs are maintained simultaneously for each pair of nodes and for each SAID. The two TEK lifetimes overlap so that there is always at least one active set of TEKs when the other set of TEKs expires. As a result, one set of TEKs is generated at the half lifetime of the other set of TEKs.

8.3 RRM in Tree Topology Multihop 802.16 Networks

In this chapter two RRM frameworks for 802.16 multihop relay networks are discussed. In both RRM frameworks, radio resources are managed through multihop route selection and time slot scheduling. In both schemes, the key design concept is to create an RRM framework with a cross-layer optimized perspective. The RRM framework is aware of PHY layer interference, resource spatial reuse in a multihop network, and end-user traffic demand. In the first RRM scheme, the 802.16 multihop relay network forms multihop routes in a tree topology and allocates radio resources in this context. The second RRM scheme utilizes multipath route construction and creates multipath scheduling in the context of a mesh topology multihop relay network.

RM can be achieved with route selection and time slot allocation in multihop 802.16 networks. A spectrally efficient IEEE 802.16 mesh network RRM scheme with cross-layer design approach is proposed in Wei et al. [4]. The system performance in this scheme has improved significantly based on the joint design and optimization that relies on application layer load demand information, PHY layer interference information, as well as the scheduling and route selection mechanism in the data link layer. The RRM utilizes route selection and scheduling to achieve better system performance.

The objective of this scheme is to provide an efficient approach for increasing the utilization of the 802.16 mesh network through appropriate design of multihop routing and scheduling. The simulation study in Wei et al. [4] shows that this scheme effectively improves the network throughput performance in IEEE 802.16 mesh networks and achieves high spectral utilization. Two efficient wireless radio resource allocation extensions in route selection and concurrent scheduling are described below.

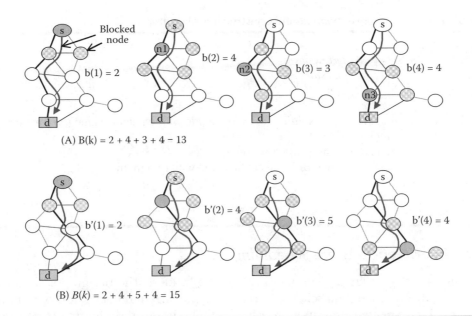

(A) $B(k) = 2 + 4 + 3 + 4 - 13$

(B) $B(k) = 2 + 4 + 5 + 4 = 15$

Figure 8.4   Blocking metric calculations.

8.3.1   Interference-Aware Route Selection

To achieve efficient spectral utilization in 802.16 mesh networks, the route construction is aware of PHY layer interference. The blocking metric $B(k)$ of a multihop route indicates the number of blocked (interfered with) nodes by all the intermediate nodes along the route from the mesh BS toward the destination SS node k. We also define the blocking value $b(\eta)$ of a node $b(\eta)$ as the number of blocked (interfered with) nodes when node η is transmitting. Therefore the blocking metric of a route is the summation of the blocking values of nodes that transmit or forward packets along the route. A simple example of a blocking metric computation is shown in Figure 8.4. In this example, the blocking metric computation shows a simplified case where only nodes within the transmission range of a transmitting node are blocked. In various scenarios, a transmitting node can interfere with nodes that are a greater distance away. Other types of blocking metrics (such as a detailed propagation model or measurement with receiver sensitivity) can be defined based on information availability and system design trade-offs.

The basic design idea of this interference-aware scheme is to select the routes with less interference. As shown in Table 8.4, the interference-aware route construction algorithm begins with a single mesh BS node (i.e., node 0 in the algorithm). Later mesh nodes join this mesh one by one in each iteration. The time order of node η joining the mesh is given as $\sigma(\eta)$. When the SS node η joins the mesh, it selects the sponsoring node with the minimum blocking from all possible sponsoring nodes (i.e., $W(\eta)$).

To implement this interference-aware scheme, the blocking metric information is incorporated into the network descriptor of an MSH-NCFG message. When a CN scans for an active network during the network entry process, the CN chooses the potential SN with minimum blocking metric to reduce the interference of the multihop route and hence to improve the throughput. This route construction algorithm constructs a tree topology multihop route to limit the interference condition.

Table 8.4   Interference-Aware Route Construction Algorithm

Algorithm:
1. $S \leftarrow \{0\}$ / *set of selected nodes*
2. $N_s \leftarrow \{1, 2, \ldots, n\}$ *nodestobeselected*
3. $P(i) \leftarrow \phi$ $i \in 1, 2, \ldots, n$
4. While $N_s \neq \phi$ do
5. $\eta \leftarrow argmax$ $i \in N_s \cap Neighbor(S)$ $\sigma(i)$ // Node with greater σ value joins earlier
6. $C(\eta) \leftarrow Neighbor(\eta)$
7. $W(\eta) \leftarrow C(\eta) \cap S$ // candidate sponsoring nodes
8. $p(\eta) \leftarrow argmin$ $i \in W(\eta) B(i)$ // select the node with minimum blocking
9.   Add η to S
10.   Remove η from N_s
11. End While

8.3.2   Interference-Aware Scheduling

Scheduling is another key component in IEEE 802.16 RRM. The design goal of the proposed centralized scheduling scheme is to exploit concurrent transmission opportunities to achieve high spectral utilization and hence high system throughput. Moreover, the RRM framework takes the user traffic demand into allocation consideration. Meanwhile, it also limits the interference level. Given the user traffic demand at each end node, the scheduling algorithm allocates time slots for better spectral utilization.

The scheduling design seeks to maximize the number of concurrent transmissions without creating interference. The simultaneous transmission scheduling in the multihop environment also takes bandwidth requests from the mesh SS into consideration. The scheduling algorithm is shown in Table 8.5. We use $Y(k)$ to denote the capacity request of an SS node k from the mesh BS. The radio resource allocation conducts iteratively until all $Y(k)$ are satisfied. The resource allocation iteration begins with allocating the highest unfulfilled traffic demand $Y(j)$. Inside the iteration, the scheduler seeks to find possible simultaneous transmission from the nonactive and nonblocked link set (i.e., links not within $A \cup B$). The $BlockedNeighbor(k)$ indicates the neighboring nodes that should be blocked from simultaneous transmission when node k is active. Each iteration ends when there is no more possible transmission to be assigned due to blocking constraints.

Table 8.5   Interference-Aware Scheduling Algorithm

Algorithm:
1. $t \leftarrow 1$ // initialize time
2. While $\exists Y_j > 0$ do
3. $k \leftarrow argmax_{\forall j} Y(j)$ // select link k
4. $B \leftarrow \emptyset$ // initialize the set of block links
5. $A \leftarrow \emptyset$ // initialize the set of selected active links
6.   While $k \neq \emptyset$ do
7.   Add k to A
8.   Add $BlockedNeighbor(k)$ to B
9. $k \leftarrow argmax_{\{j \nsubseteq A \cap B; Y(j) > 0\}} Y(j)$
10.   End While
11. $ActiveLinks(t) \leftarrow A$
12. $t \leftarrow t + 1$
13. $Y(j) \leftarrow Y(j) - 1$ $\forall j \in A$
14. End While

The mesh BS grants radio resource according to the capacity requests of all SS nodes and the route information of the mesh network. With the obtained route information from network entry and the initialization process, the node capacity request can also be equivalently represented in terms of link demands for every link. The scheduling algorithm iteratively determines the set of active links at time t. In each allocation iteration t, a link with the highest unallocated traffic demand is selected for the next allocation of a unit of traffic. The scheduling algorithm is designed to find the maximum number of concurrent transmissions that satisfy the signal-to-interference plus noise ratio (SINR) constraints. The iterative allocation continues until there is no unallocated capacity request.

8.4 RRM in Mesh Topology Multihop 802.16 Backhaul Networks

In the second RRM framework, the resource management issues in 802.16 mesh backhaul networks are investigated. To effectively utilize wireless radio resources, the route construction scheme exploits the multipath routing opportunities. Compared to the scheduling algorithm in the scheme, the scheduling algorithm allows dispatching data blocks dynamically, based on current buffer condition and route condition, without the a priori knowledge of traffic demand. The scheduling algorithm is also more suitable for the multipath routing scenario. Moreover, this RRM scheme is more suitable for application in scenarios where the 802.16 mesh is used as wide-area backhaul connections [5].

8.4.1 Route Construction in Mesh Backhaul

The aim of this route construction algorithm is to create maximum disjoint routes from the source node S toward the destination node D. Figure 8.5 illustrates a route construction procedure with a simplified example. We try to establish multiple routes from A to D. The number shown next to a link is the normalized radio resource required to transmit a fixed data block through the link. We show two interference links by circling them together. In the first route construction step, shown in Figure 8.6, the first route A-C-D is constructed and updates the bandwidth usage. Both residual radio resources of selected links and interfered links are updated. Although A-B-D was originally a higher bandwidth (also a shorter route) route than A-E-F-G-D, it is not a better route selection due to the interference. In Figure 8.7, the second route is constructed. After these two routes are constructed, the route selection algorithm determines which route will be used for a given traffic flow.

As shown in Table 8.6, the route construction algorithm first adds the source node S, which is also the mesh BS, which serves as the gateway of the wireless mesh backbone. Until there are no more resources to create a route, a route Γ_D to destination node D is computed in each route construction iteration, and the wireless resource usage from the available wireless resource is deducted. A directional link from node i to node j is denoted as L_{ij} with link rate r_{ij}. Radio resources are calculated in normalized time. Residual radio resource capacity is represented as t_{ij}, which is the normalized residual usage time of link L_{ij}.

During a route construction iteration, there are several node addition iterations (lines 8 through 22) that add one node to a routing tree in each iteration. First, the tree Υ is initialized and one node is added each time to the tree Υ and the set of already added nodes *InRoute*. The mesh BS node, which is the source node S, is the first node to be added.

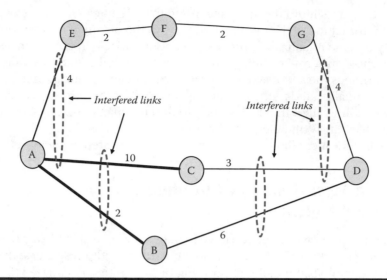

Figure 8.5   Interference-aware multiple route construction—network configuration.

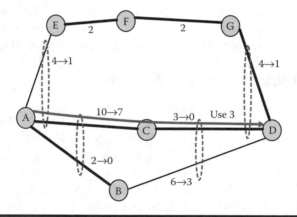

Figure 8.6   Interference-aware multiple route construction—first route.

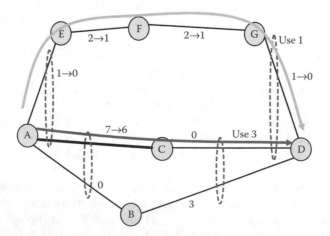

Figure 8.7   Interference-aware multiple route construction—second route.

Table 8.6   Algorithm to Create Multiple Routes

Input:
N - Set of Nodes - S is the source; D is the destination
r_{ij} - physical link rate of link L_{ij} $i, j \in N$

Output:
T - Set of Routes

Algorithm:
1. $T \leftarrow \phi$
2. $t_{ij} \leftarrow 1 \; \forall_{i,j}$
3. Do
4. $\Upsilon \leftarrow \phi$ // Initialize tree Υ
5. $InRoute \leftarrow \{S\}$
6. $U_\Upsilon(S) \leftarrow 0$
7.   While $D \notin InRoute$ do
8.   For each $n \in InRoute$ do
9. $M_n \leftarrow min_j\{1/r_{nj}\} \; s.t. \; j \notin InRoute$
10. $n \leftarrow$ Node with M_n
11.   End For
12. $x \leftarrow$ Node with $min_n\{U_\Upsilon(n) + M_n\}$
13. $y \leftarrow$ Node with $min_n\{1/r_{xn}\}$
14. $\Upsilon \leftarrow \Upsilon \cup L_{xy}$ // We'll attach node y to node x
15.   For each node $v \in InRoute$ do
16. $\Gamma_v \leftarrow$ Route from S to node v in tree Υ
17. $U_\Upsilon(v) \leftarrow U_\Upsilon(v) + \sum_{l \in \Gamma_v} \Gamma_{L_{xy} \succ l}(1/r_{xy})$
18.   End For
19. $InRoute \leftarrow InRoute \cup y$
20.   End While
21. $\Gamma_D \leftarrow$ Route from S to D in tree Υ
22.   For each link $L_{uv} \in \Gamma_D$ do
23. $\tau_{uv} \leftarrow \sum_{L_{ij} \succ L_{uv}, L_{ij} \in \Gamma_D}\{1/r_{ij}\}$
24.   End For
25.   For each link $L_{ij} \in \Gamma_D$ do
26. $\phi_{ij} \leftarrow t_{ij}/\tau_{ij}$
27.   End For
28. $\Phi \leftarrow min_{(ij)}\{\phi_{ij}\}$
29. $t_{ij} \leftarrow t_{ij} - \Phi \cdot \tau_{ij} \; \forall L_{ij} \in \Gamma_D$
30. $T \leftarrow T \cup \Gamma_D$
31. End Do
32. Return T

One important metric, $U_\Upsilon(n)$, is the temporary radio resource usage along the multi-hop route from S toward node n on Υ. Since our wireless mesh network model considers interference and has different link rates at different wireless links, it is difficult to directly measure radio resource usage in bits per second across spatial locations. We opt for calculating radio resources in normalized time (t_{ij} and τ_{ij}) while considering interfered links, and then converting to link capacity in bits per second (R_{ij}) at different spatial locations.

The criteria for route selection is to minimize radio resource usage (in terms of normalized time) along the multihop route. The notation $U_\Upsilon(n)$ represents the multihop radio resource cost from S to node n using routing tree Υ. The notation $\succ$ is used to indicate interference between two links. $L_{ij} \succ L_{uw}$ implies that link L_{ij} interferes with link L_{uw} (when $L_u v$ is active, $L_i j$ cannot be active). We also define $L_{ij} \succ L_{ij}$. The time of a bit transmitted over a link L_{ij} is $1/R_{ij}$. As $L_{ij} \succ L_{uw}$, L_{uw} cannot be active during this $1/R_{ij}$ time duration. The wireless radio resource cost of transmitting a bit over L_{ij} is $1/R_{ij}$ on L_{ij} plus $1/R_{ij}$ on the interfered link L_{uw}.

At the beginning of the node addition iteration, nodes already in the route examine the neighboring nodes and pick a candidate child node with the minimum one-hop transmission cost (lines 9 through 12). The candidate child node, which has the minimum multihop radio resource transmission cost, is added to the routing tree. The multihop transmission cost values along the route will be updated (lines 16 through 19). The node addition iterations continue until Γ_D, a route from S to D, is derived.

After new route Γ_D is given, the cost at L_{uw} for transmitting one bit over route Γ_D, τ_{uw}, is computed (lines 24 through 26). The ratios of available link resources and transmission cost, ϕ_{ij}, are calculated to determine the bottleneck link, which is the link with the minimum ϕ_{ij} value. Then the maximum possible wireless resources Φ along the route are allocated. After updating the available resources and adding the current route Γ_D to the set of routes, the next available route is calculated.

8.4.2 Mesh BS Scheduling

The mesh BS, which acts as a gateway, is the coordinator of radio resource allocation by scheduling and dispatching data blocks. In this centralized scheduling scheme, per-access point (AP) first-in first-out (FIFO) buffers are employed for in-sequence data delivery. The admitted traffic load of destination node y is L_y, while the backlog to AP_y is D_y. Therefore the scheduler selects the buffer with the maximum weighted buffer length (D_y/L_y) and schedules the dispatch to block data from the buffer.

The goal of the scheduler design at mesh BS is to maximize the overall data throughput. To improve the radio resource utilization efficiency when serving multiple interfering routes, the route interference can be resolved if two routes are served with a time gap Δ. Figure 8.8 shows a simple example of dispatching data blocks on multiple routes. We consider a single mesh backbone with three APs. For clarity of illustration, all links in the paths have the same transmission rates and data blocks are a fixed size. Let T denote the time to completely transmit a single block over a link. The current block is sent along path $P1$, and the second block is sent along path $P2$. In the multihop mesh, the block delivery is sequenced as the following transmission events in $P1$: $GW \rightarrow 1$, $1 \rightarrow 4$, $4 \rightarrow AP1$, each taking T units of time. As can be seen, the event mesh BS $\rightarrow 2$ can only take place at the same time as the event $4 \rightarrow AP1$. The mesh BS has to wait $\Delta = 2T$ units to send on $P2$ after it has finished dispatching data for AP1 on link $GW \rightarrow 1$. As can be noted for the same figure, $\Delta = T$ if the second block is sent along path $P3$. Without this waiting, there will be a collision at the receiver side leading to a decrease in the throughput. Next we present the algorithm to find Δ for any two paths in the mesh network. Based on finding the Δ, we subsequently propose pipeline scheduling in dispatching data blocks from the mesh BS. Note that the minimum value of Δ is T since the mesh BS cannot transmit simultaneously to more than one destination.

8.4.3 Computing Data Dispatching Delay Between Two Routes

To manage the interfering multihop radio resource utilization, an algorithm to compute the minimum delay to resolve interference between multihop transmission is shown

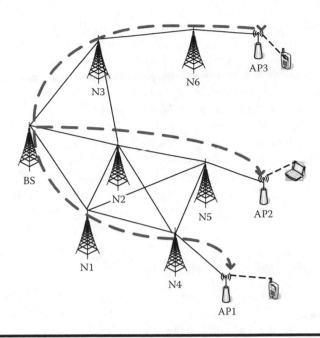

Figure 8.8   Interference between multiple routes in an IEEE 802.16 mesh backhaul network.

in Table 8.7. This algorithm serves as the foundation of a parallel pipeline dispatching scheduler, which is described in the next subsection.

With two routes P_0 and P_1, our objective is to find the minimum delay value Δ such that dispatching a block on P_1 does not lead to interference with the block sent on P_0. We denote a conflict node pair $\xi(n_0, n_1)$ as one relay node n_0 on P_0 and the other relay node n_1, which will interfere with each other when they are both transmitting.

Table 8.7   Algorithm to Compute Δ for a Conflicting Path Pair

Input:
ξ - the set of conflicting nodes along path P_0 and P_1
$t_{l_o}^k$ - time that n_0 start transmission; $\xi_k(n_0, n_1) \in \xi$; $n_0 \in P_0$
$t_{u_o}^k$ - time that n_0 stop transmission; $\xi_k(n_0, n_1) \in \xi$; $n_0 \in P_0$
$t_{l_1}^k$ - time that n_1 start transmission; $\xi_k(n_0, n_1) \in \xi$; $n_1 \in P_1$
$t_{u_1}^k$ - time that n_1 stop transmission; $\xi_k(n_0, n_1) \in \xi$; $n_1 \in P_1$

Output:
Δ - Minimum delay to dispatch on conflicting path P_0, P_1

Algorithm:
1. $\Delta \leftarrow 0$
2. For each conflicting node pair $\xi_k(n_0, n_1) \in \xi$
3.   If $(t_{u_1}^k + \Delta) \in [t_{l_o}^k, t_{u_o}^k])$ or $(t_{l_1}^k + \Delta) \in [t_{l_o}^k, t_{i_0}^k])$
4. $\delta = t_{u_0}^k - (t_{l_1}^k) + \Delta)$
5. $\Delta = \Delta + \delta$
6.   End If
7. End For
8. Return Δ

Consider a conflict node pair $\xi(n_0, n_1)$ for route pair P_0 and P_1. For block b_0 sent on P_0, let the transmission from relay node n_0 start at t_{l_0} and end at t_{u_0}. Similarly, on path P_1, the block b_1 will be transmitted between t_{l_1} and t_{u_1} at relay node n_1. Nodes n_0 and n_1 are located within the interference range. We say the conflict node pair $\xi_k(n_0, n_1)$ is active when two time intervals $[t_{l_0}^k, t_{u_0}^k]$ and $[t_{l_1}^k, t_{u_1}^k]$ overlap, which implies the transmissions at n_0 and n_1 are actually interfering with each other and we need to impose a time delay δ_k to avoid this interference.

We compute Δ with iterative steps. The scheduling of P_0 is fixed and the scheduling of P_1 is delayed with Δ to avoid interference. The initial value of Δ is set to zero. When an active conflict pair is found, we compute the necessary delay value δ_k to resolve the conflict of ξ_k, and then update $\Delta = \Delta + \delta_k$. After completion of iterative steps on all conflict pairs we get the *minimum* Δ, which resolves all possible interference between P_0 and P_1.

8.4.4 Pipeline Dispatch at the Mesh BS

To increase wireless system performance, managing radio resource efficiently is critical. In pipeline dispatch scheduling at the mesh BS, after dispatching current block b_i, the scheduler selects the next block b_{i+1} from the buffer that has the maximum level. It computes the waiting time, $\Delta(b_i, b_{i+1})$, and assigns the dispatch time as $T_c + \Delta(b_i, b_{i+1})$ to the block b_{i+1}, where T_c is the current time. However, this dispatching rule may result in a void (a period of no transmission from the mesh BS) between two dispatches, leading to low utilization. To improve the radio resource utilization, the scheduler applies the pipeline dispatch process to fill the void.

The pipeline dispatch process is illustrated in Table 8.8. A dispatch of a block refers to the event corresponding to initiating transmission of a block at the mesh BS. A data block from the main schedule is denoted as b^m and the block from the void filling schedule

Table 8.8   Algorithm to Schedule Data Block Dispatch

Algorithm:
1. On dispatch of a block b_i^m at current time T_c
2. Select block b_{i+1}^m from buffer with max weighted backlog
3. $\Delta = \Delta(b_i^m, b_{i+1}^m)$
4. Schedule b_{i+1}^m at $T_c + \Delta$

/ Start void filling
5. $\delta \leftarrow 0$
6. Last block $b_l \leftarrow b_i^m$
7. While $\delta \leq \Delta$
8.   For each non-empty buffer B_k
9. $b_k^v \leftarrow$ data block from buffer B_k
10.   End For
11. $b_k \leftarrow \text{argmin}_k \, \Delta(b_l, b_k^v)$
        s.t. $\Delta(b_l, b_k) + \Delta(b_k, b_{i+1}^m) \leq \Delta - \delta \, \forall k$
12. $\delta \leftarrow \delta + \Delta(b_l, b_k)$
13.   Schedule b_k at $T_c + \delta$
14. $b_l \leftarrow b_k$
15. End While

Table 8.9   Algorithm for Access Point and Route Selection

Input:

$\Gamma_y(k)$ - The route k from Gateway to AP y

$A(x)$ - set of access points that mobile node x can connect to

$\tau_y(x)$ - physical link rate of from AP_y to MN_x, $y \in A(x)$

ϕ_y - available radio resource at AP_y

Output:

$B_s(x)$ - Serving access point of mobile node x

$P_s(x)$ - Serving route of mobile node x

Algorithm:

1. For each mobile node x do
2.   Select AP y and route $\Gamma_y(k)$ with minimal
       $\{\Delta(\Gamma_y(k), \Gamma_y(k)) + 1/(r_y(x)\dot{\varphi}_y)\}$
3. $B_s(x) \leftarrow y$
4. $P_s(x) \leftarrow \Gamma_y(k)$
5. End For
6. For each mobile node x do
7.   Select AP y and route $\Gamma_y(k)$ with minimal
       $\{\Delta(\Gamma_y(k), \Gamma_y(k)) + 1/(r_y(x)\dot{\phi}_y) + \overline{\Delta(\Gamma_s, \Gamma_y(k)} + \overline{\Delta(\Gamma_y(k), \Gamma_s)}\}$, where $\forall \Gamma_s \in P_s(x)$
8. $B_s(x) \leftarrow y$
9. $P_s(x) \leftarrow \Gamma_y(k)$
10. End For
11. Return $B_s(x)$, $P_s(x)$

---

is denoted as b^v. The time void is filled iteratively by first selecting block b_k from the buffer B_k which has a minimum $\Delta(b_i, b_k)$ such that $\Delta(b_i, b_k) + \Delta(b_k, b_{i+1}) < \Delta(b_i, b_{i+1})$, where $\Delta(b_i, b_k)$ and $\Delta(b_k, b_{i+1})$ is the waiting time required to avoid collision with the blocks b_i and b_{i+1}, respectively. If there is more than one buffer B_k with a minimum $\Delta(b_i, b_k)$ and satisfying the above constraint, one with a more weighted buffer level is selected. The above process is repeated until no more buffer can be served in the void.

8.4.5   WiMAX Mesh as Backhaul: Access Point and Route Selection

An interesting application for the 802.16 mesh network is as backhaul connections for wireless APs, as shown in Figure 8.8. Mobile nodes (MNs) attach to APs and are unaware of the mesh backhaul. The route construction algorithm computes multiple routes from the mesh BS to an AP. There might be multiple candidate APs to which MN x can attach. The AP selection algorithm determines which available AP y should serve this MN. Within the mesh backhaul, the route selection algorithm determines the multihop route for packets sent through the mesh BS, which is sometimes known as the gateway (GW), to AP y. As illustrated in Table 8.9, the AP and route selection algorithm consider load condition and packet delivery in both the mesh backbone and wireless APs.

Since an MN can connect to multiple APs, the set of candidate serving APs for MN x is denoted as $A(x)$. We denote the physical link rate between x and AP y as $r_y(x)$. The available radio resources of AP y is given in normalized time φ_y, which is a real number between zero and one. When no MN is served by y, φ_y is set to one. When no radio resources are available at access point y, φ_y is zero. The residual capacity concept is applied to APs. Since the link rates from an AP to different MNs depend on distance and radio propagation conditions, the residual capacity value is different for different MNs and is given as $C_y(x) = \varphi_y \cdot r_y(x)$. The expected delay for a bit to be delivered from AP y to MN x is $1/C_y(x)$.

In the mesh backbone, the route selection procedure is based on the concept of the estimated dispatching delay from the mesh BS to the serving AP as shown in Table 8.8. Given the fixed wireless mesh topology, Δ between a pair of routes can be computed by the algorithm in the previous section. When there is currently no active serving route and AP, we initialize the serving routes and APs (lines 1 through 5). Without considering background traffic, the expected dispatching delay from the mesh BS to AP y on route $\Gamma_y(k)$ is $\Delta(\Gamma_y(k), \Gamma_y(k))$. The expected delay from the AP to the MN is $1/(r_y(x)\varphi_y)$.

The second part of the selection process (lines 6 through 10) estimates the delay with coexisting traffic. In addition to the expected delay of transmitting two packets sequentially over the same route and the delay from the AP to the MN, the $\overline{\Delta(\Gamma_s, \Gamma_y(k)}$ term represents the average delay of first transmitting a packet from an active route Γ_s before transmitting on $\Gamma_y(k)$, while the $\overline{\Delta(\Gamma_y(k), \Gamma_s)}$ term represents the average delay of first transmitting a packet from $\Gamma_y(k)$ before transmitting from an active route. Notice that the actual dispatching delay with pipeline dispatching is unknown during the stage of route and access point selection. The delay metric for selection is a heuristic estimation.

8.5 Conclusion

In this chapter, IEEE 802.16 mesh mode operations were introduced and two radio resource allocation frameworks were discussed. One RRM framework considers the tree topology multihop 802.16 scenario, while the other RRM framework can be applied to mesh topology multihop 802.16 networks. In both RRM schemes, high radio resource utilization efficiency is the design goal. A cross-layer approach for RRM is applied to avoid multihop route interference through proper route selection and scheduling. The RRM schemes also consider user bandwidth demand for enhanced QoS. Cross-layer design in RRM shows improved system performance. There are several open issues in the RRM and cross-layer design context that should be investigated to achieve better IEEE 802.16 mesh network performance:

- RRM with multiple antennas and directional antennas.
- RRM with space-time coding.
- Transmission power control and spectral reuse.
- QoS requirements with different applications (real-time and non-real-time applications).

References

1. IEEE, IEEE Standard for Local and Metropolitan Area Networks—Part 16: Air Interface for Fixed Broadband Wireless Access Systems—Amendment 2: Medium Access Control Modifications and Additional Physical Layer Specifications for 2–11 GHz, Standard 802.16a-2003, IEEE, Washington, DC, 2003.
2. IEEE, IEEE Standard for Local and Metropolitan Area Networks—Part 16: Air Interface for Fixed Broadband Wireless Access Systems, Standard 802.16-2004 (Revision of IEEE Std 802.16-2001), IEEE, Washington, DC, 2004.

3. IEEE, IEEE Standard for Local and Metropolitan Area Networks—Part 16: Air Interface for Fixed Broadband Wireless Access Systems—Amendment 2: Physical and Medium Access Control Layers for Combined Fixed and Mobile Operation in Licensed Bands and Corrigendum 1, Standard 802.16e-2005, IEEE, Washington, DC, 2006.
4. H.-Y. Wei, S. Ganguly, R. Izmailov, and Z. Haas, Interference- Aware IEEE 802.16 WiMax Mesh Networks, *61st IEEE Vehicular Technology Conference*, IEEE, Washington, DC, 2005.
5. H.-Y. Wei and S. Ganguly, Design of 802.16 WiMAX-Based Radio Access Network, *17th IEEE International Symposium on Personal, Indoor and Mobile Radio Communications* IEEE, Washington, DC, 2006.

Chapter 9

Cross-Layer Design in WirelessMAN

Taesoo Kwon and Dong-Ho Cho

Contents

The wireless metropolitan area network (MAN) is a highly promising technology for fourth-generation (4G) systems. The IEEE 802.16e standard has been released for wireless MAN supporting mobile access. Cross-layer optimization is the most important concept for next-generation wireless communication systems, and the IEEE 802.16 standard also has a protocol architecture for efficient support of cross-layer operation, such as the adaptive modulation and coding (AMC) scheme, channel-aware resource allocation, and quality of service (QoS) handling mechanisms for data transport. These cross-layer mechanisms require vertical coupling between layers and need tightly coupled interlayer interaction. In this chapter we introduce the general concept of and approaches to the design of cross-layer optimization in wireless communication systems and present various cross-layer schemes in IEEE 802.16e systems for capacity enhancement, efficient QoS support, and power conservation. We also discuss future cross-layer issues for wireless MAN systems that may arise in emerging wireless communication systems.

9.1 Introduction

The IEEE 802.16e standard specifies the air interface for mobile wireless MANs that support multimedia services [1,2]. WirelessMAN technology, such as the IEEE 802.16e system, has the potential to replace third-generation (3G) systems and become a highly promising candidate for fourth-generation (4G) systems. Many technologies (e.g., AMC, channel-aware resource allocation) used in recent wireless communication systems are based on interlayer operation, and the IEEE 802.16e standard also has an efficient structure for cross-layer operation [3,4].

First, the IEEE 802.16 system has a flexible frame structure using variable subchannelizations. It enhances link quality through frequency diversity and also improves system throughput through frequency-selective scheduling. Second, using a MAP based signaling scheme, it provides a flexible mechanism for the optimal and dynamic scheduling of time, frequency, and space over the air interface on a frame-by-frame basis. Third, it enhances system performance by supporting a variety of advanced antenna technologies, which are applied adaptively according to users' environments. Fourth, QoS-aware operation is possible because the system applies appropriate data handling mechanisms for data transport based on the service type and its QoS parameters. This flexible and adaptive structure enables system performance to be optimized through a careful cross-layer design.

The remainder of this chapter is organized as follows. In Section 9.2 we define the general concept of cross-layer design in wireless communication systems and provide a survey of various approaches to cross-layer design. In Section 9.3 we review system characteristics to support interlayer operation in the IEEE 802.16e system, which is representative of the wireless MAN system. In Section 9.4 we introduce and analyze various cross-layer schemes in IEEE 802.16e systems for capacity enhancement, efficient

QoS support, and power conservation. In Section 9.5 we highlight cross-layer issues for future wireless MAN systems by presenting the principle technologies in emerging systems, such as mobile multihop relay in wireless MAN and beyond 3G systems. Section 9.6 concludes.

9.2   Cross-Layer Design in Wireless Communication Systems

A novel design paradigm, the so-called cross-layer optimization, is one of the most promising avenues of research for the improvement of wireless communication systems [5]. Traditional communication systems, such as wired networks and early wireless networks, were designed on the basis of a layered protocol stack, which simplifies protocol design and treatment through layer independence (modularity). However, close cooperation among layers was needed, due to variations in wireless link channels and the various QoS requirements of multimedia and data services: hence the requirements for a strict layered architecture were violated. A cross-layer design can be defined as a protocol design by the violation of a reference-layered communications architecture and includes the creation of new interfaces between layers, the redefinition of layer boundaries, a layer protocol design that depends on another layer, and the joint tuning of parameters across layers [6]. The vertical coupling between layers and tightly coupled interlayer interaction can improve system capacity, meet QoS requirements efficiently, and minimize power consumption. However, the coexistence of conflicting cross-layer protocols negatively affects system performance [7]. Thus how the various cross-layer protocols depend on each other has to be considered carefully.

In this section we review several cross-layer design paradigms and approaches proposed in the literature and analyze their advantages and disadvantages by comparing various solutions. Moreover, we examine cross-layer feedback information utilized for efficient interlayer operation and present examples of its utilization.

9.2.1   Paradigms and Approaches for Cross-Layer Design

Cross-layer operation can be formulated conceptually as the selection of strategies across multiple layers such that the resultant interlayer operation is optimized, as shown in Figure 9.1 [8]. Each layer has optimal schemes under given states, such as channel condition and QoS parameters, and the combination of schemes selected in all layers results in optimized interlayer operation. For example, strategies in the physical (PHY) layer may implement variable modulation and coding schemes and multiple antenna technologies (e.g., beamforming and space-time coding), and strategies in the media access control (MAC) layer may implement different automatic repeat request (ARQ), scheduling, and resource allocation schemes. Subject to station and system constraints, such as delay requirements, total power, and fairness, the strategies in each layer are selected cooperatively so that the utility function, $Q(S(\mathbf{x}))$, is maximized. A variety of utility functions can be designed to maximize system throughput, meet QoS requirements of multimedia services as well as data services, and minimize power consumption. This optimization problem is just a conceptual formulation. In fact, even analytical expressions for strategies and constraints may be unavailable.

Many proposals for the design of cross-layer protocols have been submitted, which may be grouped into four categories according to how layers are coupled, as shown in Figure 9.2 [6].

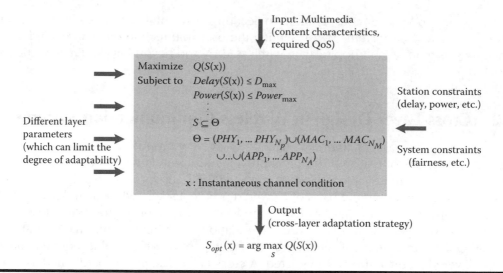

Figure 9.1 Conceptual formulation for cross-layer optimization.

9.2.1.1 Creation of New Interfaces

The exchange of information between layers at runtime needs a new interface between those layers. The creation of such a new interface violates the requirements for a strict layered architecture (hence it is a cross-layer protocol design), but can be accommodated within conventional architecture without causing significant problems. This approach to cross-layer design is one of the most representative of those used in practical system. Channel-aware scheduling, such as proportional fair (PF) scheduling, utilizes instantaneous channel quality information, and this is a good example for the MAC layer to require an upward information flow from the PHY layer. In addition, QoS parameters from the application layer can be utilized on the transport, network, and MAC layers for the support of multimedia services (e.g., transmission control protocol/user datagram protocol/real-time transfer protocol [TCP/UDP/RTP], Internet protocol/integrated services/differentiated services[IP/IntServ/DiffServ], QoS-aware scheduling, etc.). This is an example of downward information flow for cross-layer operation. Moreover, joint scheduling and power control (e.g., the water-filling algorithm [9,10]) exemplifies a cross-layer design that requires information to flow back and forth.

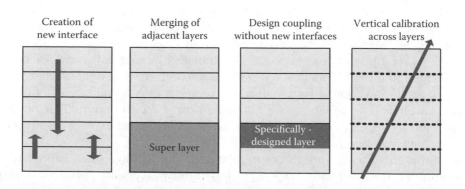

Figure 9.2 Layer-coupling methods for cross-layer operation.

9.2.1.2 Merging of Adjacent Layers

A new layer is made by combining two or more existing layers, and such a layer is called a superlayer. By using superlayers, the system can use conventional layered architecture without any modification. Though there exists no system that defines superlayers explicitly, tightly coupled algorithms between two layers have been introduced, such as hybrid automatic request (HARQ). The HARQ scheme is based on the tight combination of ARQ and forward error correction (FEC) algorithms that originally operated on the MAC layer (or link layer) and the PHY layer, respectively.

9.2.1.3 Design Coupling Without New Interfaces

One layer may be designed specifically with the operation of another layer in mind, without a new interface being created. For example, if the PHY layer can receive more that one packet simultaneously, the functions of the MAC layer may be altered considerably [11].

9.2.1.4 Vertical Calibration Across Layers

Parameters are adjusted jointly across all layers. An example of cross-layer design based on vertical calibration is determination of the HARQ retransmission number according to the delay requirements of real-time service and channel condition. Vertical calibration may be achieved in two ways: static and dynamic. In the former, parameters are adjusted once at the design stage, while in the latter, parameters are tuned dynamically at runtime. Dynamic vertical calibration has the advantage of flexible operation, but it may cause severe overheads.

9.2.2 Cross-Layer Feedback Information and Its Utilization

Unless adjacent layers are merged or one layer is designed specifically with the operation of another layer in mind (see Section 9.2.1), cross-layer design requires that information be exchanged or shared between layers. This information may flow from upper to lower layers or vice versa. For example, the MAC layer schedules packets according to channel condition (using information from the PHY layer; i.e., lower to upper) and QoS parameters (using information from application layer; i.e., upper to lower) in order to maximize system throughput and meet QoS requirements. Table 9.1 shows examples of feedback

Table 9.1   Examples of Cross-Layer Feedback Information

Layer	Examples of Feedback Information
PHY	SINR, received signal power
Link/MAC	Current FEC scheme, number of frames retransmitted, frame length, time when wireless medium is available for transmission, hand-off-related events
Network	Mobile IP hand-off initiation/completion events, network interface currently in use
Transport	RTT, RTO, MTU size, receiver window size, congestion window size, number of lost packets, actual throughput (or goodput)
Application	QoS requirements (i.e., delay tolerance, acceptable delay variation, required throughput, acceptable packet loss rate)
User	User-perceived QoS

Table 9.2 Cross-Layer Feedback Information and Its Utilization

Layer	Information	Operation	Objective
PHY	Channel condition	Link adaptation	Throughput increase
Link/MAC	Number of retransmission	Reestimation of TCP retransmission timer	Throughput increase
Network	Mobile IP hand-off initiation/ completion events	Manipulation of TCP retransmission timer	QoS support
Transport	Packet loss at TCP	Adaptation of application source rate	QoS support
Application	QoS parameters	Link layer scheduling	Throughput increase QoS support
User	User-perceived QoS	Trade-off between information and battery saving	QoS support Power saving

information that can be utilized for cross-layer operation [12]. Raisinghani and Iyer [12] consider the user as the uppermost layer of the protocol stack and utilize user requirements (i.e., user-perceived QoS) for cross-layer information. For example, even though the system may decide to conserve battery power by not receiving data, the user may think that receiving the information (e.g., stock) is more important than saving the battery. Table 9.2 shows variable cross-layer operations for throughput enhancement, QoS support, and power conservation by the exchange of the appropriate information.

For implementation, it is very important to consider how cross-layer information is exchanged between layers, and a number of methods for sharing such information are available. Figure 9.3 shows three approaches to cross-layer interaction [13] and they all allow information to be shared at runtime. (i) Direct signaling between nonadjacent as well as adjacent layers [14] follows the protocols for a conventional layered architecture, except for the creation of new interfaces, and it makes system design easy. (ii) A shared database can easily be accessed by all layers, but the database and access mechanisms need to be designed carefully [12]. This shared database approach fits well with vertical calibration. (iii) Finally, a completely new approach introduces a novel protocol architecture, based on heaps [15]. This may provide optimized performance, but requires completely new system-level design.

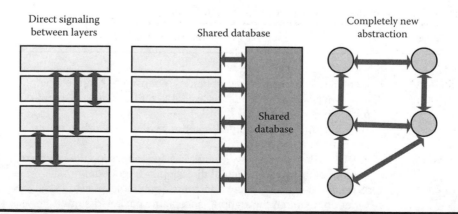

Figure 9.3 Architectures for cross-layer interaction.

9.3 WirelessMAN System and Its Cross-Layer Protocol

The IEEE 802.16e standard was designed efficiently to apply cross-layer techniques for on-demand data services as well as multimedia services. In this section we discuss the characteristics of IEEE 802.16e systems with respect to cross-layer design. First, we introduce its frame structure, which can support cross-layer operation through time, frequency, and space diversity. Then, we review a number of uplink control channels for the fast exchange of information under cross-layer operation.

9.3.1 Frame Structure of 802.16e OFDMA Systems

IEEE 802.16e systems can support time division duplex (TDD) and frequency division duplex (FDD). The frame can be composed of several zones that are divided according to subcarrier allocation methods or multiple-input multiple-output (MIMO) modes. Figure 9.4 shows an example of an IEEE 802.16 TDD frame structure, which is also a model frequently mentioned when constructing wireless broad band (WiBro) systems that support only TDD mode. (WiBro is the Korean wireless broadband service based on Mobile Worldwide Interoperability of Microwave Access [WiMAX] technology and is compatible with the IEEE 802.16e orthogonal frequency division multiple access [OFDMA]/TDD system with a 1024 fast fourier transform [FFT] size.) In the case of an FDD frame structure, the downlink (DL) and uplink (UL) subframes are allocated in a different frequency band without guard time such as the transmit/receive transmission gap (TTG) and receive/transmit transmission gap (RTG).

The frame structure consists of the following fields: a preamble using the first symbol; frame control header (FCH) with fixed size for resource allocation of the DL partial usage of subchannels (PUSC) zone and DL-MAP; DL-MAP and UL-MAP messages for resource allocation of DL and UL data bursts; DL/UL data bursts for data or control messages; and UL control channels for ranging, UL acknowledge (ACK), and channel quality indication (CQI) feedback.

The DL-MAP can present resource allocation information for each burst or each user. Either method causes considerable overhead. Presenting the MAP information for each burst reduces the number of DL-MAP information element (IE) messages, but causes processing overhead because a mobile station (MS) needs to find its own packet among many packets concatenated into a burst. Presenting the MAP information for each user can allocate resources to each user effectively, but causes considerable overhead due to the transmission of many DL-MAP IE messages. So the IEEE 802.16e system defines various MAP messages, such as compressed MAP and compact MAP, that reduce the size of MAP messages or allocate resources effectively for each user.

To support various types of physical channel conditions, IEEE 802.16e OFDMA systems define two types of methods for building subchannels: distributed subcarrier permutation mode (partial usage of subchannels [PUSC], optional PUSC [OPUSC], full usage of subchannels [FUSC], or optional FUSC [OFUSC] modes in Figure 9.4) and adjacent subcarrier permutation mode (AMC mode in Figure 9.4). The ratio of these modes can be flexible in the IEEE 802.16e standard. However, one burst for data transmission consists of several slots and one slot is the minimum possible data allocation unit. In addition, the definition of this slot depends on the OFDMA symbol structure, which varies for DL and UL, for FUSC and PUSC, and for distributed subcarrier permutations and adjacent subcarrier permutations, as shown in Figure 9.4.

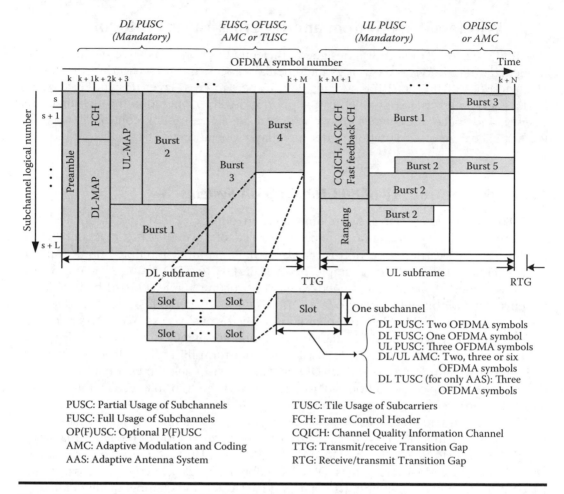

Figure 9.4   Example of IEEE 802.16e OFDMA/TDD frame structure.

9.3.1.1   Diversity Subchannels

Distributed subcarrier permutation mode is a very useful scheme for averaging inter-cell interference and avoiding deep fading by selecting subcarriers pseudo-randomly. Therefore it is expected to be suitable for users with high velocity or low signal-to-interference-plus-noise ratio (SINR). Basic resource units in the frequency domain of this mode are called diversity subchannels.

9.3.1.2   Band AMC Subchannels

In adjacent subcarrier permutation mode, adjacent subcarriers are grouped into clusters and are allocated to users. In this channel structure, the channel response can be seen as a flat fading channel. Thus the frequency selectivity of the channel cannot be exploited. Due to the flat fading nature of this subchannel, the system can make better use of multiuser diversity as long as the channel state does not change significantly during the scheduling process. Therefore it is expected to be suitable for users with low velocity or high SINR. Basic resource units in the frequency domain of this mode are called band AMC subchannels.

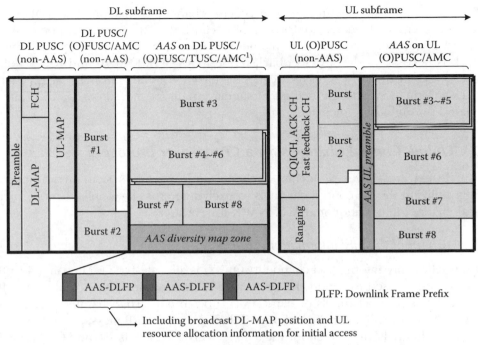

In AMC mode, the first and last numbered subchannels of AAS DL zone may contain AAS diversity map zone

Figure 9.5   IEEE 802.16 TDD frame structure for AAS support.

9.3.1.3 Support of Advanced Antenna Technologies

The IEEE 802.16 standard supports various multiple antenna technologies (such as space-time coding [STC] for space diversity, spatial multiplexing [SM] for higher throughput, and adaptive antenna system [AAS] for beam forming), which are applied in different zones within a frame. In particular, AAS can be used for space division multiple access (SDMA), which transmits data bursts concurrently to spatially separated MSs by applying different beam patterns to them, as well as a coverage extension. When AAS is used for SDMA, it requires close cooperation between the MAC and PHY layers. Figure 9.5 shows a logical frame structure for AAS support. DL and UL AAS zones are defined by a broadcast MAP message (i.e., UL and DL extended AAS-IE), and the DL AAS zone includes an AAS diversity map zone, which occupies two subchannels within the DL AAS zone. An MS receives AAS downlink frame prefix (DLFP) within the AAS diversity MAP zone by scanning known AAS DL preamble patterns. AAS-DLFPs within an AAS diversity MAP zone can be transmitted using different beams from each other and the MS selects the AAS-DLFP that has the best beam. The AAS-DLFP includes the position of the broadcast DL-MAP, which is beam formed or can be used to page a specific MS that cannot receive the normal DL-MAP, and UL resource allocation for initial UL access. Once an MS obtains initial DL resource allocation information through a broadcast DL-MAP pointed to by the AAS-DLFP, subsequent allocations can be managed with private DL-MAP and UL-MAP messages that are unicast and beam formed with high modulation and coding scheme (MCS) levels.

However, this beam steering needs a precise channel estimation. The TDD system may not need channel feedback information, because the DL channel can be estimated from the UL channel state using its channel reciprocity; but, in FDD systems, channel feedback is necessary. The AAS feedback request (AAS-FBCK-REQ) and AAS feedback response (AAS-FBCK-RSP) messages can be utilized for the transmission of channel feedback information, and the real and imaginary parts of measured channel gain are transmitted through these messages.

9.3.2 Uplink Control Channels for a Cross-Layer Protocol

We can design cross-layer protocols efficiently and improve system performance by carefully utilizing the UL control channels to exchange cross-layer information, such as physical channel information and ACK/NACK for HARQ.

9.3.2.1 Channel Quality Information Channel

The channel quality information channel (CQICH) is allocated to an MS using a CQICH control IE and is used to report the DL carrier-to-interference-plus-noise ratio (CINR) for either diversity subchannels or band AMC subchannels. This channel occupies one UL slot in the fast feedback region allocated through a UL-MAP message. In the case of diversity subchannels, the MS reports the average CINR of the base station (BS) preamble from which the BS is able to determine the DL MCS level. Here, a CINR measurement is quantized into 32 levels and encoded into five information bits. In the case of band AMC subchannels, an MS can report the difference in CINR values for five selected frequency bands (increment 1 and decrement 0 with a step of 1 dB) on this CQICH after reporting the CINR measurements of the five best bands using a MAC management message such as REP-RSP.

9.3.2.2 Fast Feedback Channels

Fast feedback channels may be allocated individually to MSs for the transmission of PHY-related information that requires a fast response from the MS. One fast feedback channel occupies one UL slot in the fast feedback region allocated through a UL-MAP message.

Using these fast feedback channels, the MS can report the following: (i) variable information for MAC operation, such as information for selecting the anchor BS in the event of macrodiversity handover, or a request for adapting the UL rate for voice over Internet protocol (VoIP) services, and (ii) PHY-related information, such as information about DL channel measurement for MIMO operation, the MIMO-related coefficients for the best DL reception (e.g., antenna weights), and MIMO mode selection (e.g., space-time transmit diversity [STTD], SM, and beam forming).

9.3.2.3 UL ACK Channel

A resource region for HARQ ACK channels is allocated using the HARQ ACK region allocation IE, and this resource region can include one or more ACK channels for HARQ-support MSs. The UL ACK channel occupies half a slot in this HARQ ACK channel region, which may override the fast feedback region. This UL ACK channel is assigned implicitly to each HARQ-enabled burst, according to the order of the HARQ-enabled DL bursts in the DL-MAP. So using this UL ACK channel, MSs can quickly transmit ACK or NACK feedback for DL HARQ-enabled packet data.

9.3.2.4 UL Sounding

In 802.16e OFDMA systems, UL sounding is defined to support smart antenna or MIMO, and this UL sounding is a kind of UL pilot signal. The BS measures the UL channel response from UL sounding waveforms transmitted by each MS and translates the measured UL channel response to an estimated DL channel response under the assumption of TDD reciprocity. In order to allocate resources for the transmission of UL channel sounding, the BS allocates a sounding zone through a UL-MAP message. In this sounding zone, each MS can transmit its UL sounding signal while maintaining signal orthogonality between multiple multiplexed MS sounding transmissions.

9.4 Cross-Layer Design in IEEE 802.16e OFDMA Systems

In this section we present various cross-layer design issues by classifying cross-layer operations in IEEE 802.16 systems into three categories according to their objectives.

- For capacity enhancement: IEEE 802.16e OFDMA systems provide many adaptive and flexible mechanisms for optimal and dynamic radio resource management, such as frequency-selective resource allocation, advanced multiple antenna technologies, and HARQ. They are based on cautionary cross-layer operations and we present these cross-layer design issues.
- For QoS support: IEEE 802.16e systems support five QoS scheduling types: unsolicited grant service (UGS) for constant bit rate service; real-time polling service (rtPS) for variable bit rate service; extended real-time polling service (ertPS) for VoIP service with silence suppression; non-real-time polling service (nrtPS) for non-real-time variable bit rates; and best effort (BE) for service with no rate or delay requirements. We discuss cross-layer operations to meet the QoS of each type.
- For power saving: The conservation of battery power in MSs is extremely important. A power-aware protocol can be optimally designed through cross-layer optimization, such as sleep mode operation, according to QoS requirements. We present various cross-layer design issues for power saving.

9.4.1 Cross-Layer Design for Capacity Enhancement

The IEEE 802.16 standard has enhanced its capacity by utilizing smart technologies, such as band AMC subchannel allocation, MIMO diversity, enhanced AAS, and HARQ. These technologies need careful interlayer operations, so we discuss various cross-layer issues for capacity enhancement with respect to such matters as adaptive resource allocation, advanced antenna technique, and HARQ. Moreover, we present an example of the design of primitives to exchange PHY layer information for the operation of cross-layer protocols.

9.4.1.1 Cross-Layer Adaptation for Efficient Resource Allocation

IEEE 802.16 OFDMA systems support two main modes for subcarrier permutation: distributed and adjacent. Distributed subcarrier permutation can employ frequency diversity effectively by distributing the allocated subcarriers to subchannels using such permutation

Table 9.3   Comparison of Diversity and Band AMC Subchannels

Items	Diversity Subchannels	Band AMC Subchannels
Principle	Frequency diversity	Frequency selectivity
Effect	Avoidance of deep fading and intercell interference averaging	Capacity enhancement
Interference robustness	Robust	Sensitive
Channel condition	Not limited (appropriate for mobile MSs)	Slowly varying channel (appropriate for stationary MSs)
Control overhead	Low or moderate	Large
Scheduling complexity	Low or moderate	High

mechanisms as FUSC, PUSC, OFUSC, and OPUSC (these subchannels are called diversity subchannels). This frequency diversity effect reduces performance degradation due to the fast fading characteristics of mobile environments. In addition, these permutation mechanisms minimize the probability that the same physical subcarriers will be used in adjacent cells and sectors, so this operation results in averaging intercell/sector interference. Adjacent subcarrier permutation can maximize throughput by adaptively allocating adjacent subcarriers according to the characteristics of users' frequency-selective fading (the sets of these adjacent subcarriers are called band AMC subchannels). Table 9.3 summarizes and compares the characteristics of diversity subchannels and band AMC subchannels [16]. Diversity subchannels have average intercell interference because each cell or sector has a different subcarrier permutation pattern, so one-cell frequency reuse can be implemented. However, cell or sector boundary users using band AMC subchannels may experience severe intercell interference if adjacent cells or sectors use the same subcarriers, so resources need to be allocated carefully, according to the level of interference. Band AMC mode achieves an improvement in capacity of up to 30% over diversity mode [17,18], but requires more overhead and is more complex than diversity mode because it is necessary to estimate channel quality correctly and a lot of feedback is required in order to report the channel quality for each band. Therefore the use of band AMC mode requires careful cross-layer design, and it is expected that system performance can be improved if diversity mode and band AMC mode are utilized appropriately according to such conditions of MSs as SINR, mobility, and QoS.

Figure 9.6 shows MAC-PHY cross-layer adaptation functions and corresponding information flows for efficient resource allocation using diversity and band AMC modes [4,19]. Though there are many other possible architectures, this architecture provides a good model of a cross-layer adaptation framework for interlayer operation between the MAC and PHY layers.

In this architecture, the MAC layer contains an airlink scheduler, a packet scheduler, and a resource controller, while the PHY layer contains a diversity mode PPDU controller, a band AMC mode PPDU controller, entities for HARQ operation, and a PHY layer control information controller. The airlink scheduler runs on the MAC layer of the BS and determines the DL and UL modes of each frame, such as diversity mode and band AMC mode. In addition, it classifies users who utilize those modes. As shown in Table 9.3, since the properties and usages of diversity mode and band AMC mode are very different, an airlink scheduler needs to determine the transmission modes carefully, taking into account the channel condition and status of each MS so that system throughput increases. Hence

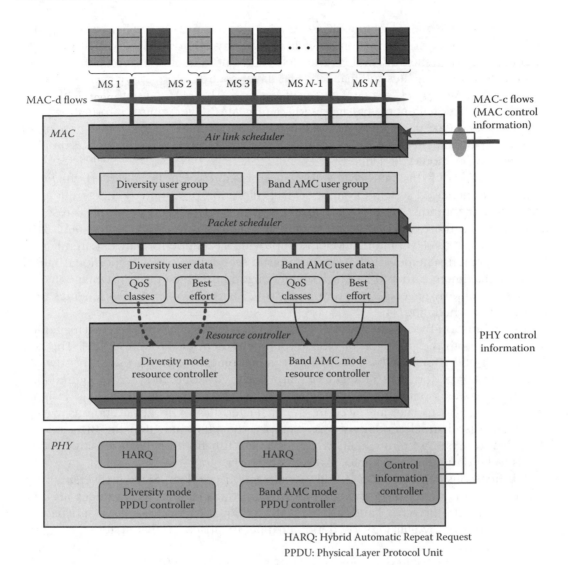

Figure 9.6 Cross-layer adaptation scheme for efficient resource allocation.

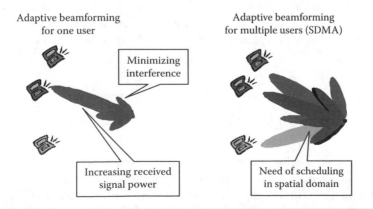

Figure 9.7 Adaptive beam-forming technology.

these MAC layer entities, such as the scheduler and resource controller, should gather and utilize PHY layer information as well as MAC layer information so that interlayer operation between the MAC and PHY layers is efficient. A control information controller in the PHY layer generates and manages PHY layer control information, such as channel matrix, SINR, MCS level, velocity, and user location. The MAC-c controller, which is a MAC layer functional entity, manages and controls MAC layer control information, such as fairness and QoS parameters gathered from upper layers (e.g., the application layer). In order to implement the exchange of this cross-layer information efficiently, primitives between layers should be defined.

On the basis of this cross-layer information, resources can be allocated efficiently, resulting in increased system throughput. This allocation of resources is determined by scheduling algorithms. Scheduling algorithms are closely related to physical resource allocation and the two together are widely accepted key functions in the MAC layer for exploiting cross-layer information and improving system performance. Careful cross-layer adaptation schemes, such as opportunistic scheduling (e.g., proportional fair algorithm, maximum carrier-to-interference ratio algorithm), that utilize instantaneous link quality can improve average cell throughput by 25% to 60% over coarse schemes (e.g., round-robin scheduling) [4].

Utility function-based scheduling is a representative cross-layer resource management method for both the provision of QoS and enhancement of throughput [20]. The central idea of utility function-based scheduling is mapping of resource usage (bandwidth, power, etc.) or performance criteria (data rate, delay, etc.) onto the corresponding utility function and then optimizing these utility functions. Thus this utility function is designed in such a way as to take into account the characteristics of an application as well as instantaneous channel conditions. For example, the utility function for BE service is designed to maximize throughput, while the utility function for delay-sensitive service needs to be designed to guarantee packet delay.

We first consider a rate-based utility optimization problem. Since a diversity mode utilizes only frequency diversity to achieve robust transmission, its resources are one-dimensional. On the other hand, a band AMC mode can allocate optimal subchannels to each end user, by utilizing closed-loop channel feedback, so that system throughput is maximized.

When the achievable data rate of user i on subchannel k ($c_{i,k}$) is a function of the SINR ($\rho_{i,k}$) and the transmission power (p_k) on subchannel k, the data transmission rate of user i (r_i) is obtained as $r_i = \sum_{k \in D_i} c_{i,k}(\rho_{i,k}, p_k)$, where D_i is a set of subchannels assigned to user i. (The diversity mode can be also modeled by using the same notations, and the value of $c_{i,k}$ for this mode is independent on subchannel index k.) Thus, the rate-based utility optimization problem assuming a fixed allocation of power, is formulated as follows [21]:

$$\max_{D_i, i \in \mathcal{M}} \quad \sum_{i \in \mathcal{M}} U_i(r_i)$$
$$\text{subject to} \quad \bigcup_{i \in \mathcal{M}} D_i \subseteq \mathcal{K}, \ D_i \cap D_j = \emptyset, \ i \neq j, \quad \text{for} \quad \forall \, i, j, \in \mathcal{M}, \tag{9.1}$$

where $\mathcal{M}$ and $\mathcal{K}$ denote the sets of users and subchannels, respectively, and the constraints mean that each subchannel is only assigned to one user. Moreover, by adding the adaptive transmission power allocation scheme in Equation (9.1), this resource management scheme can be extended to joint subchannel and power allocation, as follows [21]:

$$\max_{D_i, i \in \mathcal{M}, \mathbf{p}} \quad \sum_{i \in \mathcal{M}} U_i(r_i)$$
$$\text{subject to} \quad \bigcup_{i \in \mathcal{M}} D_i \subseteq \mathcal{K}, \; D_i \cap D_j = \emptyset, \; i \neq j, \; \text{for} \quad \forall \, i, j \in \mathcal{M}$$
$$\sum_{k \in \mathcal{K}} p_k \leq \bar{P} \tag{9.2}$$
$$p_k \geq 0,$$

where $\mathbf{p}$ denotes a transmission power allocation vector and $\bar{P}$ is a total transmit power constraint. If the utility functions are concave, the solutions of these optimization problems can be obtained effectively with low complexity [21]. In other words, for concave utility functions (a local maximum of a concave function is also its global maximum), an original optimization problem under a fixed allocation of power can be simplified into a gradient scheduling algorithm in which a scheduler maximizes the projection onto a gradient of total utility [22]. Thus the utility function-based scheduler can select the user, $m^*(k, n)$, to be served on the ith subchannel at the nth frame based on the following rule [20, 21, 23]:

$$m^*(k, n) = \arg\max_{i \in \mathcal{M}} \left\{ U_i'(\bar{r}_i(n)) \cdot c_{i,k}(n) \right\}, \tag{9.3}$$

where U_i' denotes marginal function (i.e., the first-order derivative) of U_i and $\bar{r}_i(n)$ is the average data rate up to the nth frame.

The following utility function shows one example of a concave utility function [22, 24]:

$$U_i(\bar{r}_i(n)) = \begin{cases} \frac{q_i}{\alpha}(\bar{r}_i(n))^{\alpha}, & \alpha \leq 1, \alpha \neq 0 \\ q_i \log(\bar{r}_i(n)), & \alpha = 0 \end{cases}, \tag{9.4}$$

where $\alpha \leq 1$ and q_i denote a fairness parameter and a QoS weight, respectively. So, from Equation (9.3),

$$m^*(k, n) = \arg\max_i \left\{ q_i(\bar{r}_i(n))^{\alpha - 1} \cdot c_{i,k}(n) \right\}. \tag{9.5}$$

In the rule of Equation (9.5), $\alpha = 1$ results in a scheduling rule that maximizes total throughput (e.g., maximum carrier-to-interference [max C/I] scheduling) and $\alpha = 0$ yields a PF rule.

So far we have discussed the rate-based resource allocation problem. This rate-based resource management scheme provides no mechanism for the provision of QoS. Thus we also need to consider a utility function that depends on other parameters, such as the queue size, $Q_i(n)$, and the delay of the head-on-line (HOL) packet, W_i. Assuming a fixed allocation of power, this problem can be formulated as

$$\max_{D_i^{(n)}, i \in \mathcal{A}^n} \quad \sum_{i \in \mathcal{A}^n} w_i(n) r_i(n)$$
$$\text{subject to} \quad \bigcup_{i \in \mathcal{A}^n} D_i^{(n)} \subseteq \mathcal{K}, \tag{9.6}$$
$$D_i \cap D_j = \emptyset, \; i \neq j, \; \text{for} \quad \forall \, i, j \in \mathcal{A}^n,$$

where $w_i(n) \geq 0$ is a time-varying weight assigned to the ith user at the nth frame, and $\mathcal{A}^n = \{i : Q_i(n) > 0\}$ denotes the set in which each queue is not empty at the nth frame

[22,25]. From Equation (9.3), the scheduler can select the user based on the rule given as

$$m^*(k, n) = \arg\max_{i \in \mathcal{A}^n} \{w_i(n) \cdot c_{i,k}(n)\}. \tag{9.7}$$

In addition, a number of scheduling rules have been studied for delay-sensitive traffic in single-carrier systems, and these can also be applied in OFDMA systems [25].

9.4.1.1.1 Modified Largest Weighted Delay First (M-LWDF) Rule

The QoS requirement for delay-sensitive traffic can be modeled using a packet delay bound and the allowed probability of violating this delay bound for each service flow. That is,

$$\Pr\{W_i > T_i\} < \delta_i. \tag{9.8}$$

The M-LWDF rule not only increases system throughput using multiuser diversity, but also compensates for delayed traffic by taking into account the QoS constraint of Equation (9.8) [26]. This scheduler selects the user transmitting data on the kth subchannel at the nth frame, $m^*(k, n)$, on the basis of the following rule:

$$m^*(k, n) = \arg\max_{i \in \mathcal{A}^n} \left\{ a_i \frac{c_{i,k}(n)}{\bar{r}_i(n)} W_i(n) \right\}, \tag{9.9}$$

where a_i is set to $-\log \frac{\delta_i}{T_i}$ and this weight provides the priority according to the QoS requirement. Using the simple rule of Equation (9.9), the M-LWDF scheduler for the provision of QoS can be easily implemented.

9.4.1.1.2 Exponential (EXP) Rule

In order to reduce the dependency of the M-LWDF scheduling rule on the weight a_i for each service flow, the EXP scheduling rule was designed for single-carrier systems [27]. This scheduler is very similar to M-LWDF, except that it has different time-varying weights. The EXP scheduling algorithm selects a user for data transmission on the kth subchannel at the nth frame using the following rule:

$$m^*(k, n) = \arg\max_{i \in \mathcal{A}^n} \left\{ a_i \frac{c_{i,k}(n)}{\bar{r}_i(n)} \exp\left(\frac{a_i W_i(n) - \overline{aW}}{1 + \sqrt{\overline{aW}}} \right) \right\}, \tag{9.10}$$

where $\overline{aW} = \frac{1}{N} \sum_{i=1}^{N} a_i W_i(n)$ and $a_i > 0$. This rule tries to equalize the weighted delays $a_i W_i(n)$ of queues. In other words, if there is a queue with a large value of $W_i(n)$, its priority metric becomes very large due to the exponential term, so this queue is selected by the scheduler. If the difference of weighted delay values of all queues is small, the exponential term is close to unity, so the scheduler behaves as the PF rule. Since the $\overline{aW}$ term in the exponent is common for all queues, it may be dropped.

9.4.1.1.3 Maximum Delay Utility (MDU) Rule

Since a user with a long delay has a low level of satisfaction or utility, the optimization problem can be formulated by defining a decreasing utility function for HOL packet delay. Thus the decreasing utility function for $W_i(n)$ can be substituted for the objective function in Equation (9.7). Song et al. [28] showed that the long-term optimization objective with respect to average waiting times leads to an instantaneous optimization objective given as

$$m^*(k, n) = \arg\max_{i \in \mathcal{A}^n} \left\{ \frac{|U_i'(W_i(n))|}{\bar{r}_i(n)} \min\left(r_i(n), \frac{Q_i(n)}{T_f}\right) \right\}, \qquad (9.11)$$

where T_f denotes the frame length and the min(x, y) function ensures that the service bits of each user should be less than or equal to the accumulated bits in its queue to avoid bandwidth wastage. The scheduling that uses this rule is called maximum delay utility (MDU) scheduling.

As we have seen, according to utility functions, that a variety of schedulers can be designed to meet the QoS requirement of the application layer and optimize MAC-PHY cross-layer operation. However, the optimization problems for band AMC resource allocation are generally formulated as integer programming or combinatorial programming, and these are mostly nondeterministic polynomial (NP)–hard or NP-complete. In addition, the design of a scheduling and resource allocation algorithm for band AMC mode is complex and difficult because each user's unique channel condition is considered and calculated. Thus the design of band AMC scheduling for bursty traffic is a new and interesting problem. Even though a variety of solutions for band AMC resource allocation have been proposed, practical problems, such as complexity and feedback overhead, remain.

9.4.1.2 Cross-Layer Design for Advanced Antenna Techniques

Multiple-antenna technology is considered to be one of the most promising technologies for future wireless communications systems. It offers increased spectral efficiency through spatial multiplexing gain and improved link reliability due to antenna diversity gain. Further, by combining this multiple-antenna technique with orthogonal frequency division multiplexing (OFDM) or OFDMA, dividing the frequency-selective channel into a set of parallel flat fading channels, each element of the channel matrix has a scalar value and the system will be less complex. Utilizing these benefits, the IEEE 802.16e standard supports variable options for multiple-antenna technologies, such as STC based on Alamouti code, SM using vertical encoding and two-user collaborative SM, and an AAS for beam forming. Table 9.4 summarizes and compares these advanced antenna technologies in the IEEE 802.16e standard [29]. STC reduces the fade margin by providing spatial diversity through the transmission of structured redundancy, such as Alamouti transmission, in which it sends information on two transmit antennas during two consecutive transmission times. SM improves capacity because multiple streams can be transmitted over multiple antennas of an MS with good SINR and low spatial correlation. In other words, STC overcomes severe fading environments through spatial diversity, while SM increases the peak data rate by transmitting multiple data streams simultaneously. Thus the system may optimize its performance and coverage by applying SM in the near field and STC in the far field. The IEEE 802.16e standard also supports an adaptive MIMO switch (AMS), and the STC or SM can be optimally selected to adapt to channel conditions. However, in the UL, two users can transmit their data streams collaboratively, as if the two streams are spatially multiplexed from two antennas of the same user. This operation allows each MS to use only one UL transmit antenna, and is called UL collaborative SM. Because these techniques provide improved capacity and coverage, the MAC layer has to determine the scheduled users and the number of transmitted packets by carefully considering the physical state of the network, such as each user's available data rate and the estimated packet error rate.

Table 9.4   Advanced Antenna Technologies in IEEE 802.16 Systems

	Space-Time Coding	*Spatial Multiplexing*	*AAS (Beam Forming)*
Application Benefit	Spatial diversity Fade margin reduction	Spatial multiplexing Higher peak rate, throughput increase	Adaptive beam forming Link budget improvement Interference reduction SDMA
Feedback	Open loop	Open loop	Open loop (TDD), Closed loop
Antenna	DL: $N_t = 2$, $N_r \geq 1$ UL: N/A	DL: $N_t = 2$, $N_r \geq 2$ UL: $N_t = 1$, $N_r \geq 2$ (two-user collaborative spatial multiplexing)	DL: $N_t = 2$, $N_r \geq 1$ UL: $N_t \geq 1$, $N_r \geq 2$
Effect on MAC layer	Improvement of link reliability	Throughput enhancement	Outage probability reduction, scheduling in spatial domain

The IEEE 802.16e standard supports adaptive beam forming, called AAS. In a system with multiple antennas, by applying a weight to each antenna element, a beam can be steered, and this adaptive beam-forming technique can be utilized with two objectives, as shown in Figure 9.7: (i) coverage extension or interference reduction, and (ii) SDMA. First, by focusing the transmit power into desired directions, the received SINR and link budget are improved. Since this beam focusing minimizes the energy that is emitted in other directions, the interference for other users or other cells can be reduced. The improvement in link budget and reduction in interference lead to a reduction in the probability that there will be an outage. Second, multiple beams can be transmitted to spatially uncorrelated users at the same time, on the same frequency. This behavior enables the space to be exploited as new scheduling resources. Since the spatial resources have different data transmission capabilities, depending on how spatially separable users are clustered, the design of a scheduler and resource allocator that utilizes spatial resources is very difficult and requires a sophisticated cross-layer design. The adaptive allocation of subchannels (e.g., in band AMC mode) and power with beam forming improves system performance more and is also an important cross-layer issue. In addition, scheduling and resource allocation in the space domain cause large computational complexity, and it is necessary to design spatial resource management techniques with low complexity.

Figure 9.8 shows an example of cross-layer resource management mechanisms supporting SDMA, which are based on an architecture proposed in Riato et al. [30]. The PHY layer is responsible for clustering spatially separable users and collecting channel-state information for each user. A clustering entity classifies users into groups of spatially separable users and reduces the computational load related to the allocation method. It is an issue of implementation whether this clustering should be separated with SDMA resource allocation or joined tightly with resource allocation mechanisms. The MAC layer is responsible for queue management and the provision of QoS. The algorithm for SDMA resource allocation operates on the basis of QoS requirements received from the MAC layer and information about spatial channels received from the PHY layer and provides the scheduler with sets of possible solutions for scheduling and the allocation of resources. The algorithm may be based on objective functions discussed in Section 9.4.1. The MAC scheduler can select an optimal or suboptimal solution from the sets of possible solutions. Some schemes for scheduling and resource allocation have been proposed for low complexity, and they all are based on interlayer operation.

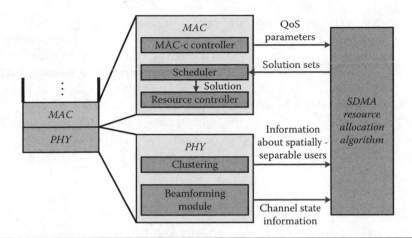

Figure 9.8   Cross-layer SDMA resource management.

9.4.1.3   Cross-Layer Design for HARQ

The 802.16e standard provides the option to combine gain by incremental redundancy. HARQ is a very important technique for link adaptation and makes aggressive MCS-level decisions possible. Thus the use of HARQ can result in considerably increased throughput [31]. However, it is a critical issue how to decide the MCS level and packet size for the original transmission of a HARQ-enabled connection. Extensive retransmissions cause considerable overhead because HARQ retransmission requires control messages, such as compact DL-/UL-MAP IE in 802.16e systems. Thus it is necessary to consider a trade-off between the overhead caused by control messages and the efficiency of link adaptation. Recently various schemes have been studied that attempt to optimize ARQ performance by applying a channel-aware scheduling algorithm to the HARQ retransmission packet. System performance can be improved through an HARQ-aware scheduler design [32,33]. However, when HARQ is applied to real-time services, it is desirable to design retransmission strategies that consider service delay bound, as well as channel quality.

9.4.1.4   Other Cross-Layer Design Issues for Capacity Enhancement

There are many other cross-layer issues. The IEEE 802.16e standard supports frequency reuse-1 to maximize spectral efficiency, which means that all sectors or cells use the same frequency channel and cell planning related to frequency reuse is not required. Users at the boundary of a cell may experience severe interference because adjacent sectors or cells use the same frequency channel. To overcome this problem, fractional frequency reuse at cell boundaries can be also used. This frequency reuse can be reconfigured dynamically on a frame-by-frame basis, taking into consideration the network load and level of interference [34]. The reuse of frequencies generates a number of issues related to interference-aware resource management and how to divide users into cell-center and cell-boundary groups. In addition, several header compression schemes, such as robust header compression (ROHC) and enhanced compressed real-time protocol (ECRTP), are widely applied to utilize resources in the air interface efficiently, and the IEEE 802.16e standard also provides architecture for header compression. The contents of the upper-layer header, such as the sequence number and time stamp in the real-time transport protocol (RTP), are constant or vary sequentially over a long period compared with the time required for link-layer operation. Thus once these contents are transmitted successfully, then they can be suppressed in the next transmission to reduce redundancy. Also, if packet error occurs, a full header or a part of the header is transmitted. Although this suppression operates in the convergence sublayer, the scheduler in the MAC layer needs to consider their effects because their packet sizes, including headers, depend on channel status. In addition, packet aggregation within a delay bound results in a better forward error correction (FEC) effect as well as the reduction of control overhead, such as resource allocation information, so the scheduler can utilize this packet aggregation to enhance capacity.

9.4.1.5 Exchange of Cross-Layer Information

In OFDMA-based systems, the condition of the UL and DL channels should be considered when scheduling, for the purpose of increasing throughput. For this reason, the IEEE 802.16 standard defines variable UL control channels, as described in Section 9.3.2. We present a cross-layer protocol for CQI feedback using these channels as a basis, as shown in Figure 9.9. This figure shows MAC-PHY primitives, the cross-layer protocol sequence of CQI feedback for DL channel measurement, and the UL sounding signal for UL channel measurement. All AMC subchannel users that have a transport connection identifier (CID) should periodically transmit a DL channel measurement report on CQICH. To construct a CQI feedback message, the MAC layer needs to receive channel measurement results from the PHY layer. Primitives such as CQI-MSG.request and CQI-MSG.response in the DL CQI feedback shown in Figure 9.9 are used for this purpose. Once a CQI feedback message is constructed in an MS MAC layer, it is transmitted to a BS MAC layer through CQICH. This information is exploited during scheduling and resource allocation. The UL sounding shown in Figure 9.9 shows the transmission sequence of the UL sounding signal. UL sounding is a kind of UL pilot signal and is defined to support smart antenna or MIMO in 802.16e. If an MS confirms its sounding channel allocation in an UL-MAP message, an MS MAC layer sends a SOUNDING.request primitive to an MS PHY. Then an MS PHY layer sends a sounding signal on the allocated UL sounding region. A BS can use the received sounding signal to measure the quality of the UL channel and translate the measured UL channel quality to an estimated DL channel quality under the assumption of TDD reciprocity.

9.4.2 Cross-Layer Design for QoS Support

The IEEE 802.16e standard supports five types of data delivery services: (i) UGS, which supports real-time applications that generate fixed-rate data; (ii) real-time variable rate service (RT-VR), which supports real-time data applications with variable bit rates that require a guaranteed data rate and delay; (iii) extended real-time variable rate service (ERT-VR), which supports real-time applications with variable data rates that require a guaranteed data rate and delay (e.g., VoIP with silence suppression); (iv) non-real-time variable rate service (NRT-VR), which supports applications that require a guaranteed data rate but are insensitive to delays; and (v) BE service, which supports applications that have no requirements for data rate or delay. Each type of data delivery service is associated with a certain predefined set of QoS-related service flow parameters, as summarized in Table 9.5, and the MAC layer can support the appropriate data handling mechanisms for data transport, using those service types and QoS parameters as

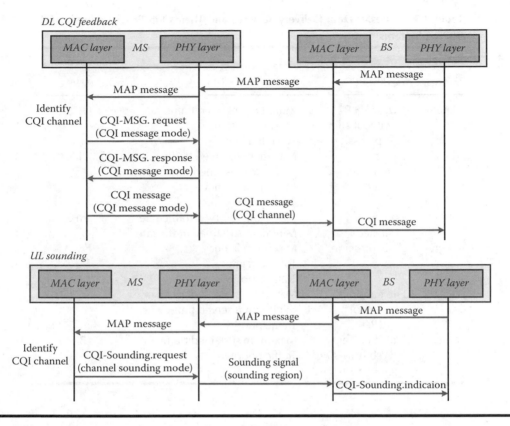

Figure 9.9 Cross-layer protocol for DL CQI feedback and UL sounding.

a basis. In other words, the MAC layer exploits these service type and QoS parameters to improve system performance. The definitions of these parameters are illustrated in Table 9.6. These parameters can be exchanged between upper layers and the MAC layer, through primitives, as discussed in Section 9.4.1, and managed dynamically through MAC management messages to accommodate the demand for dynamic service. The parameters are also utilized in the UL grant/scheduling mechanisms.

The IEEE 802.16e standard provides the various UL bandwidth (BW) request mechanisms, such as unsolicited grants, unicast poll, and piggyback. On the basis of these BW request mechanisms, five types of UL grant/scheduling are supported according to service types, as shown in Table 9.5. First, the UGS offers fixed-size grants for data transport, with the base period equal to the unsolicited grant interval and the offset upper-bounded by tolerated jitter. In addition, the grant size is decided on the basis of the minimum reserved traffic rate. Second, the rtPS offers periodic unicast request opportunities. The minimum reserved traffic rate and the maximum latency are the key QoS parameters of the rtPS scheduling type, and the polling period is optimized by the BS, based on the maximum latency requirement. In addition, the polling period may be specified explicitly as an optional QoS parameter. Third, the ertPS offers a mechanism for periodic UL allocations, which may be used for requesting BW as well as for data transfer, taking into consideration the traffic characteristics of VoIP with silence suppression. Like the rtPS, the minimum reserved traffic rate and the maximum latency are the key QoS parameters. Fourth, the nrtPS offers unicast polls on a regular basis; typically in an interval of 1 s or less. The BS may consider its minimum reserved traffic rate for the nrtPS. Fifth, the

Table 9.5 Types of Data Delivery Services and Their QoS Provisioning in IEEE 802.16 Systems

Service Type	Application	QoS Parameters	UL Scheduling Type
UGS	T1/E1, VoIP without silence suppression	Minimum reserved traffic rate Maximum latency Tolerated jitter	UGS
RT-VR	Streaming audio or video	Maximum reserved traffic rate Minimum sustained traffic rate Maximum latency Traffic priority	rtPS
ERT-VR	VoIP with silence suppression	Maximum reserved traffic rate Minimum sustained traffic rate Maximum latency Tolerated jitter Traffic priority	ertPS
NRT-VR	File transfer protocol (FTP)	Maximum sustained traffic rate Minimum reserved traffic rate Traffic priority	nrtPS
BE	Data transfer, Web browsing, etc.	Maximum sustained traffic rate Traffic priority	BE

BE may offer contention-request opportunities. These UL grant/scheduling mechanisms themselves are based on cross-layer operation and enable UL wireless resources to be used efficiently through the careful utilization of QoS parameters.

Figure 9.10 shows the conceptual scheduling architecture of the BS and MSs for the support of QoS for multimedia services in IEEE 802.16 systems, and it also shows the control information flows for cross-layer operation. The DL packets from the upper layer are classified into service flows by a packet classifier within the BS and these classified packets are transmitted once they are selected and their resources are allocated by the DL scheduler. The design of the scheduler needs to consider cross-layer information, such as DL channel quality information reported from the PHY layer and QoS parameters noticed from the upper layer, in order to maximize system throughput and meet the requirements

Table 9.6 Definition of QoS Parameters

Parameter	Definition
Maximum reserved traffic rate	Peak information rate of service
Maximum sustained traffic rate	Minimum amount of data to be transported when averaged over time
Maximum latency	Maximum allowable time between ingress of a packet to convergence sublayer and the forwarding of SDU to its air interface
Tolerated jitter	Maximum delay variation (jitter) for the connection
Traffic priority	Priority assigned to a service flow

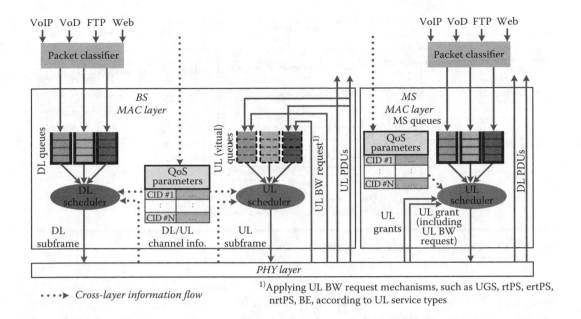

Figure 9.10   QoS support for multimedia services in IEEE 802.16 systems.

for QoS for multimedia services. For example, the utility-based cross-layer schemes for resource management discussed in Section 9.4.1 provide good solutions. The UL packets from the upper layer are classified into service flows by a packet classifier within the MS, like the DL, and the MS requests BW according to the UL grant/scheduling type. From the amount of BW requested, the BS estimates the queue status information of each MS. In IEEE 802.16 systems, all resources are managed by the BS, thus the BS performs channel- and QoS-aware scheduling, on the basis of measured UL channel information, the negotiated QoS parameter and estimated queue status. As shown in Figure 9.10, the BS builds UL virtual queues on the basis of estimated queue status, and allocates UL grants to MSs. Here, the MSs request BW for each connection and (i.e., service flow), virtual queues are managed for each connection, and the BS allocates a UL grant not for each connection, but for each MS. If an MS has more than one connection, the MS has to perform QoS-aware packet scheduling on the basis of the UL grant allocated by the BS. If the amount of allocated resources is insufficient, the MS can request additional BW. Thus both the BS and MSs have to perform a precise exchange of cross-layer information between layers or entities and a careful cross-layer scheduling operation by utilizing that information so that the system meets QoS requirements using the minimum amount of wireless resources.

So far we have discussed cross-layer operation related to resource allocation and scheduling. Figure 9.11 shows examples of other cross-layer adaptation schemes for QoS support [14,35]. IntServ or DiffServ can be used for network-level QoS management, and a variety of transport layers, such as TCP, UDP, and RTP, are available according to QoS requirements. The number of retransmissions can be determined by the delay requirement of services, and the link layer error parameters can be utilized for adaptive operation in the transport layer, which can deal with jitter as well as error-related parameters such as packet loss ratio. The application layer may control multimedia qualities (providing better quality multimedia services under good channel conditions and minimum quality

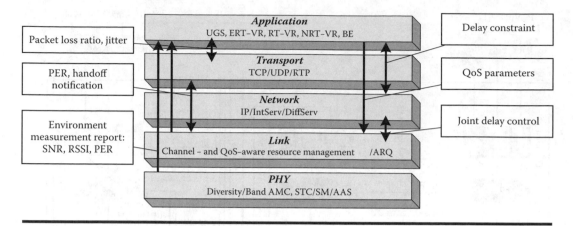

Figure 9.11   Cross-layer protocol adaptation for QoS support.

services under poor channel conditions) using information about channel conditions from the PHY layer, link quality from the link layer (or the MAC layer), packet loss ratio and jitter from the transport layer, and so on. Mobility also has a significant impact on perceived QoS for multimedia services. Maintaining an acceptable service quality during hand-off is another important and challenging research issue.

9.4.3   Cross-Layer Design for Power Saving

Efficient mechanisms for managing energy are needed to support the mobility capability of MSs, which are battery powered. The IEEE 802.16e standard defines sleep and idle mode power management functions, which preserve battery life for MSs by stopping some of the communication operations for a period of time. While in sleep mode, an MS does not have to monitor the serving BS and receives only a traffic indication message (MOB-TRF-IND) from the BS during listening intervals in order to check whether the BS has any DL traffic buffered to transmit to the MS. If there is buffered data, the MS wakes up; and otherwise it remains in sleep mode. If an MS in sleep mode has UL data to transmit, it can issue a BW request through random access. The determination of sleep mode parameters, such as the sleep interval and start time of the sleep mode, has an important effect on system performance. For example, a long sleep interval minimizes power consumption, but it may violate the delay requirement of real-time service because it takes longer to receive a traffic indication message that notifies of the existence of DL buffered data. Thus there is a trade-off between power consumption and system performance.

The IEEE 802.16e standard defines three types of power-saving classes for sleep mode, which differ by operation procedure according to service characteristics, as shown in Table 9.7. Non-real-time services are not delay sensitive and allow buffered transmission. So BE and NRT-VR are grouped into power-saving class (PSC) I, and in order to minimize their power consumption, the sleep interval is determined on the basis of a binary truncated exponent algorithm. In other words, the sleep interval is initiated to a negotiated initial window size, and if there is no buffered traffic, it is doubled continuously, but cannot exceed the final window size. In the case of real-time services, such as UGS and RT-VR, the sleep interval needs to be smaller than the period of maximum latency because buffered data should not experience delays greater than the maximum latency. PSC II has a fixed-size sleep interval, so it is suitable for use in real-time services.

Table 9.7 Power-Saving Classes for Sleep Mode Operations in IEEE 802.16 Systems

	Class I	*Class II*	*Class III*
Application	BE, NRT-VR	UGS, RT-VR	Management operation, multicast connections
Sleep interval	Binary truncated exponent	Fixed	Once (no listening interval)

However, PSC III is recommended for management operation and multicast connections. In PSC III, an MS enters sleep mode for a specified period and then automatically enters awake mode without any listening interval.

Each sleep mode operation, specified according to service characteristics, has sleep mode parameters, such as initial sleep interval, final sleep mode interval, and sleep mode triggering time, the values of which can be configured adaptively by considering QoS requirements and upper-layer operation. For example, the fixed sleep interval of PSC I should be smaller than the maximum latency required by the application layer. In the case of PSC II, the sleep interval can be determined on the basis of queue-empty time, taking the operation of transmission control protocol (TCP) into account [36]. Moreover, sleep mode triggering time is an important design factor that has an effect on system performance and has to be determined cautiously considering QoS requirements.

Adaptive resource management, such as link adaptation [37] and scheduling [38,39], also helps to minimize the average power consumed. Link dynamics of the wireless channel and multiuser diversity enable opportunistic scheduling and transmission, so this channel-adaptive resource allocation allows the system to transmit more bits using the same or lower power. Moreover, reliable transmission results in reduced retransmission and collision. In other words, channel-aware operation through careful cross-layer design reduces the number of redundant operations for transmission and reception and minimizes power consumption. However, QoS requirements, such as delay bound, may require a short sleep interval or limited system adaptability. Thus optimal energy efficiency requires a trade-off between QoS and energy due to varying user expectations and changes in the environment, and needs careful cross-layer design [40].

9.5 Future Cross-Layer Issues for WirelessMAN Systems

Many technologies and new system concepts have been proposed for the design of advanced wireless communication systems. OFDM, multiple antennas or MIMO, and adaptive radio resource management are buzzwords for the design of novel systems, and the cross-layer design utilizing them. The group responsible for the IEEE 802.20 standard has been developing the specification for mobile broadband wireless access (MBWA), which supports various vehicular mobility classes up to 250 km/hr, and has been trying to apply the latest technologies such as MIMO. Meanwhile, the Third-Generation Partnership Project (3GPP) has been specifying the long-term evolution (LTE) for systems beyond 3G, applying OFDM and MIMO. Moreover, the Wireless World Initiative New Radio (WINNER) project has been developing a ubiquitous scalable radio access system based on common radio access technologies. In addition, the IEEE 802.16j Mobile Multihop Relay (MMR) Group is newly established for coverage extension and throughput enhancement through multihop paths. The development of these systems presents new issues for cross-layer design.

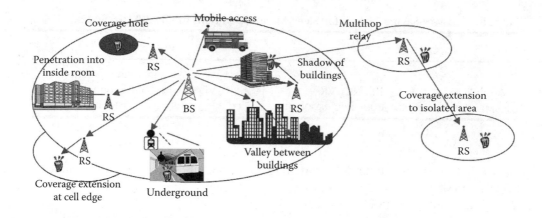

Figure 9.12 Deployment of relay stations for throughput enhancement and coverage extension.

9.5.1 Cross-Layer Design for the Multihop Relay System

A new concept of deployment, as exemplified by the relay system shown in Figure 9.12, was introduced in order to enlarge the area of coverage, increase capacity at cell boundaries, and reduce transmit power/interference (IEEE 802.16j). This requires the modification of conventional protocols for relaying. Links and protocols between the BS and relay station (RS), as well as between RSs, have to be defined. Figure 9.13 shows the IEEE 802.16j MMR protocol stack. New MAC and PHY layers for relay operation should be developed, which raises many new cross-layer issues, such as scheduling, ARQ, radio resource management, power control, call admission policies, and QoS support mechanisms. These can be designed efficiently by cross-layer operation.

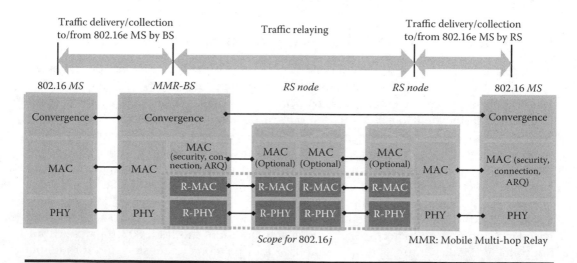

Figure 9.13 IEEE 802.16j protocol stack.

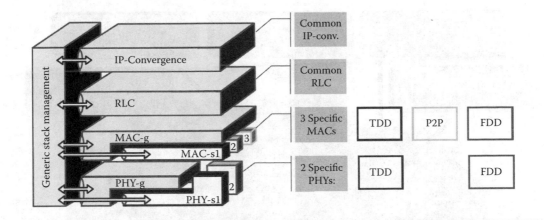

Figure 9.14   Multimode protocol architecture.

9.5.2   *Adaptive Cross-Layer Protocol Design*

The implementation of protocol operations in wireless environments that differ with respect to application and user scenarios requires different radio interface solutions to optimize the use of available radio resources. In order to meet these different requirements, adaptive protocol architecture can be considered, and these multimode protocol architectures allow the integrated use of relay, multiple access, and advanced antenna technologies. Each layer entity is selected optimally according to channel conditions and usage scenarios, and the optimal combination among entities is possible through tightly coupled interlayer operation (i.e., careful cross-layer design). The WINNER project has been developing a multimode protocol tailored to different application and usage scenario requirements; Figure 9.14 shows its architecture. Furthermore, adaptive resource management techniques can be applied according to such factors as load, propagation conditions, terminal capabilities, number of receivers, antenna configurations, and user needs, and they also need smart cross-layer design. "Adaptation" is destined to become another buzzword for future wireless MAN systems.

9.5.3   *Cross-Layer Design for Multiple-Antenna Systems*

Different multiple-antenna solutions exploit different properties of the radio channel and have different performance objectives (e.g., optimization of cell throughput, interference reduction, improvement of link reliability, and SDMA). Thus different multiple-antenna solutions, such as diversity, spatial multiplexing, beam forming, and SDMA, can be applied adaptively according to channel conditions and usage scenario (with respect to such factors as the number of users, terminal capabilities, user velocity, and service type). For adaptive spatial resource management, spatial processing is required in the MAC layer as well as the PHY layer and needs tightly coupled resource management techniques between the two layers, as shown in Figure 9.15 [41]. Currently the IEEE 802.16e standard supports a number of advanced antenna techniques, but does not provide a complete coupling scenario due to the overhead and complexity of these techniques. So, in future wireless MAN systems, a cross-layer protocol design is required to effectively manage spatial resources.

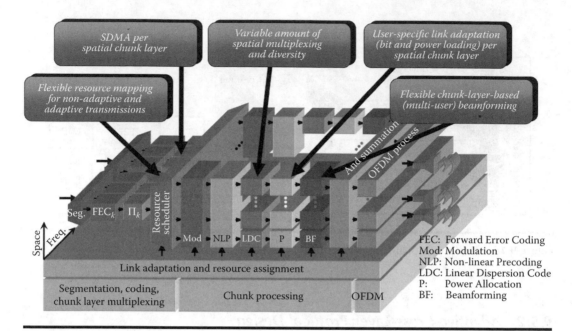

Figure 9.15 Adaptive spatial processing mechanisms.

9.6   Conclusion

Careful vertical coupling between layers and tightly coupled interlayer operation can improve system capacity, meet QoS requirements efficiently, and minimize power consumption. We have presented various cross-layer design issues for these three objectives in wireless MAN systems, such as IEEE 802.16 systems: (i) for capacity enhancement: adaptive channel- and QoS-aware resource management schemes based on utility functions, and interlayer operation for advanced antenna techniques and HARQ; (ii) for QoS support: DL/UL QoS handling mechanisms and cross-layer protocol adaptation schemes for QoS support; and (iii) for power saving: QoS-aware sleep mode operation and energy-aware resource management. The IEEE 802.16 standard supports many cross-layer operations, but does not provide completely coupled operation between layers due to the overhead and complexity of such operation. Therefore future wireless MAN systems will have to be designed to support more adaptive and flexible cross-layer operation by taking into account such emerging technologies as multihop relay, multimode protocols, and advanced multiple-antenna systems.

References

1. IEEE, IEEE Standard for Local and Metropolitan Area Network—Part 16: Air Interface for Fixed Broadband Wireless Access Systems, Standard 802.16-2004, IEEE, Washington, DC, 2004.
2. IEEE, IEEE Standard for Local and Metropolitan Area Network—Part 16: Air Interface for Fixed and Mobile Broadband Wireless Access Systems Amendment 2: Physical and Medium Access Control Layers for Combined Fixed and Mobile Operation in Licensed Bands and Corrigendum to 802.16-2004. Standard 802.16e-2005 and IEEE Standard 802.16-2004/Cor1-2005, IEEE, Washington, DC, 2006.

3. A. Ghosh, D.R. Wolter, J.G. Andrews, and R. Chen, Broadband wireless access with WiMAX/802.16: current performance benchmarks and future potential, *IEEE Commun. Mag.*, 43, 129, 2005.

4. T. Kwon et al., Design and implementation of a simulator based on a cross-layer protocol between MAC and PHY layers in a WiBro compatible IEEE 802.16e OFDMA system, *IEEE Commun. Mag.*, 43, 136, 2005.

5. S. Shakkottai, T.S. Rappaport, and Karlsson, P.C., Cross-layer design for wireless networks, *IEEE Commun. Mag.*, 41, 74, 2003.

6. V. Srivastava and M. Motani, Cross-layer design: a survey and the road ahead, *IEEE Commun. Mag.*, 43(12), 112, 2005.

7. V. Kawadia and P.R. Kumar, A cautionary perspective on cross layer design, *IEEE Wireless Commun.*, 12(1), 3, 2005.

8. M.V.D. Schaar and S. Shankar, Cross-layer wireless multimedia transmission: challenges, principles, and new paradigms, *IEEE Wireless Commun.*, 12(4), 50, 2005.

9. C.Y. Wong, R.S. Cheng, K.B. Lataief, and R.D. Murch, Multicarrier OFDM with adaptive subcarrier, bit, and power allocation, *IEEE J. Select. Areas Commun.*, 17(10), 1747, 1999.

10. Z. Shen, J.G. Andrews, and B.L. Evans, Optimal power allocation for multiuser OFDM, *IEEE Global Telecommunications Conference*, p. 337, IEEE, Washington, DC, 2003.

11. L. Tong, V. Naware and P. Venkitasubramaniam, Signal processing in random access, *IEEE Signal Process. Mag.*, 21(5), 29, 2004.

12. V.T. Raisinghani and S. Iyer, Cross-layer design optimizations in wireless protocol stacks, *Computer Commun.*, 27(720), 2004.

13. V.T. Raisinghani and S. Iyer, Cross-layer feedback architecture for mobile device protocol stacks, *IEEE Wireless Commun.*, 44(1), 85, 2006.

14. Q. Wang and M.A. Abu-Rgheff, Cross-layer signalling for next-generation wireless systems, *IEEE Wireless Communications and Networking*, vol. 2, p. 1084, IEEE, Washington, DC, 2003.

15. R. Braden, T. Faber, and M. Handley, From protocol stack to protocol heap—role-based architecture, *First Workshop on Hot Topics in Networking*, International Computer Science Institute, Berkeley, CA, 2002.

16. H. Yaghoobi, Scalable OFDMA physical layer in IEEE 802.16 wirelessMAN, *Intel Technol. J.*, 8, 201, 2004.

17. WiMAX: E vs. D—The Advantages of 802.16e over 802.16d, White Paper Motorola, Inc., 2005, available at http://www.motorola.com/networkoperators/pdfs/new/WiMAX_E_vs_D.pdf

18. S. Sung, I.S. Hwang, and S. Yoon, On the gain of data rate control in OFDMA systems, *1st International Workshop on Broadband Convergence Networks*, p. 1, IEEE, Washington, DC, 2006.

19. G. Nair, J. Chou, T. Madejski, K. Perycz, P. Putzolu, and J. Sydir, IEEE 802.16 medium access control and serving provisioning, *Intel Technol. J.*, 8, 213, 2004.

20. G. Song and Y. Li, Utility-based resource allocation and scheduling in OFDM-based wireless broadband networks, *IEEE Commun. Mag.*, 43(12), 127, 2005.

21. G. Song and Y. Li, Cross-layer optimization for OFDM wireless network—part I and part II, *IEEE Trans. Wireless Commun.*, 4(2), 614, 2005.

22. J. Huang, V. Subramanian, R. Agrawal, and R. Berry, Downlink scheduling and resource allocation for OFDM systems, *40th Annual Conference on Information Sciences and Systems*, IEEE, Washington, DC, 2006.

23. A.L. Stolyar, On the asymptotic optimality of the grandient scheduling algorithm for multiuser throughput allocation, *Opers. Res.*, 53(1), 12, 2005.

24. R. Agrawal, A. Bedekar, R. La, and V. Subramanian, A class and channel-condition based weighted proportionally fair scheduler, *Proceeding of the International Teletraffic Congress*, ITC-17, 2001.

25. G. Song, Cross-layer resource allocation and scheduling in wireless multicarrier networks, *Ph.D. dissertation, Georgia Institute of Technology*, Atlanta 2005.

26. M. Andrews, K. Kumaran, K. Ramanan, A.L. Stolyar, R. Vijayakumar, and P. Whiting, CDMA data QoS scheduling on the forward link with variable channel conditions, *Bell Laboratories Technical Memo*, 10009626-00404-05TM, 2000.

27. S. Shakkottai and A.L. Stolyar, Scheduling for multiple flows sharing a time-varying channel: the exponential rule, *Anal. Meth. Appl. Probability*, 207, 185, 2002.

28. G. Song Y. Li, L.J. Cimini, Jr., and H. Zheng, Joint channel-aware and queue-aware data scheduling in multiple shared wireless channels, *IEEE Wireless Communications and Networking Conference*, vol. 3, p. 1939, IEEE, Washington, DC, 2004.

29. A. Salvekar, S. Sandhu, Q. Li, M.-A. Vuong, and X. Qian, Multiple-antenna technology in WiMAX systems, *Intel Technol. J.*, 8, 229, 2004.

30. N. Riato, G. Primolevo, U. Spagnolini, and T. Baudone, A cross-layer architecture for SDMA, *IST Mobile & Wireless Communication Summit*, 2006.

31. S. Kallel, Analysis of a type-II hybrid ARQ scheme with code combining, *IEEE Trans. Commun.*, 38, 1133, 1990.

32. H. Zheng and H. Viswanathan, Optimizing the ARQ performance in downlink packet data systems with scheduling, *IEEE Trans. Commun.*, 4, 495, 2005.

33. J. Huang, R.A. Berry, and M.L. Honig, Wireless scheduling with hybrid ARQ, *IEEE Trans. Wireless Commun.*, 4, 2801, 2005.

34. J. Yun and M. Kavehrad, PHY/MAC cross-layer issues in mobile WiMAX, *Bechtel Telecomm. Tech. J.*, 4(1), 45, 2006.

35. Q. Zhang, F. Yang, and W. Zhu, Cross-layer QoS support for multimedia delivery over wireless Internet, *EURASIP J. Appl. Signal Process.*, 2, 207, 2005.

36. J. Jang, S. Choi, and K. Han, Adaptive power saving strategies for IEEE 802.16e mobile broadband wireless access, *Asia-Pacific Conference on Communications*, p. 1, IEEE, Washington, DC, 2006.

37. C. Schurgers, O. Berthorne, and M. Srivastava, Modulation scaling for energy aware communication systems, *International Symposium on how Power Electronics and Design*, p. 96, IEEE, Washington, DC, 2001.

38. E. Uysal-Biyikoglu, B. Prabhakar, and A. El Gamal, Energy-efficient packet transmission over a wireless link, *IEEE/ACM Trans. Network*, 10(4), 487, 2002.

39. Y. Zhang and K.B. Letaief, Energy-efficient MAC-PHY resource management with guaranteed QoS in wireless OFDM networks, *IEEE Conference on Communications*, vol. 5, p. 16, IEEE, Washington, DC, 2005.

40. W. Eberle, B. Bougard, S. Pollin, and F. Catthoor, From myth to methodology: cross-layer and system-level mixed-signal design concepts in actual designs, *42nd ACM/IEEE Design Automation Conference*, p. 303, IEEE, Washington, DC, 2005.

41. Wireless world initiative new radio (WINNER): https:/www.1st-winner.org

Chapter 10

Mobility Management in Mobile WiMAX

Shiao-Li Tsao and You-Lin Chen

Contents

The IEEE 802.16 standard, the Worldwide Interoperability for Microwave Access (WiMAX), is a broadband wireless technology that offers all packet-switched services for fixed, nomadic, portable, and mobile accesses [10]. The first specification (i.e., IEEE 802.16-2004) [1] that was ratified by IEEE in 2004 targets fixed and nomadic accesses in line-of-sight (LOS) and non-line-of-sight (NLOS) environment. With the amendment of IEEE 802.16e-2005 [2], IEEE 802.16e, also called Mobile WiMAX [5,6], further provides handover, sleep mode, idle mode, robust security, and roaming functions for mobiles. Similar to other IEEE 802 standards, IEEE 802.16 only specifies media access control (MAC) and physical (PHY) layer functions and lacks networking support. To address the demands for establishing an interoperable WiMAX network, the WiMAX Forum was formed to promote WiMAX and certify WiMAX products, and also proposed an end-to-end network architecture and service operations for WiMAX and Mobile WiMAX [7,8]. With these efforts from IEEE 802.16 and other organizations, WiMAX and Mobile WiMAX are not only PHY and MAC layer technologies, but also a complete network solution for a broadband wireless access system beyond third generation (3G) [4].

Mobility management is one of the essential functions for a mobile network. In Mobile WiMAX, mobility management schemes that handle link and network layer handover have been jointly developed by IEEE 802.16e and the Network Working Group (NWG) of the WiMAX Forum. This article provides an overview of mobility management in a Mobile WiMAX network. First, the system architecture of a Mobile WiMAX network is introduced. Based on this network architecture, location management of a mobile station (MS) in idle mode, link layer mobility management (also called access service network [ASN]-anchored mobility management), and network layer mobility management (i.e., connectivity service network [CSN]-anchored mobility management) are presented. Finally, mobility management in Mobile WiMAX is summarized.

10.1 Introduction to Mobile WiMAX Network

Based on IEEE 802.16 and IEEE 802.16e specifications, the NWG of the WiMAX Forum defines functionalities and protocols for network entities in a WiMAX network and reference points between these entities. These network entities are logical components and may be integrated together on a physical network node. The reference point is a conceptual interface between network entities and is associated with a set of protocols. While logical entities colocate on a network node, reference points between entities may be implicit. Figure 10.1 illustrates the system architecture of a Mobile WiMAX network.

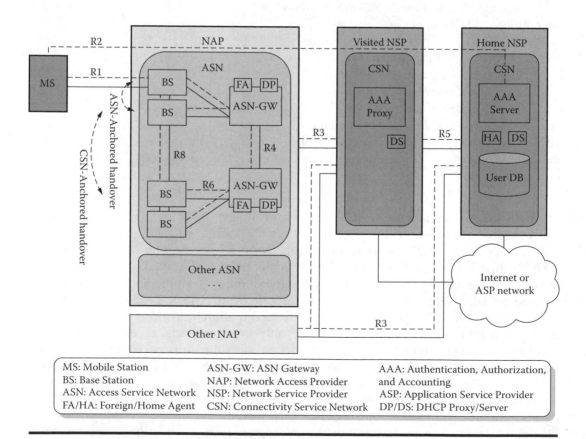

Figure 10.1 System architecture of a mobile WiMAX network.

A Mobile WiMAX network consists of three major components: mobile stations (MSs), network access providers (NAPs), and network service providers (NSPs). The R1 reference point is the interface between the MSs and BSs, and it implements control and data protocols conforming to IEEE 802.16 and IEEE 802.16e standards [3, 9]. Moreover, an MS further needs to establish the R2 logical connection with its home authentication, authorization, and accounting (AAA) server for authentication and authorization purposes.

A NAP that establishes, operates, and maintains WiMAX networks is an operator for access networks. A NAP may own several ASNs that are deployed in different geographical locations. An ASN is an access network infrastructure to which an MS attaches, and it consists of a number of base stations (BSs) that are controlled by one or more ASN-gateways (ASN-GWs). An ASN-GW is a gateway between an ASN and a CSN for a NSP. It serves as a relay node to tunnel MS packets between ASNs and specific CSNs that MS associate with. An ASN plays an important role to process MS packets from and to CSNs, handles mobility management functions such as ASN-anchored handover and mobile Internet protocol (IP) foreign agent (FA), implements security functions such as an authenticator and key distributions, and manages radio resources of the BSs in an ASN. The R4, R6, R7, and R8 reference points are defined in an ASN. The R4 is an interface between ASN-GWs and it is mainly used for transferring control plane messages for mobility management and forwarding data packets between ASN-GWs during handover. The R6 reference point defines control and data plane protocols between BSs and an ASN-GW. The R8 is an interface for transferring control plane packets and, optionally, data packets between BSs. The R8 facilitates fast and seamless handover between BSs.

The ASNs of an NAP further connect to an NSP which owns WiMAX service contracts with end customers. An NSP operates a CSN, which manages subscribers' information such as service policies, AAA records, etc. To provide services and applications to MSs, an NSP can either establish its own service networks in a CSN or forwards MS requests to other application service providers (ASPs). From a subscriber's point of view, a user initially subscribes to the service and has a contract agreement with an NSP. An NSP establishes contact agreements with one or more NAPs that offer WiMAX access services. Also, an NSP might have roaming agreements with other NSPs so that an MS may attach to its home NSP via visited NSPs. In such a case, an MS first associates with an NAP that only has a contact agreement with a visited NSP. Then the visited NSP relays the MS's authentication messages to the MS's home NSP, and finally the home NSP authenticates and authorizes the MS. To further access the Internet or services provided by ASP networks, Internet protocol (IP) addresses should be assigned to MSs. An MS must associate with an ASN first, and attaches to the home CSN. An ASN-GW implements a dynamic host configuration protocol (DHCP) relay to forward IP acquisition requests to either a visited NSP or home NSP for the IP assignment. The R3 and R5 interfaces are defined between an ASN and a CSN, and between NSPs, respectively. The R3 reference point defines control plane protocols such as AAA, policy enforcement, and mobility management between ASNs and CSNs. Data plane packets, which are tunneled packets, are also transferred between an ASN and a CSN over the R3 interface. The R5 reference point consists of control and data plane protocols for interconnecting a visited NSP and home NSP.

Mobility-related procedures for MSs in a Mobile WiMAX network can be categorized into location management for idle-mode MSs and mobility management for active-mode MSs. MSs without any radio and network connection may enter idle mode in order to save power, and they have to perform idle-mode procedures such as entering and leaving idle mode, listening to paging messages, and updating their locations. Idle-mode procedures also specify BSs and ASN-GWs to page a particular idle MS. On the other hand, MSs with active connections have to perform handover while they move from one BS to another BS. Handover in a Mobile WiMAX network can be further classified into ASN-anchored and CSN-anchored handover. ASN-anchored handover, also called micromobility, implies that an MS moves from one BS and another BS without a need to update its care-of address (CoA). The handover can be a hard handover, a hard handover with fast base station switching (FBSS) support, or a microdiversity handover (MDHO). CSN-anchored handover, on the other hand, defines macro mobility, which involves MSs to change its serving ASN-GW/FA and their CoAs. CSN-anchored handover facilitates network-layer mobility in both IPv4 and IPv6 networks. To support handover in an IPv4 network, proxy mobile IP (PMIP) and client mobile IPv4 (CMIPv4) are specified. CMIPv4, which inherits from the conventional mobile IPv4, simply applies mobile IP to MSs, ASG-GWs, and home networks. Different from CMIP, PMIP suggests running a PMIP client on the ASN-GW to perform mobile IP functions for an MS during CSN-anchored handover. Thus message exchanges over the air interface and the development efforts on MSs are

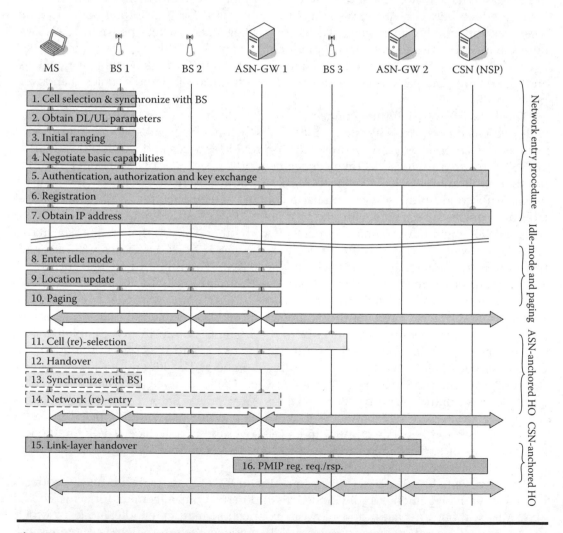

Figure 10.2 Overview of mobility management in a mobile WiMAX network.

both minimized. Based on the PMIP approach, network-layer handover is transparent to MSs and there is no mobile IP message exchanged over the air interface. CMIPv6, which applies mobile IPv6 to a Mobile WiMAX network, is introduced to support the network-layer mobility in an IPv6 network.

Figure 10.2 provides an overview of mobility management in a Mobile WiMAX network. In this figure, each procedure block represents a number of message exchanges between nodes. Procedures 1 through 7 are the initial steps for an MS to attach to a WiMAX network. An MS must first synchronize with a BS, obtain downlink (DL)/uplink (UL) parameters, and enter the ranging phase with the BS. After ranging procedures, an MS communicates with the serving BS to exchange basic capabilities, performs authentication and authorization, registers to the network, acquires an IP address from the home or visited network, and then can access the Internet. In this example, an MS attaches to a WiMAX network via BS 1 and ASN-GW 1 and accesses the Internet. The MS may enter idle mode and turns off its radio interface to save the power. During idle-mode periods, the MS still has to wake up periodically to update its location in order to be reached by BSs. Procedure 9 in Figure 10.2 shows that the MS moves from BS 1 to BS 2 and completes a location update. While there is an incoming packet sent to the MS, ASN-GW 1 receives the packet and pages the MS via BS 2. Then, a connection is established. The MS may move from one BS to another BS. Procedures 11 to 14 illustrate that the MS performs an ASN-anchored handover from BS 2 and BS 1. An ASN-anchored handover comprises cell (re)selection, handover, and network (re)entry steps. Also, the MS may perform a CSN-anchored handover while the MS moves from one BS to another BS when the two BSs are managed by different ASN-GWs. Figure 10.2 also shows an example of a CSN-anchored handover based on the PMIP approach. The network-layer handover is triggered by a link-layer handover. After a link-layer handover, a PMIP client performs network-layer handover procedures for the MS.

10.2 Idle-Mode Management

While an MS does not have any connection for a period, an MS might want to turn off its WiMAX interface to save power and switches to idle mode. Similar to other mobile communication systems, Mobile WiMAX also defines its own idle-mode operations and a paging network architecture. Four logical entities for idle-mode and paging operations are defined in a Mobile WiMAX network. First, a paging controller (PC) is associated with a paging group (PG) which comprises one or several paging agents (PAs) in the same NAP. The major task of a PC is to administer the activities of all idle-mode MSs situated in the PG managed by the PC. A PC can function as an anchor PC, which is in charge of the paging and idle-mode management, or a relay PC, which only forwards paging-related messages between PAs and an anchor PC. A PC could either colocate with a PA (i.e., a PC is implemented on a BS) or a PC can be implemented on a network node such as an ASN-GW and uses the R6 interface to communicate with its PAs. PAs which are normally implemented on BSs interact with the PC to perform paging functions. Finally, a PC can access a distributed database, called a location register (LR), which contains information such as paging parameters for idle-mode MSs.

Figure 10.3 illustrates examples for an MS to enter idle mode, update its location, and to be paged by the network. The examples in Figure 10.3 assume that an LR and PC colocate on an ASN-GW and PAs are implemented on BSs. PA 1 and PA 2 are in the same PG associated with PC 1. When an MS decides to switch to the idle mode, it first

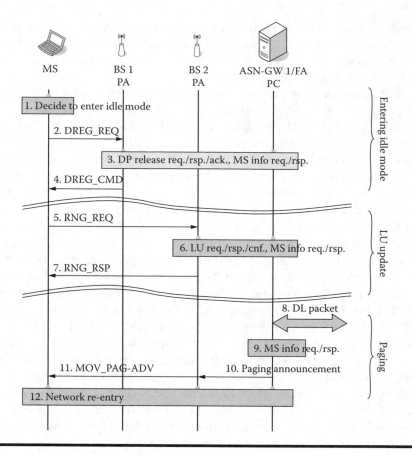

Figure 10.3   Idle-mode management and paging procedures.

sends a deregistration message (DREG-REQ) to the ASN-GW. The serving BS/PA and ASN-GW/PC release resources such as the data path occupied by the MS and update the information of the MS to the LR. Meanwhile, the PA and PC negotiate, configure, and inform the paging parameters, such as paging cycle, paging offset, paging interval length, anchor PC identifier, and paging group identifier for the MS. Based on the paging cycle (PAGING_CYCLE), paging offset (PAGING_OFFSET), and paging interval length, the MS derives the BS paging listening interval. A BS paging listening interval begins from the PAGING_OFFSET frame in every paging cycle and each paging listening interval lasts for the paging interval length. The MS has to stay awake during the entire BS paging listening interval in order to receive BS broadcasting paging messages (MOV_PAG-ADV).

An MS performs a location update (LU) based on four LU evaluation conditions (i.e., paging group update, timer update, power-down update, and MAC hash skip threshold update). The paging group update is activated when an MS detects a change in the paging group. The timer update is a periodic LU, and an MS performs an LU when the idle-mode timer expires. When an MS turns off or the MS MAC hash skip counter exceeds the MAC hash skip threshold, the MS also has to perform LUs. After a BS receives LU messages, the BS/PA updates the MS information to the PC/LR. While receiving an incoming packet sent to an idle MS, the ASN-GW/FA first obtains information of the MS from the LR and informs the PC to page the MS. Then the PC generates a paging announcement message and sends it to the relay PC or PA. Based on the paging parameters of the MS, PAs/BSs send BS broadcasting paging messages (MOV_PAG-ADV) to the MS. Once an MS is paged, the MS exits idle mode, performs ranging with the serving BS, and completes network (re)entry procedures.

10.3 ASN-Anchored Mobility Management

While an MS detects a poor signal quality from the serving BS, an MS may have to find a target BS to hand over. In a Mobile WiMAX network, a handover may be a link-layer or network-layer handover, and can be triggered by an MS or initiated by the network. Handover from one BS to another BS without a CoA update is a link-layer handover, also called an ASN-anchored handover. In general, an ASN-anchored handover involves the following steps. First, an MS performs a cell (re)selection that comprises scanning and association procedures to locate candidate BSs to handover. Second, an MS is informed or decides to hand over to the target BS. Finally, an MS completes network (re)entry procedures. Three possible approaches are specified in the IEEE 802.16e standard to implement ASN-anchored handover. Hard handover, which requires an MS to disconnect from the serving BS first and then connect to the target BS, is mandated. FBSS and MDHO are two options that improve the handover latency and reduce packet losses during handover.

10.3.1 Scan Procedures

Scan procedures can be activated by an MS or the serving BS according to the trigger criteria. The purpose of the scan is to measure the signal qualities of the neighboring BSs of an MS, and the measurement reports are used for MSs or BSs to select the target BS during handover. Initially the serving BS may indicate the scanning trigger conditions in downlink channel descriptor (DCD) messages or neighbor advertisement messages (MOB_NBR-ADV). The MOB_NBR-ADV is a broadcasting message that contains a list of suggested BSs for scanning and DCD, uplink channel descriptor (UCD), and other parameters of the BSs. Therefore an MS can synchronize with the neighbor BSs. After receiving DCD or MOB_NBR-ADV messages, MSs should measure signal qualities such as carrier-to-interference plus noise ratio (CINR), received signal strength indicator (RSSI), or round-trip delay (RTD) of the serving BS and other BSs, and check if the measurement results satisfy the trigger criteria. If a scan is triggered, the MS performs one of three actions. First, an MS can actively report the measurement results to the serving BS via a scan report message (MOB_SCN-REP). Second, an MS can initiate a handover by sending a handover request message (MOB_MSHO-REQ) with a list of BSs to which the MS prefers to connect. Third, an MS may send a scan request message (MOB_SCN-REQ) to the serving BS to allocate scanning intervals for scanning procedures. The MOB_SCAN-REQ message contains a list of BSs that are selected from the neighbor BSs in the MOB_NBR-ADV message or other BSs that are not on the neighbor BS list. While the serving BS responds to the MS scanning request, the serving BS can ask the MS to report the scanning results after the scans. These measurement results assist the serving BS in judging if a handover for this MS is required.

The serving BS may allocate a scanning period to an MS in order to scan neighboring BSs or find other candidate BSs to hand over. Figure 10.4 shows an example of the scanning process. In this figure, the scanning process is triggered by an MS according

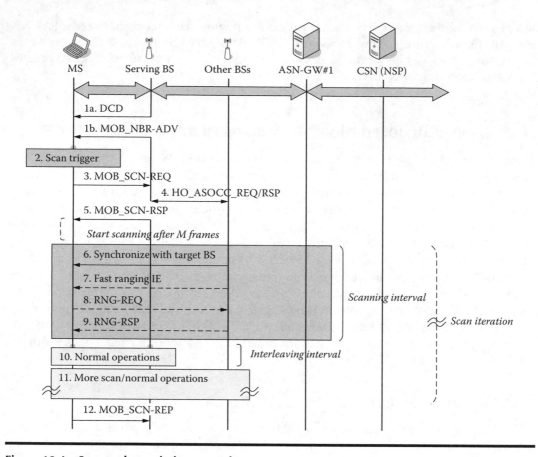

Figure 10.4 Scan and association procedures.

to the trigger conditions specified in DCD or MOB_NBR-ADV messages. An MS sends a MOB_SCN-REQ message to the serving BS with the MS's preferred scanning and interleaving intervals. Then the serving BS responds to the scan request using a scan response message (MOB_SCN-RSP) that contains the final list of BSs to scan, the start frame of the scan, the length of the scanning and interleaving interval, and the scan iteration. The start frame of the scan indicates to the MS the exact frame for performing the scan, and the scan and interleaving interval are used to determine the length of the scan and normal operation periods. The scanning and interleaving intervals are scheduled on a round-robin basis, and the scan iteration controls the number of iterating scanning intervals. An MS can scan other BSs in scan intervals and transmits/receives packets to/from the serving BS during normal operation periods. During a scan interval, the serving BS sees the MS as a sleep-mode MS and needs to buffer incoming packets to the MS. During normal operation periods, the serving BS then sends the queued packets to the MS.

10.3.2   Association Procedures

An MS may perform associations with neighbor BSs during scanning intervals. Association helps an MS to establish basic relationships, such as ranging, for these BSs, which may be potential target BSs for the MS. With associations before handover, MSs can reduce the time to synchronize and register with the target BS. The scan type in a MOB_SCN-RSP message indicates whether an MS should perform an association with a neighbor BS or not and what association type an MS and a BS should establish. The scan type may be one of the following configurations:

- Without association: This scan type indicates that an MS does not have to perform associations during scanning intervals. In this case, an MS only needs to monitor signal quality, report to the serving BS, or initiate a handover if the trigger condition is satisfied.
- Association level 0: scan/association without coordination. Scanning with association level 0 implies an MS should perform an association during scanning intervals, but the neighbor BSs do not allocate dedicated ranging regions for the MS. Thus, in this case, MSs must perform ranging procedures such as an initial ranging on a contention basis.
- Association level 1: association with coordination. While the serving BS indicates an MS to perform this association level, the serving BS coordinates ranging parameters of the neighbor BS for an MS. The serving BS sends an association request over the backbone to notify the neighbor BSs, and the neighbor BSs allocate ranging opportunities for the MS and reply to the serving BS. Then, the serving BS sends the MS the association parameters via a MOB_SCN-RSP message that contains rendezvous time and transmission opportunity offset. This information helps the MS send a ranging request to the neighbor BSs in the reserved ranging slots.
- Association level 2: network-assisted association reporting. Association level 2 is similar to association level 1. The serving BS needs to negotiate association parameters with the neighbor BSs for an MS over the backbone network. Different from association level 1, association level 2 does not require an MS to wait for ranging response message replies from the neighbor BSs. The ranging response messages are sent to the serving BS over the backbone network and then the serving BS forwards the ranging responses to the MS using a MOB_ASC-REP message instead of a RNG-RSP message. Therefore an MS does not have to wait for ranging responses, and this design reduces the delay of associations.

10.3.3 Handover Procedures

A handover followed by scanning and association procedures can be initiated by an MS or the network. Figure 10.5 shows an example of an MS-initiated handover. First, an MS sends a handover request message (MOB_MSHO-REQ) to the serving BS. The handover request message contains a list of candidate BSs and a measurement report of the BSs. Based on this report and other information on the serving BS, the serving BS sends a handover request (HO request) message to one or several neighbor BSs over the backbone network to identify the possible target BSs. Once the neighbor BSs receive handover requests from the serving BS, the BSs may send a context request to the context server to gather information such as the quality of service (QoS) of current connections of the MS and check whether they have sufficient resources to support this handover. After the context transfer and data path preregistration, the neighbor BSs send handover response (HO response) messages to the serving BS. The serving BS summarizes the results from the neighbor BSs and finally decides on a new list of recommended BSs and replies with an MOB_BSHO-RSP message to the MS. Meanwhile, buffering schemes

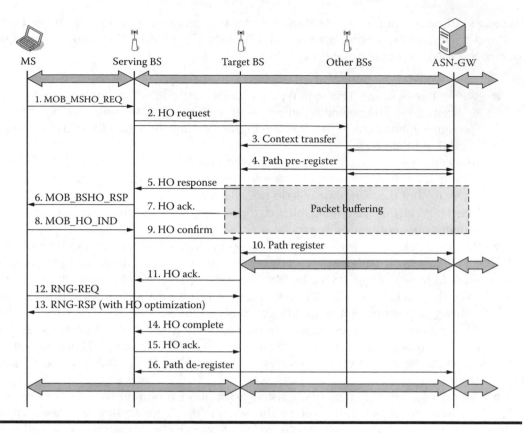

Figure 10.5   Handover procedures.

should be activated on an ASN-GW or BS that start to buffer the incoming packets to the MS to avoid packet losses.

After receiving a handover response message (MOB_BSHO-RSP), an MS should send a handover indication message (MOB_HO-IND) to confirm or terminate the handover process. In an MOB_HO-IND message, an MS explicitly notifies the serving BS about the target BS of the MS. Finally, the MS disconnects from the serving BS and synchronizes with the target BS. The serving BS can either release the resources for the MS or temporarily reserve the resources for the MS. This cache mechanism enables fast resumption of the services if the MS cancels the handover and returns to the serving BS.

After an MS sends a handover indication message (MOB_HO-IND) to the serving BS and switches to the target BS, the MS can either perform ranging procedures or directly access the target BS if the association has already been established during the scanning phase. Ranging from the MS to the target BS can be contention-based ranging or reserved-based ranging. If the serving BS negotiated the ranging parameters with the target BS and sent the information, such as CDMA code and reserved ranging slots, to an MS via a MOB_BSHO-RSP message, the MS can send a ranging request in these reserved ranging regions. This noncontention ranging speeds up the ranging procedure. Otherwise the MS needs to perform contention-based ranging, which is just like an initial ranging. After the ranging procedure, the MS needs to perform network (re)entry procedures. In order to accelerate the network (re)entry phase, the target BS can obtain the configurations and settings, such as service flows, state machines, and service information, of an MS from the serving BS via the context server without the involvement of an MS. This feature helps an MS and the network omit certain steps in network (re)entry. The target BS informs an MS which network (re)entry steps can be ignored in the HO process optimization field of a ranging response (RNG-RSP) message. An MS can thus skip these steps, such as session border controller, request/response (SBC-REQ/RSP), privacy key management (PKM) authentication, PKM traffic encryption key (TEK) creation, registration request/response (REG-REQ/RSP), network address acquisition, time of day acquisition, and trivial file transfer protocol (TFTP) during a network (re)entry. For example, if SBC-REQ/RSP is omitted, an MS does not have to send a SBC-REQ message after receiving a ranging response. The target BS may also generate an unsolicited SBC-RSP message or an unsolicited REG-RSP message if the target BS wants to modify the parameters of the basic capacity or the registration. The unsolicited response message is embedded in a ranging response message and sent to the MS.

10.3.4 FBSS and MDHO

During handover, an MS may have to disconnect from the serving BS and then attach the network again via the target BS. The connections and services may be disrupted during handover. To reduce the handover delay and minimize packet losses during handover, two advanced handover mechanisms (i.e., FBSS and MDHO, are proposed in the IEEE 802.16e specification. The basic ideas behind FBSS and MDHO are that an MS maintains a diversity set and an anchor BS. The diversity set is a list of candidate BSs to hand over for an MS. An anchor BS is the serving BS that transmits/receives packets to/from the MS over the air interface for FBSS. For MDHO, an MS receives the same data packets from all BSs in the diversity set and only monitors the control information from the anchored BS, which may be any BS in the set. An MS must associate with the BSs in the diversity set before handover, and should perform diversity set updates to include new neighbor BSs or remove BSs with poor signal quality from the list. The ASN-GW should multicast incoming packets for an MS to all BSs in the diversity set, and therefore the BSs in the diversity set are always ready to serve the MS for FBSS and MDHO. In packet transmission over the air interface, an MS transmits/receives packets to/from the anchored BS only for FBSS. Since packets are already in the BSs in the diversity set, the packet transmission can be resumed quickly after an MS performs an anchor BS update to change the serving BS. Packet loss and handover delay are reduced by employing FBSS. On the other hand, the BSs in the diversity set transmit the same data packets to the MS simultaneously when MDHO is applied. In this case, an MS can still receive packets from several BSs during handover, and the MDHO approach further minimizes packet loss and handover delay.

10.4 CSN-Anchored Mobility Management

CSN-anchored mobility management involves MSs moving from the current FA to another FA. This type of handover requires an MS to change its CoA. Mobile WiMAX supports network-layer handover for both IPv4 and IPv6 networks. For IPv4, CMIP and PMIP are supported. For IPv6, only client mobile IPv6 (CMIPv6) is defined, since each MS has its own IP address in an IPv6 network. CMIP fully integrates conventional mobile IP (MIP) mechanisms with the designs of an MS and a Mobile WiMAX network to handle network-layer handover. On the other hand, to minimize the development efforts on

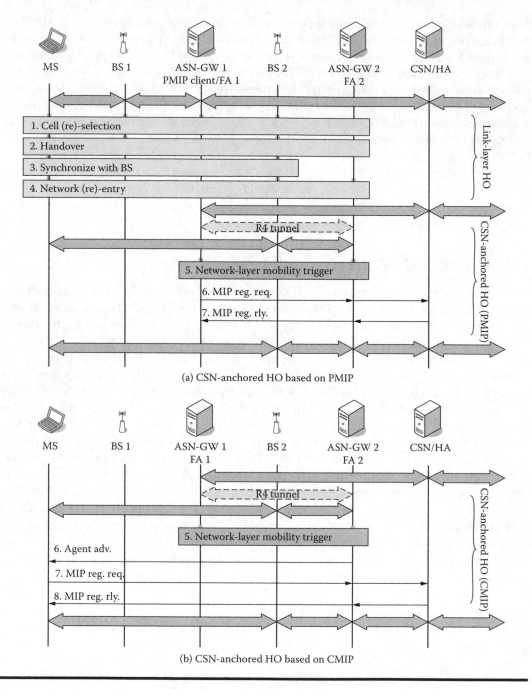

Figure 10.6   CSN-anchored handover based on (a) PMIP and (b) CMIP approaches.

MSs and also reduce MIP message exchanges over the air interface, PMIP is proposed. PMIP suggests running a PMIP client on the ASN-GW or a dedicated node in the ASN. The PMIP client serves an agent to handle network-layer handover for MSs, and thus network-layer handover is transparent to MSs.

Figure 10.6a illustrates the procedures for performing a CSN-anchored handover based on the PMIP approach. In this example, BS 1 and BS 2 are controlled by ASN-GW 1 and ASN-GW 2, respectively. Correspondent nodes (CNs) can reach an MS via the MS's home address (HoA), which is assigned by the home agent (HA) of the MS. An MS does not implement any MIP function, and all MIP functions are performed by a PMIP client running on the serving ASN-GW. Initially an MS is served by BS 1 and ASN-GW 1/FA 1. Packets from CNs to an MS first traverse to the HA, and the HA then tunnels the packets to the FA. The FA receives the packet, decapsulates it, and tunnels the packet again to the serving BS using a type I or type II tunnel. If a type I tunnel is applied, the ASN-GW uses the HoA of the MS as the key to locate the serving BS of the MS and tunnels the packets to the serving BS. If a type II tunnel is employed, the ASN-GW also uses the HoA of the MS as the key to find the connection identifiers (CIDs) or service flow identifiers (SFIDs) and the serving BS, generates layer 2 packets, and then tunnels the layer 2 packets to the serving BS. After an MS is informed by the network or decides to handover from BS 1 to BS 2, the MS performs a link-layer handover first, and attaches to BS 2 and ASN-GW 2. For a WiMAX network that implements an R4 interface and R4 relay, the packets sent to an MS via the serving ASN-GW 1 are forwarded to the target ASN-GW 2. That procedure minimizes packet losses during a CSN-anchored handover. Meanwhile, the serving and target ASN-GW detect this network-layer handover, and the PMIP client is activated to update the FA for the MS. A PMIP client registers the new FA's CoA (i.e., the CoA of FA 2) to the MS's HA. After the CoA update, the HA tunnels the MS's packets to the new ASN-GW/FA. It can be seen from the example in Figure 10.6a that an MS is unaware of a CSN-anchored handover.

CMIPv4 and CMIPv6 fully integrate the conventional MIPv4 and MIPv6 with MSs and Mobile WiMAX networks to support IPv4 and IPv6 mobility. In this case, MIPv4 and MIPv6 functions should be implemented on MSs, and MSs involve network-layer handover. Figure 10.6b shows an example of a CSN-anchored handover based on CMIPv4 approach. The new FA detects an MS joining the network and it sends an FA advertisement to the MS. The MS uses the FA's address as the CoA to perform MIP registration. Then packets for the MS sent to the HA can be tunneled to the new FA's address and then to the MS. CMIPv6 is similar to CMIPv4 except that FAs are eliminated from MIPv6, colocated CoAs for an MS are required, and CMIPv6 supports routing optimization by default. In other words, once an MS with CMIPv6 detects a network-layer handover, it acquires a new colocated CoA through autoconfiguration or DHCP in the visited network, and it sends binding updates to its HA and CNs. Then packets from CNs can be sent to the MS via the serving ASN-GW directly.

10.5 Conclusion

IEEE 802.16e and the WiMAX Forum have jointly developed location and mobility management mechanisms for a Mobile WiMAX network. To support link-layer mobility (i.e., ASN-anchored mobility), IEEE 802.16e defines complete new messages and procedures between MSs and ASNs. ASN-anchored mobility can be implemented by using three possible mechanisms. Hard handover, in which an MS must disconnect from the serving BS first before it can attach to the target BS, is mandated. FBSS, which is an enhanced version for a hard handover, suggests MSs to associated BSs that the MSs might hand over in advanced and maintain a diversity set with these candidate BSs. With these pre-association functions and negotiation procedures over an ASN network, handover latency is reduced. MDHO, which is a soft handover technology, further suggests BSs in the diversity set to send the same data to an MS. Thus the handover delay can be minimized. The above ASN-anchored handover mechanisms strongly rely on network supports, and thus the WiMAX Forum develops signaling messages and functions on BSs and ASN-GWs to support handover. Besides link-layer handover functions, the WiMAX Forum also defines network-layer handover functions for MSs, ASNs, and CSNs based on the MIP approach. CMIP and PMIP for IPv4 and CMIP for IPv6 are all supported.

Mobility management is one of the essential and crucial functions in Mobile WiMAX. Hence IEEE 802.16e, the WMAX Forum, and the Internet Engineering Task Force (IETF) have spent a lot of effort developing protocols to facilitate handover between BSs, ASNs, IPs, and administration domains. Although the infrastructure for mobility management in Mobile WiMAX has been established, challenging issues such as seamless handover for real-time applications in a Mobile WiMAX network, between a WiMAX network and other heterogeneous wireless networks, and future relay-based WiMAX networks (i.e., IEEE 802.16j) still need more research and study.

References

1. IEEE, Air Interface for Broadband Wireless Access Systems, Standard 802.16-2004, IEEE, Washington, DC, 2004.
2. IEEE, Air Interface for Fixed and Mobile Broadband Wireless Access Systems— Amendment 2: Physical and Medium Access Control Layers for Combined Fixed and Mobile Operation in Licensed Bands, Standard 802.16e-2005, IEEE, Washington, DC, 2005.
3. H. Yaghoobi, Scalable OFDMA physical layer in IEEE 802.16 WirelessMAN, *Intel Technol. J.*, 8(3), 201, 2004.
4. Understanding WiMAX and 3G for Portable/Mobile Broadband Wireless, Technical White Paper, Intel Corp., Santa Clara, CA, 2004.
5. Fixed, nomadic, portable and mobile applications for 802.16-2004 and 802.16e WiMAX networks, White Paper, WiMAX Forum, Beaverton, OR, 2005.
6. Mobile WiMAX—Part I: A Technical Overview and Performance Evaluation, White Paper, WiMAX Forum, Beaverton, OR, 2006.
7. WiMAX End-to-End Network Systems Architecture (Stage 2: Architecture Tenets, Reference Model and Reference Points, Draft Document, WiMAX Forum, Beaverton, OR, 2006.
8. WiMAX End-to-End Network Systems Architecture (Stage 3: Detailed Protocols and Procedures), Draft Document, WiMAX Forum, Beaverton, OR, 2006.
9. G. Hair, J. Chou, T. Madejski, K. Perycz, D. Putzolu, and J. Sydir, IEEE 802.16 medium access control and service provisioning, *Intel Technol. J.*, 8(3), 213, 2004.
10. A. Ghosh, D.R. Wolter, J.G. Andrews, and R. Chen, Broadband wireless access with WiMAX/802.16: current performance benchmarks and future potential, *IEEE Commun. Mag.*, 43(2), 129, 2005.

Chapter 11

Dynamic Network Selection in Wireless LAN/MAN Heterogeneous Networks

Olga Ormond, Gabriel-Miro Muntean, and John Murphy

Contents

In future generations of wireless networks, it is expected that different users with various multihomed personal wireless devices will have the option of accessing their desired services via different available radio access networks. Given the variability of the radio environment properties and user mobility, the availability and characteristics of an access network will change in time and are highly dependent on location. As a result, dynamic reselection of the access network is a necessary part of the mobility management mechanism (which maintains session connectivity as the user moves and/or the available access characteristics change). In their selection of a radio access network, customers consider cost and perceptive quality preferences for the current application and rely on intelligent network selection decision strategies to aid or automate their choice. These dynamic selection strategies need to take into account different dynamic and sometimes conflicting metrics, including price, local network capabilities and performance, application requirements, user preferences and context, and the mobile terminal capabilities. The ultimate goal is to provide a strategy to maximize the user's best interests when communicating in this multiaccess network environment.

This chapter describes the design of intelligent network selection strategies for initial call setup and for subsequent dynamic handover in wireless heterogeneous networks (which include local area network [LANs] and metropolitan area networks [MANs]) for transfer of future application data. The problem is first motivated by considering the differences between traditional cellular network selection methodologies, the changing requirements for future network selection methods, and possible architectures for implementation. The challenges involved in intelligent strategy design are outlined and the current state-of-the-art in network selection strategies is then presented. These strategies vary greatly in the parameters that they base the decision on, the way in which they discover and scale the parameters for comparison between candidate networks, and the actual decision strategies used for the final network selection decision. Each of these issues is discussed and proposed guidelines for designing such strategies are explored. A study on the design of a proposed user-centric dynamic network selection strategy is then presented, which includes a comparison between two of the existing intelligent approaches and an in-depth discussion on user utility functions for describing user preferences.

11.1 Introduction

The wireless landscape is changing. Influencing factors include deregulation, convergence of the Internet and telecommunications, and continuous evolution of network technologies and terminal capabilities. In home and office settings, wireless broadband networks are a popular alternative to wired networks, because of the ease, speed, and cost-effectiveness with which they can be deployed. At the same time, spoiled by high

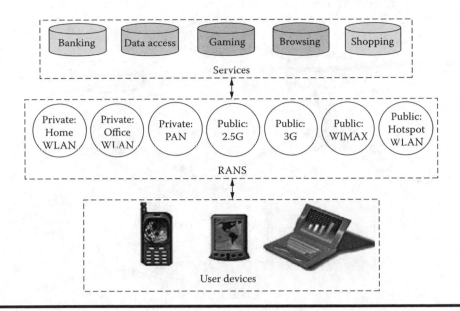

Figure 11.1   Service-oriented heterogeneous wireless network environment (SOHWNE).

bandwidth applications, users are becoming more technologically proficient and expect more services at higher quality levels on their wireless devices. The new focus is on users and providing them with quality access to a wide range of services. Offering a positive user experience is central to the success of future wireless networks. The resulting setting is envisioned to be a service-oriented heterogeneous wireless network environment (SOHWNE), as illustrated in Figure 11.1.

In SOHWNE, service provision is decoupled from the delivery mechanism, introducing service providers as a third party into the wireless arena. It is expected that an extensive collection of novel and attractive services will be produced by an array of service providers to supply the communications, entertainment, and information needs of many different customers operating on a variety of user devices. These services will need to be dynamically adaptable to the current context, such as user terminal capabilities, user preferences, and available access network characteristics.

At the network access level, both traditional and new operators will compete to offer user access to this wide range of services. Different operators will use different radio access technologies (RATs), each with different coverages and different capacities to cater to diverse service requirements. The availability of multi-mode user devices, with means to connect to different RATs [1], enables traditional operators to supplement their existing third-generation (3G) network offerings with higher bandwidth RATs, such as (WiMAX) or wireless local area network (WLAN), in certain areas (e.g., the first dual-mode WiMAX and code division multiple access [CDMA] prototype phone was recently revealed by Samsung in Singapore). Changes in regulations for communications are allowing traditional operators to sell bandwidth to mobile virtual network operators (MVNOs). MVNOs lease mobile spectrum and provide or resell different access services to consumers [2]. The development of new methods of payment, such that users do not need to have accounts with all network access providers, will be an incentive for many small businesses to offer access to customers in their locality [3]. All of these factors will change the current communications economic model and provide a greater mix of access provision for end users.

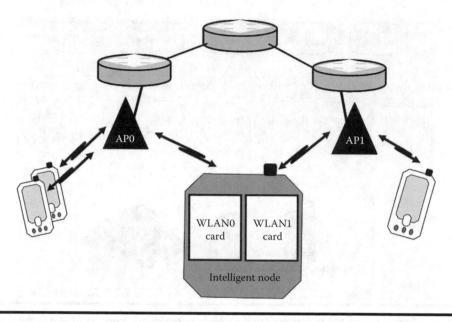

Figure 11.2   Network selection decision scenario: user device has a choice of radio access network.

Terminals are also evolving into high computing-power terminals. The popularity of portable electronic devices, such as cameras, MP3 players, and DVD players, has prompted the integration of these personal gadgets into the mobile handset. As more people are on the move, there has been an increase in the demand for 24 hr wireless access to user required services for both work and personal needs. Various users with their mixture of multiple-access devices will take advantage of provider competition when accessing a diverse range of services. Each of these users will have different expectations for the perceived quality of service (QoS) depending on their previous experiences, their current context (such as location, activity, mobility, and remaining battery power), and the capabilities of their mobile terminal. Their QoS expectations will also change with the type of applications they are currently running, the associated priority settings, and their willingness to pay.

As the Internet and telecommunications worlds merge, there are different views on how they will integrate and various visions on what will be the eventual shape of next-generation wireless networks. The final structure will depend greatly on how the communications market is regulated. As the economic model changes and more work is done on flexible payment systems, it will allow a wealth of radio access network (RAN) providers into the market to supplement the offerings of the bigger, more traditional access providers. The user-centric vision for the future is one in which a multihomed user in an area of overlapping network coverage will always have a choice of which of the existing RANs to use for their current service (the "always best connected" view). An example of such a scenario is shown in Figure 11.2. This is in stark contrast to the network-centric view, in which users are tied by contract to a sole network operator, like today. It may also be that service providers will choose on the user's behalf the most suitable transport offering from the available network providers for the application requirements. One way they might do this is to have the user download a network selection application onto their mobile device on subscription to the service.

In the case of the user-centric service-oriented heterogeneous wireless network setting, there is a need for an intelligent user-centric network selection mechanism in-built in the mobile device to aid with or automate the instantaneous best network selection decision for the user. This mechanism will rely on a dynamic decision strategy that will need to take into account the changing user preferences for current applications, current and predicted connectivity quality in the various available RANs, and traffic-related characteristics of the applications. The significance of the problem lies in the fact that a first-rate user experience is key to the success of the future communications industry. Access to the right information at the right price and at the right time is important to encourage service usage. Users will avoid services they perceive as expensive and will only spend money on services that they deem to be a good value. They will want to take advantage of the choice and diversity of the access networks' characteristics available in their current location to cater for their immediate service needs. The user-centric RAN selection strategy needs to be flexible given the varying RAN availability and characteristics, and the changing context in the user's busy day-to-day lives, whether they are at home or at the office, strolling about or in transit, on personal or business calls, in a hurry or sitting idle.

As both the technical and economic models for wireless communications change, users will require intelligent dynamic network selection strategies to provide a high quality communications experience. This chapter presents the research challenges in providing a user-centric-based network selection strategy that maximizes the user's interests in a SOHWNE, the current state-of-the-art on network selection strategies, and proposed guidelines for designing such strategies. A study of a user of non-real-time data applications using their multimode terminal in an area of overlapping RAN coverage is described, and an intelligent utility-based network selection strategy that considers user budget limitations and performance needs in terms of transfer completion time is illustrated.

11.2   Network Selection Decisions

A network selection decision is made to choose which of a number of candidate networks offers the most suitable connectivity for the current applications running on the user's mobile terminal. The decision process happens once at the beginning of a call and then subsequently during the call every time handover is triggered as part of the handover execution process. This process involves the handover of a connection from one network cell to another network cell and can happen, for example, if the provided connectivity starts to deteriorate and fails to meet the required quality standard for the current applications. During mobile communications, a number of parameters are measured periodically to monitor the performance of the connectivity. If any of these parameters exceed predefined thresholds, then a new network selection decision is triggered as part of the handover process.

There are a number of steps involved in the network selection process. The first step is to determine a set of available networks based on radio characteristics (if the radio frequency [RF] connection is poor, communication is not possible and the network is not considered as a possible candidate). From this candidate network set, the network selection strategy may choose one candidate network (or maybe multiple in the case of multihomed devices) to which to connect. In the latter process, several metrics (more than only RF-specific) can be involved.

11.2.1   Traditional Network Selection Decisions

Traditionally each mobile user was attached by contract to one sole network operator and the offered wireless applications were voice and short message service (SMS). The network selection decision was considered in the context of a cellular handover that was totally controlled by the subscriber's operator. The aims for handover were to maintain good connectivity for the voice application, to avoid diminishing link quality at the cell border, and to minimize the unstable ping-pong effect caused by continuous recurrent handovers over and back between two network cells. Handover was also performed for load balancing reasons when the network operator wanted to ensure good utilization and maximum user capacity in each cell.

At call initiation, a voice call is set up by the operator in the current serving cell. During the call, measurements are taken periodically on certain parameters, mainly to monitor the link quality. If these measurements exceed a certain threshold, a handover is triggered. The network selection decision of which candidate network to select as the target for handover is either network controlled or mobile assisted (still network controlled). The type of handover scheme employed is known as horizontal handover and is mainly caused by a user moving from the coverage of one cell to the coverage of another cell of the same RAT, and is usually within the same network operator domain.

The introduction of multihomed or multimode wireless terminals, which provides connectivity on multiple radio interfaces, together with the increase in availability of different RATs gives rise to the need for a vertical handover scheme. A vertical handover is performed when an existing connection accessing the core network via one type of RAT is switched over to a different type of RAT to maintain or improve device connectivity.

11.2.2   Future Network Selection Decisions

In future heterogeneous wireless networks it may be that either horizontal or vertical handover is within different network operator domains, in which case the decision is more likely to be terminal controlled. The multitude of choices between access network types, operators, service levels, and numerous applications adds to the number of metrics on which to base the choice, adding more complexity to the network selection decision.

As multifunctioning mobile devices begin to include MP3 players, cameras, and color video screens, the number and variety of wireless application types is increasing far beyond voice and SMS applications. These include real-time services like video streaming and conferencing, non-real-time services such as video, picture, and MP3 downloads, and interactive data services like Web browsing. To add to the complexity, some of these applications may consist of multiple flows of different traffic types (e.g., a multimedia streaming application may have synchronized audio, video, and subtitle data streams). In these cases the decision will need to establish which available connection can best satisfy all three streams.

While previously users only used one application at a time (i.e., either voice or SMS), now users may use multiple applications simultaneously (e.g., downloading emails while chatting on voice over Internet protocol [VoIP]). Some applications will have priority over others, and these priorities may change as the user selects one or the other to be in the foreground. Depending on the multihomed device and the system design, there may also be single or multiple simultaneous access network connections open. This leads to a number of possible scenarios: one connection for a single application, one connection for multiple applications, many connections for a single application (with traffic split over all of the connections), many connections with multiple applications split one connection per application, or a number of connections shared between multiple applications.

The number of network choices, traditionally limited to the operator's available cells, is continually increasing. Multihomed devices in future user-centric heterogeneous wireless environments will have the choice of many different operators, many different RATs, and many different service levels for each available access network. Network selection will no longer be controlled by the operator and may be controlled manually by the user, automatically by the user's terminal with guidance from the user, or by a service provider for their particular applications.

Given all of the choices available, the aims of network selection have changed somewhat to address user-centric goals. The goal is to always maintain the "best" connection possible for the duration of the current call, satisfying the application requirements and the user's expectations. As mentioned above, the number of criteria or metrics used to compare the different candidate networks is increasing. Some of these metrics are outlined in the sections below; they are a mix of dynamic and some static parameters, some of which are conflicting and involve the need for trade-offs.

One important difference between traditional and future network selections is the level of knowledge or information available to the decision maker. In traditional networks, where the operator makes the network selection decision, they have a lot of information at their disposal, including current and predicted cell traffic trends and loads for the current and the candidate cell, user profile and behavior trends, terminal characteristics, and call duration probabilities for voice traffic. Also, once the initial connection is made, the operator sets the network parameters so as not to lose the call (i.e., some guard band may be reserved for handover). The pricing scheme is also set at a fixed rate per time unit, depending on the user's subscription type, and known to the user in advance of making a call.

Information available to the user terminal includes the current reachable networks and the recent radio link quality for each, and also specific user trends for each application they use. However, devices do not have the wealth of knowledge that each of the operators have on the current and predicted cell traffic trends and loads for the current and the candidate cells, nor do they have much information on other users' profiles and behaviors. No information exists on whether the surrounding networks have better or worse quality or price. If global positioning system (GPS) or location-finding technology is available, then they can perhaps learn trends for networks in locations that are frequently visited. However, in this case the user may not be sure of the price of the call if they switch between different network operators and different service levels. Possible pricing schemes in future networks are discussed later.

All in all, the network selection decision is becoming more complex. Increased choice means more access network services to assess and more metrics on which to compare them. There are a large number of parameters involved, with complex relationships, many of which are dynamic in nature. Some conflicting parameter settings require trade-offs. This is further complicated by the lack of information available to the decision maker and the fact that given the limitations of the terminal processor power and battery, and the need for a fast and seamless handover, it is impossible to collect and interpret all information. As a result, some decision metrics may be enhanced or substituted with predicted information. A highly flexible strategy is required to suit the user's changing personal and business needs, their high mobility, and changing context and expectations. The strategy should also incur minimal loss of time, energy, money, and user inconvenience. Some user information or trends can be learned by the terminal over time.

11.2.3 Handover Execution Process

Network selection decision is part of the handover execution process. Handover execution requires functionality to dynamically manage network access for the user device. This management functionality will make use of three major subsystems: an access discovery function, a network selection decision scheme, and a mobility protocol responsible for tracking current subscriber access locations, and executing the connection setup and teardown.

The network selection decision is based on information available in the user device and on inputs from both the network and the user. The network metrics are established by the multihomed terminal during access discovery. This is an on-going phase during which the terminal communicates with the surrounding RANs to determine network availability, link characteristics, and pricing information. The user preference metrics are based on inputs from the user and stored in the terminal.

The network selection mechanism makes the decision of which RANs can best provide for the user's current communications needs. This is a dynamic decision and is triggered on call initiation and subsequently when any of the network selection triggers occur.

The mobility protocol covers two tasks: location management and connection management. Connection management involves signaling for the setup of connections to the selected RAN service at call initiation and subsequent handover execution. Location management involves tracking the mobile node and the current interfaces in use; this is done through location updates. Devices seeking to communicate with the user can contact this facility to get the current direct paths to the user terminal. In the case of a home server style protocol, a method for redirecting packets destined for the mobile node is also required.

Other elements required include the measurements and predefined thresholds that monitor the current link and other possible network connections in order to determine when a fresh network selection decision is needed. It is important to avoid network instability. The stability can be improved by monitoring the frequency of handover executions.

11.2.4 Network Selection Triggers

In recent research, much of the work on RAN selection policy design considers the access network selection problem in the context of seamless vertical and horizontal handovers. However, the network selection decision is now also relevant to the initial RAN choice at the beginning of a call. The network selection decision may be invoked at the initiation of service connectivity or as part of the handover execution during the service call when

- The current RAN characteristics are worse than expected or degrade considerably.
- Any of the other available RAN's characteristics improve.
- A new better RAN becomes available.
- The user changes their preferences or the service requirements.
- An imminent handover situation is predicted and triggered by any predictive algorithms running on the terminal. An example of a predictive algorithm is one that uses the terminal location, speed, and trajectory together with a map of local network coverage to help determine the approach of the current serving cell's border area in advance of arriving, to allow more time to choose a candidate cell for a seamless handover.

11.2.5   Metrics and the Importance of User Preferences

During traditional handover, network selection is based on received signal strength (RSS) thresholds and hysteresis values to reduce the ping-pong effect. Subsequent strategies consider the RSS along with other link quality metrics. More recent work considers other metrics, such as remaining battery life, power output limitations, the terminal speed, and candidate cell diameters, in addition to previous metrics. In these strategies the user preferences for the current services in the current context are not considered. Because providing a positive user experience is crucial, it is important to take user preferences into account in the access selection decision.

Where network selection strategies consider user preferences, most are presented in the context of network-centric vision. These solutions describe each user's preferences and then maximize the social welfare or overall satisfaction for all the users in a sole operator's network. The user preference information is used to the operator's advantage in order to maximize revenue and network utilization. For example, Chan et al. [4] use their understanding of user behavior to maximize network gains. They do not consider a user-centric approach, but rather look at congestion-based pricing to influence user behavior, with the goal of optimal resource allocation.

A smaller portion of the existing literature considers the always best connected, user-centric scenario where users take advantage of operator competition and select the best available RAN for current service in the existing context. Examples include Song and Jamalipour [5] and Ormond et al. [6], which both propose network selection techniques to provide the user with the current best available network, taking into account user preferences, current available connectivity, and application requirements.

11.3   Challenges Arising from the State-of-the-Art

This section looks at the state-of-the-art for each of the research issues associated with network selection and highlights the aspects that need more work. The most significant research issues that need to be addressed in accommodating a user-centric network selection decision include

- Decision architecture design. This determines where the network selection-based information and intelligence is located, what components are involved in the network decision and execution, and the signaling necessary between them. The location of the intelligence and the signaling involved in making a decision are crucial to the speed of that decision. The speed is also impacted by what information is readily available and what information is missing or imprecise and needs to be sought or predicted.
- Access discovery. The purpose of this phase is to assess available RANs and the associated network information for each of them. In this phase, the information discovered depends on the antenna system used, the kind of advertised information that is sent out by the different operators on different RATs, and whether the terminal passively listens for available information or actively seeks it. The challenge is to gain the right information that will lead to a correct characterization of each network's current connectivity offerings, with minimal interruption to the current connection, minimal impact on battery consumption, in a speedy manner, and with minimal need for data collection, storage, and interpretation.

■ User preference discovery. The goal of this phase is to collect, trade off, and model conflicting user metrics. The challenge is to do this as accurately as possible with minimal user interruption.

■ Network selection decision methodology. The object of this step is to compare candidate RANs and decide which is the best access network to use in the current situation. This is a challenge because of the extended complexity of the network selection problem and the need for a fast, efficient, and stable decision algorithm that dynamically chooses the best available network combination for the duration of the call that will provide the highest user satisfaction.

Another issue that is important to network selection is mobility management functionality, to dynamically manage access network connectivity on the selected RANs and to manage packet buffering, routing, and forwarding in the networks. The type of mobility management system in use may affect the network selection decision in terms of the latency involved in setting up a target connection, the mobility management architecture, and any security issues of the system (which should be accounted for in decisions for applications that require high security connectivity). Any delay in switching from one network cell to another is a metric for consideration, especially for applications that require seamless handover. The architecture design will impact the amount of signaling required and thus the speed of the entire handover execution process. It will also affect the security of the system.

Current research to find solutions to all of these issues spans many research fields. Examples include next-generation wireless networks, beyond third generation (3G), fourth generation (4G), mobility protocols, multihomed devices and adaptive antenna configurations, fast handover mechanisms, context awareness and intelligent adaptive agents, predictive techniques, economic incentives, autonomic computing, and modeling consumer behavior. The focus of this chapter is on the network selection decision methodology for user-centric benefit and the associated decision metrics. This section discusses these issues in detail, outlining their associated challenges, and also takes a quick look at architecture design and mobility management issues, as both will have some impact on the network selection decision method.

11.3.1 Architecture Design

The architecture for implementing network selection decision functionality is addressed in some recent work and is discussed in this section. The architecture design for mobility management is discussed separately, in the section on mobility management (11.3.1). There are a number of common issues between the two architectures. One such issue is the seamlessness of the network selection. The speed of the mobility management to execute the network selection decision and change from the current to the selected connection, and the penalties incurred while doing this (e.g., switching time and the power required on the mobile device, data loss, etc.), will impact the seamlessness. The network selection architecture impacts the speed at which decision metrics are obtained, and thus the speed in making the network selection decision, which also impacts the seamlessness of the entire process. In designing new architectures, it is important to limit the modifications required to existing wireless systems and to minimize the amount of additional network traffic needed.

The selection decision may be network controlled, mobile-assisted (still network controlled) or terminal controlled. Arguments in favor of a terminal controlled decision include the fact that most of the decision factors are based on information already available in the device and reduced signaling in the network. In the case of a multihomed terminal, a network controlled decision is difficult to implement. It is hard to determine which network should be the controller and then trust that network to opt for the access that is best for the terminal user and not for itself. Also, in the case that the decision is performed by some agent in the network, there is a lot of signaling required to relate the information from the terminal to the agent and then the decision back from the agent to the terminal. Therefore it is plausible that for a user-centric network selection, the decision will be terminal controlled and the actual decision function will be part of the user's terminal.

While the decision will most likely be terminal controlled, the layer at which the decision sits will depend on the mobility protocol in use and the mobility management architecture. The decision may be classified as being made at the link, network, or application layer or maybe at the level of a cross-layer decision manager. Each layer is closer to different information on the link quality, the available networks, or the application requirements. For the non-cross-layer case, there is a need to change the existing interlayer signaling to include information required by the network selection decision manager.

The user devices can operate with a single network access connection at a time, or with multiple simultaneous connections. These devices run various applications. Some applications may be integrated services that generate multiple data flows. For example, a video conferencing application could generate parallel video, audio, and whiteboard or file-sharing flows. As the traffic characteristics of different flows favor different radio access technologies, a terminal capable of maintaining multiple simultaneous traffic flows via different access networks may do so, whereas a terminal sending multiple flows on the one network access has to compromise the favored RATs of some flows. Different flows may also be charged differently. Decision policy design is outlined by McNair and Zhu [7] in the context of vertical hand-off for a single mobile user running multiple communications sessions. Adamopoulou et al. [8], use multiple simultaneous applications with different networks selected for each application. Gazis et al. [9] map the problem of identifying the network, or a combination of access networks, from the available candidates that will best satisfy the current user requirements in their current circumstance to an nondeterministic polynomial-timer (NP)-hard optimization problem. Heuristics may solve this complex combinatorial optimization problem to produce optimal or near-optimal solutions for some input instances.

Other works, such as Chebrolu and Rao [10], have proposed using multiple interfaces to extend the bandwidth available to a single traffic flow. This involves selecting the right combination of network interfaces and then scheduling the packets to be transmitted on different interfaces based on the estimated data transfer time.

In the literature on network selection decision management, many propose architectures that focus on the mobile terminal. The design of some of these architectures is based on more formal methods of decision management implementation, such as policy-based design, policy-based with fuzzy logic design, reinforcement learning mechanisms, agent-based design, or some combination of these.

Pérez-Romero et al. [11] discuss a framework for the development of policy-based initial RAT selection algorithms within the context of common radio resource management (CRRM). They outline a basic set of policies for the initial RAT selection decision for a specific service type, such as always allocate voice users to Universal Mobile Telecommunications System (UMTS) Terrestrial Radio Access Network (UTRAN), or always allocate indoor users to GERAN. These basic policies are inflexible, having no defaulting mechanism built in. Consequently, if the UTRAN or the Global System of Mobile Communications (GSM)/Edge Radio Access Network (GERAN) in the example is not available to the user device, then another network is not used, severely affecting user connectivity. More complex policies are defined by combining some of the basic policies. Each of the policies results in a prioritized list of RATs to use.

Murray et al. [12] use a selection policy to control the selection decision. The proposed policy system architecture is divided into a network policy engine and a mobile policy engine. Each engine is used to make decisions based on locally available information, stored in the associated policy repository. The network selection decisions may be triggered in the network policy engine by a congested network wishing to force a handover or an access request from the user. Triggers on the mobile policy engine are reduced levels of QoS or network unavailability. Both engines contain functionality for a policy decision point (PDP), a policy enforcement point (PEP), and a policy repository (which contains possible decision metrics). The network engine also has a network health monitor, which continuously updates the network condition information.

McNair and Zhu [7] also propose a policy-based networking architecture for user-centric handover. This architecture includes the two main elements for policy control: PEP and PDP. Two scenarios are presented: in the first, the vertical hand-off policy is optimized for all the user's active sessions being collectively handled and handed off to the same target network at the same time. In the second scenario, each service is handled separately, is prioritized, and may be independently handed off to different target networks.

Ylitalo et al. [13] look at how to facilitate a user making a network interface selection decision. They concentrate on a policy-based architecture for the end terminal and do not concentrate on any particular network selection strategy. However, they do mention the always cheapest (AC) network selection strategy, where the terminal always selects the cheapest of the available RANs. The policy driver module within the terminal kernel interprets the stored user (or operator) preferences and creates the necessary policies, which are then stored in the policy database. The policy driver makes a new network selection decision if any changes in the air interface start to violate the policy rules. The authors employ simultaneous multiaccess mobile-IPv6 as the mobility protocol, and thus can cater to multiple simultaneous connections to a correspondent node.

Bircher and Braun [14] propose an agent-based architecture with a user agent decision function. The agent-based marketplace architecture consists of three agents: a user agent (split into three entities: the travel assistant, the negotiation agent, and the configuration tool), an Internet service provider (ISP) agent, and a marketplace agent. The ISP agent has access services to sell and the marketplace agent is where the buyer and seller meet. The user agent acts as a buyer on behalf of the customer. It interacts with the user via a graphical user interface (GUI), compares and selects services with the best performance/price ratio, negotiates with providers for offered services, and prereserves the resources for an agreed price. Details of the exact negotiation terms are not covered, but they are based on differently weighted QoS parameters such as delay, bandwidth and packet loss, and SecureMobile-IP (SecMIP) is used for mobility management.

Mohyeldin et al. [15] describe a generic architecture for context awareness and adaptation behavior in 4G networks. It consists of the following functions: profile and context management, adaptation decisions, and technical deployment. This system uses its cognitive abilities and background knowledge to learn the recurring scenarios and user preferences and behavior. A profile is kept for the user, the terminal, the network, and the service (can be implemented as an extensible markup language [XML] document). An interpreter makes the decision based on the context information. It has the intelligence to adapt to new scenarios and can use applied decision logic, simple rules, fuzzy logic, or neutral networks for decision making. However, few details on the interpreter are supplied.

An example of a cross-layer architecture for network selection decision is the universal convergence layer (UCL). The UCL, proposed by Lanza et al. [16], isolates the upper layers from the underlying radio access technologies. The proposed UCL enables cross-layer optimization, using the information from the lower layers to enrich the network selection decision process in the upper layers. It also informs the lower layers of upper-layer requirements or limitations. Inserted between the network layer and the lower layers, the UCL contains a network resource collector, which collects and translates relevant data from different layers, and a path optimization module, in which the optimal access interface is chosen for each individual packet based on the information available in the network resource collector. Details of the actual decision-making algorithm applied are based on a simple approach using appropriate signal-to-noise-ratio (SNR) threshold levels, although more complicated approaches can be easily implemented.

A different approach is proposed by the emerging IEEE 802.21 standard, which supports seamless vertical handovers. This approach introduces a vertical layer (often denoted as layer 2.5) between the data link and network layers to ensure that multiple RANs and RATs are not visible in the upper layers.

Adamopoulou et al. [8] also present a mobile terminal architecture. Their terminal management system consists of a network interface adaptation module, which is responsible for access discovery and monitoring and the associated information collecting, a user preference module with a GUI, a mobility management module that employs Session Initiation Protocol (SIP) with Mobile IP, and the intelligent access selection function.

Stemm and Katz [17] propose an architecture for vertical handover for multihomed nodes. This is based on a modified version of Mobile IP. A handoff controller is implemented in the mobile terminal to make the decision of which network and which base station (BS) to connect to. The selection decision is based on network availability, as indicated by an increase or decrease in received beacons from the surrounding BSs, but it may be influenced by network constraints outlined by the user in the user control panel (on the terminal) or by heuristic advice from a subnet manager. The focus of their paper is reducing hand-off latency; details on the user-specified network constraints are not outlined.

11.3.2 Decision Metrics

Decision metrics and the need for decision policy design are outlined by McNair and Zhu [7] in the context of vertical hand-off. The context information or decision metrics can be either static or dynamic during the communications session. Figure 11.3 describes the network selection process at block level. The decision maker takes in inputs from the user, the applications, the network and the user terminal, interprets these inputs into decision metrics, uses some criteria or a scale of comparison to compare the list of available RAN services offered, and based on a decision algorithm, outputs a network service or a set of network services that can best cater to the user's current communication needs. Feedback and outcomes of previous decisions stored in the device can be used to supplement the inputs for the decision and help with the interpretation of inputted data. Theoretically, any change in the environment can be captured through constant

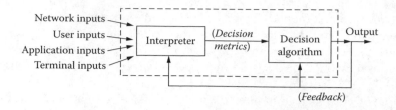

Figure 11.3   Block-level description of the network selection decision.

monitoring and can then be filtered and interpreted into actual decision metrics. In reality, it is not possible to gather and interpret all the information available, and some input data supplied may be enhanced or substituted with predicted information.

Interpreting the user inputs involves resolving conflicting metrics to accurately model user behavior. Collecting user preferences should incur minimal user interaction. Resultant decision metrics for each application running on the terminal include priority level, delay restrictions, and budget restrictions. Gathering user requirements is further discussed in Section 11.3.2.3.

Interpreting the application inputs involves collecting the information from the current application, or group of applications, requiring connectivity. One application may comprise multiple traffic flows, each with very different requirements. Resultant decision metrics from each application traffic flow include the application flow type, the application's current status (running in the background or foreground), the application flow's bandwidth, and delay restrictions.

Interpreting the terminal inputs involves using information gathered on the fly or stored on the terminal which could affect the selection decision. A significant resultant decision metric is the battery power status.

Gathering and interpreting the network inputs is sometimes referred to as the access discovery phase, and involves scanning and receiving information messages from each of the available networks and determining other information from the radio interface. This process may be repeated on a periodic basis or if some predictor or sensor detects a major change in the radio settings and triggers an access discovery phase. The procedure can consume battery power, and the frequency of running it should be minimized whenever possible. Resultant decision metrics for each access service available on each RAN include the RSS, RAT type, operator name, pricing information, current and predicted network conditions, terms of the service level agreement, and location information. Access discovery mechanisms are further discussed in Section 11.3.2.1.

There are various metrics used to compare different RANs. Adamopoulou et al. [8] consider quality, preferred network operator, preferred technology type and cost, while Wang et al. [18] and Chen et al. [19] use bandwidth, power consumption, and cost as metrics for comparison. The latter's proposed scheme relies on a score function that is based on functions of allocated bandwidth (transfer completion time), battery power consumption, and cost of the available networks. Zhang [20] uses a decision matrix with the following criteria: price, bandwidth, SNR, sojourn time, seamlessness, and battery consumption. Song and Jamalipour [5] consider a large number of parameters, including RSS, coverage area, throughput, delay, response time, jitter, bit error rate (BER), burst error, average number of retransmissions per packet, packet loss ratio, security, and cost.

The more metrics considered in the decision, the more complex—thus requiring more processor power and time—the resulting decision will be. There is a need for a trade-off between the amount of information processed and decision speed. The question of when to switch networks is an important one. Network selection may be done on different time scales depending on the trigger—if the current access link is not in immediate danger of being severely disrupted, then time is not as critical as if the node is on a cell border or in a highly congested cell. The number and type of decision metrics used may change depending on the need for speedy handover, the particular application involved, and user preferences for that application.

11.3.2.1 Access Discovery and Associated Information

During the access discovery phase, the wireless terminal continuously seeks information on all available access networks. The amount of information supplied should be minimal, as receiving current network condition information may be resource consuming and wasteful in the ever-changing unreliable wireless environment. In the case of an existing connection, the purpose of this information is to monitor the current access links and other available network interfaces to detect degradation of the current service or availability of a better service and then trigger a possible handover situation to the network selection decision module. This information will be used as decision metrics in order to determine which available RAN service is best for the current traffic flow.

It may be that a network operator offers a number of different access services over its RAN. Different networks may advertise their network availability in different ways. The network advertisement message should carry information on the operator identity and the pricing information and current network condition or service level available (may be a simple one-bit flag for congested or not congested, or may be a minimum bit rate level that can be obtained or an average bit rate for a particular timeframe) for each access service type it is offering. Other information that might be advertised is the location information (exact coordinates or a rough area); this information can be used to determine the speed and trajectory of travel of a mobile user and as another factor in the decision process. Also, having this information together with information on previous network selections for that location can provide some good predictive information on user movement patterns and imminent handover situations. Proactive handover triggers like this will help to improve the latency for the handover mechanism. Feng and Reeves [21] present the advantages and limitations of this approach. In the case of a highly mobile user, it may also rule out the use of certain RANs with small cell diameters or reduced capability to handle calls for nodes traveling at high speed.

The information sought on each RAN may depend on the RAT employed by that network. The main information determined by the terminal is the RSS. From the frequency used it may also determine the RAT type, and by monitoring the gaps between beacon signals it may be able to estimate the available throughput [22]. If the location information is not supplied by the network, it may also estimate the current location through prior knowledge of each RAN's location together with the difference in RSS for each network. Other information that may be sought includes terminal proximity to the BS and recent trends for network congestion conditions.

In the case of multihomed wireless terminals, the access discovery procedure depends on the antenna configuration. If it is a software-defined radio (SDR) where only one transceiver is used, then some kind of compressed mode needs to be used to temporarily suspend the current connection and allow the receiver to scan other frequencies [23]. Research in this area aims to find a fast and efficient method to determine the selection of the best available networks without interrupting the current connectivity or using up precious battery power.

In Wang et al. [18], bandwidth is determined either by use of an agent in each RAN which estimates and broadcasts the current network load, or in the case of commercial networks, by the "typical" value of bandwidth advertised by these networks. Chen et al. [19] measure bandwidth by using a probing tool. The main disadvantage of the probing technique is that it generates extra traffic, adding overhead to the different RANs tested. Stemm and Katz [17] consider increases or decreases in received beacons from the surrounding BSs indicating network availability. All these works—Wang et al. [18], Chen et al. [19], and Stemm and Katz [17]—collect current information on available bandwidth on all local networks. This may require heavy power consumption, a factor that should possibly be added to the cost of implementing the suggested strategy. It is possible that any network becomes more, or less, congested just after the information has been collected. Ormond et al. [24] use a predicted throughput which is based on previous throughput rates in each specific network.

11.3.2.2 Pricing Information

Other information that may be communicated in the access discovery phase is the pricing scheme employed by each available network and current charges. The user network selection decision strategy is influenced by the pricing scheme employed in the available networks. Many new pricing schemes that are being proposed for RANs are discussed by Falkner et al. [25]; the majority focus on network-centric benefit. Le Bodic et al. [26] look at networks with an auction-based pricing scheme and employ two different network selection strategies based on user preference for low service charge or user preference for networks with a good reputation.

The most popular charging schemes to date are fixed pricing schemes together with dynamic schemes like peak/off-peak pricing and usage-based pricing, with the user being charged on a per-unit-time, per-byte-transferred, or flat-rate basis. With service differentiation and different levels of QoS, operators can now begin to charge and differentiate between customers (e.g., on a per-traffic-type, per-service-class, or user-class basis).

Congestion-based pricing is being suggested as a convenient means of radio resource management. The implication of this is that a stationary user will not know in advance if the current call prices will remain the same for the duration of the current connection, as the price may increase or decrease. Users may not be keen to use such a system, and they may require some guarantees that the price will not increase beyond their maximum budget limit during communication before they agree to commence a call under this pricing regime.

Mobile users who move within different network operator domains will not know the exact price of their call in advance of call initiation; that is, unless they have a map of all the networks in their path, together with exact pricing and congestion information for each of them, and timing information for their journey and call. In this situation they may be able to calculate the maximum and minimum prices the call could possibly cost. Consequently users could set their expectations against these values.

11.3.2.3 User Inputs and Preferences

Modeling user preferences is a challenging task. Naturally users want quality connectivity for their current application requirements at as low a price as possible. Often the user's preferences conflict for different metrics. It is important to model the complex trade-off between different user preferences to determine which service the user really wants, if any, or what service can best serve the user's current needs. Often user preferences change with time, so it is important to enable dynamic update of the user profile. This can be done implicitly by monitoring user interaction (e.g., monitoring the user's frustration through observing repeated clicking on a button) or explicitly by prompting for user input.

In order to determine the user preference relation on different metrics, some works probe the user for their required settings for a number of user metrics and convert these to weightings for the network metrics. This may be invasive or annoying to the user. For example, Adamopoulou et al. [8] incorporate a GUI on the terminal to collect and weight four parameters (quality, preferred network operator, preferred technology type, and cost) in order of the indicated user preference for each service in use. Some intelligent learning mechanisms may be used to predict user behavior and preferences and to minimize interaction with the user. These mechanisms may even share information with other learning agents they meet in the networks and learn information for predicting the user model for new services encountered by the user.

Utility-based functions are commonly used to describe the user preference rating relationship for a number of metrics. User utility is defined as a function that captures the consumer's level of satisfaction with a service. Once the user satisfaction outweighs the respective charges, users will continue consuming the service [7]. Wang and Schulzrinne [27] consider user preferences to be represented quantitatively through a utility function when comparing two congestion pricing schemes. Das et al. [28] consider users to choose a pricing plan based on their data delay considerations, described by a user utility function. Wu and Yin [29] use a utility function to describe the user's bandwidth preferences and then perform admission control, path selection, and load balancing by maximizing the sum of the satisfaction (utility) of all users. This utility maximization is known as maximizing social welfare.

11.3.3   Network Selection Decision Methodology

The network selection strategy must use some scale on which to compare candidate RAN services, each of which have different characteristics, and then decide which is the best access service or combination of access services to use in the current situation. User-centric network selection strategies are the topic of Section 11.4.

11.3.4   Mobility Management Functionality

After a network selection decision is made for either call initiation or handover procedures, the mobility protocol communicates with the selected network to dynamically set up the connection. The aim of mobility management is to provide seamless connectivity in a multiaccess scenario, such that handover latency and packet loss are both kept to a minimum to avoid a noticeable interruption in the connectivity as the device is switched from one RAN to another. Other aims are to avoid network instability, to keep signaling and packet duplication to a minimum in order to avoid wasting precious bandwidth, and to provide scalability and reliability. It is also important to avoid designing a mobility management architecture that involves too many changes in the existing network's infrastructure, for backward compliance with existing communications systems and ease of deployment, which encourage general acceptance and widespread penetration of the new solution.

A number of different mobility management protocols have been proposed. Although most of these protocols do not include any standardized functionality for either the management of the network selection decision or for dynamically updating the current RAN selection mechanism with the latest user preferences or application requirements, it may be possible to adapt the protocols to include some time-varying metrics that are critical to making an appropriate user-centric RAN selection decision.

Challenges include the mobility management architecture, addressing, security, and the impacts of wireless network link characteristics, mobility, and multihoming. The Internet architecture was not designed with wireless networking characteristics in mind. The original design was intended for machines in a fixed location with only one network interface to the wired network. Wireless network link characteristics, mobility and multihoming concerns, and associated security issues were not considered. Nikander et al. [30] discuss many of these problems when introducing their host identity protocol (HIP) architecture proposal. Some networking functions that are provided for are addressed by multiple layers of the Internet suite, which can result in replication of responsibility and contradicting outcomes. For example, transmission control protocol (TCP) always assumes that packet loss is due to congestion and will take congestion avoidance measures. It is unaware of wireless channel conditions, where a packet may be lost due to a poor radio link, disconnection, or handover execution. The modular functionality of the protocol stack is seen as inflexible for adding the many new interlayer communications links that are required for collecting information, making the network selection decision, and informing the appropriate layers on the latest network connection and the current available access characteristics. Eddy [31] addresses the problem of which layer of the stack mobility belongs to. He analyses the disadvantages of Mobile IP and concludes that while the transport layer is the most likely place for a mobility protocol, a cross-layer approach, where interlayer communication is used, may be the best approach.

The cross-layer approach to wireless network design is a current hot topic [32]. The aim of the approach is to overcome the inflexibilities of the strict layered protocol stack and provide interlayer communications within the stack to improve the overall performance. Various architectures for cross-layer feedback in the mobile terminal protocol stack have been proposed; a summary of some of these architecture proposals can be found in Raisinghani and Iyer [33].

One main issue is addressing. IP addresses, once considered as location-bound identifiers for the mobile device, can now be dynamically assigned during a call session as the mobile device hands over from the original operator's network to other RANs. On a multihomed terminal where multiple wireless interfaces are possible, a set of addresses is necessary to describe each of the possible access paths. These addresses may also change, requiring a means for dynamically mapping the addresses to the changing availability in RANs. There is a need to keep track of the address of the current wireless interfaces in use and to convey this to the communicating peer nodes. The double jump problem [34] happens if two communicating nodes both change their current connections at the same time and cannot inform each other of the changes. The eventual architecture should include measures to avoid this problem.

Another major concern is security. To avoid denial of service and other malicious attacks, there is a need to determine that all the received messages are from whom they say they are from and are not originated by some bogus attacker. Address updates also need to be verified as being genuine.

There are many existing protocols and current research works on proposed approaches for mobility management, with many protocol extension proposals to address different problems found in each. These include Mobile-IP (MIP) [35,36], a network layer protocol, and its various extensions including Mobile IPv6, multihoming Mobile IP

(M-MIP) [22], secure Mobile IP (SecMIP) [37], and a combination of Session Initiation Protocol (SIP) and MIP [38], stream control transmission protocol (SCTP) [39], a transport protocol that was proposed by the IEFT, and its extensions including the proposed mobility extension, mSCTP [40], TCP migrate [41] a secure end-to-end architecture for host mobility, and (HIP) [42], a recent IETF draft protocol proposal that solves the security issues currently surrounding mobility and multihoming solutions.

11.4   Access Network Selection Strategies

Previous sections have described the network selection decision problem, architectural design, mobility functions, and decision metrics required to support this decision for a next-generation heterogeneous wireless environment. The focus of this section is the actual decision-making strategy used by the terminal to select a network or a set of networks on the user's behalf. This network selection should be optimal, although given the large number of parameters involved and their complex relationships and dynamicity, it may be hard to make an optimal decision. The unreliability of the wireless environment, computational effort required in the terminal, battery power and user budget limitations, different data priorities, and the need for minimum connection decision and setup latency (especially for the real-time applications) all further complicate the design of the decision-making process. In this context there is a need for a fast and efficient decision strategy that involves minimal user interruption and application connectivity disruption and makes the best choice for the user's set preferences. The solution should avoid causing network instability or the ping-pong effect by switching too often between the same networks. Also, the selection may be for an instantaneous connection setup or for a reservation-based procedure (in which case the user inputs and predicted information ratio may change considerably).

11.4.1   Types of Network Selection Decision Algorithms

Network selection decision policies may be simple, with straightforward objectives; for example, always stick with one particular network regardless of its current characteristics (always select the cheapest RAN service available, always select the preferred operator's network), or always select a WLAN if one is available. Another possible strategy is random network selection. However, none of these strategies are responsive to changes in the dynamic radio environment, and in the case that the chosen network is heavily congested, the user will be stuck with a poor network choice. The user node needs a more sophisticated, intelligent approach to break down the options and decide on an optimal or suboptimal choice. Section 11.4.2 describes a number of proposed intelligent strategies.

11.4.2   State-of-the-Art Network Selection Decision Algorithms

This section examines some of the proposed access network selection strategies, what inputs they choose to use, how the inputs are interpreted into decision metrics, and the actual decision algorithm used. Song and Jamalipour [5] have their user-centric network selection module implemented in the link layer, with cross-layer signaling messages delivering the QoS information from different layers in the IP stack, including the application layer, where users describe their desired attributes in a QoS context. The scheme is based on a large number of parameters that describe availability, throughput, timeliness, reliability, security, and cost. Two mathematical techniques—analytic hierarchy process (AHP) and grey relational analysis—are used to perform the analysis and trade-off between the parameters. The decision maker then chooses the network with the best score.

Adamopoulou et al. [8] propose a user-centric network selection strategy and the associated mobile terminal architecture. A GUI on the terminal prompts the user to set the weight (wc, wq, wt, wop) for each of four user input parameters: cost, quality, preferred technology type, and preferred network operator. The cost information is received from the networks in advance of network selection and is not expected to change often, and especially not during the call session. Each network operator charges per data volume, or per unit time, for access at a specific quality level. Multiple simultaneous applications are provided, with different networks selected for each application.

Lee et al. [43] use a cognitive agent to determine the user's particular needs and preferences with minimal user interaction and use the resulting user profile, together with network service profiles for each access provider, to make a decision on which access service to request. The network service profiles contain network condition information in the form of a two-bucket profile and pricing information. The total access cost for the user is made up of a price per unit time plus a price per kilobyte sent. The user profile is described by a utility function, which is a function of quality and cost of the access service in a given context, and a weight (range zero to one), which represents the trade-off between the cost and quality objective. Feedback buttons are included so the user can indicate if they are not satisfied with the current network choice. The strategy has an built-in risk function which may make a selection to an estimated lesser quality network based on the fact that it may be better than estimated.

Zhang [20] uses fuzzy logic to deal with the imprecise information of some decision criteria and user preferences, he then compares the use of two classical multiple attribute decision-making (MADM) methods to get a final ranking of candidate networks. The methods compared are simple additive weighting (SAW) and technique for order preference by similarity to ideal solution (TOPSIS). A decision matrix is defined containing the following criteria: price, bandwidth, SNR, sojourn time, seamlessness, and battery consumption. The user preferences for each application are modeled as a matrix of weights assigned by the user for each of the six decision criteria. Two applications are considered: voice and file download. The available networks are ranked separately for each application using the two methods. TOPSIS is found to be more sensitive to criteria with high user preference weights, whereas SAW provides a more conservative ranking.

Wang et al. [18] describe a handover "policy" for heterogeneous wireless networks, which is used to select the "best" available network and time for handover initiation. They consider the cost of using a particular network in terms of the sum of weighted functions of cost, bandwidth, and power consumption. The weights are supplied by the user and may be modified at runtime by either the user or the system. Power consumption is limited by battery life and cost is limited by the maximum sum of money a user is willing to spend for a period of time. Network inputs include the available bandwidth, as reported by the network, and throughput that is observed at the terminal. The network that is consistently calculated to have the lowest cost is chosen as the target network. A randomized waiting scheme based on the impact of the estimated handover delay is used to achieve stability and load balancing in the system and to avoid handover synchronization.

Many papers, all of which reference Wang et al. [18], use similar cost functions. A smart decision model for vertical hand-off is the focus of Chen et al. [19]. Their proposed scheme relies on a score function that is based on functions of allocated bandwidth (transfer completion time), battery power consumption, and cost charged by the available networks. The hypertext transfer protocol (HTTP) hand-off decision model presented by Angermann et al. [44] and the work in McNair and Zhu [7] also use a cost function that compares available networks from the list of options and establish the network to handoff to according to the importance weightings associated with different metrics.

11.4.3 Analysis and Comparison of Network Selection Decision Algorithms

The strategies presented above vary greatly, in the parameters that they base the decision on, the way in which they discover and scale the parameters for comparison between candidate networks, and the actual decision strategies used for the final network selection decision. While some of these proposed strategies are placed specifically at various layers (e.g., link, network, or application) or involve a cross-layer design, others do not specify or leave this as an open issue, which may be better addressed in the ongoing work on mobility management protocols. As discussed previously in both the network selection decision and mobility management architecture sections, placement at each layer has its advantages and disadvantages, with the current work on cross-layer design being a probable winner.

Network decisions may be based on either end-to-end link conditions or the local link conditions. Monitoring local network conditions as opposed to the end-to-end network conditions is easier and faster. Discovery of the end-to-end network conditions for each possible access network solution may be time consuming and add to the traffic load on each of the investigated networks. This is especially true in the case where there are two access networks involved, one on each end of the link, as opposed to the link when a mobile user connects to a high-powered server in a high-speed wired network. With the millions of miles of high-speed fiber backbones the main bottlenecks are envisioned as being on the access networks. In most cases the local conditions on the wireless access links are the basis of the decisions.

Some of the work—Song and Jamalipour [5] and Zhang [20]—uses formal methods from multiple attribute decision making (i.e., AHP, SAW, TOPSIS). Most of the work looks at the sensitivity of the decision strategies to the user preference weightings. A few look at user satisfaction with the network selections. It is important to consider other performance metrics in the analysis of any proposed network selection strategy. Some of these metrics are outlined in Section 11.4.4.

11.4.4 Performance of the Network Selection Decision Algorithm

It is important to find the right balance of decision metrics. Too many parameters can make the decision very complex and cumbersome to process, adding to processor load and decision duration. The number and type of metrics used may change depending on how soon the handover needs to happen, the requirements of the particular application involved, and the current user preferences for the application. The frequency of remaking network selection decisions is important, and whether it should be one decision made at call initiation for the duration of the connection, a periodic decision made at set time intervals, or a decision only made when handover conditions are triggered. If the decisions are too frequent, this can cause network instability and will also impact on battery power and the device processor load (causing interruption of user applications, with which it shares the processor). If the decision is periodic, then the chosen time intervals are important and may depend on the type or priority of the applications running. Other factors to be considered are when to switch and not to switch between networks or network cells.

When considering any network selection decision algorithm, a number of performance metrics should be considered:

- Additional processor load and battery consumption for access discovery.
- Additional network traffic load involved in access discovery, the decision, and the handover execution phase.
- Required user interaction and the annoyance factor involved.
- Additional processor load and battery consumption in the device for the actual decision making.
- Speed and efficiency of the decision-making process.
- How accurate the decisions are compared to the ideal.
- The frequency of decisions made for the call session duration.
- User satisfaction rating with the decisions made.
- Other penalties incurred in the process (i.e., amount of information lost and resent, handover latency, connections lost).
- How the decision process learns from feedback on previous decisions and evolves over time.
- How the decision strategy deals with multiple competing intelligent users, especially in the case where the majority employ similar schemes to the device user.

11.5   Consumer Utility-Based Network Selection Strategy

This section presents a study on the proposal for an intelligent utility-based network selection strategy for initial call setup, and for subsequent dynamic handover, in wireless LAN/MAN heterogeneous networks for transfer of non-real-time data files. The focus is on the user-centric decision-making problem of which available network to choose for data transfer and concentrates on serving one application's communications needs at a time. The strategy will be implemented in the user device. This device uses a single connection at a time (as opposed to multiple simultaneous links).

The solution involves describing the user's preference relations for conflicting service metrics and predicting the throughput in each of the available RANs. Then, by applying economic principles, network selection decisions are taken that maximize the user's interests in an ever-changing radio environment.

These decisions will rely heavily on the charging schemes implemented by the candidate networks, as users make an effort to get the best value for their money. The study considers the case where a user wishing to send non-real-time data (a large file) has a choice of several available networks, each of which employ a set price-per-byte pricing scheme, but each charging different prices. It is assumed that the price advertised by the network will not change for the duration of the call. Every user wants timely delivery of their data at the lowest price. While the strict time restrictions imposed by real-time traffic are not used, it is assumed that every user has a patience limit and will only be willing to wait so long for the completion of their data transfer before they become dissatisfied.

At the time of access network selection, the user's terminal surveys the radio interfaces and determines a list of current available access networks and their associated prices. The terminal employs an algorithm to predict the current transfer rates in each of the listed RANs. The work described here uses a simple weighted average of previous rates achieved on the candidate networks. It will then apply user time and budget preferences for the current service and choose the most suitable RAN.

A user utility function is employed to describe the user time and budget preferences for non-real-time data (i.e., the user's willingness to pay in relation to their willingness to wait for a particular service transfer). The more delay a user experiences the less the user may be willing to pay. The shape of the user utility function is related to the user's priorities for delay or cost savings. The difference between the value of the data to the user (willingness to pay) and the actual price they are charged is known in microeconomic terms as the consumer surplus (CS). We propose a utility-based algorithm that accounts for user time constraints, estimates complete delivery time (for each of the available access networks), and then selects the most promising access network based on consumer surplus difference.

11.5.1 Throughput Prediction Scheme

The actual average transfer rate determines the transfer completion time and therefore the value of the data to the user. In the access discovery phase, the wireless terminal seeks information on available access networks. Since receiving current network condition information may be resource consuming and wasteful in the changeable, unreliable wireless environment, the amount of information supplied should be minimal.

The model employed here is a rate prediction scheme that forecasts the available rates offered in the available wireless links. The method used considers available information on the networks' recent history, using a weighted average of the previous N rates ($N = 5$ in the tests presented here) experienced for the end-to-end connection in a particular network to predict the transfer completion time for that network. If previous rates are unavailable, a default rate is used. This rate is the minimum acceptable rate that will meet the user's transfer completion time deadline for the current application. If the prediction produces a bad rate, we only consider this rate for a number of decisions before we decide that the information is stale and replace it with the default rate. A drawback in this approach is that if all networks suffered congestion a number of times in succession, the user continues to select other networks until the same (lower) throughput is obtained or the information is deemed stale and is replaced. One of the networks may well have recovered from its slump in transfer rate before this time.

11.5.2 User Utility Function

Choosing an appropriate utility function to model human user preferences under the uncertainty of the radio environment is a challenge. A reasonable set of assumptions and conditions helped to determine possible shapes and threshold values for the utility function. User satisfaction is expressed in terms of the user's willingness to pay and willingness to wait for data transfer. The users must always be reasonable in their expectations and consider current local prices and RATs available to them. From the typical rates associated with RATs in a heterogeneous environment, a range of possible throughput rates were determined. Corroborated with the file sizes to be transferred, this gives users realistic expectations for transfer completion time. The transfer costs depend on the pricing schemes employed by the candidate networks. In this work it is assumed that all networks employ a fixed cost per byte pricing scheme (i.e., the price is per byte, is known to the user in advance of network selection, and will not change for the session duration).

11.5.2.1 The Shape

Utility theory from the field of economics provides a means to qualitatively capture the user's perception of, or satisfaction with, application performance in relation to network performance metrics. Utility functions, which have already been used in many areas of telecommunications, are mostly determined in one of two ways: by a user subjective survey, as in Jiang et al. [45], or by examining and modeling typical user behavior, as in Wu and Yin [29]. One of the first works to use utility functions to describe user perceived quality with respect to effective serving rate was by Shenker [46].

Application data are mainly categorized into elastic or best effort data, and delay or rate-sensitive data. Rate-sensitive applications depend greatly on effective or instantaneous bandwidth for user-perceived quality. While the strict time restrictions imposed by real-time traffic do not apply to elastic non-real-time data, it is assumed that every user has a patience limit, a threshold value for the duration that they are willing to wait for completed transfer of their data. Beyond this value, they become dissatisfied and less willing to pay. The shape of the utility function depends on the type of application considered. Elastic applications are described with a concave curve that approaches the maximum utility value as the bandwidth increases. The value or benefit function of offered link capacity is a concave increasing function following the economic assumption of diminishing marginal utility (i.e., value will increase for each unit of added bandwidth up to a certain point, after which the gain in value for extra capacity is marginal). Delay or rate-sensitive applications have a convex portion to their curve which describes the sharp decrease in utility once the rate is below a certain threshold.

The utility function shapes in much of the literature are based on this model. Some utility functions are defined as ranging from zero to one, where zero describes a totally unsatisfactory service and one describes a completely satisfactory one. Others consider utility rating related to the mean opinion score (MOS), which was described by an International Telecommunication Union Telecommunication Standardization Sector (ITU-T) recommendation [47] to grade user perception of voice quality. The range of MOS scores is zero to five, where five represents the highest achievable satisfaction with the service.

Most of these utility functions are continuous, but a piecewise utility function built on a zone-based structure may be more appropriate to reflect human behavior. The zone-based structure introduced by Sevcik [48] classifies users' patience with file delivery completion time in three zone-like classes: satisfaction, tolerance, and frustration. This structure easily incorporates two basic concepts on the application's bandwidth requirements. The first idea is that there is a basic minimum bandwidth requirement for each application type. The user will be frustrated with any bandwidth values below this minimum bandwidth threshold, with the resulting correspondent utility value of zero. Once the user's application receives the minimum bandwidth, they will be willing to pay a basic amount for this tolerant situation. The amount is expressed as the minimum utility value for this application and is denoted as U_{min}. The second concept is that there exists a maximum bandwidth threshold at which the utility value will be at its maximum U_{max} and will increase no further for bandwidths received beyond this maximum bandwidth threshold. Wu and Yin [29] propose using similar thresholds for their utility functions.

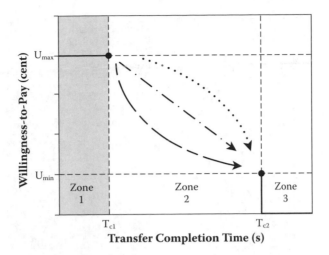

Figure 11.4 Designing the user utility function.

For non-real-time data delivery, the user's willingness to pay or utility $U_i(T_c)$ for a file i depends on the actual transfer completion time (T_c). The utility function used here is a plot of the user's willingness to pay (in cents) versus the estimated file transfer completion time (in seconds). The three zones described above correspond to zones 1, 2, and 3 in Figure 11.4; that is, the satisfaction zone, tolerance zone, and frustration zone, respectively. Timely arrival of the file anywhere within zone 1 is worth the highest value to the user, U_{max}. This corresponds to data transfers that meet or exceed the maximum bandwidth threshold. We denote completion time T_{c1} as the threshold of the satisfaction zone. After waiting T_{c1} seconds, the value of completed transfer decreases with an increase in delay, until the time exceeds the user-defined maximum tolerated delay, T_{c2}. Delivery of the complete file by this specified maximum completion time corresponds to receiving the minimum bandwidth threshold, and the user is willing to pay a minimum amount, U_{min}, for the data transfer. Any file arriving after T_{c2} does not meet the minimum threshold and is worth 0 cents to the frustrated user.

There are a number of possible shapes to describe the utility function for zone 2 depending on the user's flexibility to delay and how the user expects the cost to decrease with an increase in delay for the transfer. A number of economics publications on user attitude to risk taking (e.g., Keeney [49]) describe three different attitudes to risk and their associated utility function shapes. Work by Ormond et al. [50] adopted these three shapes in an exploration of a number of possible utility functions based on different user's attitudes to risk for money and delay preferences. The attitudes in terms of their function metrics are

- Risk neutral: translates to a neutral utility function where the user equally prefers paying less to experiencing less delay ($U_{1i}(T_c)$).
- Risk seeking: translates to a delay-sensitive utility function where the user prefers the alternative of less delay to assured money saving ($U_{2i}(T_c)$).
- Risk averse: translates to a price-sensitive utility function where the user prefers to be certain of paying less ($U_{3i}(T_c)$).

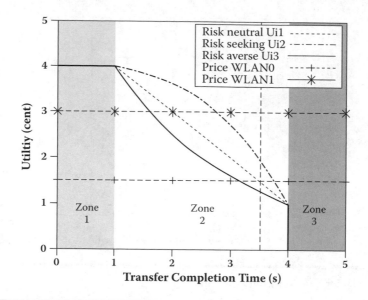

Figure 11.5   User utility functions for 100 kbyte payload.

These utility functions are graphically represented in Figure 11.5. These three possible utility functions were tested and their comparative results correspond with their approach to risk taking:

- Risk-neutral users see their delay and cost results sit between the other two utility results.
- Delay-sensitive users end up paying more, but enjoying less delay.
- Price-sensitive users pay less money, but experience more delay and money waste (some of the files they paid for were worth 0 cents to them because they exceeded T_{c2}).

The results suggest that the difference between these user types increases with the file size. That is, as the file size increases, the associated delays and costs are larger, and therefore the value of employing an appropriate utility function for a given user's preferences increases with an increase in file size. In order to facilitate the case when the user attitude towards risk is not known in advance, the neutral utility function is used.

The neutral utility function, $U_{1i}(T_c)$, is described by Equation (11.1). It is a piece-wise linear user utility function that describes the relationship between the user's time and budget limitations:

$$U_{1i}(T_c) = \begin{cases} U_{\max} & T_c \leq T_{c1} & \text{(zone 1)} \\ \frac{(U_{\max} - U_{\min}) \cdot (T_c - T_{c1})}{T_{c2} - T_{c1}} & T_{c1} < T_c \leq T_{c2} & \text{(zone 2)} \\ 0 & T_c > T_{c2} & \text{(zone 3)} \end{cases} \qquad (11.1)$$

11.5.2.2   The Thresholds

There is a need to determine reasonable threshold values for the user utility function (i.e., T_{c1}, T_{c2}, $U_{\max}$, and $U_{\min}$). Table 11.1 shows the association between expected user

Table 11.1   Rates, File Sizes, and Corresponding Delays

File Size (kbytes)	File Size (kbits)	Mbps 5 (S)	Mbps 2 (S)	Mbps 1 (S)	kbps 500 (S)	kbps 200 (S)	kbps 100 (S)	kbps 50 (S)	kbps 20 (S)
10	80	0.02	0.04	0.08	0.16	0.40	0.80	1.60	4.00
20	160	0.03	0.08	0.16	0.32	0.80	1.60	3.20	8.00
50	400	0.08	0.20	0.40	0.80	2.00	4.00	8.00	20.00
80	640	0.13	0.32	0.64	1.28	3.20	6.40	12.80	32.00
100	800	0.16	0.40	0.80	1.60	4.00	8.00	16.00	40.00
200	1600	0.32	0.80	1.60	3.20	8.00	16.00	32.00	80.00
500	4000	0.80	2.00	4.00	8.00	20.00	40.00	80.00	200.00

delays and the bit rate as seen at the application level. It takes into account a large range of possible throughput rates available in existing and emerging RANs (10 kbps to 5 Mbps), and a range of likely file sizes (10 kbytes to 500 kbytes) to give the range of likely transfer completion times. T_c, in seconds, is related to the size of the file and depends on the average end-to-end throughput rate according to Equation (11.2),

$$T_c = F_i / r \qquad (11.2)$$

where F_i is the size of file i in bits and r is the average throughput rate seen by the application. Depending on the RAN type available, and its configuration and congestion policy, the bit rates will vary, and what is presented here is only one set of rates. In this table, any delay of less than 1 s is considered negligible, whereas delays greater than 30 s breach the end user's expectation for good service. Different users can have different expectations for file completion times, and the same user can change their expectations, depending on the current context or over time, as higher bandwidth RATs become available or prices decrease. Consequently the bolded values in the table can vary.

In order to provide good transport services, a bit rate of 200 kbps is considered as the threshold for user acceptability. This value was chosen from the table of possible rates as one which currently provides a medium transport service rate. The associated user utility function for a file transfer protocol (FTP) application transferring a payload of 100 kbytes is shown in Figure 11.5. The graph indicates how any completion time up to $T_{c1} = 1$ s is worth the highest price to the user, whereas files arriving at $T_{c2} = 4$ s are worth the minimum price to the user, and any files arriving after $T_{c2} = 4$ s are worth 0 cents. In between T_{c1} and T_{c2}, the value decreases, with time dependent on the user's attitude to delay and money conservation. These transfer completion time thresholds scale according to file size.

To determine reasonable values for the maximum and minimum utility values the user is prepared to pay, the current asking price for transferring data on local networks must be considered alongside the user's budget limitations. The utility function in Figure 11.5 is described for an environment where the cost range is 1 to 5 cents per 100 kbytes and the user's budget limitation is 4 cents per 100 kbytes. The user's maximum willingness to pay for 100 kbytes arriving within the satisfied zone is $U_i(T_{c2}) = U_{max} = 4$ cent, and the minimum value per 100 kbytes of data arriving by the transfer completion time is $U_i(T_{c2}) = U_{min} = 1$ cent. These threshold values scale with file size.

11.5.3 The Consumer Surplus-Based Network Selection Strategy

The network selection strategy proposed here is a utility-based consumer surplus strategy. Consumer surplus (CS) is a term used in microeconomics to describe the difference between what the user is willing to pay and what the user actually pays. The available access network options are considered, with the goal of maximizing the consumer surplus, subject to user-imposed time constraints related to complete file transfer. The greater the CS, the more satisfied the user is, provided that the transaction meets the transfer completion time deadline set by the user.

If CS_i is the consumer surplus (in cents) for transporting file i, the predicted CS_i can be calculated using Equation (11.3).

$$CS_i = U_i(T_c) - C_i$$

$$\text{subject to } T_c \leq T_{c2} \tag{11.3}$$

where $U_i(T_c)$ is the monetary value (in cents) that the user places on the transfer of file i in the given transfer completion time (T_c) and C_i is the cost charged by the network in question, also in cents, for the completed file transfer. Figure 11.6 shows the fixed prices charged by the two available RANs in the scenario described below. A few examples of CS can be seen for files sent on WLAN1 and WLAN0.

The proposed approach involves the user terminal surveying the radio interface and determining a list of current available access networks at the time of or, in the case of proactive discovery function, just before a network selection decision. For each available RAN, the user terminal must predict the current offered rate and use these estimated rates, together with the provided utility function for the current application, to select the network that will give the greatest consumer surplus while meeting the transfer completion time deadline. This strategy was first outlined by Ormond et al. [51] and further expanded and tested (Ormond et al. [6]).

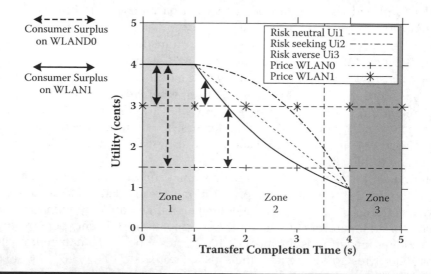

Figure 11.6   Examples of consumer surplus.

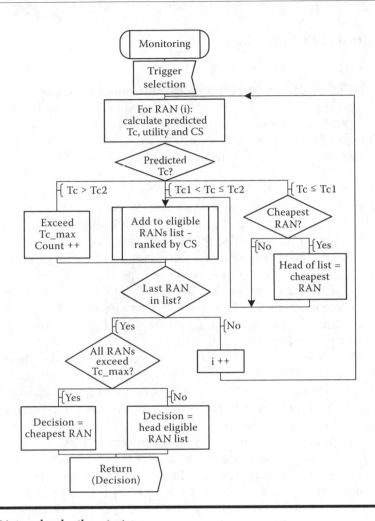

Figure 11.7   Network selection strategy.

Figure 11.7 presents the proposed CS-based network selection algorithm in detail. First, the predicted completion time (T_c) is calculated, and from that the predicted utility (U_i) and CS_i for each candidate network are determined. The network with the best predicted CS_i, which is also predicted to meet the completion deadline, is then selected as the best network for data transfer.

In order to find an intelligent solution to the selection decision problem for non-real-time data services for large data transfers, Ormond et al. [24] proposed a renegotiation of the selection during transfer of large files with varying background traffic mixes and patterns. To handle the large file, it is fragmented into smaller chunks and a network selection decision is made for each chunk. The objective of the decision is to transfer the entire file (chunk by chunk) over the current best of the available access networks. It uses the throughput rate information from previous chunks to predict the available rates in each network. There is an assumption made that there is no interface switching time required for changing between the RANs and that there are no other penalties associated with switching either. The proposal is for a utility-based algorithm that accounts for user time constraints, estimates complete chunk delivery time (for each of the available access networks), and then selects the most promising access network based on consumer surplus difference.

11.5.4 Performance Evaluation

11.5.4.1 Testing Setup and Scenarios

A simulation model was developed using Network Simulator (NS2) version 2.27 (http://www.isi.edu/nsnam/ns/) with IEEE 802.11b wireless LAN parameter settings (data rate 11 Mbps), the no ad-hoc routing (NOAH) extension, and a new application that simulates a simple multihomed terminal with a set of built in network selection strategies. Three decision strategies are supported: consumer surplus (CS), always cheapest (AC), and network centric (NC).

The CS network selection strategy conforms to the description presented in the previous section. The AC network selection strategy aims to minimize the cost of data transfer, regardless of the performance of the transmission, always selecting the cheapest network available (in this situation WLAN0). The NC network selection strategy considers the case when the transfer is always performed over the network owned by the user's preferred network operator (in this case WLAN1—the more expensive network).

The scenario for testing is one where an intelligent mobile user's FTP application requires the transfer of a large file (size range 100 kbytes to 5 Mbytes) in a SOHWNE. This is a typical range for the size of an FTP transfer. The user terminal has built-in intelligent features for network detection and selection. The file is broken into smaller chunks of data (length 10 kbytes). Each data chunk is to be sent uplink from the terminal through the access point (AP) of the selected network to a server in the wired network.

The wired network is connected via point-to-point links so that delay in the wired network is negligible and any significant delays experienced are dependent on the selected radio access network. For example, the user may be uploading a photo slideshow from a laptop, emailing a large document from their mobile phone, or uploading vending machine data from their personal digital assistant (PDA) to a central server. The user is faced with a scenario like the one shown in Figure 11.8, where the terminal employing the intelligent network selection strategy has a choice of two IEEE 802.11b networks: WLAN0 or WLAN1. Both networks employ a fixed price per byte pricing scheme, but each charges a different price: WLAN1 at 3 cents per 100 kbytes is twice the price of WLAN0. Also each network has varying capability to transfer data with varying background traffic mixes and patterns. The background traffic pattern on WLAN0 changes from high traffic to medium, whereas WLAN1 starts with a medium amount of traffic which increases. As WLAN1 is twice as expensive as WLAN0, it is assumed that it will get slightly less traffic; this was allowed for in the background traffic patterns.

A repeated decision is required for which RAN to use for transporting each chunk of user's application data. Naturally the user wants timely delivery of their data at the lowest

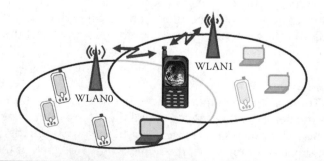

Figure 11.8   Choice of access network: WLAN0 or WLAN1.

possible price. The problem is to select the current network that has the greatest chance to achieve this goal. The neutral utility function, $U_{1i}(T_c)$, is employed to represent the perceived user preferences. In the case of each network selection decision, the intelligent node will short-list the networks, based on delivery time prediction, to meet T_{c2}. The file is then transmitted over the short-listed RAN which is predicted to maximize the CS.

The quality of the intelligent node connectivity in each RAN is affected by the mix and patterns of the background traffic, generated by the other nodes. In IEEE 802.11b networks, if the bulk of data traffic is on the downlink, an FTP session transferring data uplink would have a clear advantage, as the bulk of the traffic would create a bottleneck at the AP, leaving less contention for uplink traffic. The background traffic mix used in this simulation was a mix of HTTP, wireless application protocol (WAP), FTP, and two-way non-real-time video sessions. The connectivity quality is affected mainly by the background FTP and video traffic (both with uplink streams). Although WAP is not a common application in current WLANs, in the future when smart-phone users will be taking advantage of the cheaper rates on a WLAN, they may only be interested in WAP sessions on their smaller screens. During successive tests, file size is varied from 100 kbytes to 5 Mbytes. All three network selection strategies—CS, AC, and NC—are employed in turn. If a chunk transfer exceeds the transfer completion deadline, the transmission is stopped and the chunk is resent.

11.5.4.2 Performance Analysis

The entire file transfer completion time, cost of transfer, and the number of chunk resends required are compared against the other tested network selection strategies. The transfer completion time is important to the non-real-time users, as they are not willing to wait too long for the file transfer to finish. The sooner the file transfer ends, the more satisfied the user will be. A transfer that exceeds the user's time deadline is worth 0 cents to them, although the user will end up paying for the data that is transferred. The user wants to take advantage of the competition and pay as little as possible to achieve the expected perceived quality for the current service. The cost performance trade-off is important to them. One measurement of transmission efficiency is in terms of how many network resources are used in order to transfer a given amount of data. The number of chunk resends indicates the number of chunks that exceed the time deadline and have to be resent, costing customers more in time and money in the long run. As the file size increases, the differences in the strategies may become more apparent.

Analyzing the results shown in Figures 11.9 through 11.11, it can be seen that the average duration for files (Figure 11.9) sent with the CS-based strategy is shorter than for files sent with either of the other two methods. As the file size increases, the difference between the CS average transfer time and the others increases.

Figure 11.10 shows the comparison of average cost (in cents) per file transfer using each of the three strategies considered in this chapter. As expected, the NC strategy users who always choose their network operator's WLAN (WLAN1) end up paying the most, as this is the most expensive network of the two. Although the AC strategy achieves the least costly transfers, the difference between the average costs paid per file for the CS and AC strategy users is not very much considering the difference in average delay and the number of chunk resends.

Looking at the number of chunk resends counted when each of the three network selection strategies were employed (Figure 11.11), it is obvious that while customers employing the AC strategy are paying less per chunk sent, their chunks exceed the time deadline more often and have to be resent, costing these customers more in time and

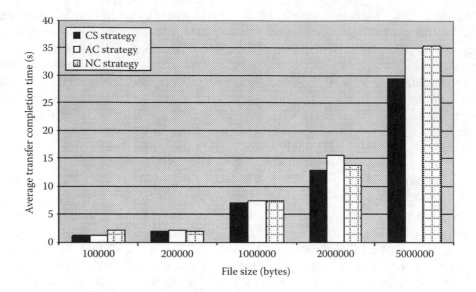

Figure 11.9 Average file transfer completion time.

money in the long run. As file size increases, the advantage of using the CS strategy becomes more apparent; chunk resends are far less than those recorded in both AC and NC cases.

The CS strategy was then tested against one of the intelligent strategies (from Section 11.4). The results of which are published in Ormond et al. [24]. More rigorous testing of the strategy in comparison to other intelligent network selection strategies is currently under way, including scenarios with different competing traffic mixes and different chunk sizes. Future models will use more advanced throughput prediction methods that consider the time of day, day of week, previous history, congestion recovery time, and other possible influential metrics.

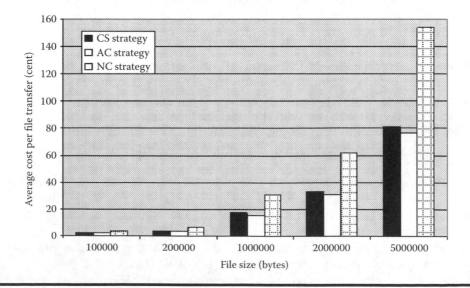

Figure 11.10 Average cost per file transfer.

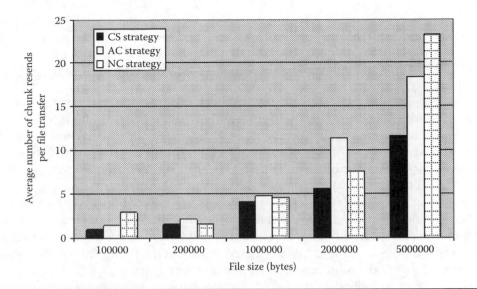

Figure 11.11   Average number of chunk resends per file transfer.

11.6   Conclusion

Thanks to deregulation and advancements in technology, users are becoming more powerful in their everyday choices as consumers. Users in a user-centric service-oriented heterogeneous wireless network environment (SOHWNE) will be free to choose the available access network that can best satisfy their needs and objectives in the current connectivity situation. This chapter discussed the network selection decision problem and presented current research carried out on the issues surrounding a fast, efficient, and intelligent solution to this problem.

There are many open issues to be solved regarding the access network selection problem, and this chapter details some of them, including access discovery and the associated network metrics that can be determined from this phase, gathering and interpreting user inputs with minimal user inconvenience, providing a decision strategy that takes the most relevant inputs into consideration in a fast, computationally efficient algorithm that does not overburden the battery, and a mobility protocol that takes into account mobility with a choice of RAT and operator, multihoming, and security prevention against malicious attacks.

Some of the existing proposals for network selection were presented. More traditional selection strategies were designed in the context of network-centric seamless handover for a connection that required an immediate handover to prevent severe interruption or loss of connectivity. New triggers for the network selection process take into account multihoming and the user-centric facility to be always best connected. These include the initial connection setup at the beginning of a call and when the local RAN conditions change. Many researchers have concentrated on the architecture and the type of decision mechanism for the network selection, with fewer works concentrating on the actual decision.

Works on the actual network selection decision consider many possible parameters and objectives on which to base the decision, some of which are conflicting. There is a need for a technique to convey the user preference trade-off to help interpret which of the available RANs can best service current needs. Given the need for a fast and energy-efficient decision with minimal processor loading, and the variety and dynamicity of the decision metrics, there is a need to take only the most relevant metrics into account to dynamically provide the user with the best access possible within their delay and budget restrictions.

A short study was presented in which the ultimate goal is to find the best user-centric network selection strategy for non-real-time data transfer in the next-generation service-oriented heterogeneous wireless network environment. This section of the chapter introduced a proposal for a novel solution, a consumer utility-based strategy for intelligent cost-oriented and performance-aware selection between available networks. The proposed user-centric network selection strategy is based on maximizing consumer surplus subject to meeting user-defined constraints in terms of transfer completion time. The strategy includes a user utility function to describe the user preference relation for the metrics important to the users and means to determine or predict the transfer completion time of non-real-time data without too much overhead. It is a dynamic strategy, as it reconsiders the decision during larger file transfers.

Testing involved a simulation model of an intelligent terminal employing a network selection strategy with a choice of two IEEE 802.11b networks. The proposed CS-based strategy was evaluated against an AC and an NC strategy which were each deployed in turn on the intelligent terminal. Test results showed that CS was the best strategy of the three in terms of combined average cost and duration of transfer per file as well as transfer efficiency. The proposed CS strategy was further tested against another intelligent strategy, and future plans are outlined for testing and comparison with other intelligent strategies.

Future goals are to experiment with the threshold values of the utility function and the throughput prediction scheme to further improve the network selection strategy (taking the performance metrics outlined in Section 11.4.4 into account), and to examine the impact of multiple intelligent users operating in the same region. Especially in the case when the user is in a particular location with homogeneous users who all have similar cost performance preferences (e.g., students on a university campus, or business users in a business park).

References

1. X. Gelabertt, J. Pérez-Romero, O. Sallent, and R. Agusti, On the impact of multi-mode terminals in heterogeneous wireless access networks, *2nd International Symposium on Wireless Communication Systems*, IEEE, Washington, DC, 2005.
2. J.S. Park and K.S. Rye, Developing MVNO market scenarios and strategies through a scenario planning approach, *7th International Conference on Advanced Communication Technology*, IEEE, Washington, DC, 2005.
3. M. Koutsopoulou, A. Kaloxylos, A. Alonistioti, L. Merakos, and P. Philippopoulos, An integrated charging, accounting and billing management platform for the support of innovative business models in mobile networks, *Int. J. Mobile Commun.*, 2(4), 418, 2004.
4. H. Chan, P. Fan, and Z. Cao, A utility-based network selection scheme for multiple services in heterogeneous networks, *International Conference on Wireless Networks, Communications, and Mobile Computing*, IEEE, Washington, DC, 2005.
5. Q. Song and A. Jamalipour, An adaptive quality-of-service network selection mechanism for heterogeneous mobile networks, *Wireless Commun. Mobile Comput.*, 5(6), 697, 2005.

6. O. Ormond, G.-M. Muntean, and J. Murphy, Network selection strategy in heterogeneous wireless networks, *Information Technology and Telecommunications Conference*, p. 175, October 2005.

7. J. McNair and F. Zhu, Vertical handoffs in fourth-generation multi-network environments, *IEEE Wireless Commun.*, 11(3), 8, 2004.

8. E. Adamopoulou, K. Demestichas, A. Koutsorodi, and M. Theologou, Intelligent access network selection in heterogeneous networks—simulation results, *Second International Symposium on Wireless Communication Systems*, p. 279, IEEE, Washington, DC, 2005.

9. V. Gazis, N. Houssos, N. Alonistioti, and L. Merakos, On the complexity of always best connected in 4G mobile networks, *Vehicular Technology Conference*, vol. 4, p. 2312, IEEE, Washington, DC, 2003.

10. K. Chebrolu and R. Rao, Communication using multiple wireless interfaces, *Wireless Communications and Networking Conference*, vol. 1, p. 327, IEEE, Washington, DC, 2002.

11. J. Prez-Romero, O. Sallent, and R. Agust, Policy-based initial RAT selection algorithms in heterogeneous networks, *7th IFIP International Conference on Mobile and Wireless Communications Networks*, September 2005.

12. K. Murray, R. Mathur, and D. Pesch, Intelligent access and mobility management in heterogeneous wireless networks using policy, *1st International Symposium on Information and Communication Technologies*, vol. 49, p. 181, ACM, New York, 2003.

13. J. Ylitalo, T. Jokikyyny, T. Kauppinen, A.J. Tuominen, and J. Laine, Dynamic network interface selection in multihomed mobile hosts, *36th Hawaii International Conference on System Sciences*, vol. 9, p. 315. IEEE, Washington, DC, 2003.

14. E. Bircher and T. Braun, An agent-based architecture for service discovery and negotiations in wireless networks, *Wired/Wireless Internet Communications*, p. 295, Springer, Berlin, 2004.

15. E. Mohyeldin, M. Fahrmair, W. Sitou, and B. Spanfelner, A generic framework for context aware and adaptation behaviour of reconfigurable systems, *16th International Symposium on Personal, Indoor and Mobile Radio Communications*, IEEE, Washington, DC, 2005.

16. J. Lanza, L. Snchez, and L. Muoz, Performance evaluation of a cross-layer based wireless interface dynamic selection on WPAN/WLAN heterogeneous environments: an experimental approach, *Sixth International Workshop on Applications and Services in Wireless Networks*, 2006.

17. M. Stemm and R.H. Katz, Vertical handoffs in wireless overlay networks, *Mobile Networks Applic.*, 3(4), 335, 1998.

18. H.J. Wang, R.H. Katz, and J. Giese, Policy-enabled handoffs across heterogeneous wireless networks, *Second IEEE Workshop on Mobile Computing Systems and Applications*, p. 51, IEEE, Washington, DC, 1999.

19. L.-J. Chen, T. Sun, B. Chen, V. Rajendran, and M. Gerla. A smart decision model for vertical handoff, *4th ANWIRE International Workshop on Wireless Internet and Reconfigurability*, Athens, Greece, 2004.

20. W. Zhang, Handover decision using fuzzy MADM in heterogeneous networks, *Wireless Communications and Networking Conference*, vol. 2, p. 653, IEEE, Washington, DC, 2004

21. F. Feng and D. Reeves, Explicit proactive handoff with motion prediction for Mobile IP, *Wireless Communications and Networking Conference*, vol. 2, p. 855, IEEE, Washington, DC, 2004.

22. C. Ahlund, R. Brännström, and A. Zaslavsky, Agent selection strategies in wireless networks with multihomed Mobile IP, *Service Assurance with Partial and Intermittent Resources*, Springer, Berlin, p. 197, 2004.

23. Z. Zhang, WCDMA compressed mode triggering method for IRAT handover, *Wireless Communications and Networking Conference*, vol. 2, p. 849, IEEE, Washington, DC, 2004.

24. O. Ormond, G.-M. Muntean, and J. Murphy, Evaluation of an intelligent utility-based strategy for dynamic wireless network selection, *IFIP/IEEE International Conference on Management of Multimedia Networks and Services*, p. 158, IEEE, Washington, DC, 2006.

25. M. Falkner, M. Devetsikiotis, and I. Lambadaris, An overview of pricing concepts for broadband IP networks, *IEEE Communication Surveys and Tutorials* Q3, 2000.

26. G. Le Bodic, J. Irvine, D. Girma, and J. Dunlop, Dynamic 3G network selection for increasing the competition in the mobile communications market, *Vehicular Technology Conference*, vol. 3, p. 1064, IEEE, Washington, DC, 2000.

27. X. Wang and H. Schulzrinne, An integrated resource negotiation, pricing, and QoS adaptation framework for multimedia applications, *IEEE J. Select. Areas Commun.*, 18(12), 2514, 2000.

28. S. Das, H. Lin, and M. Chatterjee, An econometric model for resource management in competitive wireless data networks, *IEEE Network Mag.*, 18(6), 20, 2004.

29. Z. Wu and Q. Yin, A heuristic for bandwidth allocation and management to maximise user satisfaction degree on multiple MPLS paths, *3rd IEEE Consumer Communications and Networking Conference*, vol. I, p. 35, IEEE, Washington, DC, 2006.

30. P. Nikander, J. Yliatalo, and J. Wall, Integrating security, mobility, and multi-homing in a HIP way, *Network and Distributed Systems Security Symposium*, Internet Society, Reston, UA, 2003.

31. W.M. Eddy, At what layer does mobility belong?, *IEEE Commun. Mag.*, 42(10), 155, 2004.

32. V. Kawadia and P.R. Kumar, A cautionary perspective on cross-layer design, *IEEE Wireless Commun.*, 12(1), 3, 2005.

33. V.T. Raisinghani and S. Iyer, Cross-layer feedback architecture for mobile device protocol stacks, *IEEE Commun. Mag.*, 44(1), 85, 2006.

34. C. Huitema, B. Zhang, and L. Zhang, Multi-homed TCP, Internet Draft, Network Working Group, 1998.

35. C. Perkins, IP mobility support for IPv4, IEFT RFC 3220, Network Working Group, 2002.

36. D. Johnson, C. Perkins, and J. Arkko, Mobility support in IPv6, IETF RFC 3775, Network Working Group, 2004.

37. T. Braun and M. Danzeisen, Secure mobile IP communication, *26th Annual IEEE Conference on Local Computer Networks*, IEEE, Washington, DC, 2001.

38. S.M. Faccin, P. Lalwaney, and B. Patil, IP multimedia services: analysis of Mobile IP and SIP interactions in 3G networks, *IEEE Commun. Mag.*, January 2004.

39. R. Stewart, Q. Xie, K. Morneault, C. Sharp, H. Schwarzbauer, T. Taylor, et al., Stream control transmission protocol, IETF RFC 2960, Internet Society, Reston, VA, 2000.

40. M. Riegel and M. Tuexen, Mobile SCTP, IETF Internet Draft, Network Working Group, 2006.

41. A.C. Snoeren and H. Balakrishnan, An end-to-end approach to host mobility, *6th ACM/IEEE International Conference on Mobile Computing and Networking*, IEEE, Washington, DC, 2000.

42. Host identity protocol (HIP), IETF Working Group, 2005.

43. G. Lee, P. Faratin, S. Bauer, and J. Wroclawski, A user-guided cognitive agent for network service selection in pervasive computing environments, *2nd IEEE Annual Conference on Pervasive Computing and Communications*, p. 219, IEEE, Washington, DC, 2004.

44. M. Angermann and J. Kammann, Cost metrics for decision problem in wireless ad hoc networking, *IEEE CAS Workshop on Wireless Communications and Networking*, 2002. http://citeseer.1st.psu.edu/angermanne2cost.html

45. Z. Jiang, H. Mason, B.J. Kim, N.K. Shankaranarayanan, and P. Henry, A subjective survey of user experience for data applications for future cellular wireless networks, *Symposium on Applications and the Internet*, p. 167, IEEE, Washington, DC, 2001.

46. S. Shenker, Fundamental design issues for the future Internet, *IEEE J. Select. Areas Commun.*, 13(7), 1176, 1995.

47. ITU-T Rec. P.800, Methods for subjective determination of transmission quality, Technical Report, ITU-T, 1996.

48. P.J. Sevcik, Understanding how users view application performance, *Business Commun. Rev.*, 32(7), 8, 2002.

49. R.L. Keeney, *Value-Focused Thinking*, Harvard University Press, Cambridge, MA, 1992.

50. O. Ormond, G.-M. Muntean, and J. Murphy, Utility-based intelligent network selection in beyond 3G systems, *IEEE International Conference on Communications*, vol. 4, p. 1831, IEEE, Washington, DC, 2006.

51. O. Ormond, P. Perry, and J. Murphy, Network selection decision in wireless heterogeneous networks, *16th International Symposium on Personal Indoor and Mobile Radio Communications*, vol. 4, p. 2680, IEEE, Washington, DC, 2005.

Chapter 12

Mobility Support for Wireless PANs, LANs, and MANs

Marc Emmelmann, Berthold Rathke, and Adam Wolisz

Contents

Mobility support in IEEE 802 wireless networks may be defined as the superset of all necessary functionalities providing uninterrupted user data communication services in the presence of moving mobile devices. Using this definition, an interruption of communication on a wireless channel is acceptable if the application's behavior (e.g., interruption of a telephone call) is not significantly affected.

These mobility supporting functions can be divided into two classes. The first one includes handover functions which delegate communication from one device to another, possibly including a transfer of all relevant information (i.e., stored data or link states and parameters) from the old device to the new one. The second class includes all functionality expediting or enabling the handover process (e.g., neighborhood detection; measurement functions gathering, such as information about link quality; or media access control [MAC] layer link control).

This chapter illustrates mobility requirements from the user perspective and how IEEE 802 wireless networks fit specific user needs. An overview of required technology-independent mobility functions (namely, generic handover mechanisms) is given. The different approaches providing these functionalities are illustrated for each of the following technologies: wirless personal area networks (PANs) (IEEE 802.15.1 and IEEE 802.15.3), wireless local area networks (LANs) (IEEE 802.11), and metropolitan area networks (MANs) (IEEE 802.16). Next, a brief overview of technology-independent mobility functionality (IETF) and IEEE 802.21 is presented. The chapter concludes by describing the authors' view of open challenges and future trends.

12.1 Mobility Aspects in IEEE 802

Technologies are driven by the expectations of users regarding the support of their communication needs. In general, the increased coverage area of today's wireless networks requires increased handover support as users' mobility increases in both their current velocity and how frequently they communicate on the move. Figure 12.1 illustrates the supported mobility in terms of user's velocity and offered bandwidth.

At first guess, high velocity cases seem to be most challenging for providing seamless handover. But indoor environments supporting users at low or pedestrian speed might impose even harsher system requirements regarding the support of seamless, low-latency handover. Table 12.1 illustrates the reason: the number of handovers per time for a given user is one magnitude larger than for high velocity, outdoor environments.

In the following, we present a historical review of how mobility influences the technologies in MANs, LANs, and PANs:

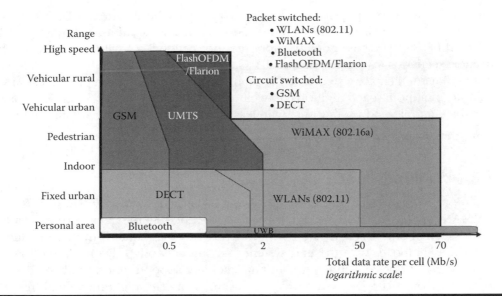

Figure 12.1   Mobility support in various RAN technologies.

12.1.1   Mobility in LANs

With respect to mobility support (e.g., the first wireless communication in the 1980s was a mere cable replacement), cordless telephones like digital enhanced cordless telecommunications (DECT) were introduced. Similar to voice communication, one of the first data application scenarios of IEEE 802.11 was wireless Ethernet. Mobility aspects were not of primary interest because of the size of former end systems: heavy desktop computers were equipped with WLAN interface cards. Later on, the increasing number of laptop computers shifted the application scenario toward providing nomadic mobility within limited areas, such as buildings, hot spots, or campuses. Today, even smaller end systems paired with multimedia applications (e.g., personal digital assistants [PDAs] and Internet protocol [IP] telephones) enable users to communicate continuously on the move. Thus users expect seamless mobility support and high bandwidth communication even beyond the coverage area of wireless LANs. Today, cellular networks enable continuous communication, while in the near future, WirelessMANs will offer higher bandwidth alternatives to these cellular networks.

12.1.2   Mobility in MANs

Wireless MANs (IEEE 802.16) were originally designed as a wireless local loop. They provide high bandwidth and enable easy, low-cost cable replacement for rural and

Table 12.1   Coverage Area and User Velocity of Wireless PAN, LANs, and MANs

RAN Technology	Coverage Area	User Velocity	Handover Frequency
IEEE 802.15.1	10 m	$\leq 1\ m/s$	0.100 Hz
IEEE 802.11 (indoor)	15 m	$\approx 1.5\ m/s$	0.100 Hz
IEEE 802.11 (outdoor)	250 m	$\approx 20\ m/s$	0.080 Hz
IEEE 802.16e	2500 m	$\approx 70\ m/s$	0.028 Hz

outlying areas with low population densities. Typically these wireless point-to-point links connect the wired backbone with premises having a local infrastructure (e.g., IEEE 802.11 or IEEE 802.3). Thus mobility support is not needed. As wireless MANs enable application scenarios similar to wireless LANs, with the same stringent service requirements, it is a natural choice to enhance the point-to-point configuration of wireless MANs by point-to-multipoint configurations as well as mobility support for single end users. This enables mobile end users to use wireless LANs or wireless MANs without service degradation.

12.1.3   Mobility in PANs

Typical representatives of PANs are IEEE 802.15.1 (Bluetooth) and IEEE 802.15.3. In contrast to MANs and LANs, PANs are designed to connect peripherial devices (e.g., wireless keyboards, telephone headsets, etc.) with the end systems. Thus, if the end system is mobile, the PAN moves with the end system (moving networks). Even though this application scenario does not inherently require mobility support among PAN devices, in some cases mobility support is desirable. For example, a user might move with his telephone headset outside the coverage area of an iPod and still want to listen to music via other PAN devices.

12.1.4   Technology-Independent Mobility

All of the technologies above have their specific radio cell coverage area and their own technology-specific mobility support. With the increasing deployment of each technology, the offered overall coverage area increases as well. This enables seamless communication if end systems use several different technologies while on the move. This heterogeneous mobility support is targeted at a technology-independent layer and consequently is covered, at least for Internet-based deployments, by IETF using various protocols (e.g., mobile IP, fast mobile IP, etc.). Again, the IEEE provides the bridge between each technology and the IETF's technology-independent approaches: IEEE's Media Independent Handover Group (IEEE 802.21) specifies a technology-independent interface to signal mobility relevant information.

In summary, today's and future applications require higher data rates as well as quality of service (QoS) support. TripplePlay will converge applications (i.e., Internet, television, and telephony). In addition, not only end systems are moving, but entire networks. For example, LANs may be mobile when installed in vehicular environments like trains, buses, or cars. As a result, advanced mobility support has to ensure ongoing real-time communication while moving among heterogeneous networks, possibly at high speeds.

12.2   Generic Mobility Functions

The handover process, in general, requires the following four functionalities: detection of available radio cells, the handover decision and involved decision criteria, reestablishment of the link layer, and if necessary, higher layer procedures (i.e., network layer aspects and above) that finalize the handover. All these functions do not occur in a strict sequence, but may also overlap or even happen in parallel. Security aspects are usually handled both at link layer and higher protocol layer levels.

12.2.1 Detection of Available Radio Cells

Detection functionality aims at finding available radio cells. This involves a scanning process in which the mobile device may either passively listen on the wireless channel for potential communication partners or actively send probe requests via the wireless channel awaiting a response from present interlocutors. (In principle, apart from scanning, wireless systems may also employ a dedicated signaling channel announcing available radio resources.)

Normally the wireless channel is divided into several subchannels (e.g., using frequency, time, or code division multiplexing). This implies a trade-off between the number of offered subchannels and the time spent in the detection phase, and in the worst case, the wireless device has to scan each subchannel independently. Thus communication may interrupt if scanning takes longer than the time a mobile device spends in a radio cell. In addition, the mobile devices' capabilities may determine if scanning is possible while having an active communication with the serving radio cell. This feature implies additional hardware—and costs—as compared to accepting an interruption of the communication during the scanning process. Finally, the scanning interval influences the handover performance: if it is too long, the gained information on available channels might be outdated; if it is too short, signaling overhead may increase dramatically.

The following phases are in general involved in detection functionality: In a first step, the mobile device listens in order to determine if a transmitter is operating on that subchannel. This can be either a noncommunication device (e.g., a microwave oven or radar system) or another communication device (e.g., wireless fidelity [WiFi] or Worldwide Interoperability for Microwave Access [WiMAX]). This step merely tracks the radio signal level and its derivations for a given time. Thus at this point the device has only knowledge of the presence of a transmitter, but not the kind of underlying technology. Now the mobile device has to determine if the observed signal represents a technology of its own kind. This is primarily achieved at the physical (PHY) level. In principle, the device could either try to decode the received analog or digital signal. The former case might compare the received signal with typical technology-specific radio frequency (RF) patterns (e.g., energy change or employed spectrum), while the latter case usually tries to detect bit patterns, like technology-specific preambles or start frame delimiters. If the recognized pattern matches with the mobile device's technology, the received information is forwarded to the MAC layer. Finally, the latter tries to retrieve wireless network-specific parameters such as network identification (ID) or operation mode. Employing these parameters, the mobile device evaluates if the detected link can be used to establish a further MAC layer connection.

The mobile device can reduce the time spent in the scanning phase by means of actively probing the wireless channel if the channel is idle for a long period of time. Probing should be initiated carefully, as a presumably idle channel may still be occupied by an alien technology.

12.2.2 Handover Decision and Criteria

The general goal of this functionality is to decide when to initiate the handover and to which neighbor. The underlying assumption is that the mobile device has knowledge of the existence of neighboring radio cells acquired by the detection phase. The mobile device must take care that the information obtained during the detection phase is not outdated. One way to ensure this is to initiate a reduced detection phase once again before the decision for handover is made. The detection phase just before the handover can be shortened if information of former detections is used (e.g., to verify that the selected handover candidate is really available).

Regarding the decision of when to initiate the handover, several parameters may be considered. Traditionally only a characterization of the wireless link was considered: the main parameter included the signal strength. If the latter fell below a certain threshold, a handover was initiated. Along with increasing service requirements, apart from the signal level, other parameters are considered for a handover decision. These include loss, delay, jitter, and throughput considered at various levels (e.g., at the mobile device, at the application level, at the current and target radio cell, or even network-wide). Consequently other parameters describing the interaction of neighboring cells—interference, location, and usage costs—are of interest.

The main question is how to measure these parameters. First, the mobile itself may measure them directly at the PHY, MAC, or application level. In addition, the access points of the current and neighboring radio cells are also able to measure these parameters. As each involved entity can conduct the measurement, each of them may potentially initiate the handover process if the criteria merely depends on locally retrieved information. If the handover decision is not feasible based on local measurements, means to signal values of considered parameters have to be provided. This may yield to an additional signaling communication between two end systems, between an access point and end system, as well as between two access points. Thus the handover decision can be regarded as a collaborative process.

Another problem is how to ensure the correctness of the measured parameter. The values of the parameters are oscillating in time caused by fluctuations of the wireless channel. To avoid misinterpretation of parameter values, a proper sampling rate has to be ensured and an additional low-pass filter may leverage the effects of short time fluctuations. If parameters have to be signaled between distributed systems, the introduced delay also has an effect on the validity of the measured parameter values. For high signaling delays, the value could be outdated.

After measuring, the handover decision has to be made depending on the instantaneous parameter's value or its history. Normally not all discussed parameters are part of a specific decision criteria. Policies determine the type and number of parameters included in the criteria. In general, policies reflect the desired behavior of the preferred network operation (e.g., optimized radio cell utilization, high throughput for a single mobile device, stabilization of application-level QoS, or reduction of interference phenomena). As discussed previously, the whole time-consuming measurement process has to be finished before the mobile device leaves the radio cell.

12.2.3 Link Layer Reestablishment

After the decision for a handover has been made, communication with the new access point has to be established. This involves synchronization on the PHY level and establishing link layer communication. The latter may include a renewal or a renegotiation of resources at the target access point.

Depending on the method used to establish this new link, packets may be lost, duplicated, or reordered. Releasing the old link before establishing the new link may involve packet losses (hard handover or break-before-make approach), whereas releasing the old link after establishment of the new link may produce duplicated and reordered packets (soft handover or make-before-break approach). The handover process has to deal with these aspects during the link layer establishment phase if the employed technology offers services that do not accept any loss, duplication, or reordering.

Depending on the applied security and authentication scheme, a time-consuming message exchange between the mobile terminal, the new access point, and possibly other network instances (e.g., if IEEE 802.1X is employed) may be part of the link establishment as well. Initiation of the handover may be terminal (end system) or network driven. Resource allocation can be conducted more efficiently if the handover process is initiated and also controlled by the network (i.e., the network decides at what time a handover takes place and to which neighboring radio cell a handover should be made). This is because a network controlled handover entity is more likely to have a global view regarding the utilization of each radio cell and possibly has information regarding the QoS experienced by each mobile device. The same applies if a provider offers a heterogeneous network infrastructure. Nevertheless, this approach is mostly limited to cases when a handover takes place across provider boundaries and no efficient signaling channel is available between providers. Depending on the policy of the network provider, a handover can be initiated to a radio cell with minimum utilization, to the best signal propagation conditions on the wireless channel (always best connected), or to radio cells with the largest size. The policies are not limited to the described parameters. Even a mixture of these parameters or various economic cost parameters may be included.

If the handover process is initiated and controlled by the terminal, only local knowledge about the network (i.e., about the neighboring cells) is available. The handover decision may not include information regarding current conditions in the second or third neighboring radio cell. To avoid poor network performance, sometimes an admission control is included in the network, which is able to deny handovers to desired radio cells. Nevertheless, if the handover process is controlled and initiated by the terminal, the handover process becomes quite simple compared to network-controlled handover. There is no need to store the state of all terminals in the network. This is done by each terminal separately. Another advantage is that handover across provider barriers is easier. In this case, the terminal acts as a middleman between providers. It is also possible that multiple streams of a single application can be distributed over multiple radio cells using different technologies.

12.2.4 *Higher Layer Aspects*

Higher layer mobility concepts—at the IP layer and above—aim to provide technology-independent communication while moving. Today, end systems are mainly identified by their IP address. The end system's IP address belongs to a subnet for routing purposes. If a packet is directed to the end system, it will first be routed to the subnet and then to the end system. If the end system changes its location to a new subnet, routing will not direct the packets to the new location because the packets will still be routed to the home network.

In order to ensure that the packet is routed to the new location, two approaches are used: either the network changes its routing paradigm from a subnet-oriented to a host address-based routing scheme, or the end system has to retrieve a new IP address whenever changing its location to a new subnet. In the latter case, the end system has to make its new IP address available network-wide. Both approaches require location change signaling. For scalability reasons, this signaling should be kept as efficient and small as possible.

Nevertheless, if the IP address changes, only nomadic mobility is possible because transport layer connection will abruptly terminate whenever the IP address changes. This is due to the fact that the IP address is part of the transport layer connection identifier. Decoupling the transport layer identifier from the network address used for routing purposes is one possible solution.

12.3 Handover Support Mechanisms in IEEE Wireless Networks

Before describing the mobility aspects, a brief overview of the selected technologies (IEEE 802.11, IEEE 802.15.1, and IEEE 802.15.3) is provided. For an introduction regarding IEEE 802.16, the reader is referred to Chapter 1.

12.3.1 IEEE 802.11

IEEE 802.11 is a wireless MAC and PHY layer specification for LANs. It employs a decentralized, stochastic MAC, namely, carrier sense multiple access/collision detection (CSMA/CD). Devices may either operate in ad-hoc mode, forming an independent basic service set (IBSS), or in infrastructure mode forming an infrastructure basic service set (BSS). In the latter, a distribution system may connect several BSSs into an extended service set (ESS) forming a mobility domain. Devices may operate in the 2.4 GHz or 5 GHz industrial, scientific and medical (ISM) bands and provide a bandwidth of up to 54 Mbps. To establish communication, devices have to scan the channel for possible interlocutors. Then, an authentication and association procedure employing a four-way handshake (for unencrypted communication) establishes communication between devices. Handover mechanisms are not included in the baseline standard, which only offers mobility supporting functions (e.g., scanning, establishing, and releasing a link) [1].

12.3.2 IEEE 802.15.1

IEEE 802.15.1 is a wireless MAC specification including a PHY layer specification for wireless PANs. This specification is used for Bluetooth devices building an ad hoc network. The ad hoc network is of short range (a few meters) and operates in the 2.4 GHz band. IEEE 802.15.1 employs a fast frequency hopping scheme and offers bit rates up to 700 kbps for asynchronous and up to three 64 kbps channels for isochronous data transfer. The communication is based on a master slave principle—the master polls the slaves. The latter two form a so-called piconet. To establish communication between two devices, first an inquiry process has to be performed that tries to find other devices in the device's proximity. Then, a paging process is conducted that synchronizes the probing device with those detected in the inquiry phase. Finally, this procedure yields to the establishment of a link layer connection. A handover mechanism is not included in the specification. Nevertheless, IEEE 802.15.1 offers needed services that can be used by higher layers to initiate a handover. These services include detection of other devices as well as establishment and release of link layer connectivity [6].

12.3.3 IEEE 802.15.3

IEEE 802.15.3 is a wireless MAC and PHY layer specification for high-rate wireless PANs. High data rates range from 11 Mbps to 55 Mbps in the 2.4 GHz band. The wireless

PAN is built by piconets. A piconet is formed by a number of devices that are controlled by a piconet controller (PNC). In contrast to IEEE 802.15.1, the access mechanism is based on a synchronous slotted time division multiple access (TDMA) control where the PNC coordinates access of the devices to the wireless media. Joining and leaving a piconet is also controlled by the PNC. Thus the PNC is necessary for operation of the network. On the other hand, because of the wireless nature of the network, the PNC can move outside the proximity of its devices. In this case, the PNC has to hand over its functionality to another device. Thus IEEE 802.15.3 introduces PNC handover, where a PNC can delegate its operation to another device [7].

12.3.4   Detection of Available Radio Cells

The detection of available radio cells requires receiving messages according to the mobile devices communication protocol. As the following considered protocols all differ in their PHY specification—either in the used PHY-level synchronization patterns, modulation, or physical layer convergence procedure (PLCP) frame formats—messages not belonging to the mobile device's technology are discarded at the PHY level. As soon as messages from available radio cells are received at the MAC level, the latter may retrieve additional information (e.g., network/provider ID or location information) and decide based on these parameters if the detected radio cell is actually a possible handover candidate. Retrieving this information is possible by simply listening (passive scanning) or actively probing (active scanning).

12.3.4.1   IEEE 802.11

In IEEE 802.11 networks, access points frequently announce the existence of an infrastructure basic service set using beacons. (The corresponding use of beacons in independent basic service sets [i.e., ad hoc mode] is not further elaborated as our focus is on handovers occurring in the infrastructure mode.) Thus, in order to detect an access point, mobile clients may scan for this beacon iterating through all possible frequencies an IEEE 802.11 device operates on. Depending on the regulatory domain, this process includes up to 13 possible channels in the 2.4 GHz band. It should be noted that the standard does not define the order in which a station (STA) should iterate through all the frequencies.

Regarding IEEE 802.11, the scanning process is the most time-consuming phase during a handover. The standard defines passive as well as active scanning. Both support either looking for all available networks or limiting the result of the scanning process to those being identified by a specific service set ID (SSID).

When employing passive scanning, a STA must listen to each channel. User-defined parameters specify the minimum and maximum time spent on each channel. As the STA has to receive at least one beacon to recognize an access point, this period should be longer than the target beacon transmission time (TBTT), which in a typical deployment is set to produce a beacon transmission every 100 msec. Obviously the time spent in this detection phase can be reduced, using smaller TBTTs at the cost of an additional channel load caused by beacon transmissions, as illustrated in Figure 12.2.

Active scanning is another mechanism defined by the standard to reduce the duration of the detection phase. Figure 12.3 shows how an STA has to listen for a minimum duration—the probe delay—on the channel and then sends a probe request frame. If it does not receive a response within the minimum probe response time (not illustrated), it may scan another channel. Otherwise, if it receives a probe response within this time

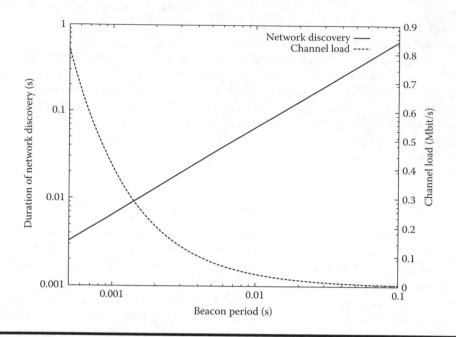

Figure 12.2 Effect of beacon period on network discovery duration and imposed channel load (IEEE 802.11 employing 1 Mbps base transmission rate).

interval, the STA waits for further responses up to the maximum probe response time. This gives existing access points a reasonable amount of time to compete for channel access [1].

The amendment for radio resource management—IEEE 802.11k [2]—introduces several approaches that reduce the detection phase duration, including measurement pilots and neighbor information. The new measurement pilot management frame is relatively small compared to a beacon. It is transmitted at a smaller time interval. Thus it allows rapid discovery of available networks using passive scanning without imposing such a high channel load as for a reduced TBTT interval.

Another approach to expedite the detection phase is to scan only for networks that are known to be available. IEEE 802.11k introduces neighbor reports, which may be requested by an STA. The report element contains a list of known access points in the proximity of the reporting STA and provides information on the measurement pilot interval and the channel number on which an access point operates. The reported information is retrieved from a corresponding MIB table (dot11RRMNeighborReportTable in Annex D of IEEE 802.11k), but the standard does not specify how these entries are obtained. Rather, it indicates two possible approaches: an access point may request STAs in its

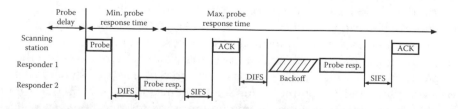

Figure 12.3 Network discovery in 802.11 using active scanning.

proximity to report their local neighbor list, which in turn may be obtained using local scanning, or it can receive the information via an external management interface [2].

In general, the detection phase should not interfere with any ongoing communication. If a STA switches to another channel for network discovery, interlocutors communicating on the former frequency should refrain from sending data to the STA. Actually this requirement is not explicitly stated in the IEEE 802.11 standard. Nevertheless, mechanisms exist (i.e., STA power management modes) that can be used for vendor-specific solutions. For example, a STA can switch from the "awake" to the "doze" state during scanning. The STA indicates this state change to its current access point, which in turn does not immediately transmit data addressed to the STA but holds them in a buffer. The access point signal to the STA buffered data pending for transmission via beacons (traffic information map [TIM] information element) upon whose reception the STA explicitly polls the pending data (PS-poll frame) [1].

12.3.4.2   IEEE 802.16

IEEE 802.16 employs passive scanning for network discovery. The subscriber station (SS) listens consecutively on downlink frequencies until it detects a base station (BS). (Originally IEEE 802.16 considered only point-to-point links and made it mandatory to stop scanning after finding the first available BS.) To expedite this procedure, SSs are required to store connection parameters, including channel and synchronization information, of the last operational channel. They start trying to acquire a downlink channel using these parameters first [8].

To support network discovery while having an active communication between SSs and BSs, IEEE 802.16e amends schemes for continuous network discovery for mobile SSs. Two variants are standardized: SS-initiated scanning and BS-initiated scanning.

For the former, the SS may request a time interval from its BS which may be used for scanning and thus is not used for communication between the SS and the serving BS. Apart from this SS-initiated scanning, the BS may reserve and advertise time slots to be used by all current SSs for scanning purposes. The results of the scanning process are reported to the BS and may be used to build a local neighbor list, including synchronization and timing information, as well as the most recently seen uplink and downlink slot assignment [9].

12.3.4.3   IEEE 802.15.1

IEEE 802.15.1 devices have to discover the available devices in their proximity. To detect other devices by just listening on the wireless channel is not suitable because devices change their frequency every 625 μs because of the fast frequency hopping scheme implemented in the PHY layer. If only listening, the discovery phase has to estimate the hopping sequence of any device. This sequence is determined by 77 frequencies and derived from the MAC address. In addition, a discovering device has to guess the currently used frequency that is derived from the native clock (CLKN) of the device. Both parameters, the MAC address and the native clock, cannot be estimated.

Therefore, IEEE 802.15.1 has introduced a separate signaling channel that is called the "inquiry channel." Devices may use this channel to search for other devices by means of active probing. The inquiry channel is defined by an inquiry hopping sequence that is known by every device. In contrast to the hopping sequence for user communication, which has a length of 77 frequencies, the inquiry sequence repeats after 32 frequencies. The discovery of devices now consists of two steps: first, synchronization of the discovering device with other devices, and second, information transfer between the devices.

First, we will focus on the synchronization phase. Even though all devices know the hopping sequence of the inquiry channel, they do not know which frequency of the sequence each device will use at the moment. The frequency to use is selected dependent from the native clock of each device. Due to the fact that the native clocks of two devices are not synchronized, every device is using another frequency, but of the same sequence. To solve this problem, the discovering device sends "inquiry" messages of a series of frequencies (called frequency train) on a trial-and-error basis. The inquiry message has a duration of 68 μs and is sent twice using two different frequencies within 625 μs. The inquiry message contains an identity (ID) packet. The content of the ID packet is constructed to allow an easy reception via a correlation receiver at the receiving device. It contains a so-called inquiry access code (IAC), which is unique and provides a certain power level. If a device has received that message—meaning two devices are using the same frequency of the inquiry sequence—it will transmit an inquiry response message 625 μs after the inquiry message. At this stage the discovering device knows the offset between its own used frequency and the frequency used by the responding device. As stated earlier, the offset of the frequencies is proportional to the offset of the native clocks of both devices. In fact, both devices are now synchronized. All they have to do now is exchange information about themselves. This is done inherently by the inquiry response message, which contains a frequency hop synchronization (FHS) packet. The FHS packet containing, the device address of the discovered device and its native clock value. (The native clock is represented by a 2^2 eight bit value.)

From the above description, one major question arises: What happens if two devices access the inquiry channel at the same time using the same frequency? The answer is quite simple: collisions occur. This unlikely event may happen if two or more devices are synchronized, as they have the same native clock value. To avoid collisions, additionally a back-off mechanism is integrated by IEEE 802.15.1. The time to start scanning or to listen for scan requests is stochastically chosen [6].

12.3.4.4 IEEE 802.15.3

Searching of available radio cells—called piconets in the IEEE 802.15.3 context—will be initiated by the station (device) management entity by means of the MAC sublayer management entity (MLME) scan request. The device now tunes its receivers to a specific radio channel (passive scanning) and starts searching for beacons which are periodical transmitted by PNCs. If a beacon is detected by the device, additional information about the found piconet will be gathered and stored in a piconet description. The piconet description contains piconet identifiers (the base station identification [BSID], a text string identifier of the piconet; and PNC ID, a numerical identifier of the piconet), the MAC address of the piconet controller, and the used channel number of the piconet.

If the piconet is controlled by another piconet (called a dependent piconet), the piconet description also contains needed information about the parent piconet, which is a subset of the dependent piconet description. At a minimum, the piconet description contains the BSID and the MAC address of the parent piconet.

For the whole scan process, the scanning time is limited (the standard does not specify the scanning time) and the management entity can instruct the scan process to scan for a specified piconet by its BSID and PNC ID or can initiate an open scan where the scan process searches for all piconets in all channels.

After the specified scanning time has been reached, the scan process stops scanning and starts ranking each found piconet. Ranking is done by determining the signal quality of the wireless channel to each found piconet. The signal quality is a measure of interference from other piconets. The result of the ranking is stored in a channel rating list—an ordered list of channel numbers starting from the best channel to the worst channel. Each channel represents a found piconet [7].

12.3.5 *Criteria for Handover Decision*

Neither of the considered IEEE standards specifies when to initiate a handover or to which neighbor a mobile device should switch. To support vendor-specific criteria, they offer numerous measurement capabilities to assess the quality of the wireless channel. Probes of the measured parameters may either be obtained locally or distributed by employing signaling schemes requesting remote measurements. The latter case enables channel evaluation from the uplink and downlink perspective, whereas local measurements assess the wireless channel characteristics in only one direction.

12.3.5.1 *IEEE 802.11*

The radio resource management amendment (IEEE 802.11k) provides several means to either obtain relevant measurements on a regular basis or by explicitly requesting them. For example, the previously described measurement pilot frame includes information on the used transmission power and the noise floor experienced at the sender which can be used for vendor specific link budget calculations on the receiver side [1].

In addition, an STA may explicitly request remote measurements from another STA using action request management frames [1]. Specified measurement reports refer either to statistics gained from the entire radio cell (i.e., BSS) or statistics representing the status of an individual STA. The former includes the service load of the BSS expressed as the average medium access delay, whereas the latter includes reports on channel load, current location, experienced noise in the form of a histogram, and reports on STA counters such as experienced frame check sequence (FCS) errors, number of acknowledgement (ACK) failures, channel utilization, or access delays for various QoS classes [2].

12.3.5.2 *IEEE 802.16*

IEEE 802.16 devices are capable of continuously measuring the radio signal strength based on messages received via the downlink channel. Depending on the employed PHY, RSSI value and also the carrier-to-interference plus noise ratio (CINR) are reported. The BS may request the SS report the mean or standard deviation of the measurements using a prioritized feedback channel and thus may forward them as an input into vendor-specific handover decision algorithms [9].

12.3.5.3 *IEEE 802.15.1*

IEEE 802.15.1 defines an interface to configure, control, and gather information about an IEEE 802.15.1 device. This interface is called the host controller interface (HCI). Even though IEEE 802.15.1 does not contain any handover, higher layer handover processes can obtain RSSI values about the link quality between devices, which along with device information obtained during the inquiry phase, can help decide if a handover should be initiated. It should be noted that the standard itself does not introduce the RSSI for handover decisions, but for power adjustment reasons. Nevertheless, the RSSI value can

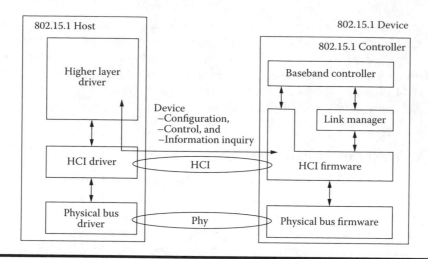

Figure 12.4   Information inquiry via the human-computer interaction (HCI) interface in 802.15.1.

also be used for mobility support. Figure 12.4 illustrates the interfaces between IEEE 802.15.1 functional components including the HCI and PHY drivers [6].

12.3.5.4   IEEE 802.15.3

The ad hoc nature of IEEE 802.15.3 devices does not differentiate between different types of stations like end systems, access points, and base stations. Thus it is necessary to include not only link characteristics like bit error rate and corrupted and lost packets, but also capabilities of the devices itself into a handover decision. This capability decides if a station can act as a special device, like a controller, for a piconet. This capability includes an indication of whether the device is battery powered, because externally powered devices have a longer lifetime and thus may be preferred as a PNC. The IEEE 802.15.3 standard defines the following ranking to determine possible handover candidates: the ability to act as an exponent device, capable of supporting security operations, connectivity to an external power supply, maximum number of associated devices (which can be supported), maximum number of channels (that can be managed), maximum transmission power, and the device address (higher addresses are preferred). Whenever a device that joins a piconet has greater capabilities compared to the current controller of the piconet, the piconet controller will give up its controlling function and delegate it to the new device [7].

12.3.6   Reestablishment of Link Layer Connectivity

12.3.6.1   IEEE 802.11

Originally IEEE 802.11 did not provide advanced mechanisms for handover support: an STA moving from the coverage area of one access point to another can simply reestablish a link to the new access point which includes authentication, association, and possibly a renegotiation of resources. The authentication process consumes a rather long time if advanced authentication and encryption schemes are employed. Since the protocol message exchange for all of these phases takes place via the air interface, application-level communication is interrupted [1].

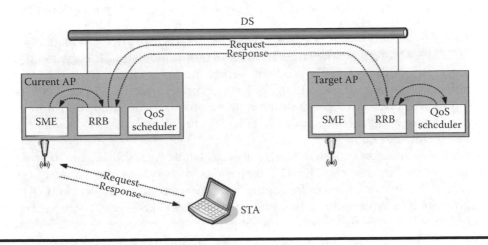

Figure 12.5   Function of the 802.11 remote request broker.

The fast BSS transition amendment (IEEE 802.11r) reduces this time period by moving the authentication, and optionally the resource negotiation, prior to actually switching the permanent connection from the old access point to the new one. IEEE 802.11r supports two methods to conduct the exchange of protocol messages: over the air and via the backbone (i.e., the distribution system). For the latter case, a new component—the remote request broker (RBB)—is added to the architecture of an access point as illustrated in Figure 12.5. The RBB, as part of the station management entity, can communicate with the target access point on behalf of the STA. Thus an STA can authenticate itself at the new access point prior to the actual transition. In addition, radio resources can be requested at the target access point and the STA may decide not to switch to the target access point if this resource allocation fails. IEEE 802.11r does not specify any mechanisms to transfer the actual state of an ongoing connection (e.g., buffered packets or connection relevant timer) from the old access point to the new one, but only outlines ways to preestablish a possibly secure connection in between the STA and the target access point via the distribution system [4].

This actual state transfer was partially covered by IEEE 802.11F. (802.11F has expired. Its withdrawal has been voted on by the IEEE 802.11 Working Group in November 2005 and was approved by the IEEE SA in March 2006.) The recommended practice defines an interaccess point protocol that allows the new access point to trigger the old access point, thereby forcing the latter to forward packets buffered for the STA [5].

In all situations, the current IEEE 802.11 standard only foresees a mobile-initiated BSS transition, whereas upcoming standardization efforts target providing access point triggered transitions to balance traffic load among several access points.

12.3.6.2   IEEE 802.16

The general approach of IEEE 802.16 is to conduct as many phases of link reestablishment prior to switching from one BS to another. During these phases, the SS synchronizes on the downlink channel of the BS. While in scanning mode, the mobile device obtains transmission parameters, adjusts its transmission power level (ranging), negotiates capabilities between itself and the BS, and conducts authentication [8]. This process may occur between the SS and the BS via the wireless channel, called "association without coordination," or may happen via the backbone as "association with coordination" or "network-assisted association."

In case of association with coordination, the serving BS forwards information (e.g., the SS's MAC address) directly to the target BS. As synchronization and PHY layer information of the target BS were previously acquired during the network discovery phase, the SS may immediately start with the ranging process. Even the duration of the latter process may be further reduced as a result of the ranging process. The latter in turn provides a condensed answer to the SS. In addition, IEEE 802.16e allows an SS to maintain several associations simultaneously.

Besides the approach to shift many phases of link reestablishment to before the occurrence of the handover, IEEE 802.16e provides two optional mechanisms to enable seamless handover support even for higher open systems interconnection (OSI) layers: macrodiversity handover and fast base station switching (FBSS). The former is based on a synchronous bi-cast of data on the downlink, as shown in Figure 12.6. This allows the mobile device to switch its uplink from one BS to another on a frame-by-frame basis. In the latter case, the SS establishes associations as previously described, with several BSs forming a so-called diversity set, but uplink and downlink data are only exchanged via a single BS called the anchor BS. In order to conduct a handover, the SS has only to send an anchor update to conduct a time slot–synchronized switch from one BS to another. Thus the two approaches generically differ in the required uplink and downlink capacity on the wireless channel.

It should be noted that both the macrodiversity handover as well as the FBSS mechanisms require time synchronization between the involved BSs. The standards do not specify how this synchronization is achieved [9].

12.3.6.3   IEEE 802.15.1

IEEE 802.15.1 only offers handover supporting functions, namely the establishment and release of a connection between a master and slave. Link establishment and release is controlled by a link manager protocol (LMP). The LMP is responsible for handling synchronous and asynchronous links [6].

12.3.6.4   IEEE 802.15.3

IEEE 802.15.3 offers two mechanisms to support mobility. First, an IEEE 802.15.3 device offers the mechanisms to leave a piconet and connect to another piconet to support a handover process located in higher layers. Second, a PNC can delegate its functionality

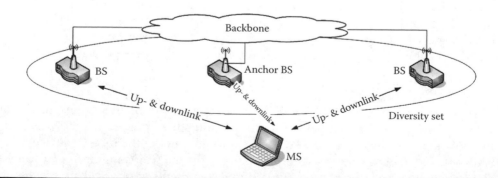

Figure 12.6   Macrodiversity handover in 802.16.

of controlling the piconet to another device. This mechanism is called PNC handover. There are two types of PNC handover. PNC handover can be done in an independent picocell or in a dependent picocell, which is called dependent PNC handover. Conceptually both types are similar, except that the new dependent PNC has to inform its parent PNC about the handover.

12.3.6.5   Handover Support for Higher Layer Mobility

IEEE 802.15.3 offers the service of association and disassociation. Association describes the process of joining a piconet, whereas disassociation describes the process of leaving a piconet.

12.3.6.6   PNC Handover

A PNC leaving the piconet because of movement or switching itself off should initiate a handover to give the responsibility of the piconet to another device. To begin the handover, the PNC selects a candidate as successor. The information is acquired from status reports where devices indicate their ability to act as a PNC.

After issuing the handover request, information transfer between the PNC and the nominated device is initiated. The PNC then allocates channel time with the chosen device. After the channel time has been reserved, the PNC transmits all needed information to the new PNC. The PNC sends all information about itself—the PNC personality. In addition, it transmits all necessary information about the associated devices, the current received requests for channel allocation from devices, and power save information about devices. Power save information includes all devices that are in power-saving mode. After all relevant information is sent to the new PNC, the current PNC awaits a response from the new PNC indicating that it is ready to take over PNC functionality. The new PNC cannot refuse the handover request unless it is controlling a dependent piconet. If the new PNC acknowledges the request with a handover response, the latter is acknowledged for reliability reasons by the current PNC (three-way handshake). The current PNC then informs all devices that a PNC handover will occur.

This is done by inserting a handover flag in the beacon, together with the device ID of the new PNC and the number of superframes remaining before the new PNC starts to act as the new PNC. After the remaining number of superframes, the current PNC stops sending beacons and the new PNC starts sending beacons. At this point the handover is completed.

Even though associated devices do not need to reassociate, resources for asynchronous transmission time have to be renewed at the new PNC, whereas resources for synchronous transmissions remain. Nevertheless, the number and perhaps the size of the slots allocated for synchronous transmission can change. This is signaled by the PNC. It should be noted that the entice handover process is monitored by a timer. If the time expires, the handover is canceled. The handover is also canceled if the selected new PNC recognizes that the current PNC is shutting down [7].

12.4   Handover Between Different Technologies

12.4.1   IETF Mobility Support

As described in Section 12.2.4, IETF mobility support has to deal with the problem that IP addresses are used for routing purposes as well as for host identification. If end systems move to another IP subnet, the routing will not adapt to the new location. In addition, all current transport connections are abruptly interrupted with the change in IP address because the IP address is part of the transport-level identifier. With the introduction of mobile IP, this problem has been partially solved: Mobile IP attaches two IP addresses to a single end system. One IP address, the permanent home address, is used to identify the end system, whereas the other temporary IP address is used for routing the care of address. This concept enables nomadic mobility between different IP networks [11].

True mobility cannot be supported because of handover latencies due to prior link layer establishment and signaling between the home network and the network where the mobile end system is currently attached. To reduce the handover latency, hierarchical mobile IP (HMIP) was introduced. The concept of HMIP introduces anchor points within the router hierarchy of the Internet. In case of movement to another IP network, these anchor points redirect packets to the mobile node without interaction of the home network [12].

Another improvement in mobility support is the fast Mobile IP (FMIP). The idea behind FMIP is to establish a tunnel between the old and new IP network of the moving end system. In case of a handover, all outstanding data received at the old IP subnet are forwarded through the tunnel to the end system. It is assumed that this mechanism reduces the packet loss rate during a handover [13].

Although these basic concepts of mobile IP, hierarchical mobile IP (HMIP), and FMIP, as well as combinations of the latter two have been established, certain issues remain open: IP subnet detection and host alerting of multihomed operation. Several study groups had been established to focus these problems. The Internet Engineering Task Force (IETF) Detecting Network Attachment (DNA) Working Group deals with the problem of detecting a change in the IP network in case of link change [14]. The Network-based Localized Mobility Management (NETLMM) Working Group is working on solutions so a mobile node can keep its IP address even if it moves to another IP network near its location (micromobility) in the same wireless access network [15]. When a mobile endsystem has connectivity to more than one wireless network (multihomed endsystem), a decision has to be made as to which interfaces should be used. Today all flows of an end-system are routed through the same interface, called terminal mobility. The Mobile Nodes and Multiple Interfaces in IPv6 (MONAMI6) Working Group is working on distributing multiple flows to multiple interfaces of a single end system [16].

12.4.2 Media Independent Handover

The previously discussed higher layer IETF mobility schemes implicitly anticipate layer-2 constructs providing status information about the underlying link. Link status information should be technology independent and therefore should reflect only triggers (e.g., a link goes down or is coming up) and abstract descriptions of the link (e.g., link quality is good or bad).

IEEE's Media Independent Handover Group (MIH, IEEE 802.21) provides such an interface description for existing and future IEEE 802 standards. The concept is also extendable and may apply to non-IEEE technologies, such as cellular networks. Media independent handovers do not limit IEEE 802.21 to heterogeneous technologies, but also support homogeneous technologies. It offers the possibility to adapt a common handover process to link layer technologies. Therefore it is not necessarily important that a specific technology support mobility functions inherently enabling a true seamless handover. On the other hand, every link layer technology has to provide handover supporting functions like network discovery, network and link measurements, and link status information. IEEE 802.21 provides an interface specification so mobility supporting functions can communicate technology independently with the handover process.

The key services of the media independent handover functionality include link layer triggers, handover commands, and primitives to exchange network information among the involved MIH entities. For example, link layer triggers include the indication of link state changes: a handover process placed on top of the MIH interface (e.g., employing make-before-break concepts) may utilize predictive IEEE 802.21 events (i.e., link going down) as well as events indicating a link state change (link up/down or a change in link parameters, such as signal strength or packet error rate). If the underlying link layer already incorporates local handover mechanisms as found in cellular networks or IEEE 802.16e, IEEE 802.21 provides commands allowing higher processes to make use of them (e.g., calling the handover-initiate service primitive). Information on the mobile device's neighborhood may be obtained by requesting the available layer-2 technologies to scan for new links or networks or may be retrieved via the media independent information service (MIIS), which may be represented by a provider-specific database. At the moment, IEEE 802.21 defines the structure of the contained information and how to retrieve it; how to build this database or add information to it is not considered [10].

12.5 Future Trends and Challenges

The ongoing trend in wireless communication is continuously moving toward providing higher and higher bit rates either by employing advanced modulation schemes or protocol optimization [3] or by simply shifting the operation into higher frequency bands. The latter inherently leads to reduced radio cell sizes which in turn leads to an increased number of handovers (per time unit) for a given mobile device's velocity. In combination with higher bit rates, future approaches providing seamless handovers have to reduce today's handover latency by several magnitudes. These systems may offer bit rates of several gigabits per second. For these and other "microcellular" systems, the overlap of adjacent radio cells is extremely small due to reduced deployment costs. Thus handover algorithms have only an extremely small time interval to decide on switching from on access point to another. (This phenomena is even present in today's systems: For example, a high-availability and robust wireless LAN system for high-speed train communication faces the same challenges [17].) The aspect of the mobile device's velocity influencing the handover performance of IEEE 802.11 systems is by now a well-acknowledged aspect of ongoing standardization efforts [18,19].

Another challenge for providing seamless handover in a heterogeneous environment, especially if it consists of IEEE-based wireless networks as well as cellular networks, is how to handle the different handover philosophies: mobile initiated and possibly controlled handovers versus network initiated or controlled handovers in cellular networks. The latter inherently have state information about the end system and can therefore decide when and to which radio cell a handover should occur. In contrast, this decision is mostly left to the end system in IEEE wireless networks. Apart from the mere technical aspects, the provider view may be neglected. The provider may not wish to leave the handover decision to the mobile device.

Third, a large number of access points are deployed in private households. The unused capacity of these access points could be used to extend the access network of providers or could form a separate wireless network. Unfortunately these access points are connected via long-delay lines (digital subscriber line [DSL]), and the gateway to the Internet for two different households might be geographically far apart. Thus, since the handover delay is influenced by the signaling delay between the entities involved in the handover process (especially the access points), the handover latency increases as well. Advanced handover schemes should increasingly use available location information as well as mobile devices trajectories in order to predictively prepare handovers.

References

1. IEEE, Wireless LAN Medium Access Control (MAC) and Physical Layer (PHY) specifications, Standard 802.11—REVma/D7.0, IEEE, Washington, DC, 2006.
2. IEEE, Amendment 9 to IEEE 802.11—REVma/D5.2: Radio Resource Measurement, Standard 802.11k/D4.1, IEEE, Washington, DC, 2006.
3. IEEE, Draft amendment to IEEE 802.11 Wireless LAN Medium Access Control (MAC) and Physical Layer (PHY) specifications: Enhancements for Higher Throughput, Standard 802.11n/D1.0, IEEE, Washington, DC, 2006.
4. IEEE, Amandment 2 to IEEE 802.11—REVma/D5.2: Fast BSS Transition, Standard 802.11r/D2.0, IEEE, Washington, DC, 2006.
5. IEEE, Trial-use Recommended Practice for Multi-Vendor Access Point Interoperability via an Inter-access Point Protocol Across Distribution Systems supporting, IEEE 802.11, Standard 802.11F, IEEE, Washington, DC, 2002.
6. IEEE, Wireless medium access control (MAC) and physical layer (PHY) specifications for wireless personal area networks (WPANs), Standard 802.15.1–2005, IEEE, Washington, DC, 2005.
7. IEEE, Wireless Medium Access Control (MAC) and Physical Layer (PHY) Specifications for High Rate Wireless Personal Area Networks (WPANs), Standard 802.15.3–2003, IEEE, Washington, DC, 2003.
8. IEEE, Air Interface for Fixed Broadband Wireless Access Systems, Standard 802.16–2004, IEEE, Washington, DC, 2004.
9. IEEE, Amendment 2 to IEEE 802.16-2004/Cor1-2005: Physical and Medium Access Control Layers for Combined Fixed and Mobile Operation in Licensed Bands, Standard 802.16e–2006, IEEE, Washington, DC, 2006.
10. IEEE, Draft IEEE Standard for Local and Metropolitan Area Networks: Media Independent Handover Services, Standard 802.21/D2.00, IEEE, Washington, DC, 2006.
11. D. Johnson, C. Perkins, and J. Arkko, Mobility Support in IPv6, RFC 3775, IETF, June 2004.
12. H. Soliman, Hierarchical Mobile IPv6 Mobility Management (HMIPv6), RFC 4140, IETF, August 2005.
13. R. Koodli, Fast Handovers for Mobile IPv6, RFC 4068, IETF, July 2005.
14. J. Kempf, S. Narayanan, E. Nordmark, and B.P.J. Choi, Detecting Network Attachment in IPv6 Networks (DNAv6), Internet Draft, IETF, May 2006.
15. I. Akiyoshi and M. Liebsch, NETLMM protocol, Internet Draft, IETF, October 2005.
16. T. Ernst, N. Montavont, R. Wakikawa, C. Ng, and K. Kuladinithi Motivations and Scenarios for Using Multiple Interfaces and Global Addresses, Internet Draft, IETF, February 2006.
17. M. Emmelmann, Influence of Velocity on the Handover Delay Associated with a Radio-Signal-Measurement-based Handover Decision, *62nd Vehicular Technology Conference*, vol. 4, p. 2282, IEEE, Washington, DC, 2005.
18. S. Bangolae, C. Wright, C. Trecker, M. Emmelmann, and F. Mlinarsky, Test Methodology proposal for measuring fast BSS/BSS transition time. IEEE, doc. 05/0537r3, 802.11 TGt Wireless Performance Prediction Task Group, Vancouver, Canada, November 14–18, 2005, accepted into the IEEE P802.11.2 Draft Recommended Practice.
19. M. Emmelmann, T. Langgärtner, C. Wright, F. Mlinarsky, B. Rathke, and P. Egner, Methodology for employing variable attenuators in a conducted test environment (Substantive Standard Draft Text), IEEE doc. 05/0702r3, 802.11 TGt Wireless Performance Prediction Task Group, Vancouver, Canada, November 14–18, 2005.

Chapter 13

Energy Management in the IEEE 802.16e WirelessMAN

Yan Zhang, Jianhua Ma, Laurence T. Yang,
Yifan Chen, and Jun Zheng

Contents

The IEEE 802.16 Wireless Metropolitan Area Networks (WirelessMANs) standard is emerging as a highly promising technology for the future generation of broadband wireless access (BWA). To provide multimedia services and satisfy various scenario requirements, the standard specifies point-to-multipoint (PMP) and mesh working modes. Since a mobile subscriber station (MSS) is normally powered by a rechargeable battery, an efficient energy management strategy is an indispensable component to lengthen the MSS and network lifetime.

In this chapter we characterize the sleep mode in the PMP working mode in the IEEE 802.16e WirelessMAN and perform an extensive analysis of the specified energy-saving scheme in terms of energy consumption and packet delay. Illustrative examples are presented to show the interactions between the key parameters and the aforementioned metrics. This chapter should be helpful for understanding the energy management schemes in the IEEE 802.16 WiMAX PMP mesh networking modes. Also, it can serve as guidance for efficiently managing limited energy in IEEE 802.16 WirelessMANs.

13.1   Introduction

WirelessMANs provide broadband wireless access as an excellent alternative to cabled access networks, such as fiber-optic links, coaxial systems, and digital subscriber line (DSL) links. This system has the advantages of low-cost construction and maintenance. IEEE 802.16 WirelessMAN is emerging as a highly promising technology for the future BWA. It was originally organized to develop standards and recommend practices to support the development and deployment of fixed BWA [1–4]. Presently, by supporting mobility and seamless hand-off, the IEEE 802.16 Working Group is designing a high-speed, high-bandwidth, and high-capacity standard for both fixed and mobile BWA [1,3]. The BWA alternative from the European Telecommunications Standards Institute (ETSI), is the high-performance radio metropolitan area network (HiperMAN). Another competing standard from Korea is wireless broadband (WiBro). WiBro defines the specifications in the licensed radio spectrum, offering data throughput of 30 to 50 Mbp/s and covering a service radius of 1 to 5 km for portable Internet usage over the 2.3 GHz spectrum. In addition, WiBro is designed to provide all-Internet protocol (IP) services and offer quality-of-service (QoS) schemes to differentiate loss-sensitive data and real-time streaming video multimedia services.

Worldwide Interoperability for Microwave Access (WiMAX) is a certification mark for products that pass conformity and interoperability tests for the IEEE 802.16 standard. To ensure interoperability while reducing deployment cost, the WiMAX Forum is working on approaches that enable 802.16, HiperMAN, and WiBro to seamlessly interconnect.

13.1.1   Standardization Activities

The first version, IEEE 802.16-2001, defined the PMP mode and addressed the line-of-sight (LOS) problem employing the orthogonal frequency division multiplexing (OFDM) technique with the 10 to 66 GHz range and a data rate of up to 134 Mbps. Because of the characteristics of this radio spectrum, the system is inapplicable in non-line-of-sight (NLOS) environments. This version was completed in December 2001 and can only be used in fixed scenarios. The next version, 802.16.2, attempted to minimize the interference between coexisting WirelessMAN systems. The next version, 802.16c, defined detailed system profiles in the 10 to 66 GHz range.

An emerging wireless network architecture, wireless mesh networking (WMN) can be traced back to multihop ad hoc networks. WMN refers to a network architecture in which the nodes can communicate with each other via multihop routing or forwarding. It is characterized by dynamic self-organization, self-configuration, and self-correction to enable quick deployment, easy maintenance, low cost, high scalability, and reliable service. Because of its inherent advantages, WMN is believed to be a highly promising technology for future wireless mobile networks. IEEE 802.16a introduced and defined the key operation procedures for the mesh networking mode. This version, approved in January 2003, supports the NLOS capability, operational in both the licensed and unlicensed spectrum ranging from 2 to 11 GHz. It can support a data rate of up to 75 Mbps and a maximum range of 50 km. However, the mobility capability is still absent in this version. All the aforementioned versions serving fixed BWA were incorporated into the finalized standard, 802.16d-2004.

In December 2005, the latest version, IEEE 802.16e, was approved by adding the mobility capability, including components supporting the PMP and mesh modes, seamless handover operation, and location management schemes. It can achieve a data rate of up to 15 Mbps in the 5 MHz channel bandwidth.

To develop the standard into a mature and complete specifications, there are several projects under development. Project 802.16i, from in the IEEE 802.16 Network Management Task Group (NetMan), is developing the mobile management information base. Project 802.16j is developing mobile multihop relay networking. The scope includes enhanced frame structure in the physical (PHY) layer and new protocols in the media access control (MAC) layer to expand service coverage and improve network throughput. To achieve this, the hand-off, protocol performance, and backward compatibility are key challenges. Project 802.16g is still being developed. Its goal is to develop management plane procedures and services. Project 802.16h is also under development, with the objective of improving coexistence mechanisms for license-exempt operation. Project 802.16k is also under development. The scope of this project is to support the bridging function of the IEEE 802.16 MAC. Specifically, it will describe the service interface between the internal sublayer service and the IEEE 802.16 MAC, and add schemes for priority mapping.

In this chapter we will concentrate on the energy management mechanism in IEEE 802.16e Mobile WiMAX. Normally MSSs are powered by rechargeable batteries with very limited capacity. Radio transmitters and receivers consume large amounts of power, hence efficient power-saving mechanisms can significantly increase MSS operating time as well as life span. In particular, we will explain the operating mechanisms in the framework of the energy management in IEEE 802.16e Mobile WiMAX. Furthermore, the performance of the standardized sleep strategy will be evaluated under various scenarios. This chapter is organized as follows: Section 13.2 introduces PHY layer and MAC layer fundamentals. The orthogonal frequency division multiple access (OFDMA) modulation scheme is explained. In Section 13.3, the frame structure and basics of the PMP and mesh modes are presented. Following the explanation of the frame structure and the functionalities of each subframe, we elaborate on the distributed scheduling algorithms for determining the transmission opportunities of the control messages and the data subframe. In Section 13.4 to Section 13.7, we introduce the sleep mode and then extensively analyze the performance of the power-saving mechanism. Section 13.8 concludes.

13.2   WiMAX Overview

13.2.1   OFDMA

IEEE 802.16 defines three different PHY specifications for the PMP mode in the 2 to 11 GHz frequency band. Each of the following air interfaces is able to work together with the MAC layer to provide a reliable end-to-end link:

- WirelessMAN-SCa: a single carrier (SC) modulation technology.
- WirelessMAN-OFDM: a 256-carrier OFDM modulation. Time division multiple access (TDMA) is utilized for multiple access. This air interface is mandatory for license-exempt bands.
- WirelessMAN-OFDMA: a 2048-carrier OFDM modulation. Multiple access is achieved by allocating a subset of the available carriers to an individual.

The latest version, IEEE 802.16e-2005, employes the OFDMA approach. Essentially OFDMA is a combination of two multiple access strategies: TDMA and frequency division multiple access (FDMA). TDMA refers to the mechanism where each user is allocated a unique time slot to transmit. FDMA refers to the access policy when each user receives a unique carrier frequency and bandwidth. In OFDMA, each user is dynamically assigned subcarriers in different time slots.

Multiuser diversity and adaptive modulation and coding (AMC) are the two most important techniques in OFDMA to achieve higher performance. The objective of multiuser diversity is to select subcarriers with the highest signal-to-noise ratio (SNR). The objective of AMC is to select the appropriate modulation scheme under a specific environment.

13.2.2   MAC Layer Overview

The MAC layer is comprised of three sublayers: the service-specific convergence sublayer (SSCS), the common part sublayer (CPS), and the privacy sublayer (PS). The SSCS is an interface to higher layers and provides mapping functions from the transport layer. The CPS provides core MAC functions including access control, collision resolution, control or data scheduling, bandwidth request, and allocation. The PS ensures secure connection establishment, provides network access authentication, and generates key exchange and encryption for data privacy [5].

13.3   PMP and Mesh Modes in IEEE 802.16 WiMAX

13.3.1   PMP and Mesh Networking Modes

Figure 13.1 compares the PMP and mesh topologies. In PMP mode, a base station (BS) coordinates and relays all communications. The subscriber station (SS), under the management of the BS, communicates with the BS first before transmitting data to other SSs. This architecture is similar to that of cellular networks.

Under certain situations (e.g., emergency or disaster scenarios), the PMP mode may not be suitable for timely and efficient deployment, while the mesh mode is an excellent candidate for temporary network construction. Unlike the PMP mode, there are no clearly separate downlinks and uplinks in the mesh mode. Every SS can directly communicate

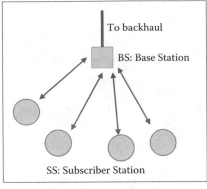
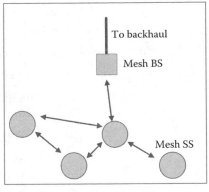

Point to MultiPoint (PMP) mode   Mesh mode

Figure 13.1   PMP mode and mesh mode.

with its neighbors without the help of a BS. In a typical installation, one or several nodes play the role of BS to connect the mesh network to the external backhaul link (e.g., Internet or telecommunications networks). Such nodes are called mesh BSs, while the other nodes are called mesh SSs.

13.3.2   Frame Format in PMP Mode

The OFDM PHY layer of the system supports frame-based transmission with a frame length of 0.5, 1, or 2 ms. Figure 13.2 illustrates the frame structure for the OFDM PHY layer operating in time division duplex (TDD) mode. Each frame consists of a downlink (DL) subframe and an uplink (UL) subframe. The transmit/receive transition gap (TTG) is used to separate the DL subframe and UL subframe and allows the terminal to operate from reception to transmission. Similarly the receive/transmit transition gap (RTG) is used to separate the UL and DL subframes and enables the terminal to operate from transmission to reception.

The DL subframe consists of one DL PHY packet data unit (PDU) starting with a long preamble (two OFDM symbols), which is used for PHY synchronization. The frame control header (FCH) follows the preamble and has a length of one OFDM symbol. The FCH is modulated by binary phase-shift keying (BPSK) with coding rate 1/2. The FCH is followed by one or multiple DL bursts. The first DL burst contains the broadcast MAC management messages (i.e., DL and UL mobile application parts [DL-MAP and UL-MAP, respectively] as well as the DL and UL channel descriptors [DCD and UCD, respectively]). The DL-MAP defines the access strategy to the DL channel, while the UL-MAP specifies the access scheme to the UL channel. The DCD and UCD define the physical channel characteristics. Each of the other DL bursts starts with an optional preamble to enhance synchronization and channel estimation. Following the preamble, a number of MAC PDUs are scheduled to transmit in a DL burst. These MAC PDUs may be associated with different service flows/connections or SSs, but all of these PDUs are encoded and modulated using the same PHY mode. In either the DL or UL direction, the size of the burst is an integer number of the OFDM symbol length to exactly match the OFDM symbol and burst boundaries. To form an integer number of OFDM symbols, unused bytes in the burst payload might be padded by the bytes 0xFF.

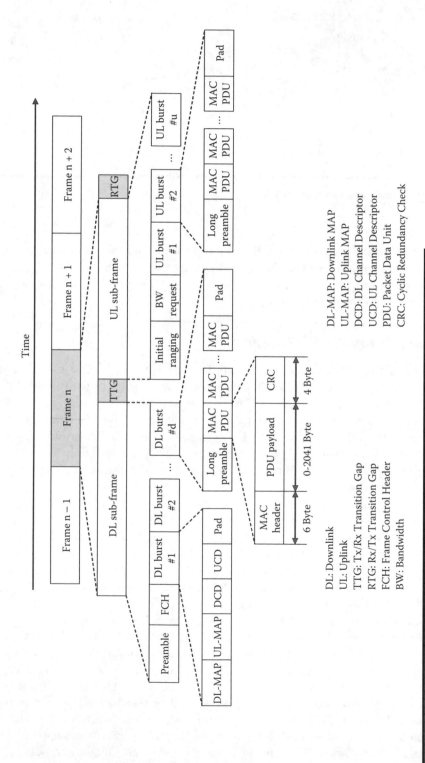

Figure 13.2   PMP frame format with TDD in IEEE 802.16.

The UL subframe consists of contention slots for initial ranging, contention slots for bandwidth requests, and one or a multiple number of UL PHY transmission bursts. The purpose of the initial ranging is for entry of the SSs into the system, including power control, frequency offset adjustment, time offset correction, and basic management request. The bandwidth request interval is used for the SSs to transit the bandwidth request message. The UL burst structure is similar to the DL burst.

13.3.3 Frame Format in Mesh Mode

In the mesh mode, traffic can be routed through other SSs and can occur directly between SSs without being routed through a mesh BS. Mesh mode defines the direct communication between SSs in the MAC layer and allows multihop communication. Here the setup can be achieved using two scheduling schemes: centralized scheduling and distributed scheduling.

Figure 13.3 shows the frame format in the mesh mode. A frame consists of a control subframe and a data subframe. The length of the control subframe is fixed as mesh control length (MSH-CTRL-LEN) × 7 OFDM symbols, where the parameter MSH-CTRL-LEN has 4 bits (i.e., values range between 0 and 15) and is advertised in the `network descriptor information element (IE)`. The data subframe is divided into minislots.

Figure 13.3 illustrates two control subframes, the network control subframe in case (a) and the schedule control subframe in case (b). The network control subframe occurs

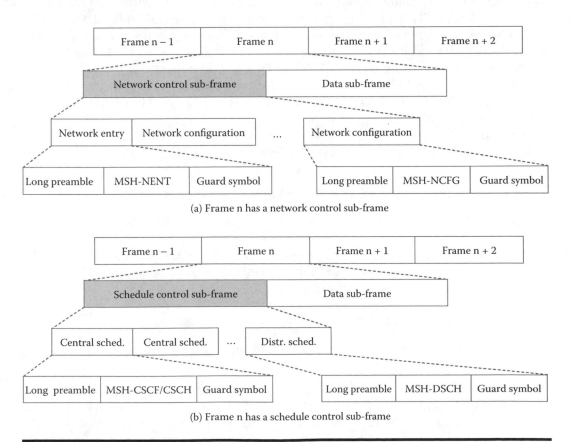

(a) Frame n has a network control sub-frame

(b) Frame n has a schedule control sub-frame

Figure 13.3   Frame structure in the mesh mode in IEEE 802.16.

periodically, with the period indicated in the `network descriptor IE`. The schedule control subframe occurs in all other frames without a network control subframe. In particular, the `scheduling frame` in the `network descriptor IE` defines the number of frames having a schedule control subframe between two frames with a network control subframe in multiples of four frames. For example, if `scheduling frame = 4`, then after a frame with network control subframe, the following 4×4 frames have a schedule control subframe, which is again followed by the next frame with a network control subframe.

The network control subframe is defined primarily for new nodes gaining synchronization and joining a mesh network. The first transmission opportunity is the network entry component carrying the mesh network entry message (MSH-NENT). This part is reserved for new nodes that expect to enter the mesh network system. The remaining (MSH-CTRL-LEN − 1) transmission opportunities are the network configuration components carrying the mesh network configuration message (MSH-NCFG). The portion for network configuration is defined to broadcast network configuration information to all nodes. The length of each transmission opportunity accounts for 7 OFDM symbols. Hence, the length of the transmission opportunities carrying the MSH-NENT and MSH-NCFG is equal to 7 OFDM symbols and (MSH-CTRL-LEN − 1) × 7 OFDM symbols, respectively.

The schedule control subframe is defined for centralized or distributed scheduling of the sharing nodes in a common medium. In `network descriptor IE`, there are mesh distributed scheduling numbers (MSH-DSCH-NUM) of mesh distributed scheduling messages (MSH-DSCH). This suggests that the first (MSH-CTRL-LEN − MSH-DSCH-NUM) × 7 OFDM symbols are allocated for transmitting the mesh centralized scheduling message (MSH-CSCH) and the mesh centralized configuration message (MSH-CSCF).

The data subframe serves the PHY layer transmission bursts. The PHY layer bursts start with a long preamble (2 OFDM symbols) for synchronization, immediately followed by several MAC PDUs. Each MAC PDU is comprised of a 6-byte MAC header, a 2-byte mesh subheader with identification node (ID), a variable length MAC payload (0 to 2039 bytes), and a 4-byte optional cyclic redundancy check (CRC). Consequently, the length of a MAC PDU varies between 12 and 2051 bytes.

13.3.4 Distributed Scheduling in WiMAX Mesh Networks

Distributed election scheduling determines the next transmission time (`NextXmtTime`) of a node's MSH-NCFG during its current transmission time (`XmtTime`). There are two fields—`NextXmtMx` and `XmtHoldoffExponent`—in the MSH-NCFG to determine the next eligibility interval. Here the eligibility interval refers to the time during which the node can transmit in any slot, and is given by

$$2^{\text{XmtHoldoffExponent}} \cdot \text{NextXmtMx} < \text{NextXmtTime}$$
$$\leq 2^{\text{XmtHoldoffExponent}} \cdot (\text{NextXmtMx} + 1). \quad (13.1)$$

The length of the eligibility interval is equal to the difference between the upper bound and the lower bound (i.e., $2^{\text{XmtHoldoffExponent}}$). After the eligibility interval, the node has to wait a hold-off time (`XmtHoldoffTime`) before a new transmission with

$$\text{XmtHoldoffTime} = 2^{\text{XmtHoldoffExponent}+4}. \quad (13.2)$$

For example, if NextXmtMx = 2 and XmtHoldoffExponent = 4, the node is eligible for the next MSH-NCFG transmission between the 33rd and 48th transmission opportunity. After the eligibility interval with length 16 transmission opportunities, the node waits for 256 transmission opportunities before the next transmission.

The node chooses the temporary transmission opportunity (TempXmtTime) equal to the first transmission slot after the hold-off time (XmtHoldoffTime). Then the node determines the set of all eligible nodes competing for this slot (TempXmtTime). The set of eligible competing nodes includes all nodes in the extended neighborhood satisfying the following properties:

- The NextXmtTime includes TempXmtTime.
- The NextXmtTime is unknown.
- The EarliestSubsequentXmtTime occurs no later than the TempXmtTime, where

$$EarliestSubsequentXmtTime$$

$$= NextXmtTime + XmtHoldoffTime$$

$$= NextXmtTime + 2^{XmtHoldoffExponent+4}. \qquad (13.3)$$

After building a set for the specific node, a pseudorandom mixing function calculates a pseudorandom MIX value for each node. If the specific node generates the biggest MIX value, it wins the competition and the next transmission time (NextXmtTime) is set as TempXmtTime. Then the node broadcasts to its neighbors in the MSH-NCFG message. Otherwise, the specific node fails in competing for this slot. The node sets the TempXmtTime as the next transmission slot and repeats the competing procedure until it wins.

The design of the distributed election scheduling algorithm takes into account distribution, fairness, and robustness. In terms of the distributed algorithm, this protocol requires no centralized control for coordinating transmission opportunity allocation. Fairness refers to the strategy whereby the algorithm treats all nodes equally in competing for transmission opportunities. For robustness, the seed in the pseudo-random algorithm varies in each frame and this mechanism is able to resolve the persisting collision.

13.4 Sleep Mode in the IEEE 802.16e

Amendment 802.16e [3] adds the mobility component for WiMAX and defines both the PHY and MAC layers for combined fixed and mobile operations in the licensed bands. Due to the promising mobility capability in IEEE 802.16e, the mechanism for efficiently managing limited energy is becoming very significant, since an MSS is generally powered by a battery. For this, sleep mode operation has recently been specified in the MAC protocol [3,6].

Figure 13.4 shows the sleep mode message sequence between the BS and MSS. Before entering the sleep mode, the MSS sends a request message MOB-SLP-REQ to the BS for the permission to transit into sleep mode. Upon receiving the request, the BS replies with the response message MOB-SLP-RSP. This response message indicates the parameters initial-sleep window (T_{min}), final-sleep window (T_{max}), and listening window (L). Upon receiving the MOB-SLP-RSP, the MSS enters into the sleep mode.

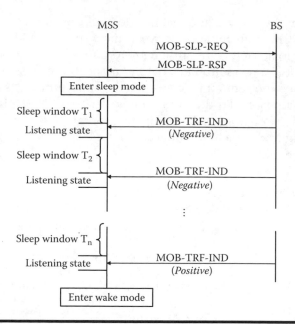

Figure 13.4   Sleep mode operation sequence in IEEE 802.16e.

Now we focus on the mechanism in the sleep mode. The duration of the first sleep interval T_1 is equal to the initial-sleep window T_{min}. After the first sleep interval, the MSS transits into listening state and listens to the traffic indication message MOB-TRF-IND broadcasting from the BS. The duration of the listening state is fixed as the parameter L indicated in the message MOB-SLP-RSP. The message MOB-TRF-IND indicates whether there has been traffic addressed to the MSS during its sleep interval T_1. If MOB-TRF-IND shows a negative indication, then the MSS continues its sleep mode after the listening interval L. Otherwise (i.e., the message MOB-TRF-IND indicates a positive indication) the MSS will return to the wake mode. We call the sleep interval and its subsequent listening interval a "cycle."

If the MSS continues in the sleep mode, the next sleep window starts from the end of the previous listening window, and it doubles the preceding sleep interval. This process is repeated as long as the sleep interval does not exceed the final-sleep window T_{max}. When the MSS reaches T_{max}, it fixes the sleep interval at T_{max}. That is, the duration of the sleep interval in the nth cycle is given by

$$T_n = \begin{cases} T_{min}, & n = 1 \\ \min(2^{n-1}T_{min}, T_{max}), & n > 1. \end{cases} \tag{13.4}$$

Alternatively, Figure 13.5 shows the wake mode and sleep mode of an MSS. The MSS alternately stays in wake mode and sleep mode during its lifetime. In case there are packets from the BS to a sleeping MSS, the packets are buffered in the BS and the MSS exits the sleep mode in the next listening interval. Different sleeping mechanisms may lead to substantially different consumed energy. Hence comprehensively evaluating the performance trade-off in a sleeping mode becomes significant. In the following three sections we characterize the standardized sleep mechanism and also perform an extensive analysis of the energy-saving scheme in terms of energy consumption and packet delay.

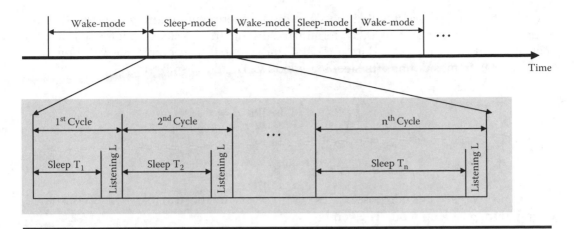

Figure 13.5   Wake mode and sleep mode transition in IEEE 802.16e.

13.5   Energy Consumption Analysis with Downlink Traffic

In this section we analyze the energy consumption during the MSS sleeping mode when only considering DL traffic. Here, DL traffic refers to the packets/frames from the BS to the MSS; UL traffic refers to the packets/frames from the MSS to the BS.

We assume that the packets addressed to the MSS follow Poisson processes with rate λ. Then, the interarrival time of the DL frame follows the exponential distribution with mean $1/\lambda$. Let e_j denote the event that there is at least one DL frame during the jth sleep window plus its preceding listening window. Then, $\Pr(e_j = \text{false})$ represents the event that there is no DL frame during the jth sleep window plus its preceding listening window:

$$\Pr(e_j = \text{false}) = e^{-\lambda(T_j+L)}; \quad j = 1, 2, \cdots \tag{13.5}$$

Note that, in principle, the expression for the first sleep window $\Pr(e_1 = \text{false})$ is a little different since there is no preceding listening window before the first sleeping window. However, since the listening window L is sufficiently small, it is acceptable to express the probability $\Pr(e_j = \text{false})$ for all sleeping windows in a similar form.

Let E_S and E_L denote the consumed energy units per unit time in the sleep interval and listening interval, respectively. As shown in Figure 13.5, we suppose that the sleep mode is terminated during the nth sleeping interval. This supposition implies that there are DL packets during the nth sleeping interval and temporay buffering in the BS. Upon the nth sleeping completion, the MSS transits into the listening state and receives the broadcast message MOB-TRF-IND with positive indication, and consequently terminates the sleep mode. Thus the energy consumption is the sum of the consumed energy during all sleeping intervals and the consumed energy during all listening windows. Denote E_n as the consumed energy provided that the sleeping mode is terminated during the nth cycle. Then, E_n is given by

$$E_n = \sum_{j=1}^{n} T_j E_S + nL\,E_L. \tag{13.6}$$

For the sake of presentation, we denote

$$W_n = \sum_{j=1}^{n} (T_j + L). \tag{13.7}$$

Let ϕ_n denote the probability that the sleep mode is terminated during the nth sleeping interval. The situation that an MSS is terminated during the nth sleeping interval indicates that there are no DL frames during the 1st, 2nd, $\cdots$, $n-1$ sleeping interval; however, there are DL frames during nth sleeping window. Then, ϕ_n is given by

$$\phi_n = \prod_{j=1}^{n-1} \Pr(e_j = \text{false})\Pr(e_n = \text{true})$$

$$= \left[1 - e^{-\lambda(T_n+L)}\right]e^{-\lambda W_{n-1}}. \tag{13.8}$$

Taking into account all possibilities with respect to the variable n, the consumed energy during a sleep mode is given by

$$\text{Energy} = \sum_{n=1}^{\infty} E_n \phi_n$$

$$= \sum_{n=1}^{\infty} \left[\sum_{j=1}^{n} T_j E_S + nL E_L\right] \left[1 - e^{-\lambda(T_n+L)}\right]e^{-\lambda W_{n-1}}. \tag{13.9}$$

13.6 Energy Consumption Analysis with Downlink and Uplink Traffic

In case there are downlink frames to a sleeping MSS, the MSS exits the sleep mode in the next listening interval. In contrast, if there are UL frames during the sleep interval, the sleep mode is terminated immediately. It is thus necessary and reasonable to differentiate traffic in determining sleep mode duration. For instance, Figure 13.6 shows a scenario with $T_{\min} = L = 1$ and $T_{\max} = 2^9 T_{\min} = 512$. No DL or UL frames arrive during the 1st, 2nd, $\cdots$, 9th cycle. During the 10th sleep interval, there are a number of downlink frames, and an UL frame arrives in the first slot. In case the frame directions are not differentiated, all these frames are temporarily stored until the sleep mode is terminated

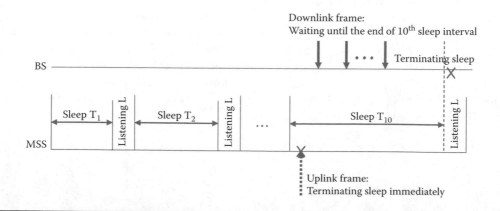

Figure 13.6   Motivation.

in the 10th listening interval. The duration of the sleep mode is $(T_1 + L) + (T_2 + L) + \cdots + (T_{10} + L) = 1033$. However, due to the UL frame, the sleep mode is actually terminated immediately at the first slot of the 10th sleep window. Hence the actual period of the sleep mode is $(T_1 + L) + (T_2 + L) + \cdots + (T_9 + L) + 1 = 521$. The overestimated error is $(1033 - 521)/521 = 98.3\%$. As a result, it is necessary to distinguish the downlink and UL traffic. This example motivates our study for differentiating traffic and performing energy management evaluation.

In this section we analyze energy consumption by considering both the downlink and uplink frames, since the instants of terminating sleep mode for downlink or uplink traffic are different.

We assume that the downlink and the uplink frames addressing the MSS follow Poisson processes with rates λ_d and λ_u, respectively. Then the interarrival time of the uplink frame, t_u, follows the exponential distribution. Let $\lambda = \lambda_d + \lambda_u$ be the total arrival rate to the MSS. Let e_j denote the event that there is at least one downlink frame during the jth sleep window plus its preceding listening window. Then we have

$$\Pr(e_j = \text{false}) = e^{-\lambda_d(T_j + L)}; \quad j = 1, 2, \cdots. \tag{13.10}$$

Under the condition that the MSS terminates the sleep mode during the nth cycle, we distinguish three possibilities, as shown in Figure 13.7. Accordingly, we denote $E_n^{(k)}$ as

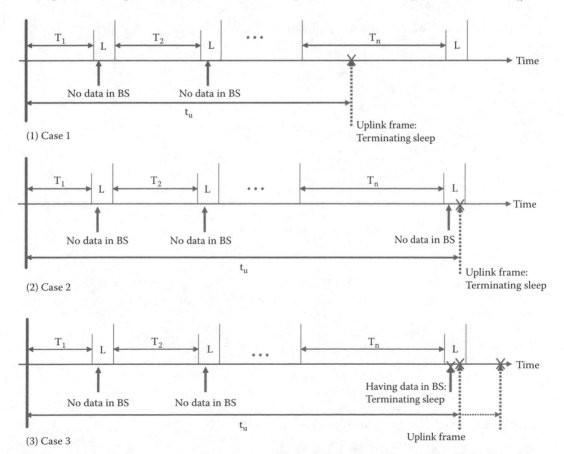

Figure 13.7   System model.

the consumed energy during the case k ($k = 1, 2, 3$), provided that the MSS terminates the sleep mode in the nth cycle. In addition, let $\phi_n^{(k)}$ be the probability for case k ($k = 1, 2, 3$) under the condition that the MSS terminates the sleep mode in the nth cycle. Considering all possibilities and employing the total probability theorem, the consumed energy during a sleep mode is thus expressed as

$$\text{Energy} = \sum_{n=1}^{\infty} \sum_{k=1}^{3} \overline{E_n^{(k)}} \phi_n^{(k)}, \tag{13.11}$$

where $\overline{X}$ denotes the average value of the random variable X.

13.6.1 Case 1 in Figure 13.7

In the first case, there is an UL frame terminating the sleep mode during the nth sleep interval. This implies that there are no packets during the jth ($j = 1, 2, \cdots, n-1$) sleep interval. In addition, t_u should be greater than $\sum_{j=1}^{n-1} T_j + (n-1)L$ and concurrently less than $\sum_{j=1}^{n} T_j + (n-1)L$. Similarly, for the sake of presentation, we denote

$$W_n = \sum_{j=1}^{n} (T_j + L). \tag{13.12}$$

In this situation, we define an alternative random variable t_u' for the variable t_u satisfying

$$t_u' = \begin{cases} t_u, & W_{n-1} < t_u < W_{n-1} + T_n \\ 0, & \text{otherwise.} \end{cases} \tag{13.13}$$

Then the average value of t_u' is given by

$$\overline{t_u'} = \frac{\int_{W_{n-1}}^{W_{n-1}+T_n} \lambda_u x e^{-\lambda_u x} dx}{\int_{W_{n-1}}^{W_{n-1}+T_n} \lambda_u e^{-\lambda_u x} dx}$$

$$= \frac{\left(W_{n-1} + \frac{1}{\lambda_u}\right)(1 - e^{-\lambda_u T_n}) - T_n e^{-\lambda_u T_n}}{1 - e^{-\lambda_u T_n}}. \tag{13.14}$$

Summarizing the energy consumption during the sleeping window and the listening window, we express the energy consumption for the first case as

$$E_n^{(1)} = \sum_{j=1}^{n-1} T_j E_S + (n-1)L E_L + \left[t_u' - \sum_{j=1}^{n-1} T_j - (n-1)L \right] E_S$$

$$= [t_u' - (n-1)L] E_S + (n-1)L E_L. \tag{13.15}$$

Hence the average energy consumption is given by

$$\overline{E_n^{(1)}} = [\overline{t_u'} - (n-1)L] E_S + (n-1)L E_L, \tag{13.16}$$

where $\overline{t_u'}$ is given in Equation (13.14).

The probability for the first situation is given by

$$\phi_n^{(1)} = \Pr(W_{n-1} < t_u < W_{n-1} + T_n) \prod_{j=1}^{n-1} \Pr(e_j = \text{false})$$

$$= \int_{W_{n-1}}^{W_{n-1}+T_n} \lambda_u e^{-\lambda_u x} dx \prod_{j=1}^{n-1} \Pr(e_j = \text{false})$$

$$= e^{-(\lambda_u + \lambda_d)W_{n-1}}(1 - e^{-\lambda_u T_n}). \tag{13.17}$$

13.6.2 Case 2 in Figure 13.7

In the second case, there is an UL frame terminating the sleep mode during the nth listening interval. This indicates that there are no packets during the jth$(j = 1, 2, \cdots, n)$ sleep interval. In addition, t_u should be greater than $W_{n-1} + T_n \,(= W_n - L)$ and also less than W_n. In this case, we denote an alternative random variable t_u' for the variable t_u satisfying

$$t_u' = \begin{cases} t_u, & W_n - L < t_u < W_n \\ 0, & \text{otherwise.} \end{cases} \tag{13.18}$$

Then the average value of t_u' is given by

$$\overline{t_u'} = \frac{\int_{W_n-L}^{W_n} \lambda_u x e^{-\lambda_u x} dx}{\int_{W_n-L}^{W_n} \lambda_u e^{-\lambda_u x} dx}$$

$$= \frac{\left(W_n + \frac{1}{\lambda_u}\right)(e^{\lambda_u L} - 1) - L e^{-\lambda_u L}}{e^{\lambda_u L} - 1}. \tag{13.19}$$

Summarizing the energy consumption during the sleep window and the listening window, we express the energy consumption for the second case as

$$E_n^{(2)} = \sum_{j=1}^{n} T_j E_S + (n-1)L E_L + \left[t_u' - \sum_{j=1}^{n} T_j - (n-1)L \right] E_L$$

$$= \sum_{j=1}^{n} T_j E_S + \left[t_u' - \sum_{j=1}^{n} T_j \right] E_L. \tag{13.20}$$

Hence the average energy consumption is given by

$$\overline{E_n^{(2)}} = \sum_{j=1}^{n} T_j E_S + \left[\overline{t_u'} - \sum_{j=1}^{n} T_j \right] E_L, \tag{13.21}$$

where $\overline{t_u'}$ is given in Equation (13.19).

The probability for the second situation is given by

$$\phi_n^{(2)} = \Pr(W_n - L < t_u < W_n) \prod_{j=1}^{n} \Pr(e_j = \text{false})$$

$$= \int_{W_n-L}^{W_n} \lambda_u e^{-\lambda_u x} dx \prod_{j=1}^{n} \Pr(e_j = \text{false})$$

$$= e^{-(\lambda_u + \lambda_d) W_n}(e^{\lambda_u L} - 1). \tag{13.22}$$

13.6.3   Case 3 in Figure 13.7

As shown in the third illustration in Figure 13.7, there are DL packets during the nth sleeping interval and temporary buffering in the BS. Moreover, during this sleep interval T_n, there are no UL frames. Upon the nth sleeping completion, the MSS transits into the listening state and receives the broadcast message MOB-TRF-IND with positive indication, and then terminates the sleep mode. Thus, from the starting moment to the stopping instant of sleeping, the sum of all sleeping windows is $\sum_{j=1}^{n} T_j$ and there are n listening windows. The consumed energy is then given by

$$E_n^{(3)} = \overline{E_n^{(3)}} = \sum_{j=1}^{n} T_j E_S + nL E_L. \tag{13.23}$$

The probability for this situation is given by

$$\phi_n^{(3)} = \Pr(t_u > \sum_{j=1}^{n} T_j + (n-1)L) \cdot \prod_{j=1}^{n-1} \Pr(e_j = \text{false})\Pr(e_n = \text{true})$$

$$= \int_{W_{n-1}+T_n}^{\infty} \lambda_u e^{-\lambda_u x} dx \cdot \prod_{j=1}^{n-1} \Pr(e_j = \text{false})\Pr(e_n = \text{true})$$

$$= e^{-\lambda_u T_n}[1 - e^{-\lambda_d(T_n+L)}]e^{-(\lambda_d+\lambda_u) W_{n-1}} \tag{13.24}$$

Substituting Equations (13.16–13.24) into Equation (13.11), we can calculate the power consumption during the sleep mode.

Figure 13.8 shows the energy consumption in terms of the ratio λ_u/λ_d with fixed $\lambda = 0.05$. The comparison shows that the consumed energy varies with different frame directions. This is because with more uplink frames (or external operations), the sleep mode is terminated instantaneously by such frames with higher probability, leading to shorter sleep mode duration and consequently less energy consumption during the sleep mode. This will eventually result in higher energy consumption during the MSS's lifetime.

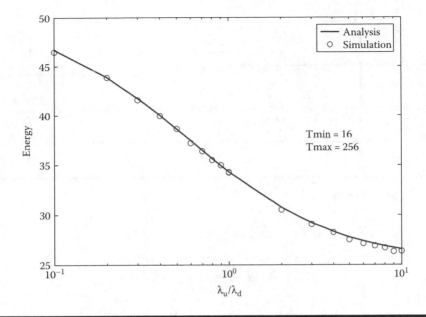

Figure 13.8   Energy consumption ($\lambda = 0.05$, $E_S = 1$, $E_L = 10$, and $L = 1$).

13.7   Energy Consumption Analysis with Generalized Traffic Process

In the previous sections, Poissons assumptions have been employed for the data packet traffic for the sake of analytical simplicity and tractability. In other words, an exponential distribution function is assumed for packet interarrival time. However, recently it has been reported that the exponential distribution may be inappropriate for data traffic [7].

In this section we will develop an analytical model and perform an extensive analysis of the specified sleep mode in the IEEE 802.16e WiMAX. The system model is derived on the basis of the generalized traffic arrival process instead of Poisson assumptions. In addition, the performance metrics are developed to investigate the trade-off in the energy management strategy with respect to energy consumption and packet delay during the sleep mode. Due to the significantly increased complexities of generalizing the traffic process, we will study the situation when only the DL traffic is considered. We leave the scenario of considering the effects of the DL and UL simultaneously, as in the previous section, as a future topic.

Figure 13.9 shows the system model for analyzing energy consumption and packet delay. We denote t_d as the packet interarrival time with mean $1/\lambda$, probability density function (pdf) $f_{t_d}(t)$, and cumulative distribution function (cdf) $F_{t_d}(t)$. To preserve universal approximation, general applicability, and also easy calculation, we suppose that t_d follows the hyper-Erlang distribution. The hyper-Erlang distribution has been proven to be able to arbitrarily closely approximate the distribution of any positive random variable as well as measured data [8]. In particular, its general approximation property has been applied in different scenarios (e.g., Marsan et al. [9]).

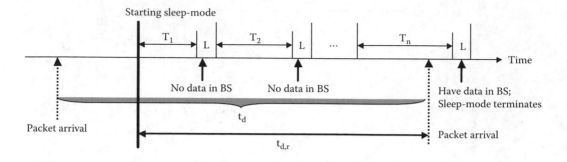

Figure 13.9   System model.

For the hyper-Erlang distributed t_d with pdf [8],

$$f_{t_d}(t) = \sum_{i=1}^{H} \alpha_i \frac{(m_i \lambda_i)^{m_i} t^{m_i-1}}{(m_i - 1)!} e^{-m_i \lambda_i t}, \tag{13.25}$$

where $\sum_{i=1}^{H} \alpha_i = 1$, H and m_i are positive integers, and α_i and λ_i are positive real constants with $0 \leq \alpha_i \leq 1, \lambda_i \geq 0$. The parameters H, α_i, and m_i in Equation (13.25) determine the shape and scale of the specific distribution.

Furthermore, for a real number α, integer m, and time t, we have the following formula:

$$\int_0^t \frac{\alpha^m x^{m-1}}{(m-1)!} e^{-\alpha x} dx = 1 - \sum_{j=0}^{m-1} \frac{(\alpha t)^j}{j!} e^{-\alpha t}. \tag{13.26}$$

Applying Equation (13.26), we calculate the cdf of hyper-Erlang t_d as

$$F_{t_d}(t) = \int_0^t f_{t_d}(x) dx$$

$$= 1 - \sum_{i=1}^{H} \alpha_i \sum_{j=0}^{m_i-1} \frac{(m_i \lambda_i t)^j}{j!} e^{-m_i \lambda_i t}. \tag{13.27}$$

We define the packet residual interarrival time as the time from an intermediate moment between two consecutive packets to the arrival time of the immediate next packet. Let $t_{d,r}$ denote the packet residual interarrival time with pdf $f_{t_{d,r}}(t)$ and cdf $F_{t_{d,r}}(t)$. Referring to the residual life theorem [8], we express the $t_{d,r}$ pdf as

$$f_{t_{d,r}}(t) = \lambda \int_t^\infty f_{t_d}(x) dx = \lambda [1 - F_{t_d}(t)]. \tag{13.28}$$

Substituting Equation (13.25) into Equation (13.28), we obtain the pdf of $t_{d,r}$ in the case of the hyper-Erlang t_d:

$$f_{t_{d,r}}(t) = \lambda \sum_{i=1}^{H} \alpha_i \sum_{j=0}^{m_i-1} \frac{(m_i \lambda_i t)^j}{j!} e^{-m_i \lambda_i t}. \tag{13.29}$$

In addition, the cdf of $t_{d,r}$ is calculated as

$$F_{t_{d,r}}(t) = \int_0^t f_{t_{d,r}}(x)dx$$

$$= 1 - \lambda \sum_{i=1}^{H} \frac{\alpha_i}{m_i \lambda_i} \times \sum_{j=0}^{m_i-1} \sum_{k=0}^{j} \frac{(m_i \lambda_i t)^k}{k!} e^{-m_i \lambda_i t}. \tag{13.30}$$

13.7.1 *Energy Consumption*

Let N denote the random variable representing the number of cycles an MSS experiences when a sleep mode terminates. In Figure 13.9, we suppose that the sleep mode is terminated during the nth cycle. This implies that there are no packets during the 1st, 2nd, $\cdots$, $n-1$ cycles, and there are packets during the nth sleep window. Under the condition that the MSS terminates the sleep mode during the nth cycle, we denote E_n as the average energy consumption provided that the MSS terminates the sleep mode in the nth cycle. Consequently the consumed energy during a sleep mode is expressed as

$$\text{Energy} = \sum_{n=1}^{\infty} \Pr(N = n)E_n. \tag{13.31}$$

From the starting moment to the stopping instant of sleeping, the sum of all sleeping windows is $\sum_{j=1}^{n} T_j$ and there are $n-1$ listening windows. The energy consumption E_n is the sum of the consumed energy in the sleep window and in the listening window:

$$E_n = \left[\sum_{j=1}^{n} T_j \right] E_S + (n-1)L E_L. \tag{13.32}$$

Now we will develop the probability distribution for the random variable N. Analogously, we denote W_n as the sum of the 1st, 2nd, $\cdots$, nth sleep window and listening window:

$$W_n = \sum_{j=1}^{n} (T_j + L); \quad n \geq 1, \tag{13.33}$$

with $W_0 = 0$.

The situation where $N = 1$ implies that there is packet arrival during T_1, or equivalently, the interval $t_{d,r}$ is shorter than the first sleep window. Hence we express

$$\Pr(N = 1) = \Pr(t_{d,r} \leq T_1)$$

$$= \int_0^{T_1} f_{t_{d,r}}(t)dt$$

$$= F_{t_{d,r}}(T_1)$$

$$= 1 - \lambda \sum_{i=1}^{H} \frac{\alpha_i}{m_i \lambda_i} \sum_{j=0}^{m_i-1} \sum_{k=0}^{j} \frac{(m_i \lambda_i T_1)^k}{k!} e^{-m_i \lambda_i T_1}. \tag{13.34}$$

For the case $N = n$ $(n \geq 2)$, there is no packet arrival during the 1st, 2nd, $\cdots$, $n-1$ cycle, but there are packets during T_n. This is equivalent with the event where interval $t_{d,r}$ is shorter than $W_n - L$ but longer than $W_{n-1} - L$. Hence we have

$$\Pr(N = n)$$

$$= \Pr(W_{n-1} - L < t_{d,r} \leq W_n - L)$$

$$= \int_{W_{n-1}-L}^{W_n-L} f_{t_{d,r}}(t)dt$$

$$= F_{t_{d,r}}(W_n - L) - F_{t_{d,r}}(W_{n-1} - L)$$

$$= \lambda \sum_{i=1}^{H} \frac{\alpha_i}{m_i \lambda_i} \sum_{j=0}^{m_i-1} \sum_{k=0}^{j} \frac{(m_i \lambda_i)^k}{k!} e^{m_i \lambda_i L}$$

$$\times \left[(W_{n-1} - L)^k e^{-m_i \lambda_i W_{n-1}} - (W_n - L)^k e^{-m_i \lambda_i W_n} \right]. \tag{13.35}$$

Substituting Equation (13.32), Equation (13.34), and Equation (13.35) into Equation (13.31), we obtain the energy consumption. The simple form of hyper-Erlang summation leads to the summation expression of performance metrics and facilitates computation complexity.

13.7.2 *Packet Delay*

For trade-off analysis, a longer sleep mode is able to save more energy and lengthen MSS serving time. However, the longer sleep mode adversely incurs greater packet waiting time in the BS. Let $\mathcal{D}$ denote the delay of a packet. In addition, let $\overline{\mathcal{D}}$ denote the average delay.

If a packet arrives at the MSS during the first sleep window, the packet has to wait in the BS until the first listening window. Alternatively, when the residual packet interarrival time $t_{d,r}$ is shorter than T_1, the delay is equal to the difference between $t_{d,r}$ and T_1. In this situation, the delay $\mathcal{D}$ is expressed as

$$\mathcal{D} = T_1 - t_{d,r}; \quad \text{if } t_{d,r} \leq T_1. \tag{13.36}$$

Under this specific condition, we define an alternative random variable v_1 for $t_{d,r}$ as

$$v_1 = \begin{cases} t_{d,r}, & t_{d,r} \leq T_1 \\ 0, & \text{otherwise.} \end{cases} \tag{13.37}$$

Then the average value of v_1 is given by

$$\overline{v_1} = \frac{\int_0^{T_1} x f_{t_{d,r}}(x)dx}{\int_0^{T_1} f_{t_{d,r}}(x)dx}$$

$$= \frac{A}{F_{t_{d,r}}(T_1)}, \tag{13.38}$$

where

$$A = \int_0^{T_1} x f_{t_{d,r}}(x) dx$$

$$= \int_0^{T_1} x \left[\lambda \sum_{i=1}^{H} \alpha_i \sum_{j=0}^{m_i-1} \frac{(m_i \lambda_i x)^j}{j!} e^{-m_i \lambda_i x} \right] dx$$

$$= \lambda \sum_{i=1}^{H} \alpha_i \sum_{j=0}^{m_i-1} \frac{(m_i \lambda_i)^j}{j!} \int_0^{T_1} x^{j+1} e^{-m_i \lambda_i x} dx. \qquad (13.39)$$

For the sake of illustration for Equation (13.39), we define

$$\Theta_n(a, b; \beta) = \int_a^b x^n e^{-\beta x} dx, \qquad (13.40)$$

where a, b, and β are real numbers and n is a nonnegative integer. The recursive algorithm for calculating $\Theta_n(a, b; \beta)$ is derived and presented as follows:

$$\Theta_n(a, b; \beta) = \frac{1}{n+1} \int_a^b e^{-\beta x} dx^{n+1}$$

$$= \frac{1}{n+1} \left[e^{-\beta x} x^{n+1} \Big|_a^b + \beta \int_a^b x^{n+1} e^{-\beta x} dx \right]$$

$$= \frac{1}{n+1} \{ [b^{n+1} e^{-\beta b} - a^{n+1} e^{-\beta a}] + \beta \Theta_{n+1}(a, b; \beta) \}. \qquad (13.41)$$

Reformatting Equation (13.41), we develop the recursive algorithm as follows:

$$\Theta_n(a, b; \beta) = \frac{1}{\beta} \left[n \Theta_{n-1}(a, b; \beta) - (b^n e^{-\beta b} - a^n e^{-\beta a}) \right],$$

with initial condition $\Theta_0(a, b; \beta) = \frac{1}{\beta}(e^{-\beta a} - e^{-\beta b})$. Then, the result for Equation (13.39) is expressed as

$$A = \lambda \sum_{i=1}^{H} \alpha_i \sum_{j=0}^{m_i-1} \frac{(m_i \lambda_i)^j}{j!} \Theta_{j+1}(0, T_1; m_i \lambda_i). \qquad (13.42)$$

If a packet arrives at the MSS during the nth sleep window, the packet has to wait in the BS until the next listening window. Equally, when the residual packet interarrival time $t_{d,r}$ is shorter than $W_n - L$, but longer than $W_{n-1} - L$, the delay is equal to the difference between $t_{d,r}$ and $W_n - L$. In this situation, the delay $\mathcal{D}$ is expressed as

$$\mathcal{D} = W_n - L - t_{d,r}; \qquad \text{if } W_{n-1} - L < t_{d,r} \leq W_n - L. \qquad (13.43)$$

For this situation, we define an alternative random variable v_n for the variable $t_{d,r}$ in the range as

$$v_n = \begin{cases} t_{d,r}, & W_{n-1} - L < t_{d,r} \leq W_n - L \\ 0, & \text{otherwise.} \end{cases} \qquad (13.44)$$

Then the average value of v_n is given by

$$\overline{v_n} = \frac{\int_{W_{n-1}-L}^{W_n-L} x f_{t_{d,r}}(x) dx}{\int_{W_{n-1}-L}^{W_n-L} f_{t_{d,r}}(x) dx}$$

$$= \frac{C_n}{F_{t_{d,r}}(W_n - L) - F_{t_{d,r}}(W_{n-1} - L)} \qquad (13.45)$$

where the C_n can be derived by following similar reasoning leading to A:

$$C_n = \int_{W_{n-1}-L}^{W_n-L} x f_{t_{d,r}}(x) dx$$

$$= \int_{W_{n-1}-L}^{W_n-L} x \left[\lambda \sum_{i=1}^{H} \alpha_i \sum_{j=0}^{m_i-1} \frac{(m_i \lambda_i x)^j}{j!} e^{-m_i \lambda_i x} \right] dx$$

$$= \lambda \sum_{i=1}^{H} \alpha_i \sum_{j=0}^{m_i-1} \frac{(m_i \lambda_i)^j}{j!} \Theta_{j+1}(W_{n-1} - L, W_n - L; m_i \lambda_i). \qquad (13.46)$$

Taking into account all possible N and the Equations (13.36) and (13.43), packet delay is given by

$$\mathcal{D} = (T_1 - v_1) \Pr(N = 1) + \sum_{n=2}^{\infty} (W_n - L - v_n) \Pr(N = n).$$

Taking the averaging operation on the two sides of the equation above, the average delay is given by

$$\overline{\mathcal{D}} = (T_1 - \overline{v_1}) \Pr(N = 1) + \sum_{n=2}^{\infty} (W_n - L - \overline{v_n}) \Pr(N = n)$$

$$= [T_1 \Pr(N = 1) - A] + \sum_{n=2}^{\infty} [(W_n - L) \Pr(N = n) - C_n], \qquad (13.47)$$

where A and C_n are given in Equation (13.42) and Equation (13.46), respectively.

Figure 13.10 shows the energy consumption and packet delay in terms of the packet arrival rate with different packet traffic processes. In this example, $T_{min} = 1$ and $T_{max} = 1024$. The various stage m represents different traffic processes, where $m = 1$ is equivalent with the Poisson arrival process. Simulation results are also presented to compare

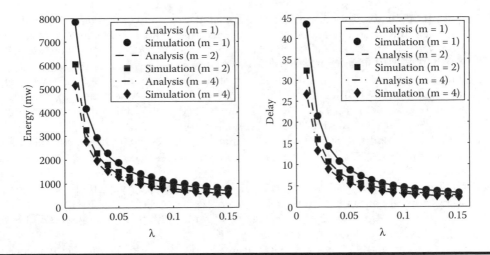

Figure 13.10   Energy consumption and packet delay in terms of packet arrival rate λ ($T_{\min} = 1$ and $T_{\max} = 1024$).

with the analytical model. It is clear that the simulation and the analysis match each other very well. The comparison shows that a smaller m leads to more energy consumption during sleep mode and longer packet delay. This can be explained as follows. A smaller m indicates larger variance in the traffic arrival process. In this case, more MSSs with longer t_d will be observed. This contributes to the longer sleeping period and consequently more energy consumption $\mathcal{E}$ and larger packet delay $\overline{\mathcal{D}}$ during sleep mode. This will eventually result in higher energy consumption during an MSS's lifetime.

13.8   Conclusion

This chapter has explained and analyzed the energy management mechanism in the IEEE 802.16e Mobile WiMAX:

- ■ WiMAX standard activities were briefly introduced and various versions were differentiated.
- ■ The standardized sleeping mode was explained in detail for efficiently managing the limited power in an MSS.
- ■ Three different scenarios were analyzed in order to evaluate energy-saving performance.

This chapter should be helpful for understanding the key procedures in the energy management framework of Mobile WiMAX. The performance evaluations in various situations provide potential approaches to efficiently utilize the very limited power in MSSs. More efficient schemes can be expected to lengthen the MSS lifetime and Mobile WiMAX resilience as well.

References

1. IEEE, IEEE Standard for Local and Metropolitan Area Networks—Part 16: Air Interface for Fixed Broadband Wireless Access Systems, Standard 802.16-2001, IEEE, Washington, DC, 2002.

2. IEEE, IEEE Standard for Local and Metropolitan Area Networks—Part 16: Air Interface for Fixed Broadband Wireless Access Systems Amendment 2: Medium Access Control Modifications and Additional Physical Layer Specifications for 2–11 GHz, Standard 802.16a-2003, IEEE, Washington, DC, 2003.

3. IEEE, IEEE Standard for Local and Metropolitan Area Networks—Part 16: Air Interface for Fixed and Mobile Broadband Wireless Access Systems—Amendment for Physical and Medium Access Control Layers for Combined Fixed and Mobile Operation in Licensed Bands, Standard 802.16e/D5-2004, IEEE, Washington, DC, 2004.

4. C. Eklund, R. Marks, K. Stanwood and S. Wang, IEEE standard 802.16: a technical overview of the WirelessMAN air interface for broadband wireless access, *IEEE Commun. Mag.*, 40(6), 98, 2002.

5. G. Nair, J. Chou, T. Madejski, K. Perycz, D. Putzolu and J. Sydir, IEEE 802.16 medium access control and service provisioning. *Intel Technol. J.*, 8(4), 213, 2004.

6. Y. Zhang and M. Fujise, Energy management in the IEEE 802.16e MAC, *IEEE Commun. Lett.*, 10(4), 311, 2006.

7. M.E. Crovella and A. Bestavros, Sel-similarity in World Wide Web traffic: evidence and possible causes, *IEEE/ACM Trans. Network.*, 5(6), 835, 1997.

8. L. Kleinrock, *Queueing Systems*, John Wiley & Sons, New York, 1975.

9. M.A. Marsan, G. Ginella, R. Maglione, and M. Meo, Performance analysis of hierarchical cellular networks with generally distributed call holding times and dwell times, *IEEE Trans. Wireless Commun.*, 3(1), 248, 2004.

Chapter 14

Link Adaptation Mechanisms in WirelessMAN

Gurkan Gur, Fatih Alagoz and Tuna Tugcu

Contents

In this chapter we introduce the link layer adaptation mechanisms currently available in wireless metropolitan area network (MAN) technologies including IEEE 802.16 (Worldwide Interoperability for Microwave Access [WiMAX]), HiperACCESS, HiperMAN, and Wireless Broadband (WiBro). First, we briefly describe the concept and the premise for link adaptation in wireless networks. Moreover, we discuss related research on link adaptation for wireless systems. Second, we present the relevant mechanisms for each broadband wireless access (BWA) system and highlight their core capabilities. Finally, we present some open issues from the perspective of link adaptation, considering the potential evolutionary directions of these systems. We mainly focus on WiMAX, compared to WiBro, HiperACCESS, and HiperMAN, since the latter specifications have adopted the former specification or a subset with the goal of global harmonization. (This work was supported in part by the State Planning Organization of Turkey [DPT] under Grant DPT03K120250, "Next Generation Satellite Networks and Applications.")

14.1   Introduction

There is an increasing need for the delivery of broadband digital communications services to individuals, households, and businesses of all sizes. While data rates supporting basic digitized voice services remain within the capability of existing wired networks, higher speeds are increasingly necessary. High-speed Internet and video are examples of services demanded by many users for a variety of business, household management, and leisure activities. Users increasingly need a combination of various services, with a flexible allocation of data rates to suit their immediate needs. Data rates can vary considerably between different applications that are traditionally delivered by a combination of several communications technologies. The adoption of WiMAX and similar broadband wireless access architectures is expected to address these issues and further push broadband penetration and services into the residential market. This goes for developed economies and especially for developing regions encumbered by legacy communications infrastructures. However, broadband wireless data transmission for these systems is a challenging task due to dynamic and adverse channel conditions in such environments. Multipath fading, shadowing, path loss, time-selective fading, and frequency-selective fading, among others, are some of the obstacles in wireless communications.

The basic rationale behind link adaptation is to optimize throughput by selecting among the set of available rates, as given by a set of modulation and coding schemes, the one that maximizes the throughput in each "short-term" channel state [1]. In other words, the communication system utilizes higher modulation levels and higher channel coding rates when the channel condition is favorable and lower modulation levels and lower channel coding rates when the channel condition is relatively harsh [2]. Therefore this capability is crucial to cope with adverse channel conditions, especially in wireless communication systems such as wirelessMANs.

In BWA systems, channel conditions can vary significantly due to propagation anomalies. It is therefore desirable to adapt the modulation and coding scheme to the channel conditions. While voice networks are designed to deliver a fixed bit rate, data services can be delivered at a variable rate. Voice networks are engineered to deliver a certain required bit rate at the edge of the cell, which constitutes the worst case. Most users, however, have more favorable channel conditions. Therefore data networks can take advantage of adaptive modulation and coding (AMC) to improve the overall throughput. In a typical adaptive modulation scheme, a dynamic variation in the modulation order (constellation size) and forward error correction (FEC) code rate is possible. In practice, the receiver feeds back information on the channel, which is then used to control the adaptation. Adaptive modulation can be used in both uplinks and downlinks. The adaptation can be performed in various ways such as user-specific only, user- and time-specific, or quality of service (QoS) dependent.

In this chapter, we present and discuss the link layer adaptation mechanisms currently available in wirelessMAN technologies including IEEE 802.16 WiMAX, Hiper ACCESS/MAN, and WiBro. This section has provided a brief introduction of wirelessMAN and link adaptation. Section 14.2 describes the link adaptation for wirelessMAN in more detail. Subsequently Section 14.3 presents a brief survey of prior work in the area of link adaptation. Section 14.4 through Section 14.7 elaborates on the link adaptation capabilities of WiMAX, HiperACCESS, HiperMAN, and WiBro, respectively. Section 14.8 focuses on crucial aspects of the link adaptation concept and provides open issues and some direction for further research on wirelessMAN link adaptation. Finally, Section 14.9 summarizes the significant phenomena and concludes the chapter.

14.2   Basic Concept of Link Adaptation

Link adaptation allows the communication system to operate around the dynamic optimal point over a time-varying channel, rather than optimally for the worstcase of a static channel model, using a trade-off between link robustness and bandwidth efficiency. Every attempt is made to sustain a connection, even if a reasonable amount of compromise has to be made with the link quality. In most systems that employ link adaptation, this ultimately results in a trade-off between link robustness and bandwidth efficiency as depicted in Figure 14.1. Link adaptation results in an ultimate increase in link capacity, in that it permits the system to make the most out of a time-varying channel instead of always operating optimally for a worst-case channel.

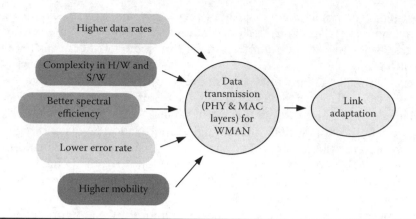

Figure 14.1 Design forces for link adaptation schemes.

There are four primary mechanisms by which a system can improve link robustness at the expense of overall bandwidth efficiency [3]:

- **Adaptive modulation:** The transmitter and receiver negotiate and decide on the most bandwidth-efficient signal constellation that can be used for the current channel condition. For M-ary quadrature amplitude modulation (MQAM) and M-ary phase shift keying (MPSK) constellations, bandwidth efficiency and robustness are inversely proportional. When the channel conditions deteriorate, bandwidth efficiency is sacrificed and a more robust, lower order modulation scheme is chosen.

- **Adaptive FEC:** The system decides to add overhead to the transmitted data in the form of FEC code words. This again buys robustness at the expense of bandwidth efficiency. Depending on the channel conditions, the strength of the FEC is determined. For instance, concatenated codes may be used to achieve fine-grain control over the level of coding applied on the data. Under excellent channel conditions, the system may choose not to use FEC and achieve maximum bandwidth efficiency.

- **Automatic repeat request (ARQ):** Some systems adopt packet retransmissions at the link layer. In this method, unacknowledged packets are assumed to be in error and are retransmitted. Retransmissions cause a reduction in overall throughput, but guarantee correct reception on subsequent trials. The system can switch the ARQ on and off, depending on the channel conditions. Many systems use an optimal combination of FEC and ARQ, called hybrid ARQ (HARQ), to achieve maximum bandwidth efficiency.

- **Optimal power and coding allocation:** Recently the multiple-input multiple-output (MIMO) technique, where both the transmitter and receiver are equipped with multiple antennas, has emerged as a cure for the multipath fading challenge [4]. The MIMO technique does not try to mitigate the adverse effect of the multipath fading channel. Instead, it tries to utilize the multipath richness environment in a smart way such that the received signal cannot be in deep fade. The reliability and spectral efficiency of the wireless communication are substantially improved. Therefore this technique brings another dimension to link adaptation schemes, which is optimal power allocation and coding (space-time coding) among parallel subchannels according to their respective conditions.

Typical link adaptation systems group the above parameters into "modes" or "profiles" and dynamically track a time-varying channel by switching between predefined profiles. In addition to defining suitable operation modes, a link adaptation algorithm needs to define the operation regions for each of these modes quantitatively. It also needs to define an appropriate metric that serves as a link quality indicator. Many practical adaptation algorithms use signal-to-noise ratio (SNR) ranges to define the operation regions for each mode. These SNR ranges are themselves determined based on the metric chosen as the quality indicator. Typical quality indicators may be bit error rate (BER), frame error rate, end-to-end delay, or throughput. In the physical layer, additional channel knowledge could be the absolute instantaneous channel information (i.e., the channel impulse response) or in an average form (i.e., the average SNR, the average channel covariance matrix, or the statistical information of the channel coefficients).

Figure 14.2 illustrates a simple example for implementing adaptive modulation studied in Proakis [5]. The system has a targeted BER of 10^{-5}. The system chooses one of three modulation schemes—64-quadrature amplitude modulation (QAM), 16-QAM, or quadrature phase-shift keying (QPSK)—according to the following rule based on the signal-to-noise plus interference ratio (SINR) value. If the received SINR is greater than or equal to the threshold for 64-QAM, the system will select 64-QAM for modulation. If the received SINR is in the region between the thresholds of 16-QAM and 64-QAM, the system will select 16-QAM for modulation. Similarly, if the received SINR is in the region between the QPSK and 16-QAM thresholds, the system will select QPSK for modulation.

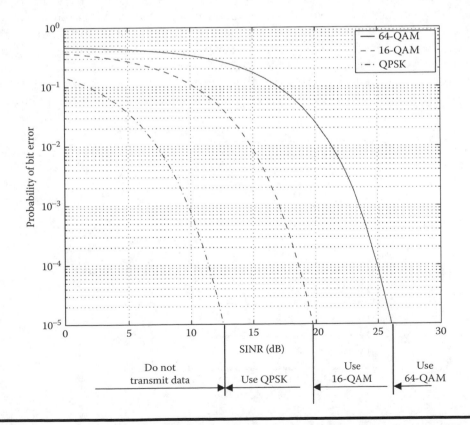

Figure 14.2   An illustration of the implementation of basic adaptive modulation based on SINR and modulation switching (assuming the noise plus the interference have Gaussian distribution).

If the received SINR is anywhere in the region below the threshold for QPSK, the system will not transmit any data until the next measured SINR becomes at larger than the threshold for QPSK (the smallest threshold). Currently, more degrees of freedom, such as error correcting code and ARQ, are added to this simple scheme for more elaborate link adaptation schemes in practical wireless systems.

14.2.1 Link Adaptation in Single-Antenna Systems

The capacity improvement in single-antenna systems offered by link adaptation (LA) over nonadaptive systems can be remarkable, as illustrated by Figure 14.3 [6]. In this figure, we represent the link-level spectral efficiency (SE) performance (b/s/Hz) versus the short-term average SNR γ in decibels, for four different uncoded modulation levels referred to as binary phase-shift keying (BPSK), QPSK, 16-QAM, and 64-QAM in a single-antenna system. The SE was obtained for each modulation by taking into account the corresponding maximum data rate and packet error rate (PER), which is a function of the short-term average SNR. The SE curves of two systems are represented by continuous and dashed lines. The first system is nonadaptive and constrained to use BPSK modulation only. The corresponding SE versus SNR is represented by the dashed curve. The second system uses adaptive modulation. Its corresponding SE is given by the envelope of the individual curves and is represented as the continuous line. Each modulation is optimal for use in different quality regions and LA selects the modulation with the highest SE for each link. The performance of the two systems is equal for SNRs up to 10 dB. However, for higher SNRs, the SE of the adaptive system is up to six times that of the nonadaptive system. When averaging the SE over the SNR range for a typical power-limited cellular scenario, the adaptive system provides a close to threefold gain over the nonadaptive system. The example in Figure 14.3 is ideal since it assumes that the modulation level is perfectly adapted to the short-term average SNR, and the probability of error as a function of the SNR is exactly known; for example, here an additive white

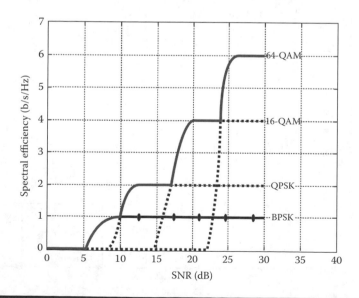

Figure 14.3   Spectral efficiency for various modulation levels as a function of short-term average SNR (extracted from Catreux et al. [6]. Reprinted with permission.).

Gaussian noise (AWGN) channel was considered, which corresponds to an instantaneous channel measurement. That assumption is true only for instantaneous feedback and is not practical due to delays in the feedback path. When there is a delay in the reverse path, more of the first- and higher-order statistics of the fading channel should be incorporated to improve the adaptation. Furthermore, other dimensions such as frequency and space (where different transmission schemes may be adapted) may yield further gains simply by providing additional degrees of freedom exploitable by LA.

Using combinations of adaptive modulation and adaptive coding is quite similar to using adaptive modulation alone. Figure 14.4 illustrates a variety of the possible combinations of adaptive modulation (QPSK, 16-QAM, and 64-QAM) and adaptive bit-interleaved coded modulation (BICM) (with code rates 1/2, 2/3, 3/4, and 7/8). When adaptive modulation is combined with adaptive coding, the network has more options to choose from than in the case of using adaptive modulation alone. An expected improvement in performance results from these combinations. These curves were obtained from simulations assuming that combinations of adaptive modulation (QPSK, 16-QAM, and 64-QAM) and adaptive BICM (code rates 1/2, 2/3, 3/4, 7/8, and 1) were used [7].

Figure 14.5 shows the effect of combining BICM with adaptive modulation throughout the sector of a base station (BS) as a function of the distance of user of interest from the BS [7]. All the interferers are placed in random positions. When BICM with code rate 1/2 is combined with adaptive modulation, the system throughput will improve in the further half of the BS. The system throughput will degrade as the user equipment is located closer to the BS. Near the BS, the combination of code rate 1/2 and adaptive

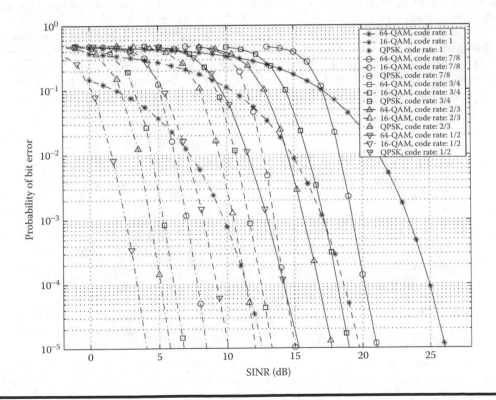

Figure 14.4 Probability of bit error for several combinations of code rates and modulations (assuming the noise plus the interference have Gaussian distribution) [7].

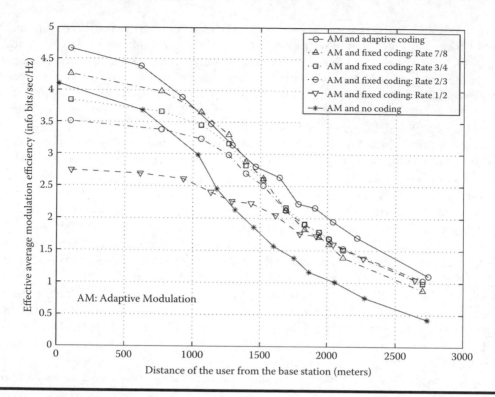

Figure 14.5   The impact of using combined adaptive modulation and adaptive coding on the transmission performance in a Rician fading channel [7]. Downlink transmission, f_c = 2.5 GHz, BER = 10^{-5}, Tx BW = 6 MHz, PTx = 200 mW, Rician parameter for the user of interest and the dominant interferer = 8, Rician parameter for the remaining interferers = 3, propagation exponent for the user and all the interferers =4, Std. Dev. of lognormal shadowing =7 dB, antenna beam width for the BS = 90°, antenna beam width for the user and all the interferers = 30°, antenna gains for the BS, the user of interest, and all the interferers: main lobe = 15 dB, side lobes = −10 dB.

modulation yields a throughput that is 34% less than adaptive modulation alone. The reason for this degradation is because code rate 1/2 reduces the received information bit by a factor of 1/2. The signal quality at this location is strong enough to be detected without losing 1/2 of the received bits for coding. A similar trend applies for BICM [8] with code rates 2/3 and 3/4. However, the amount of loss near the BS decreases for code rates 2/3 and 3/4 because of the reduction of redundant information required for them. When BICM with code rate 7/8 is combined with adaptive modulation, an improvement will be observed everywhere in the sector, with better improvement in the further half of the sector. The reason for this enhancement is because combining code rate 7/8 with adaptive modulation will reduce the required SINR to achieve a BER of 10^{-5}. The required redundant bits to implement this code (1/8) will not offset the advantage that is introduced by reducing the required SINR.

Research has shown the performance improvements and advantages provided by LA for wireless communications in single-antenna systems. However, the advent of MIMO, orthogonal frequency division multiplexing (OFDM), and space-time coding (STC) has brought new challenges and new opportunities for the LA paradigm, which constitute the subject of the next subsection.

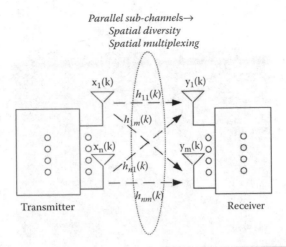

Figure 14.6 **MIMO channel. $y_i[k]$ is the kth subcarrier output for receive antenna i, $h_{ij}[k]$ is the kth subcarrier gain from the jth transmit antenna to the ith receive antenna, and $x_j[k]$ is the kth subcarrier input from antenna j.**

14.2.2 Link Adaptation in Multiple-Antenna Systems

There exist a multitude of reasons for using multiple-antenna systems. MIMO wireless systems use multiple-antenna elements at transmit and receive to offer improved capacity over single-antenna topologies in multipath channels, as shown in Figure 14.6. The most important technique in providing reliable communication over wireless channels is diversity, and of particular interest are the spatial diversity techniques. In such systems, the antenna properties as well as the multipath channel characteristics play a key role in determining communication performance. The reliability and spectral efficiency of wireless communication are therefore substantially improved.

For MIMO systems, the link adaptation can significantly enhance system peformance. In Nguyen et al. [4], several link adaptation algorithms for MIMO systems, assuming the availability of instantaneous channel coefficients (channel state information [CSI]) at both link ends, are investigated. Depending on the requirement either at the target average bit rate or the BER, the transmitted power is optimally distributed to each parallel subchannel according to its associated eigenvalue. Experimental results are illustrated in Figure 14.7. There are two important conclusions that can be drawn from the results. First, the optimum distribution of power does indeed provide better performance than the equal distribution scheme, especially at high SNR values. Second, for low SNR values, when the eigenvalues are not large enough or not evenly distributed (i.e., in a 4×4 setup), the proposed scheme gives a better BER. However, it comes at the cost of a lower number of activated subchannels. The results show that the average BER and the spectral efficiency of the spatial multiplexing MIMO system can be substantially improved using link adaptation.

In Forenza et al. [9], another adaptive transmission strategy is presented for MIMO systems, exploiting the long-term statistics of the channel. The channel model is defined as a zero-mean (i.e., K-factor of 0) correlated Rayleigh fading model, with angular spread in the range [28°, 55°] and six clusters, based on the IEEE 802.11n standard channel models. To predict the link-quality region for a given transmission, two link-quality metrics are considered: the average SNR and the relative condition number of the eigenvalues of the

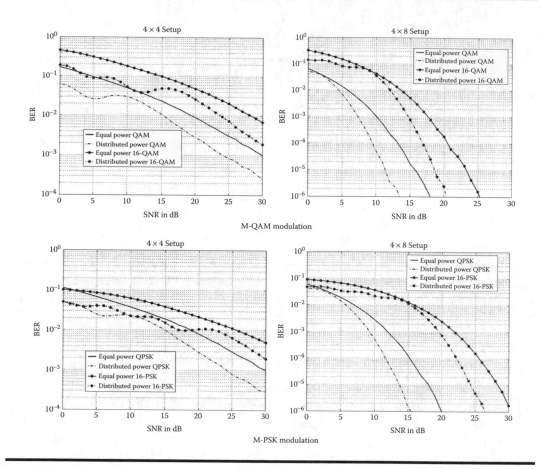

Figure 14.7   The BER of MIMO systems with and without optimum power distribution (*m* and *n* in *m* × *n* are the number of transmitter and receiver antennas, respectively) (extracted from Nguyen et al. [4]. Reprinted with permission.).

spatial correlation matrices. To enable transmissions over the wireless link, Forenza et al. [9] use a combination of modulation and coding schemes (MCSs) and practical MIMO transmit/receive techniques. They consider three common MIMO transmission schemes:

1. Beam forming (BF): with maximal ratio combining (MRC) receiver.
2. Double space-time transmit diversity (D-STTD): with minimum mean square error (MMSE) receiver
3. Spatial multiplexing (SM): with equal power allocation across the transmit antennas and MMSE receiver.

These schemes are important because they provide increasing data rates for a fixed error rate performance and for a fixed number of transmit/receive antennas. Moreover, they are being actively considered by different standards bodies such as the Third-Generation Partnership Project (3GPP) and IEEE 802.11n. Forenza et al. [9] defined eight combinations of modulation and coding schemes, according to the IEEE 802.11a standard. The combination of the three MIMO schemes with these eight MCSs results in a total of 24 different transmission modes. They selected a subset of 12 modes, according to the criterion of minimizing the SNR requirement for a given transmission rate.

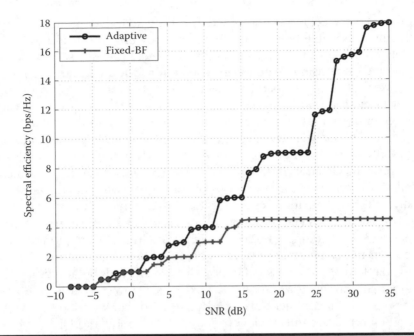

Figure 14.8 Spectral efficienc y for a single link with adaptive and fixed BF transmission techniques using channel model 1.

The proposed method adaptively selects the optimal transmission mode, which maximizes the throughput (or spectral efficiency) for a predefined target error rate, depending on the current channel condition. To enable this mode adaptation, the proposed algorithm estimates the link quality for the current transmission based on the average SNR information and the spatial selectivity indicator. These metrics are used to select the mode providing the highest throughput for the predefined target error rate. Figure 14.8 illustrates the resulting spectral efficiency for each of the transmit modes over a given range of SNRs. For high SNRs, a system employing the proposed adaptive algorithm produces a gain in spectral efficiency of 13.5 bps/Hz, compared to a system using fixed BF transmission with adaptive MCSs.

14.3   Related Research on Link Adaptation for Wireless Systems

Adapting the transmission parameters according to the variation in the channel conditions can significantly improve the overall performance of a system. The fundamental parameters that can be adapted are modulation type and index, coding depth, ARQ, and power allocation as indicated in Section 14.2. There is a plethora of research on various link adaptation strategies and their researchers have recommended efficient algorithms to improve system performance and capacity. Link adaptation has been accepted in many modern packet radio standards such as wideband code division multiple access (WCDMA), CDMA2000, and WiMAX as a key solution to increase spectral efficiency.

Recently there has been extensive research on link adaptation based on joint adaptive FEC modulation modes and OFDM systems. Adaptive modulation is described in Goldsmith and Chua [10] and Webb and Steele [11]. The presented results clearly demonstrate that adaptive modulation improves the BER performance in wireless channels that suffer from fading and shadowing. Goldsmith and Chua [10] propose a variable rate and variable power MQAM technique and prove that using adaptive modulation can provide a 5 to 10 dB gain over a fixed-rate system with only power control, and up to 20 dB gain over a completely nonadaptive system. Webb and Steele [11] use star MQAM constellations and conclude that using an adaptive scheme for SNR greater than 25 dB results in a 5 dB improvement over nonadaptive systems. Sampei et al. [12] propose an adaptive modulation scheme for a time-division multiple access (TDMA)-based cellular personal multimedia communications system using QPSK, 16-QAM, and 64-QAM. Their scheme achieves spectrum efficiencies that are 3.5 times higher than baseline nonadaptive systems using QPSK with a 10% outage probability. Goldsmith and Chua [13] apply coset codes to adaptive modulation in fading channels. The authors apply this methodology to a spectrally efficient adaptive MQAM to obtain trellis-coded adaptive MQAM. They present analytical and simulation results for this design which show an effective coding gain of 3 dB relative to uncoded adaptive MQAM for a simple four-state trellis code and an effective 3.6 dB coding gain for an eight-state trellis code. More complex trellis codes are shown to achieve higher gains. Moreover, they compare the performance of trellis-coded adaptive MQAM to that of coded modulation with built-in time diversity and fixed-rate modulation. The adaptive method exhibits a power savings of up to 20 dB.

Cameron et al. [14] recommend using adaptive modulation, adaptive FEC, and ARQ to optimize a link for maximum efficiency. They define "modulation gain" as the increase in capacity using a given modulation relative to that of binary modulation. Their results show that adaptive modulation with ARQ and no FEC performs better than the same system with FEC. For real-time services, they recommend using FEC over ARQ. Britto and Bonatti [15] analytically compare adaptive modulation, adaptive FEC, and adaptive ARQ. They conclude that adaptive FEC schemes generally perform well in all SNR ranges, but are inferior to adaptive modulation when the SNR is very low. They propose a hybrid scheme involving both adaptive FEC and adaptive modulation, which performs better in all SNR ranges with regard to throughput and delay parameters. Akyildiz et al. [16] propose an adaptive FEC scheme for wireless Asynchronous Transfer Mode (ATM) networks using Reed-Solomon coding. Their results demonstrate significant performance improvements of up to 10^{-6} BER from 10^{-2} BER prior to coding.

Although optimality can be ensured through the careful design of link adaptation for a fixed link configuration, due to changes in available power, data rate, and especially channel statistics, it is better to adapt the adaptation scheme dynamically in real time. Tang et al. [17] call this technique adaptive link adaptation (ALA) and investigate the problem of adapting the adaptation scheme online. They give the theoretic basis of ALA, and present a simple, yet working implementation of the update algorithm for ALA. Both theoretical and simulation results indicate the viability of ALA, and an 18 dB gain can be achieved for the investigated simple scenario. They state that the proposed ALA provides a promising technique to substantially increase the spectral efficiency of practical communication systems with low complexity, optimality, and generality.

Adaptive modulation and coding techniques are among the enabling technologies included in the standards for the third-generation (3G) wireless systems [18]. In particular, 3G systems support adaptive schemes on the forward link that will achieve peak data rates of up to 5 Mbps. High-speed downlink packet access (HSDPA) has been proposed for WCDMA to achieve data rates of up to 10 Mbps [2]. HSDPA recommends using two schemes for PHY layer link adaptation. One is AMC, and the other is HARQ. In Nakamura et al. [2], the authors propose a realistic and efficient method to adaptively control the threshold values for the HSDPA AMC algorithm. Blogh and Hanzo [19] propose employing adaptive QAM in conjunction with adaptive antennas and multiuser detectors to improve the average throughput in the universal terrestrial radio access (UTRA) frequency division duplex (FDD) system. Nanda et al. [20] elaborate on various rate adaptations possible in the different cellular communications standards as a result of AMC. They discuss a wide range of second-generation (2G) and 3G standards including IS-95, IS-136, general packet radio service (GPRS), enhanced GPRS (EGPRS), CDMA2000, and wideband code division (WCDMA). Yang et al. [21] address the application of AMC for 3G wireless systems. They propose a new method for selecting the appropriate MCSs according to the estimated channel condition based on a statistical decision-making approach to maximize the average throughput while maintaining an acceptable frame error rate (FER). Their AMC scheme uses a first-order, finite-state Markov model to approximate the time variations of the average channel SNR in subsequent frames. The proposed scheme is compared to the conventional techniques that use memoryless threshold-based decision making, and numerical results are presented showing that these conventional techniques are substantially outperformed.

Gesbert et al. [22] discuss the various emerging technologies in the field of fixed BWA systems. They recognize adaptive modulation as a key feature in future BWA systems and stress the need for both robust modulation modes like BPSK and QPSK, and high data rate modes such as 64-QAM. They also suggest that SINR be used to determine the switching points between the various operating modes. Catreux et al. [6] introduce link adaptation techniques in the context of BWA systems and recommend different adaptation methods based on the channel state information. They propose adaptation methods that use mean SNR, BER information, and a combination of the two to form the metrics that decide switching thresholds.

14.4   Link Adaptation in IEEE 802.16 (WiMAX)

The IEEE 802.16 family of standards specifies the air interface of fixed and mobile BWA systems that support multimedia services. The principal objective of this standardization effort has been to establish the industry standards for broadband radio access for WirelessMANs. A WiMAX system is one based on technologies of this family, sponsored by an industry consortium called the WiMAX Forum that develops test specifications and collaborates with the IEEE on implementing an 802.16 certification program.

The IEEE 802.16-2004 standard, which was also previously called 802.16d or 802.16-REVd, was published for fixed access in October 2004. The IEEE 802.16e-2005 and 802.16-2004/Cor 1-2005 standard (amendment and corrigendum to the IEEE 802.16-2004 standard) [23] has updated and expanded IEEE 802.16-2004 to allow for mobile as well as fixed (stationary) subscriber stations, called as Mobile WiMAX, as of October 2005. It has provided enhancements to IEEE 802.16-2004 to support subscriber stations (SSs) moving at vehicular speeds and thereby specifies a system for combined fixed and mobile broadband wireless access. Functions to support higher layer handover between BSs or sectors are specified. Operation is limited to licensed bands suitable for mobility below 6 GHz. Fixed IEEE 802.16 subscriber capabilities are not compromised. In addition to mobility enhancements, the corrigendum contains substantive corrections to IEEE 802.16-2004 regarding fixed operation.

The IEEE 802.16 task group standardizes only the PHY and media access control (MAC) layers for the BWA air interface [24]. To operate in a wide range of physical channel conditions, the standard defines a robust and flexible PHY layer. A wide range of MCSs (PHY schemes) enabling AMC are defined, which are denoted by burst profiles.

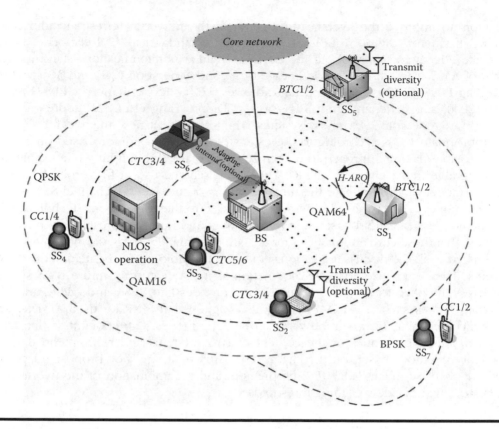

Figure 14.9 A typical Mobile WiMAX network with link adaptation in operation.

While the standard provides a framework for implementing link adaptation, it does not define how adaptation algorithms should be developed [3].

To fully utilize the flexible and robust PHY layer, the current standard defines an equally flexible radio link control (RLC) [25]. In addition to performing traditional functions, such as power control and ranging, the RLC is responsible for transitioning from one PHY scheme to another [3]. The standard defines three different modulation schemes. On the uplink, support for QPSK is mandatory, while 16-QAM and 64-QAM are optional. The downlink supports QPSK and 16-QAM, while 64-QAM is optional. In addition to these modulation schemes, the PHY layer also defines various FEC schemes on the uplink as well as the downlink. These include Reed-Solomon (RS) codes, RS concatenated with inner block convolution codes (BCC), low-density parity check codes (LDPC), and turbo codes. A typical Mobile WiMAX network in operation with these different burst profiles is shown in Figure 14.9.

14.4.1 Air Interface Nomenclature and PHY Compliance

Table 14.1 summarizes the nomenclature for the five different air interface specifications defined in WiMAX [23]. The standard makes a distinction between the licensed and license-exempt frequencies, and frequencies below and above 11 GHz. Implementations of the standard for any applicable frequencies between 10 GHz and 66 GHz shall comply with the WirelessMAN-SC PHY. Implementations of the standard for licensed frequencies below 11 GHz shall comply with the WirelessMAN-SCa PHY, the WirelessMAN-OFDM

Table 14.1   WiMAX Air Interface Nomenclature [23]

Designation	Applicability	PHY Specification	Duplexing Options	Alternative
WirelessMAN-SC	10–66 GHz	SC	—	TDD FDD
WirelessMAN-SCa	Below 11 GHz licensed bands	SCa	AAS ARQ STC mobile	TDD FDD
WirelessMAN-OFDM	Below 11 GHz licensed bands	OFDM	AAS ARQ Mesh STC mobile	TDD FDD
WirelessMAN-OFDMA	Below 11 GHz licensed bands	OFDMA	AAS ARQ HARQ STC mobile	TDD FDD
WirelessMAN-HUMAN	Below 11 GHz license-exempt bands	[SCa, OFDM or OFDMA] and HUMAN	AAS ARQ Mesh (with OFDM only) STC	TDD

PHY, the WirelessMAN-OFDMA PHY, or the WirelessMAN-SC PHY for licensed frequencies above 10 GHz. For license-exempt frequencies below 11 GHz, implementations of the standard shall comply with the WirelessMAN-SCa PHY, the WirelessMAN-OFDM PHY, or the WirelessMAN-OFDMA PHY. They shall further comply with the dynamic frequency selection (DFS) protocols (where mandated by regulation) and with the WirelessHUMAN interface.

In the following sections we will consider the PHY layer attributes of the first four of these PHY interfaces from the perspective of link adaptation, excluding the WirelessHUMAN, since it reuses the other four with some additions to enable optional support for mesh topology.

14.4.2   Monitoring Channels: Channel Quality Measurements

The received signal strength indicator (RSSI) and carrier-to-interference plus noise ratio (CINR) signal quality measurements and associated statistics can aid in such processes as BS selection and assignment and burst adaptive profile selection. As channel behavior is time variant, both the mean and standard deviation need to be defined. The process by which RSSI measurements are taken does not necessarily require receiver demodulation lock; for this reason, RSSI measurements offer reasonably reliable channel strength assessments even at low signal levels. On the other hand, although CINR measurements require receiver lock, they provide information on the actual operating condition of the receiver, including interference and noise levels, and signal strength [24].

14.4.2.1   RSSI Mean and Standard Deviation

For WirelessMAN-SC and WirelessMAN-SCa, when collection of RSSI measurements is mandated by the BS, an SS obtains an RSSI measurement from the downlink burst preambles. For WirelessMAN-OFDM and WirelessMAN-OFDMA, when collection of RSSI measurements is mandated by the BS, an SS obtains an RSSI measurement from the OFDM downlink long preambles. From a succession of RSSI measurements, the SS derives and updates estimates of the mean and the standard deviation of the RSSI, and reports them via channel measurement report response (REP-RSP) MAC layer messages. The method used to estimate the RSSI of a single message is left to individual implementation, however, relevant accuracy and range specifications are defined in the standard.

14.4.2.2   CINR Mean and Standard Deviation

When CINR measurements are mandated by the BS, an SS obtains a CINR measurement from the downlink burst preambles. From a succession of these measurements, the SS derives and updates estimates of the mean and the standard deviation of the CINR and reports them via REP-RSP messages.

For WirelessMAN-SC and WirelessMAN-SCa, the method used to estimate the CINR of a single message is left to individual implementation. Similarly, for WirelessMAN-OFDM and WirelessMAN-OFDMA, when CINR measurements are mandated by the BS, an SS obtains a CINR measurement (implementation specific). From a succession of these measurements, the SS derives and updates estimates of the mean and the standard deviation of the CINR, and reports them via REP-RSP messages or reports the estimate of the mean of the PHY CINR via the fast-feedback channel (channel quality indication channels [CQICH]).

14.4.3   Transitions among PHY Modes: MAC Functionality

To fully utilize the flexible and robust PHY layer, the WiMAX standard defines an equally flexible RLC. In addition to performing traditional functions such as power control and ranging, the RLC is responsible for transitioning from one PHY scheme to another. Combinations of PHY modulation and FEC schemes used between the BS and the SSs are called downlink or uplink burst profiles depending on the direction of flow. In WiMAX, burst profiles are identified using downlink interval usage codes (DIUCs) and uplink interval usage code (UIUCs).

The RLC is capable of switching between different PHY burst profiles on a per-frame and per-SS basis. The SSs use preset downlink burst profiles during connection setup. Thereafter the BS and the SSs continuously negotiate uplink and downlink burst profiles in an effort to optimize network performance.

The standard recommends that each SS measure $C/(N + I)$ as a metric to initiate a burst profile change when necessary. This metric defines the ratio of the received signal power (C) to the sum of the interference power (I) and the noise floor (N). Each SS is required to continuously measure $C/(N + I)$ and explicitly request a change in the downlink burst profile if the metric exceeds or falls below preset threshold levels.

An SS can request a change in the downlink burst profile by using range request (RNG-REQ) messages defined in the standard. The SS uses the initial maintenance or station maintenance intervals to transmit this message to the BS with the DIUC of the desired burst profile. The BS acknowledges the newly negotiated burst profile in the range response (RNG-RSP) message. The RLC provides the SS with the means of explicitly

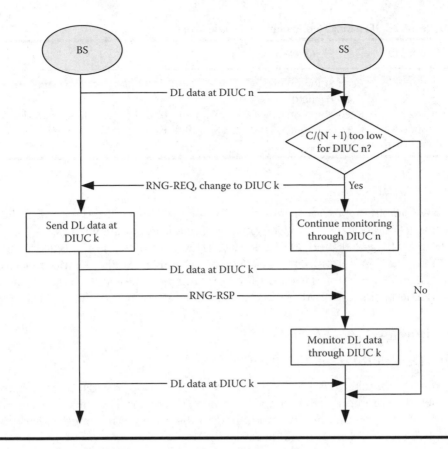

Figure 14.10   Flow for switching to a more robust burst profile [24].

requesting a change in the downlink burst profile to trade off efficiency and robustness under different channel conditions. The SS can switch to a more energy efficient burst profile under good channel conditions and switch to a more robust burst profile when the channel condition deteriorates. Transition to a more robust burst profile is shown in Figure 14.10.

It is important to understand that the standard only defines the messaging framework to be used for link adaptation. An algorithm based on the measurement of instantaneous $C/(N + I)$ is recommended. However, the standard does not specify values for the switching thresholds.

14.4.4   MAC Support for ARQ and HARQ

In WiMAX, ARQ and HARQ schemes are optional parts of the MAC layer and can be enabled on a per-terminal basis. ARQ may be supported for all the PHY specifications except the WirelessMAN-SC PHY, whereas HARQ may be supported only for the OFDMA PHY. The per-terminal HARQ and associated parameters are specified and negotiated during the initialization procedure. A burst cannot have a mixture of HARQ and non-HARQ traffic. These constraints are also valid for ARQ.

HARQ can be used to mitigate the effect of channel and interference fluctuation. HARQ provides performance improvement due to the SNR gain and time diversity

Table 14.2   IEEE 802.16 downlink FEC Code Types

Code Type	Inner Code	Outer Code
1	Reed-Solomon over Galois field (GF) (256)	None
2	Reed-Solomon over GF (256)	(24,16) Block convolutional code
3 (optional)	Reed-Solomon over GF (256)	(9,8) Parity check code
4 (optional)	BTC	—

achieved by combining previously erroneously decoded packets and retransmitted packets, and due to additional coding gain by incremental redundancy (IR).

The IEEE 802.16e-2005 and 802.16-2004/Cor1-2005 specifications have also provided optional multiple HARQ support for WirelessMAN-OFDMA PHY. Supported multiple HARQ modes may be enabled for any of the existing FEC modes. A change in the HARQ mode is signaled using the "HARQ Compact-DL-MAP IE format for Switch HARQ Mode."

14.4.5   WirelessMAN-SC PHY

14.4.5.1   Downlink FEC and Modulation

The downlink channel in WiMAX WirelessMAN-SC PHY supports adaptive burst profiling on the user data portion of the frame. Up to 12 burst profiles can be defined. The parameters of each profile are communicated to the SSs via MAC messages during the frame control section of the downlink frame.

Since there are optional modulation and FEC schemes that can be implemented at the SS, a method for identifying the capability to the BS is required (i.e., including the highest order modulation supported, the optional FEC coding schemes supported, and the minimum shortened last code word length supported). This information is communicated to the BS during the subscriber registration period. The FEC schemes are selectable from the types in Table 14.2.

To maximize utilization of the wireless link, the PHY layer uses a multilevel modulation scheme. The modulation constellation can be selected per subscriber based on the quality of the RF channel. If link conditions permit, then a more complex modulation scheme can be utilized to maximize wireless link throughput while still allowing reliable data transfer. If the wireless link quality degrades over time, possibly due to environmental factors, the system can revert to less complex constellations to allow more reliable data transfer. In the downlink the BS supports QPSK and 16-QAM modulation, and optionally 64-QAM.

14.4.5.2   Uplink FEC and Modulation

The uplink FEC schemes are as described for the downlink channel, listed in Table 14.2. The modulation used in the uplink channel is variable and set by the BS. QPSK is supported, while 16-QAM and 64-QAM are optional.

14.4.6   WirelessMAN-SCa PHY

The WirelessMAN-SCa PHY is based on single-carrier technology and designed for non-line-of-sight (NLOS) operation in frequency bands below 11 GHz.

Table 14.3   Supported Modulations and Inner (Trellis-Coded Modulation [TCM]) Code Rates

Modulation	Support Uplink	Downlink	Inner Code Rates	Bits/Symbol
Spread BPSK	M	M	(prespread) 1/2, 3/4	(postspread) 1/(2*Fs), 3/(4*Fs)
BPSK	M	M	1/2, 3/4	1/2, 3/4
QPSK	M	M	1/2, 2/3, 3/4, 5/6, 7/8	1, 4/3, 3/2, 5/3, 7/4
16-QAM	M	M	1/2, 3/4	2, 3
64-QAM	M	M	2/3, 5/6	4, 5
256-QAM	O	O	3/4, 7/8	6, 7

M = mandatory, O = optional.

14.4.6.1   Adaptive FEC and Modulation in WirelessMAN-SCa PHY

In WirelessMAN-SCa PHY, frame control header (FCH) payloads are encoded in accordance with the well-known parameters defined in the standard to avoid any compliancy problems. Adaptive modulation and concatenated FEC are supported for all other payloads. Support for convolutional turbo code (CTC) FEC and block turbo code (BTC) FEC as well as omitting the FEC and relying solely on ARQ for error control is optional for payloads carried outside the FCH.

Supported modulations and code rates for uplink and downlink transmissions are listed in Table 14.3. No-FEC operation is mandatory for QPSK but optional for other modulation types. The choice of a particular code rate and modulation is again made via burst profile parameters.

14.4.7   WirelessMAN-OFDM PHY

The WirelessMAN-OFDM PHY is based on OFDM modulation and designed for NLOS operation in the frequency bands below 11 GHz.

14.4.7.1   Adaptive FEC and Modulation in WirelessMAN-OFDM PHY

With WirelessMAN-OFDM PHY, an FEC, consisting of the concatenation of a Reed-Solomon outer code and a rate-compatible convolutional inner code, is supported on both the uplink and downlink. Support of BTC and CTC is optional. The most robust burst profile is always used as the coding mode when requesting access to the network and in the FCH burst.

Table 14.4 gives the block sizes and the code rates used for the different modulations and code rates. BPSK, Gray-mapped QPSK, 16-QAM, and 64-QAM are supported, whereas support of 64-QAM is optional for license-exempt bands. Since 64-QAM is optional for license-exempt bands, the codes for this modulation are only implemented if the modulation is implemented.

Per-allocation AMC is supported in the downlink. The uplink supports different modulation schemes for each SS based on the MAC burst configuration messages coming from the BS. The DIUC and UIUC fields are associated with the downlink and uplink burst profile and thresholds, respectively, in the MAC layer. The DIUC value is used in the

Table 14.4   Mandatory Channel Coding Per Modulation

Modulation	Uncoded Block Size (bytes)	Coded Block Size (bytes)	Overall Code Rate	RS Code	CC Code Rate
BPSK	12	24	~1/2	(12,12,0)	1/2
QPSK	24	48	~1/2	(32,24,4)	2/3
QPSK	36	48	~3/4	(40,36,2)	5/6
16-QAM	48	96	~1/2	(64,48,8)	2/3
16-QAM	72	96	~3/4	(80,72,4)	5/6
64-QAM	96	144	~2/3	(108,96,6)	3/4
64-QAM	108	144	~3/4	(120,108,6)	5/6

downlink mobile application part (DL-MAP) message and in the downlink frame prefix (DLFP) to specify the burst profile to be used for a specific downlink burst. The UIUC value is used in the uplink mobile application part (UL-MAP) message to specify the burst profile to be used for a specific uplink burst.

14.4.7.2   Transmit Diversity: STC

Space-time coding [26] (in some cases also called space-time transmit diversity [STTD]) may optionally be used on the downlink to provide higher-order (space) transmit diversity. There are two transmit antennas on the BS side and one reception antenna on the SS side. This scheme requires multiple input/single output (MISO) channel estimation. Decoding is very similar to maximum ratio combining. Both antennas transmit two different OFDM data symbols at the same time. Transmission is performed twice to decode and to get second-order diversity. Time domain (space-time) repetition is used.

14.4.7.3   Optional FEC Capabilities: BTC and CTC

Table 14.5 and Table 14.6 give the block sizes, code rates, channel efficiency, and code parameters for the different modulation and coding schemes for optional BTC and CTC, respectively. Since 64-QAM is optional for license-exempt bands, the codes for this modulation are implemented only if the modulation is implemented. These two more recently developed coding methods achieve better performance, closer to the Shannon limit, and have been integrated into the WiMAX standard.

Table 14.5   Optional BTC Channel Coding Per Modulation

Modulation	Data Block Size (bytes)	Coded Block Size (bytes)	Overall Code Rate	Efficiency (bits/s/Hz)
QPSK	23	48	~1/2	1.0
QPSK	35	48	~3/4	1.5
16-QAM	58	96	~3/5	2.4
16-QAM	77	96	~4/5	3.3
64-QAM	96	144	~2/3	3.8
64-QAM	120	144	~5/6	5.0

Table 14.6   Optional CTC Channel Coding Per Modulation

Modulation	N	Overall Code Rate	P_0
QPSK	$6 \times N_{sub}$	1/2	7
QPSK	$8 \times N_{sub}$	2/3	11
QPSK	$9 \times N_{sub}$	3/4	17
16-QAM	$12 \times N_{sub}$	1/2	11
16-QAM	$18 \times N_{sub}$	3/4	13
64-QAM	$24 \times N_{sub}$	2/3	17
64-QAM	$27 \times N_{sub}$	3/4	17

14.4.8   WirelessMAN-OFDMA PHY

The WirelessMAN-OFDMA PHY [27], based on OFDM modulation, is designed for NLOS operation in the frequency bands below 11 GHz.

The OFDMA PHY mode based on at least one of the fast Fourier transform (FFT) sizes 2048 (backward compatible to IEEE 802.16-2004), 1024, 512, and 128 are supported. This facilitates support of the various channel bandwidths. The DIUC and UIUC fields are associated with the downlink and uplink burst profile and thresholds, respectively, in the MAC layer. The DIUC value is used in the DL-MAP message and in DLFP to specify the burst profile to be used for a specific downlink burst. The UIUC value is used in the UL-MAP message to specify the burst profile to be used for a specific uplink burst.

14.4.8.1   Adaptive FEC and Modulation in WirelessMAN-OFDMA PHY

For modulation, Gray-mapped QPSK and 16-QAM are supported, whereas support of 64-QAM is optional. The FEC coding method used as the mandatory scheme is tail-biting convolutional encoding, and the optional modes of encoding in BTC, CTC, and LDPC may also be supported. Concatenation of a number of slots is performed in order to make larger blocks of coding where it is possible, with the limitation of not exceeding the largest supported block size for the applied modulation and coding. Repetition coding can be used to further increase the signal margin over the modulation and FEC mechanisms. This repetition scheme applies only to QPSK modulation; it can be applied in all coding schemes except HARQ with CTC.

The IEEE 802.16e-2005 and 802.16-2004/Cor1-2005 specifications also provide optional IR HARQ support for CC and CTC, and optional LDPC FEC support in WirelessMAN-OFDMA PHY.

- **Convolutional coding (mandatory).** Each FEC block is encoded by the binary convolutional encoder, which has a native rate of 1/2 and a constraint length equal to $K = 7$. IR HARQ implementation is optional. An IR HARQ takes the puncture pattern into account, and for each retransmission the coded block is not the same. Different puncture patterns are used to create HARQ packets identified by a subpacket ID (SPID). The puncture patterns are predefined or can be easily deducted from the original pattern, and can be selected based on SPID. At the receiver, the received signals are depunctured according to its specific puncture pattern, which is decided by the current SPID, then the combination is performed at the bit metrics level.

Table 14.7   CTC Channel Coding Per Modulation

Modulation	Data Block Size (bytes)	Encoded Data Block Size (bytes)	Code Rate
16-QAM	54	72	3/4
64-QAM	18	36	1/2

- **Block turbo coding (optional).** BTC is based on the product of two simple component codes—binary extended Hamming codes or parity check codes from the set defined in the WiMAX standard.
- **Convolutional turbo codes (optional).** Table 14.7 gives block sizes, code rates, channel efficiency, and code parameters for the different modulation and coding schemes. Since 64-QAM is optional, the codes for this modulation are only implemented if the modulation is implemented.

  IR HARQ implementation is again optional. The procedure of HARQ CTC sub-packet generation is as follows: padding, cyclic redundancy check (CRC) addition, fragmentation, randomization, and CTC encoding. The randomization, concatenation, and interleaving stages in channel coding are not applied for the encoding in this case.
- **Low-density parity check codes (optional).** The LDPC code in the WiMAX specification is based on a set of one or more fundamental LDPC codes. Each of the fundamental codes is a systematic linear block code. The LDPC code flexibly supports different block sizes for each code rate through the use of an expansion factor. Using the coding parameters, the fundamental codes can accommodate various code rates ($R = 1/2$, $2/3$, $3/4$, $5/6$) and packet sizes for modulations of QPSK, 16-QAM, and 64-QAM.

14.5   Link Adaptation in HiperACCESS

In response to trends of pervasive, personalized, and rich multimedia services based on wireless broadband connectivity, high performance radio access (HiperACCESS) systems have been specified by the Broadband Radio Access Network (BRAN) Technical Committee of the European Telecommunications Standards Institute (ETSI) for use by residential customers and by small to medium-size enterprises (SMEs) and mobile infrastructure. They aim to provide support for a wide range of voice and data services and facilities, using radio to connect the premises to other users and networks and offering "bandwidth on demand" to deliver the appropriate data rate needed for the service chosen at any time. They are expected to be a part of convergent and ubiquitous communications "ecosystems" for next-generation services.

The HiperACCESS specifications were first published in 2002, followed by ongoing fine-tuning of base and test specifications. It is an interoperable standard, allowing for point-to-multipoint (PMP) systems with BS and terminal stations from different manufacturers in order to promote mass market and low-cost products. It is based on a PMP network architecture and is intended for high-speed broadband (up to 120 Mbps) and high-QoS fixed wireless access. One of the most important applications is expected to be broadband access for residential and small business users to a wide variety of networks as a flexible and competitive alternative to wired access networks. However, HiperACCESS should not be confused with local multipoint distribution service (LMDS)–type systems. Other applications include backhauling for cellular networks like global system for mobile communications (GSM) and universal mobile telecommunications system (UMTS).

HiperACCESS standardization focuses on solutions optimized for frequency bands above 11 GHz (e.g., 26, 28, 32, and 42 GHz) with high spectral efficiency under LOS conditions. For bandwidths of 28 MHz, both FDD and time division duplex (TDD) channel arrangements as well as half-duplex fequency division duplex (HFDD) terminals are supported. The BRAN Technical Committee is cooperating closely with IEEE-SA (Working Group 802.16) to harmonize the interoperability standards for broadband fixed wireless access networks.

14.5.1 Adaptive Coding and Modulation in HiperACCESS

A HiperACCESS PHY mode includes modulation and a coding scheme (FEC). Several sets of PHY modes are specified for the downlink. The reason for specifying different sets of PHY-modes (each having different SNR gaps) is to offer greater flexibility for the HA-standard deployment, where the adequate choice of a given set of PHY-modes is determined by the deployment scenario: coverage, interference, rain zone, etc. UL PHY modes are a subset of downlink PHY modes in each set.

The modulation is based on QAM with a $2M$ point constellation, where M is the number of bits transmitted per modulated symbol. For the downlink, QPSK ($M = 2$) and 16-QAM ($M = 4$) are mandatory and 64-QAM ($M = 6$) is optional. For the uplink, QPSK is mandatory and 16-QAM is optional. The constellation mappings are based on Gray mapping.

The coding scheme is based on an outer Reed-Solomon code with $t = 8$ and a payload length of four protocol data units (PDUs) shortenable to three, two, or one; the shortening is required to avoid padding at the end of the PHY mode section (called "region"). An inner convolutional code is specified in some of the PHY modes.

For the mandatory FEC scheme in HiperACCESS, by puncturing the inner code rate for a given modulation scheme, a large number of combinations of coding and modulation are possible. However, among all these combinations, two different sets of PHY modes have been selected [28], one mandatory and one optional, where each set of PHY modes can be applied in an adaptive way for both uplink and downlink. It should be noticed that the algorithms for the selection of a given PHY mode are not specified.

In order to guarantee the interoperability, the following rules/procedures are to be applied [28]:

- The indication of the PHY mode is done on a burst-by-burst basis and the adaptation is done on a frame-by-frame basis.
- Each AT (access termination = SS) shall measure the C/N and the Rx (receiver) power and communicate these values to the AP (access point = BS). Furthermore, each AT is able to communicate the available Tx (transmitter) power margin to the AP. Then, following these parameters, the AP centrally decides to change the UL PHY mode or not.
- For the uplink, a minimum amount of traffic must be ensured in order for the AP to be able to continuously measure the C/N ratio with a given accuracy.

- HiperACCESS uses one mandatory and one optional predefined set of PHY modes. The first set of PHY modes is supported by the AT and AP; where the second set is mandatory for AT and optional for AP.
- Out of these sets of PHY modes, only one set of PHY modes is used per sector. The choice of the set of PHY modes can be determined by the network management system (NMS).

14.6 Link Adaptation in HiperMAN

HiperMAN is an air interface standard by ETSI for WirelessMANs that can be used to provide fixed applications in frequencies below 11 GHz, and nomadic and converged fixed-nomadic applications in frequencies below 6 GHz. It is confined only to the radio subsystems consisting of the PHY layer and the DLC layer—which are both core network independent—and the core network-specific convergence sublayer. For managing radio resources and connection control, the DLC protocol is applied, which uses the transmission services of the DLC layer. Convergence layers above the DLC layer handle the interworking with layers at the top of the radio subsystem [29]. HiperMAN is optimized for packet switched networks and supports fixed and nomadic applications, primarily in the residential and small business user environments.

HiperMAN is an interoperable broadband fixed wireless access system operating at radio frequencies between 2 and 11 GHz with the focus on IP-based services. The Hiper-MAN standard is designed for fixed wireless access provisioning to SMEs and residences using the basic MAC (DLC and CLs) of the IEEE 802.16-2001 standard. It has been evolving in very close cooperation with WiMAX, such that the HiperMAN standard and a subset of the IEEE 802.16 standard will interoperate seamlessly. It offers various service categories, full quality of service (QoS), fast connection control management, strong security, fast adaptation of coding, modulation, and transmit power to propagation conditions, and is capable of NLOS operation. HiperMAN enables both PMP and mesh network configurations. HiperMAN also supports both FDD and TDD frequency allocations and hybrid FDD (H-FDD) terminals. All this is achieved with a minimum number of options to simplify implementation and interoperability.

BRAN has had numerous contacts with the WiMAX Forum, the industry forum that is promoting WiMAX/HiperMAN technology and setting up a certification scheme that should ensure the interoperability of WiMAX/HiperMAN devices. The WiMAX Forum and BRAN are cooperating strongly to achieve the desired level of validation of test specifications. Subsequently, in the most recent HiperMAN PHY specification, OFDMA PHY and other various parts of the Mobile WiMAX standard have largely been adopted.

14.6.1 HiperMAN PHY Layer

In this section we present a brief description of HiperMAN PHY considering link adaptation functionalities. A more detailed description and analysis can be found in the ETSI HiperMAN Standard [29].

The HiperMAN OFDM PHY has largely adopted the OFDM subset of the IEEE 802.16 PHY specification. It uses OFDM with a 256-point transform, designed for NLOS operation in the 2 to 11 GHz frequency band, both licensed and license exempt. TDD and FDD variants are defined. Typical channel bandwidths vary from 1.5 to 28 MHz. Although optimized for the 3.4 to 4.2 GHz band, the characteristics of different frequency bands below 11 GHz have been taken into account when defining HiperMAN parameters. Currently there is a second optional air interface specification called the HiperMAN OFDMA PHY that is based on OFDMA with a 2048-point transform. This has been directly adopted from the IEEE 802.16 PHY standard, and thus is identical to the specification described in Section 14.4.8.

For HiperMAN OFDM PHY, BPSK, Gray-mapped QPSK, 16-QAM, and 64-QAM are supported. Support of 64-QAM is optional for unlicensed bands. HiperMAN's FEC scheme consists of the concatenation of a Reed-Solomon outer code and a rate-compatible convolutional inner code. The Reed-Solomon outer code may be shortened and punctured. CTC is optional for all modes. Since 64-QAM is optional for license-exempt bands, the codes for this modulation are only implemented if the modulation is implemented. The data/control plane of HiperMAN OFDM PHY is according to IEEE 802.16 [24], as modified by IEEE 802.16e [23]. Therefore the same MAC mechanisms for link adaptation are valid. The FEC options are paired with the modulation schemes listed to form burst profiles of varying robustness and efficiency and enable link adaptation.

The major differences between HiperACCESS and HiperMAN are worth noting for the sake of completeness: HiperMAN is intended to serve users in a NLOS environment and operating frequencies for HiperMAN are much lower, resulting in significantly lower cost implementations, which makes addressing the residential market economically viable [30]. To exploit the channel properties of the bands around 3.5 GHz and address the different spectrum requirements, the PHY layers of HiperMAN and HiperACCESS have to be different. HiperACCESS has a fixed bandwidth of 28/14 MHz and frame durations of 1 msec. HiperMAN, which has bandwidths in multiples of 3.5 MHz, needs to have longer frame durations. In order to maintain delay requirements (such as for voice packets), HiperMAN likely requires more stringent QoS algorithms. Since the DLC layer for HiperMAN is optimized for the IP, the DLC layer of HiperMAN is different from that of HiperACCESS, also because the PHY layer is different. The HiperACCESS standardization has focused entirely on solutions optimized for bands above 11 GHz. HiperMAN intends to complement rather than duplicate the remaining HiperACCESS efforts.

14.7   Link Adaptation in WiBro

High-speed portable Internet (HPi) is a WirelessMAN standard initiated by the Korean telecommunications industry to provide better data handling than 3G cellular systems. The standard was renamed wireless broadband (WiBro) once it adopted IEEE 802.16e as its standard for global harmonization [31].

As of 2004, WiBro began to align itself with the WiMAX Forum's implementation of IEEE 802.16-2005 using a two-phased approach. Phase I of WiBro is now based in part on the IEEE standard; however, the WiBro community has selected a different set of options which results in WiBro phase I equipment being different from WiMAX (Table 14.8). In late 2005, WiBro phase 2 standardized by the 2.3 GHz Portable Internet Project Group (PG302) of the Telecommunications Technology Association (TTA) of Korea. Over the next few years, the WiBro community will move to phase II of WiBro, which will help harmonize WiBro and Mobile WiMAX. Currently, the TTA has been trying to facilitate strong and effective global radio standards collaboration on WiBro [32].

Table 14.8   PHY Comparison of IEEE 802.16 and WiBro/HPi

Item	WiBro	IEEE 802.16e OFDMA	IEEE 802.16 OFDMA
Frequency	2.3–2.4 GHz	~ 11 GHz	~ 11 GHz
Service	Portable Internet	Fixed/mobile WirelessMAN	Fixed WirelessMAN
Terminal mobility	Vehicle in downtown	Vehicle	Fixed
Duplex	TDD	TDD/FDD	TDD/FDD
Multiple access	OFDMA	OFDMA	OFDMA
Channel bandwidth	10 MHz (occupied bandwidth: 8.65 MHz)		
Maximum data rate	30 Mbps (downlink + uplink, SISO) 50 Mbps (downlink + uplink, AAS)	36 Mbps (downlink + uplink) at 10 MHz	36 Mbps (downlink + uplink) at 10 MHz
Channel coding	CTC Subchannel concatenation	Concatenated RS-convolutional code BTC (optional), CTC (optional), subchannel concatenation	Concatenated RS-convolutional code BTC (optional), CTC (Optional)
Modulation	QPSK-16-QAM-64-QAM	QPSK/16-QAM/64-QAM	QPSK/16-QAM/64-QAM
Coverage (radius)	1–1.5 km (rural) 3–5 km (rural)	3.5–7 km	3.5–7 km

The WiBro air interface specification is targeted to be fully compatible with IEEE 802.16e and should support IEEE 802.16-2004 and IEEE 802.16e. The WiBro system deploys OFDMA/TDD in order to take an asymmetric traffic pattern into account. It is compatible with the IEEE 802.16e OFDMA system with a 1024 FFT size. The link adaptation mechanisms of IEEE 802.16e MAC and OFDMA PHY, described in Section 1.4, are also valid for WiBro. WiBro services were launched commercially at two Korean operators, Korea Telecom and SK Telecom, at the end of 2006 [33].

14.8   Open Issues

In this section we highlight some open issues for research and development in WirelessMAN systems. Although some of the topics reside in the more general context of wireless communications, they are expected to have a direct and significant impact on the standardization and design of future WirelessMAN systems.

14.8.1   Better Link Adaptation Algorithms: Smarter Decisions?

In WiMAX, the standard only defines the messaging framework to be used for link adaptation. The SS applies an algorithm to determine its optimal burst profile in accordance with the threshold parameters established in the MAC DCD message in accordance with

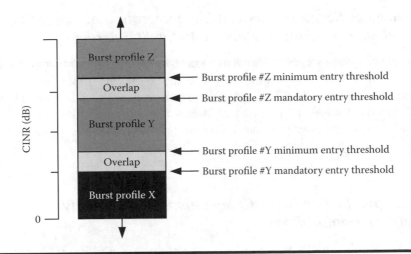

Figure 14.11   Burst profile threshold usage in WiMAX for link adaptation.

Figure 14.11. However, the standard does not specify values for the switching thresholds. Therefore novel and better link adaptation algorithms utilizing the specified framework need to be developed [3].

14.8.2   The More, the Better: More PHY Modes?

Since WirelessMAN systems attempt to address data transmission in adverse and diverse wireless propagation conditions, creating more degrees of freedom for PHY layer adaptation can mitigate the impairments, provide stronger robustness, and contribute to performance at the expense of cost and complexity. Further, given the multifaceted requirements of a global technology that can be deployed in multiple frequency bands of varying channel bandwidths, WiMAX and other WirelessMAN systems may contain more options and features, such as more PHY modes. New paradigms in wireless communications such as cognitive radio, should be taken into account more vigorously during the design of WirelessMAN standards.

14.8.3   Layers versus Monoliths: Better Cross-Layer Approaches?

Of late, there has been an avalanche of cross-layer design proposals for wireless networks [34] and the popularity of cross-layer design of communication protocols has risen considerably recently due to stricter performance requirements and ubiquitous migration to wireless communications media. Although monolithic approaches may generate "islands of performance" where systems are overtuned for specific applications and only implementable with specific hardware, a tighter and more dynamic cooperation of PHY and MAC layers may naturally boost the WirelessMAN system's performance. Moreover, some cross-layer application-optimized profiles for the most popular services, such as streaming video voice over Internet protocol (VoIP), may be integrated as optional components into upcoming standard releases.

14.8.4 Novel or Backward Compatible: Integration of Novel Channel and Source Coding Methods into the Standards?

Recent advances in channel coding such as turbo coding and MIMO systems have already been adopted by WirelessMAN standards. The development of similar novel schemes and integration of them into WirelessMAN standards is a crucial issue. However, the introduced complexity to the systems and backward compatibility are to be taken into account, and an evolutionary path has to be followed for the persistence of anticipated commercial success of WirelessMAN technologies.

14.8.5 Cooperation or Fierce Competition: Compatibility and Harmonization?

Although compatibility and harmonization among different WirelessMAN standards are a general issue, it is also paramount for the performance of link adaptation schemes since flexibility in terms of PHY modes and MAC abilities will emerge with a tighter harmonization among standards. In addition, by understanding some of these differences, it is possible to better appreciate what the migration from other standards to Mobile WiMAX will require and in what time frame that may occur.

For the PHY layer, there are some compatibility issues against the convergence of the WirelessMAN standards. For instance, we can identify a few specific examples where the Mobile WiMAX and WiBro differ: the WiMAX Forum has designated a certain number of tones for carrying data traffic via OFDMA, as well as pilot tones and null tones, which are used to limit interference in the system. Similarly WiBro has also adopted OFDMA, but the channel bandwidths and the number of associated tones, including the number of tones for carrying data, pilot, and null traffic, are not consistent with the WiMAX Forum [35]. WiBro and Mobile WiMAX both use 5 msec frames, but the number of symbols in each frame differs by technology. Since these tones, or subcarriers, serve as the basis for transporting traffic in an OFDMA-based system, it is critical for interoperability that each solution is implemented the same way. In the case of Mobile WiMAX and WiBro, it is evident that this is not the case. Further, since this is a PHY layer implementation, meaningful hardware changes (e.g., new application-specific integrated circuits [ASICs] in the devices, new channel cards in the BSs) will likely be required to bring the WiBro solution in line with Mobile WiMAX [35].

The WiMAX Forum has also mandated that mobile devices support MIMO antenna schemes, while base transceiver station support for MIMO is optional. In the case of Mobile WiMAX, the WiMAX Forum has specified that mobile devices support one transmit and two receive chains, while MIMO is optional in WiMAX BSs in order to receive WiMAX certification. Although this requirement is not mandatory until the next wave of WiMAX certification, most Mobile WiMAX solutions will support MIMO functionality by sometime in 2007. WiBro phase I, however, does not include MIMO, which implies that additional hardware changes will be required for existing solutions in order to implement the feature.

HARQ is another PHY/MAC layer feature that is supported by Mobile WiMAX and WiBro; however, with different implementations. WiBro supports incremental redundancy HARQ, while Mobile WiMAX supports Chase combine HARQ. The implementation of these two different types of HARQ technologies requires different hardware arrangements (e.g., very different memory requirements).

The IEEE 802.16e standard supports three different duplex schemes: TDD, FDD, and HFDD. Both Mobile WiMAX and WiBro presently support TDD, however, the downlink and uplink switching time gap is different. This means that the RF coexistence of Mobile WiMAX and WiBro is not possible if they are deployed in same geographical region. More specifically, since the time gaps are different, the WiBro system cannot be easily modified to Mobile WiMAX, since the two systems have different RF requirements. WiBro is only required to support TDD, so hardware elements such as a duplexer are not required. Conversely, Mobile WiMAX profiles will likely include all three duplex schemes in order to provide greater flexibility for link adaptation. In that regard, even if WiBro incorporated all of the other elements of the WiMAX PHY layer, it would still be limited to TDD [35].

14.9   Conclusion

This chapter presented the basic link layer adaptation capabilities currently available in WirlessMAN technologies including WiMAX, HiperACCESS, HiperMAN, and WiBro. We have discussed and presented these technologies in order to highlight the benefits of link adaptation. These WirelessMAN systems aim to provide broadband mobile wireless access in adverse propagation environments such as urban areas which result in formidable channel conditions (multipath propagation, shadowing, scattering, etc.). Therefore, optimization based on adaptive transmission schemes are vital for high-grade services. Novel algorithms and schemes need to be developed for smarter and more effective operation of link adaptation. Moreover, the advent of mobility support as in IEEE 802.16e incurs additional challenges. Thus link adaptation mechanisms coupled with advanced PHY layer schemes such as OFDM and MIMO are critical. However, the success of these BWA systems on a large scale relies on the trend of convergence or harmonization among these different systems. The work being done by the WiMAX Forum has resulted in the commercialization of the WiMAX standard. At the same time, WiBro, which is also based on IEEE 802.16-2005, and HiperMAN are moving to harmonize with WiMAX [35]. WiMAX is expected to be the "über-standard" where the standardization efforts will concentrate and other standards will comply with. This phenomenon will help the widespread proliferation of WirelessMAN systems in a cost-effective and rapid manner.

References

1. M. Haleem and R. Chandramouli, Adaptive transmission rate assignment for fading wireless channels with pursuit learning algorithm, *38th Annual Conference on Information Sciences and Systems*, Princeton University, Princeton, NJ, 2004.
2. M. Nakamura, Y. Awad, and S. Vadgama, Adaptive control of link adaptation for high speed downlink packet access (HSDPA) in W-CDMA, *5th International Sysmposium on Wireless Personal Multimedia Communications*, vol. 2, p. 382, IEEE Washington, DC, 2002.
3. S. Ramachandran, Link adaptation algorithm and metric for IEEE standard 802.16, Masters thesis, Virginia Polytechnic Institute and State University, Blacksburg, VA, 2004.
4. H.T. Nguyen, J.B. Andersen, and G.F. Pedersen, On the performance of link adaptation techniques in MIMO systems, *Wireless Personal Communications*, 2006 [online], available at http://www.springerlink.com.
5. J. Proakis, *Digital Communications*, 3rd ed., McGraw-Hill, New York, 1995.

6. S. Catreux, V. Erceg, D. Gesbert, and R.W. Heath, Jr., Adaptive modulation and MIMO coding for broadband wireless data networks, *IEEE Commun. Mag.*, 40(6), 108, 2002.

7. E. Armanious, Link adaptation techniques for fixed broadband wireless access systems, Masters thesis, Carleton University, Ottawa, Ontario, Canada, 2003.

8. G. Caire, G. Taricco, and E. Biglieri, Bit-interleaved coded modulation, *IEEE Trans. Info. Theory*, 44(3), 927, 1998.

9. A. Forenza, M. Airy, M. Kountouris, R.W. Heath, Jr., D. Gesbert, and S. Shakkottai, Performance of the MIMO downlink channel with multi-mode adaptation and scheduling, *6th IEEE Workshop on Signal Processing Advances in Wireless Communications*, IEEE, Washington, DC, 2005.

10. A.J. Goldsmith and S.-G. Chua, Variable-rate variable-power MQAM for fading channels, *IEEE Trans. Commun.*, 45, 1218, 1997.

11. W.T. Webb and R. Steele, Variable rate QAM for mobile radio, *IEEE Trans. Commun.*, 43, 2223, 1995.

12. S. Sampei, S. Komaki, and N. Morinaga, Adaptive modulation/TDMA scheme for personal multimedia communication systems, *IEEE Global Telecommunications Conference*, vol. 2, 989, IEEE, Washington, DC, 1994.

13. A.J. Goldsmith and S. Chua, Adaptive coded modulation for fading channels, *IEEE Trans. Commun.*, 46(5), 595, 1998.

14. F. Cameron, M. Zukerman, and M. Gitlits, Adaptive transmission parameters optimization in wireless multi-access communication, *IEEE International Conference on Networks*, 91, IEEE, Washington, DC, 1999.

15. J.M.C. Brito and I.S. Bonatti, An analytical comparison among adaptive modulation, adaptive FEC, adaptive ARQ and hybrid systems for wireless ATM networks, *5th International Symposium on Wireless Personal Multimedia Communication*, vol. 3, 1034, IEEE, Washington, DC, 2002.

16. I. Akyildiz, I. Joe, H. Driver, and Y. Ho, An adaptive FEC scheme for data traffic in wireless ATM networks, *IEEE/ACM Trans. Networking*, 9(4), 419, 2001.

17. F. Tang, L. Deneire, M. Engels, and M. Moonen, Adaptive link adaptation, in *IEEE Global Telecommunications Conference*, vol. 2, p. 1262, IEEE, Washington, DC, 2001.

18. J. Yang, N. Tin and A. Khandani, Adaptive modulation and coding in 3G wireless systems, *56th IEEE Vehicular Technology Conference*, vol. 1, p. 24, IEEE, Washington, DC, 2002.

19. J.S. Blogh and L. Hanzo, Adaptive modulation and adaptive antenna array assisted network performance of multi-user detection aided UTRA-like FDD/CDMA, *56th IEEE Vehicular Technology Conference*, vol. 3, p. 1806, IEEE, Washington, DC, 2002.

20. S. Nanda, K. Balachandran, and S. Kumar, Adaptive techniques in wireless packet data services, *IEEE Commun. Mag.*, 38(1), 54, 2000.

21. J. Yang, A.K. Khandani, and N. Tin, Statistical decision making in adaptive modulation and coding for 3G wireless systems, *IEEE Trans. Vehic. Technol.*, 54 (6), 2066, 2005.

22. D. Gesbert, L. Haumonte, H. Bolcskei, R. Krishnamoothy, and A. J. Paulraj, Technologies and performance for non-line-of-sight broadband wireless access networks, *IEEE Commun. Mag.*, 40(4), 86, 2002.

23. IEEE, IEEE Standard for Local and Metropolitan Area Networks—Part 16: Air Interface for Fixed and Mobile Broadband Wireless Access Systems Amendment 2: Physical and Medium Access Control Layers for Combined Fixed and Mobile Operation in Licensed Bands and Corrigendum 1, Standard 802.16e-2005 and Standard 802.16-2004/Cor1-2005, IEEE, Washington, DC, 2006.

24. IEEE, Standard for Local and Metropolitan Area Networks—Part 16: Air Interface for Fixed Broadband Wireless Access Systems, Standard 802.16-2004, IEEE, Washington, DC, 2004.

25. C. Eklund, R.B. Marks, K.L. Stanwood, and S. Wang, IEEE Standard 802.16: a technical overview of the WirelessMAN^{TM} air interface for broadband wireless access, *IEEE Commun. Mag.*, 40(6), 98, 2002.

26. S.M. Alamouti, A simple transmit diversity technique for wireless communications, *IEEE J. Select. Areas Commun.*, 16(8), 1451, 1998.

27. H. Sari and G. Karam, Orthogonal frequency-division multiple access and its application to CATV networks, *Eur. Trans. Telecommun.*, 9(6), 507, 1998.

28. ETSI TS 101 999, Broadband Radio Access Networks (BRAN); HIPERACCESS PHY Protocol Specification, ETSI, Sophia-Antipolis, France, 2002.

29. ETSI TS 102 177, Broadband Radio Access Networks (BRAN); HiperMAN; Physical (PHY) layer, ETSI, Sophia-Antipolis, France, 2003.

30. ETSI TR 101 856, Broadband Radio Access Networks (BRAN); Functional Requirements for Fixed Wireless Access Systems Below 11 GHz: HIPERMAN, March 2001.

31. S. Lee, N. Park, C. Cho, and H. Lee, The wireless broadband (WiBro) system for broadband wireless Internet services, *IEEE Commun. Mag.*, 44(7), 107, 2006.

32. S.Y. Yoon, WiBro technology, Technical Report, Samsung Electronics, Seoul, Korea, 2004.

33. T. Kwon, H. Lee, S. Choi, J. Kim, D. Cho, S. Cho, S. Yun, W. Park, and K. Kim, Design and implementation of a simulator based on a cross-layer protocol between MAC and PHY layers in a WiBro compatible IEEE 802.16e OFDMA system, *IEEE Commun. Mag.*, 43(12), 136, 2005.

34. V. Srivastava and M. Motani, Cross-layer design: a survey and the road ahead, *IEEE Commun. Mag.*, 43(12), 112, 2005.

35. M.W. Thelander, WiMAX or WiBro: similar names, yet dissimilar technologies, Signals Research Group, Nortel Networks April 2006, available at http://www.nortel.com/solutions/wimax/collateral/wimax_wibro_white_paper.pdf.

SECURITY, SYSTEMS, AND POLICIES

Chapter 15

Analysis of Threats to WiMAX/802.16 Security

Michel Barbeau and Christine Laurendeau

Contents

This chapter examines threats to the security of the WiMAX/802.16 broadband wireless access technology. Threats associated with the physical (PHY) layer and media access control (MAC) layer are reviewed in detail. The likelihood, impact, and risk are evaluated according to an adaptation of the threat assessment methodology proposed by the European Telecommunications Standards Institute (ETSI). Threats are listed and ranked according to the level of risk they represent. This assessment can be used to prioritize future research directions in WiMAX/802.16 security.

15.1   Introduction

This chapter pertains to an emerging broadband wireless access technology being jointly defined by the IEEE and WiMAX Forum [1]. Compared to previous wireless access technologies, WiMAX/802.16 offers wide bandwidth Internet protocol (IP)-based mobile and wireless access, handover across heterogeneous networks and management authorities, and broadband service in remote areas.

The IEEE 802.16 standard [1] defines the air interface for fixed point-to-multipoint (PMP) broadband wireless access networks. An overview of IEEE 802.16 can be found in Eklund et al. [2]. The IEEE 802.16e amendment [3] defines additional mechanisms to support mobile subscribers at vehicular speed as well as mechanisms for data authentication.

The role of the WiMAX Forum is to define profiles as subsets of the broad range of available options, to address the certification of implementations, and to define additional mechanisms for networking such as user-network mutual authentication, integration with other kinds of wireless access technologies (WiFi/802.11, 2G/3G cellular), and transfer of security and quality of service (QoS)-state information during handovers.

This chapter provides an analysis of the threats to WiMAX/802.16 security. Threats are analyzed with respect to their likelihood of occurrence, their possible impact on individual users and on the system, and the global risk they represent.

The methodology employed to conduct the threat analysis is introduced in Section 15.2. The analysis is presented in Section 15.3. Section 15.4 concludes the chapter.

15.2   Methodology

The primary goal of a threat analysis is to determine the threats inherent to a given technology and ascertain the risk posed by each identified threat. With this information, efforts to devise countermeasures can be efficiently focused solely on the more critical threats.

There are two types of threat analysis methodologies: quantitative and qualitative. Quantitative threat analysis methodologies [4,5] offer an objective means for estimating the risk posed by a threat by using the statistical probability of its occurrence. However, a significant drawback of this type of methodology lies in its reliance on historic data of past occurrences of a threat in order to predict future occurrences. In the case of emerging technologies such as WiMAX/802.16, no such history of past events exists, therefore quantitative threat analysis methodologies cannot be applied. Qualitative threat analysis methodologies [6–8], on the other hand, allow for discrete, estimated values to be assigned to a variety of risk factors, including the likelihood of occurrence and impact on the victims. Although this type of methodology has no dependence on preexisting historic data, it is more subjective than its quantitative counterpart. Risk factor values may vary according to the authors of the analysis and the information available. However, we believe that it is a nice tool for identifying security flaws and ranking them by order of importance. One example of a qualitative threat analysis methodology is the one put forth by the ETSI, which forms the basis for the method used here to assess the threats to WiMAX/802.16 security.

Table 15.1   Risk as a Function of the Likelihood and Impact

Likelihood	High	Impact Medium	Low
Likely	Critical		
Possible		Major	
Unlikely			Minor

The ETSI methodology was originally developed in 2003 to analyze the security threats to a meta protocol called the Telecommunications and Internet protocol harmonization over networks (TIPHON) [8]. This methodology allows for the risk posed by identified threats to be evaluated as critical, major, or minor, depending on estimated values for the likelihood of occurrence of the threat and its impact upon a user or a system. As a guideline, a threat ranked as minor typically requires no countermeasures, a major one needs to be dealt with, and a critical threat must be addressed with the highest priority. In the ETSI methodology, a threat is ranked as critical under the following conditions: if it is likely and has high impact, if it is likely and has medium impact, and if it is possible and has high impact. A threat is only assessed as major if it is possible and has medium impact. We have found, through experience with the ETSI methodology [9,10], that many threats are overclassified as critical when they are better ranked as major. As a result, we have adapted the ETSI methodology to assess the risks as depicted in Table 15.1. Our adaptation thus places the emphasis on the truly critical threats. The evaluation of the likelihood and impact risk factors is described in Section 15.2.1 and Section 15.2.2.

15.2.1 Likelihood of Occurrence

The likelihood risk factor denotes the possibility that attacks associated with a given threat are carried out. Three discrete levels of likelihood are defined: likely, possible, and unlikely. In order to evaluate the likelihood of a threat, two additional risk factors are taken into account: the motivation for an attacker to carry out the attack and the technical difficulties that must be resolved by the attacker in order to do so, as shown in Table 15.2.

A threat is unlikely if there is little motivation for perpetrating the specific attack or if significant technical difficulties must be overcome. A threat is possible if the motivation

Table 15.2   Likelihood as a Function of Attacker Motivation and Technical Difficulty

Motivation	Difficulty	Likelihood
High	None	Likely
	Solvable	
Moderate	None	Possible
	Solvable	
Low	Any	
Any	Strong	Unlikely

for an attacker is sufficiently high and the technical difficulties are few or solvable because the required theoretical and practical knowledge for implementing the attack is available. A threat is likely if a user or system is almost assured of being victimized, given a high attacker motivation and lack of technical hurdles. The different values for motivation and technical difficulties are further described below.

15.2.1.1 Attacker Motivation

The topic of what motivates a computer hacker to conduct attacks has been addressed in the literature in order to devise better countermeasures. In his insightful article, anthropologist Roger Blake adapts basic social stratification theory to suggest that hackers are motivated primarily by wealth, power, and prestige, although these may be sought in the acquisition of knowledge rather than money [11]. The financial gains to be made through the sale of private information or the disruption of a business rival's network, the power that private or secret information can afford an individual over others, and the prestige to be garnered before the hacker community are thus powerful motives for attacks.

Schifreen [12] proposes five possible motives for hackers:

- Opportunity: the temptation offered by systems with poor security mechanisms.
- Revenge: attacks perpetrated by a disgruntled employee, for example.
- Greed: financial gain through the sale of information, espionage, or blackmail.
- Challenge: the prestige obtained through trumping a security system.
- Boredom: the lack of better activities to occupy the hacker's time.

To these, Barber [13] adds

- Vandalism: the defacing of corporate Web sites.
- Hacktivism: attacks conducted for making political, ecological, or ethical statements.
- Information warfare: governments trying to influence foreign or domestic situations for their own interests.

In our adaptation of the ETSI methodology, we assess the possibility of gain in terms of money and power to be somewhat more motivating than prestige. We therefore associate a high motivation with an attacker reaping significant financial or power-based gains, a moderate motivation with limited gains or with creating mischief for the purpose of garnering prestige, and a low motivation with little gain for the attacker.

15.2.1.2 Technical Difficulty

Technical difficulty refers to the technological hurdles encountered by an attacker in his attempts to implement a threat. It should be noted that such difficulties are dynamic in nature. What seems like an unsurmountable obstacle today may not be so in a few years' time.

For example, WiFi implementations based on the IEEE 802.11 standard's original specifications employ the wired equivalent privacy (WEP) approach to security [14]. The standard's working group believed that WEP's security mechanisms posed strong technical difficulties to attackers. In 1999, the technical difficulty for attacks directed at WEP was strong, and attacks were unlikely to occur. However, it quickly became obvious that WEP had weaknesses. WEP's shortcomings were discovered by Fluhrer et al. [15] and made public in 2001. The technical difficulty became solvable and the likelihood of attacks rose to the level of possible. In 2002, Stubblefield et al. [16] implemented the attack. Since that time, the attack has been well documented, and the associated software has been available, reducing the technical difficulty to none and upgrading the likelihood of the attack to likely. WiFi protected access (WPA), WEP's successor, is currently believed to pose strong technical difficulties to attackers [17]. It remains to be seen whether this assumption stands the test of time.

In our WiMAX/802.16 threat analysis, we assign a strong technical difficulty to threats to security mechanisms that currently may not be defeated because some theoretical elements for perpetrating an attack upon them are missing. A solvable technical difficulty is associated with a security mechanism that may be countered or has been defeated in a related technology. A technical difficulty of none is assigned when a precedent for the attack already exists.

15.2.2 Impact on the User and System

The impact criterion evaluates the consequences for a user or a system if a given threat is carried out. Possible values for the impact are listed in Table 15.3.

15.2.2.1 Low Impact

From the single user's point of view, the impact of a threat is rated as low if an attack results in only annoyance and the consequences, if there are any, are reversible and can be repaired. From the point of view of a system serving several users, a threat is ranked as low impact if the possible outages are very limited in scope, for example, with few users affected for a short period of time.

15.2.2.2 Medium Impact

For the user, the impact is medium if a loss of service occurs for a short period of time. For a system, the consequences of a medium impact threat consist of outages that are limited in both scope and possible financial losses.

15.2.2.3 High Impact

A threat carries a high impact for a user if an attack causes a loss of service for a considerable period of time. If targeted at a system, an attack associated with a high impact threat results in outages over a long period of time with a large number of users affected, possibly accompanied by law violations or substantial financial losses.

**Table 15.3   Impact from the User or System
Point of View**

Low	Annoyance or very limited outage
Medium	Short-term loss of service or outage
High	Long-term loss of service or outage

15.3   Analysis

A WiMAX/802.16 wireless access network consists of base stations (BSs) and mobile stations (MSs). The BSs provide network attachment to the MSs. An MS selects for its serving BS the one that offers the strongest signal. In this analysis, the subscriber plays the role of the user, while a BS and the collection of MSs represent a system.

The protocol architecture of WiMAX/802.16 is structured into two main layers: the MAC layer and the PHY layer, as depicted in Figure 15.1. The diagram also indicates interfacing points where service access points (SAPs) are formally defined by the standard. The central element of the layered architecture is the common part sublayer, which is tightly integrated with the security sublayer. In this layer, MAC protocol data units (PDUs) are constructed, connections are established, and bandwidth is managed. The common part exchanges MAC service data units (SDUs) with the convergence layer, which adapts units of data (e.g., IP packets or asynchronous transfer mode [ATM] cells) from higher level protocols to the MAC SDU format, and vice versa. The convergence layer also sorts the incoming MAC SDUs by the connection to which they belong. The security sublayer addresses authentication, establishment of keys and encryption, and exchanges MAC PDUs with the PHY layer. The PHY layer is a two-way mapping between MAC PDUs and the PHY layer frames received and transmitted through coding and modulation of radio frequency (RF) signals.

In our analysis we examine the security threats first at the PHY layer, then at the MAC layer. The results of the analysis are consolidated in Table 15.4.

15.3.1   Threats to the PHY Layer

In this section we describe the frame structure used in the WiMAX/802.16 PHY layer and assess the possible threats therein.

15.3.1.1   Frame Structure

At the PHY layer, the flow of bits is structured as a sequence of frames of equal length, as shown in Figure 15.2. There is a downlink subframe and an uplink subframe, and two modes of operation: frequency division duplex (FDD) and time division duplex (TDD).

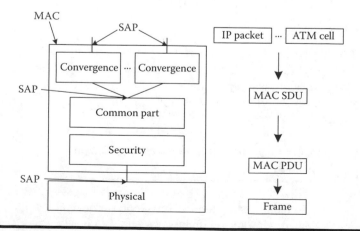

Figure 15.1   WiMAX/802.16 layered architecture.

Table 15.4   Analysis Summary from the User and System Points of View

Threat	Algorithm	Likelihood	Impact	Risk
Jamming		Possible	User: Low	Minor
			System: Medium	Major
Scrambling		Possible	Low	Minor
Eavesdropping of management messages		Likely	User: Medium	Major
			System: High	Critical
Eavesdropping of data traffic	DES-CBC, AES-CCM	Unlikely	Medium	Minor
BS or MS masquerading	Device list	Likely	User: High	Critical
			System: Medium	Major
	X.509 device authentication	User: Possible	High	Major
		System: Unlikely	Medium	Minor
	EAP	Possible	User: High	Major
			System: Medium	
Management message modification	No MAC	Possible		Major
	HMAC without counter		High	
	HMAC with counter	Unlikely		Minor
	OMAC			
Data traffic modification	Without AES	Possible	Medium	Major
	With AES	Unlikely		Minor
DoS on BS or MS		Possible	User: High	Major
			System: Medium	

Figure 15.2 depicts these two modes. The horizontal axis represents the time domain, while the vertical one represents the frequency domain. In FDD, the downlink subframe and uplink subframe are simultaneous, but don't interfere with each other because they are sent on different frequencies. In TDD, the downlink subframe and uplink subframe are consecutive in time. A frame duration of 0.5, 1, or 2 ms can be used.

All frames are of equal length. In TDD, the portion allocated for the downlink and the portion allocated to the uplink may vary. The uplink is time division multiple access (TDMA), which means that the bandwidth is divided into time slots. Each time slot is allocated to an individual MS being served by the BS.

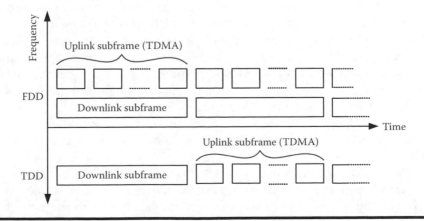

Figure 15.2   PHY layer framing.

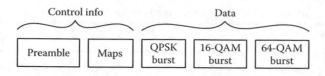

Figure 15.3   TDD downlink subframe.

A detailed representation of a TDD downlink subframe illustrates the burst nature of the transmission and can be found in Figure 15.3. A downlink subframe consists of two main parts: the control information and the data. The control information consists of a preamble and maps. The preamble is used for frame synchronization purposes. There are two maps: a downlink map describes the start position and transmission characteristics of the following data bursts, and an uplink map disseminates the allocation of the bandwidth to the MSs for their transmission. The data part consists of a sequence of bursts. Each burst is transmitted according to a profile of modulation and a kind of forward error correction. They are sent in an increasing degree of demodulation difficulty. Hence an MS may only receive the bursts while it has the capability to do so and ignores the bursts it cannot demodulate.

15.3.1.2   Threats

Since the security sublayer is above it, as represented in Figure 15.1, the PHY layer is unsecured. WiMAX/802.16 is vulnerable to PHY layer attacks such as jamming and scrambling.

Jamming is achieved by introducing a source of noise strong enough to significantly reduce the capacity of the channel. Jamming is either unintentional or malicious. Since the attacker's aim is to create mischief, the motivation can be rated as moderate. In terms of technical difficulty, the information and equipment required to perform jamming are not difficult to obtain. Poisel [18] published a book on the topic of jamming, describing how to build jamming systems and counter systems that are jamming resistant. We therefore assess the difficulty to be very low, leading us to rate a jamming attack as possible, according to Table 15.2. Resilience to jamming can be augmented by increasing the power of signals or increasing the bandwidth of signals using spreading techniques, namely frequency hopping or direct sequence spread spectrum. Note that a number of options are available to increase the power of a signal, such as using a more powerful transmitter, a high-gain transmission antenna, or a high-gain receiving antenna. Jamming is easy to detect with radio spectrum monitoring equipment. Sources are relatively easy to locate using radio direction-finding tools, and law enforcement can be involved to stop jammers. Jammed segments of bandwidth, once detected, can also be avoided in a spread spectrum scheme. Since jamming is fairly easy to detect and address, we believe that it can have a low impact on a user because of the limited loss of service, and a medium impact on a system because of the short-term outages of limited scope and number of users. According to Table 15.1, the risk associated with jamming is therefore minor for a user and major for a system.

Scrambling is a sort of jamming, with similar motivation factors, that is carried out for short intervals of time and targeted at specific frames or parts of frames. Scramblers can selectively scramble control or management information with the aim of affecting the normal operation of the network. The problem is of greater amplitude for time-sensitive messages, which are not delay tolerant, such as channel measurement report requests or responses. Slots of data traffic belonging to targeted users can be scrambled selectively, forcing them to retransmit, with the net result that they get less than their granted bandwidth. Selectively scrambling the uplink slots of other users can theoretically reduce the effective bandwidth of the victims and accelerate the processing of the data of the attacker (if it is another user). It is relatively more difficult to achieve scrambling than jamming because of the attacker's need to interpret control information and to send noise during specific intervals. There are technical difficulties for an attacker to address, but they are solvable. With the attacker motivation ranked as moderate because it is limited to creating mischief, and with the technical difficulty solvable, the likelihood of occurrence of scrambling is possible. Scrambling is more difficult to detect because of the intermittent nature of the attack and the fact that scrambling can also be due to natural sources of noise. Scrambling and scramblers can be detected by monitoring anomalies in performance criteria. This issue was studied for WiFi/802.11 systems by Raya et al. [19]. The situation for WiMAX/802.16 is much different, and further research is required for this case. The impact of scrambling is low because it results in annoyance to a limited number of users and the consequences are reversible (e.g., by retransmission). We believe that scrambling represents a minor risk at this time.

15.3.2 Threats to the MAC Layer

This section begins with an overview of the WiMAX/802.16 MAC layer, including a description of its connections, the process used by an MS for joining the network, and the MAC security model. We then proceed to discuss the threats to confidentiality and authentication.

15.3.2.1 MAC Layer Connections

The MAC layer is connection oriented. There are two kinds of connections: management connections and data transport connections. Management connections are of three types: basic, primary, and secondary. A basic connection is created for each MS when it joins the network and is used for short and urgent management messages. The primary connection is also created for each MS at the network entry time, but is used for delay-tolerant management messages. The third management connection, the secondary one, is used for IP encapsulated management messages (e.g., dynamic host configuration protocol [DHCP], simple network management protocol [SNMP], trivial file transfer protocol [TFP]).

Transport connections can be provisioned or can be established on demand. They are used for user traffic flows. Unicast or multicast can be used for transmission.

15.3.2.2 Network Entry

The network entry of an MS consists of the following steps:

- Downlink scanning and synchronization with a BS.
- Downlink and uplink description acquisition; available uplink channel discovery.
- Ranging.
- Capability negotiation.
- Authorization, authentication, and key establishment.
- Registration.

During scanning, the MS looks for downlink signals by going through the available frequencies and searches for downlink subframes. Whenever a channel is found, the MS gets the downlink and uplink description. It obtains the downlink map and uplink map in the PHY frame headers, and these maps describe the structure of the subframes in terms of bursts. The downlink/uplink channel descriptors are obtained as MAC management messages, and they describe the properties of the bursts in terms of data rate and error correction. During ranging, the MS synchronizes its clock with the BS and determines the level of power required to communicate with the BS. Ranging is done using a special channel called the ranging interval, which uses contention-based multiple access. The basic connection and primary connection are assigned during ranging. Capabilities (e.g., the supported security algorithms) are negotiated on the basic connection. Authorization and authentication can be device list based, X.509 certificate based, or EAP based. This is discussed in more detail below. The registration step results in the establishment of a secondary management connection and provisioned connections.

15.3.2.3 Security Model

The security keys and associations established between an MS and a BS during the authorization step at network entry are discussed in this section. A MAC layer PDU consists of a MAC header, a payload, and an optional cyclic redundancy check (CRC). The payload may consist of user traffic or management messages. The MAC header contains a flag, which indicates whether the payload of the PDU is encrypted or not. MAC headers themselves are not encrypted, and all MAC management messages are sent in the clear. According to the standard, this facilitates the operation of the MAC layer. A security association (SA) is a concept that captures the security parameters of a connection: keys and selected encryption algorithms. The basic and primary management connections do not have SAs, although the integrity of management messages can be secured, as discussed below. The secondary management connection can have, on an optional basis, an SA. Transport connections always have SAs. Each transport connection, a term used to refer to a MAC layer connection dedicated to user traffic, has either one SA for both the uplink and downlink, or two SAs, one for the uplink and another for the downlink. The security model is depicted in Figure 15.4: rectangles depict entities; lines represent relations with cardinalities at the termination points; preexisting elements are shown with solid lines; and dynamically established elements are shown using dashed lines.

There are three types of SAs: the primary SA, static SA, and dynamic SA. Each SA has an identifier (SAID). It also contains a cryptographic suite identifier (selected algorithms), traffic encryption keys (TEKs), and initialization vectors. There is one primary SA for each MS. The primary SA is established when the MS is initialized. The scope of the primary SA is the secondary management connection, and it is shared exclusively between an MS and its BS. Static SAs are created by the BS during the initialization of an MS. For example, there is a static SA for the basic unicast service. However, an MS may have subscribed to additional services, and there are as many additional static SAs as there are subscribed additional services. Dynamic SAs are created dynamically when new traffic flows are opened and they are destroyed when their flow is terminated. Static SAs and dynamic SAs can be shared among several MSs, for example, when multicast is used.

Core security data entities are the X.509 certificate, authorization key (AK), key encryption Key (KEK), and hashed message authentication code (HMAC) key (message authentication key). Every MS is preconfigured with an X.509 certificate. The X.509 certificate is persistent and contains the public key (PK) of the MS. The MS uses it for its authentication with the BS. All other keys are established during authorization, and they are subject to an aging process, so they must be refreshed on a periodic basis through reauthorization. The BS determines the AK, which is encrypted using the PK, and passes

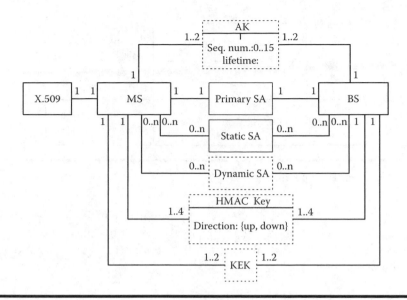

Figure 15.4 Security model.

it to the MS. The AK has a sequence number (from 0 to 15) and a lifetime. For the purpose of smooth transitions, two AKs may be simultaneously active with overlapping lifetime. The lifetime of an AK ranges from 1 to 70 days, with a default value of 7 days. The MS uses the AK to determine the KEK and HMAC key. The sequence number of the AK implicitly belongs to the HMAC keys as well. KEKs are used to encrypt TEKs during their transfer.

15.3.2.4 Threats to Confidentiality

The format of the MAC PDU payload is depicted in Figure 15.5. When applicable, before encryption, each packet is given a unique identifier as a new four-byte packet number which is increased from one data unit to another. Note that, for the sake of uniqueness, there are separate ranges of values for the uplink and downlink packets. The IEEE 802.16e standard uses Data Encryption Standard (DES) in the CBC mode or advanced encryption standard (AES) in the CCM mode to encrypt the payload of MAC PDUs. This standard introduces an integrity protection mechanism for data traffic which did not previously exist. CBC-MAC (as a component of AES-CCM) is used to protect the integrity of the payload of MAC data units.

Table 15.4 provides values for the eavesdropping threat, first for management messages, then for user traffic. Management messages, which are never encrypted, can provide valuable information to an attacker, for example, to verify the presence of a victim at his location before perpetrating a crime. This provides a high motivation for an attacker. The messages can be intercepted by a passive listener within communication range, so there are no serious technical difficulties to resolve by an attacker. The threat is therefore likely to occur. From the user perspective, eavesdropping of management messages may result in limited financial loss if a crime is committed, resulting in an attack of medium impact. From the point of view of a system, eavesdropping in itself may not create outages, but it might be used by a competitor to map the network, making it a threat of high impact. Hence eavesdropping of management messages is a major threat for users and a critical one for a system.

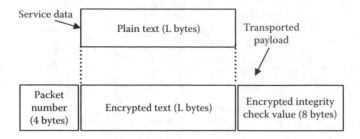

Figure 15.5   MAC layer PDU payload format.

Eavesdropping of data traffic is an unlikely threat because of the strong security measures provided by encryption, which presently pose unsurmountable technical difficulties. As a result, the threat is minor to both users and the system, and there is no need for countermeasures.

15.3.2.5   Threats to Authentication

The kinds of authentication in WiMAX/802.16 are listed in Table 15.5. There is device-level authentication, which is RSA/X.509 certificate-based and is useful for detecting stolen devices and blocking their access to the network. The certificate can be programmed in a device by its manufacturer. This type of authentication is unilateral (i.e., BSs are not authenticated).

The IEEE 802.16 standard states that identity can be verified via the X.509 digital certificate. This wording suggests that it is possible to disregard the X.509 certificate and base access control on a predetermined list of devices. In this case, a BS grants network entry only to MSs featured on a preconfigured list, while an MS is configured with its network identifier and joins a BS only if it belongs to that network.

Any weakness in authentication is an enabler for the BS or MS masquerading threat, which may result in important gains for an attacker in terms of misappropriation of resources such as air time from another user or from a system. We therefore rate the attacker's motivation as high. Specific techniques for this threat include identity theft and the rogue BS attack.

Identity theft consists of reprogramming a device with the hardware address of another device. This is a well-known problem in unlicensed services such as WiFi/802.11, but has been under control in cellular networks because it has been made illegal and more difficult to execute with subscriber ID module (SIM) cards. It is interesting to note

Table 15.5   Authentication in WiMAX/802.16

Kind	Mechanism
Device	Device list
	RSA/X.509 certificate
User level	EAP + EAP-TLS (X.509) or EAP-SIM (subscriber ID module)
Data traffic	AES-CCM CBC-MAC
Physical layer header	None
MAC layer header	None
Management messages	SHA-1-based MAC
	AES-based MAC

that a recent case of code division multiple access (CDMA) phone cloning in India has been documented [20]. The address can be stolen over the air by intercepting management messages.

A rogue BS is an attacker station that imitates a legitimate BS. The rogue BS confuses a set of MSs trying to get service through what they believe to be a legitimate BS. The exact method of attack depends on the type of network. In a WiFi/802.11 network, which uses carrier sense multiple access, the attacker has to capture the identity of a legitimate access point (AP), build a message using the legitimate AP's identity, wait until the medium is idle, and send the message. This appears to be one of the top security threats in WiFi/802.11 networks [21]. In a WiMAX/802.16 network, this is more difficult to accomplish because of the TDMA model. The attacker must transmit while the legitimate BS is transmitting. However, the signal of the attacker must arrive at the targeted receiver MSs with more strength and must put the signal of the legitimate BS in the background, relatively speaking. Again, the attacker has to capture the identity of a legitimate BS and build a message using that identity. The attacker has to wait until a time slot allocated to the legitimate BS starts. The attacker must then transmit while achieving a receive signal strength (in dBm) higher than that of the legitimate BS. The receiver MSs reduce their gain and decode the signal of the attacker instead of the one from the legitimate BS.

Mutual authentication at the user-network level has been introduced in WiMAX/802.16. Mutual authentication, when available, occurs after scanning, acquisition of channel description, ranging, and capability negotiation. It is based on the extensible authentication protocol (EAP) [22], which is a generic authentication protocol. For WiMAX/802.16, EAP can be actualized with specific authentication methods such as EAP-TLS (X.509 certificate-based) [23] or EAP-SIM [24].

There are three options for authentication: device list based, X.509 based, or EAP based. If only device list-based authentication is used, identity theft by device address reprogramming is greatly facilitated, and the likelihood of a BS or MS masquerading attack is likely because there are few technical difficulties to solve. The impact for a user is high because it can lead to a loss of service for long periods of time and the user can be billed for another user's communication fee. The impact for a system is medium because it can lead to limited financial loss or theft of resources. The risk is therefore critical for a user and major for a system, and there is the need for countermeasures.

If X.509-based authentication is used, the likelihood for a user (a MS) to be the victim of BS masquerading is possible because of the asymmetry of the mechanism and the solvable technical difficulties. The strong technical difficulties in MS masquerading render it an unlikely threat to a system. The impact is the same as for the device list-based authentication. Therefore, in the case of a user, the risk is assessed as major and countermeasures are needed. For a system, the risk is minor and there is no need for countermeasures.

If EAP-based authentication is used, we believe that at this time the likelihood of a BS or MS masquerading attack is possible. Some of the EAP methods are still being defined, and security flaws are often uncovered in unproven mechanisms. The technical difficulties in carrying out an attack are therefore best estimated as solvable. Aboba maintains a Web page about security vulnerabilities in EAP methods [25]. The impact is the same as for the device list and X.509 certificate-based authentication, so the risk is ranked as major for both the user and the system. It is a good idea to provide a second line of defense to play safe with EAP-based authentication.

The modification of MAC management messages poses a moderate motivation for an attacker because it stems from the goal to create mischief and results in limited gains.

In terms of technical difficulty, MAC management messages are never encrypted and not always authenticated. If an authentication mechanism is used for MAC layer management messages, it is negotiated at network entry. The scope of management messages to which authentication is applicable is limited in earlier versions of IEEE 802.16, but has been extended in version 802.16e. Hence, with earlier versions of the IEEE 802.16 standard, management messages are not subject to integrity protection.

Weaknesses in management message authentication open the door to aggressions such as the man in the middle attack, active attack, and replay attack. However, the following authentication mechanisms are available: the HMAC tuple and the one-key message authentication code (OMAC) tuple. The OMAC is AES-based and includes replay protection. The HMAC authentication originally specified in the IEEE 802.16 standard did not provide a counter to protect against replay attacks, but 802.16e does, so we distinguish both possibilities in our analysis. The technical difficulty in defeating the four different possibilities for authentication is as follows: none where no authentication is used, solvable for the HMAC case with no replay protection, and strong for both cases where the HMAC with replay protection or the OMAC defense is used. The likelihood of the management message modification threat is therefore possible for the first two cases and unlikely for the latter two. In all cases, the impact of an attack of that type can be high because it might affect the operation of the communications. The risk is therefore ranked as major for both cases where no authentication or HMAC without the replay counter is used, and minor for both the HMAC with the replay counter and the OMAC cases. As a result, it might be safe to provide a second line of defense against this type of attack.

Authentication of traffic messages also presents a moderate motivation for an attacker because it is an attack rooted in creating mischief. The modification of data traffic is very unlikely to occur if AES is used because of the strong technical difficulties encountered, and possible if AES is not used, given the lack of technical difficulty in carrying out an attack. We believe that such an attack has the potential to create short-term consequences for the user and system, resulting in a medium impact. If AES is not used, then this is a major threat, otherwise it is minor.

There is the potential for denial of service (DoS) attacks based on the fact that authentication operations of devices, users, and messages trigger the execution of long procedures. A DoS attack can be perpetrated by flooding a victim with a large number of messages to authenticate. With a moderate motivation on the part of the attacker bent on creating mischief, and with little technical difficulty to solve, this threat is possible. The impact is medium for a system, but could be high for a user because of lower computational resources available for handling a large influx of invalid messages. The DoS threat is therefore assessed as major for both the user and the system.

15.4 Conclusion

An analysis of the threats to the security of WiMAX/802.16 broadband wireless access networks was conducted. Critical threats consist of eavesdropping of management messages and BS masquerading. Major threats include jamming, MS masquerading, management message and data traffic modification, and DoS attacks. Countermeasures need to be devised for networks using the security options with critical or major risks. An intrusion detection system approach can be developed to address some of the threats, but more research is needed in this direction.

References

1. IEEE, Local and Metropolitan Area Networks—Part 16: Air Interface for Fixed Broadband Wireless Access Systems, Standard 802.16-2004, IEEE, Washington, DC, 2004.
2. C. Eklund, R. Marks, K. Stanwood, and S. Wang, IEEE standard 802.16: a technical overview of wirelessman air interface for broadband wireless access, *IEEE Commun. Mag.*, 40(6), 98, 2002.
3. Air Interface for Fixed and Mobile Broadband Wireless Access Systems Amendment 2: Physical and Medium Access Control Layers for Combined Fixed and Mobile Operation in Licensed Bands and Corrigendum 1, Standard 802.16e-2005 and Standard 802.16-2004/Corrigendum 1-2005, IEEE, Washington, DC, 2006.
4. F.J. Groen, C. Smidts, and A. Mosleh, QRAS—the quantitative risk assessment system, *Reliab. Eng. Syst. Saf.*, 91(3), 292, 2006.
5. E. Paté-Cornell, Finding and fixing system weaknesses: probabilistic methods and applications of engineering risk analysis, *Risk Anal.*, 22(2), 319, 2002.
6. C.J. Alberts and A.J. Dorofee, OCTAVE criteria, version 2.0, available at www.cert.org/octave/pubs.html, 2001.
7. Club de la sécurité des systèmes d'information français (CLUSIF) Methods Commission, MEHARI V3 concepts and mechanisms, available at www.clusif.asso.fr/en/clusif/present, 2002.
8. ETSI, Telecommunications and Internet protocol harmonization over networks (TIPHON) release 4; protocol framework definition; methods and protocols for security; Part 1: Threat analysis, Technical Specification ETSI TS 102 165-1 V4.1.1, ETSI, Sophia-Antipolis, France, 2003.
9. M. Barbeau, WiMAX/802.16 threat analysis, *1st ACM International Workshop on QoS and Security for Wireless and Mobile Networks*, ACM, New York, 2005.
10. C. Laurendeau and M. Barbeau, Threats to security in DSRC/WAVE, *5th International Conference on Ad Hoc Networks*, 2006.
11. R. Blake, Hackers in the mist, available at www.eff.org/Net_culture/Hackers/hackers_in_the_mist.article, 1994.
12. R. Schifreen, What motivates a hacker, *Network Security*, 1994(8), 17, 1994.
13. R. Barber, Hackers profiled—who are they and what are their motivations, *Computer Fraud Security*, 2001(2), 14, 2001.
14. IEEE, ANSI/IEEE Std 802.11—Wireless LAN Medium Access Control (MAC) and Physical Layer (PHY) Specifications, IEEE, Washington, DC, 1999.
15. S. Fluhrer, I. Mantin, and A. Shamir, Weaknesses in the key scheduling algorithm of RC4, *Selected Areas in Cryptography: 8th Annual International Workshop, SAC 2001 Toronto, Ontario, Canada, August 16–17, 2001*, P. 1, Springer, Berlin, 2001.
16. A. Stubblefield, I. Ioannidis, and A. Rubin, Using the Fluhrer, Mantin and Shamir attack to break WEP, *Proceedings of the 2002 Network and Distributed Systems Security Symposium*, ISOC, Reston, VA, 2002.
17. WiFi Alliance, Wi-Fi protected access (WPA): enhanced security implementation based on IEEE P802.11i standard, version 3.1, Wifi Alliance, Austin, TX, 2004.
18. R. Poisel, *Modern Communications Jamming Principles and Techniques*, Artech House, Norwood, MA, 2003.
19. M. Raya, J. Hubaux, and I. Aad, DOMINO: a system to detect greedy behaviour in IEEE 802.11 hotspots, *Proceedings of the 2nd International Conference on Mobile Systems, Applications and Services, Boston, MA, June 6–9, 2004*, p. 84–97, ACM Press, New York, 2004.
20. Ghosh, R. "Mobile Cloning." *The Statesman* [Calcutta, India] 15 March 2005: Science and Technology Section.
21. Ernst and Young, The necessity of rogue wireless device detection, White Paper, Ernst and Young, New York, 2004.

22. B. Aboba, L. Blunk, J. Vollbrecht, J. Carlson, and H. Levkowetz, Extensible authentication protocol (EAP), RFC 3748, Internet Engineering Task Force, ISOC, Reston, VA, June 2004.
23. B. Aboba and D. Simon, PPP EAP TLS authentication protocol, RFC 2716, Internet Engineering Task Force, ISOC, Reston, VA, October 1999.
24. H. Haverinen and J. Salowey, Extensible authentication protocol method for GSM subscriber identity modules (EAP-SIM), work in progress, December 2004.
25. B. Aboba, The unofficial 802.11 security web page—security vulnerabilities in EAP methods, available at www.drizzle.com/aboba/IEEE/, May 2005.

Chapter 16

Techno-Economic Analysis of Fixed WiMAX Networks

T. Smura, H. Hämmäinen, T. Rokkas, and D. Katsianis

Contents

Worldwide Interoperability for Microwave Access (WiMAX) radio networks offer an alternative way to provide fixed, nomadic, and mobile broadband services to businesses and consumers. In these markets, it is both complementing and competing with other access technologies such as asymmetric digital subscriber line (ADSL), wireless local area network (LAN), and third-generation (3G) mobile networks. The success of WiMAX requires the technology to reach performance and cost levels comparable or superior to these other technologies, which already have an established position in the market.

This chapter analyzes the competitive potential of fixed WiMAX networks in different market conditions. A quantitative techno-economic model is constructed for analyzing fixed WiMAX network deployments in urban, suburban, and rural areas. The model is used to assess the coverage, capacity, and cost characteristics of WiMAX systems, in contrast to the competing access technologies.

16.1 Introduction

Broadband markets have experienced significant growth during the past few years, and fixed broadband penetration exceeds 50% of households in many countries. Digital subscriber line (DSL) technologies have become the most popular broadband access method globally, constituting more than 60% of the broadband lines in Organization for Economic Cooperation and Development (OECD) countries and about 80% of broadband lines in Europe. Mobile broadband services enabled by 3G cellular technologies are also emerging, but have not yet reached maturity.

At the same time, competition between technologies and providers has intensified, as new players are entering the market. Generally, two different types of competition can be distinguished: service based and infrastructure based. Service based competition takes place when new entrant operators utilize already existing network infrastructures, offering, for example, DSL services by means of unbundled lines, or acting as mobile virtual network operators (MVNOs) in existing mobile networks. In infrastructure-based competition, the entrants build and operate their own access network infrastructure. WiMAX broadband wireless access systems induce infrastructure-based competition in both fixed and mobile broadband markets.

Initial WiMAX deployments based on the IEEE 802.16-2004 wireless metropolitan area network (MAN) air interface standard provide fixed broadband access to households and businesses, whereas the next generation of equipment (IEEE 802.16e-2005) will support mobile terminals as well. Required investments as well as expected service revenues differ between these two deployment scenarios. The profitability of the scenarios is also dependent on the characteristics of the geographical areas (e.g., subscriber density and existence of competing network infrastructure).

This chapter gives an overview of techno-economic modeling methods and tools, and how these can be utilized to analyze different types of WiMAX deployment scenarios. After a brief introduction of the techno-economic modeling methodology, we discuss the technical performance characteristics of fixed WiMAX systems and give an overview of the market- and service-related issues relevant for the analysis. Finally, a techno-economic model is constructed and used to determine the feasibility of fixed WiMAX deployments in different market settings. Sensitivity and risk analysis are carried out to recognize the most critical success factors of WiMAX and to cope with the uncertainty inherent in our analysis.

16.2 Techno-Economic Modeling: Methodology and Tools

The purpose of techno-economic modeling is to analyze the economic profitability of certain network deployment scenarios. Quantitative techno-economic models are constructed, combining market-and service-related parameters and forecasts with cost and performance-related parameters of the analyzed technologies. A number of indicators are used to determine profitability, including payback period, net present value (NPV), and internal rate of return (IRR), to get a better understanding of the economics of each examined scenario.

A typical cash balance (or cumulative cash flow) curve for a network scenario first goes deeply down to the negative side because of the high initial investments. If the scenario is profitable, the cash flow turns positive fairly soon and the cash balance curve starts to rise. The lowest point in the curve gives the amount of funding required for the project. The point in time the when the cash balance turns positive gives the payback period for the project.

Net present value is defined as the present value of future cash flows (revenues less costs), discounted using a factor that resembles the time-value of money. If the NPV is positive, the project is judged profitable. The IRR is calculated as the discount factor that gives a zero NPV. A higher IRR means higher profitability and a better return on investment. The IRR is a useful meter in a case where the scenarios to be compared are of different size and scope.

16.2.1 TONIC Tool

A tool called TONIC (TechnO-ecoNomICs of Internet protocol [IP] optimized networks and services) is used for the techno-economic modeling and evaluation of the investment scenarios described later in this chapter. TONIC is a spreadsheet-based application that has been developed within the European Union's Information Society Technologies (IST) program and is an implementation of the methodology developed by a series of projects in the field of telecommunication techno-economics. The tool has been used in several techno-economic studies among major European telecommunication organizations and academic institutes [1–4]. TONIC consists of a dimensioning model for different network architectures that is linked to a database containing information about network elements and their cost evolution over time. The tool is used to calculate revenues, investments, and operational expenditures related to the analyzed network deployment scenarios. By combining these with other economic inputs (e.g., discount rate and tax rate), TONIC calculates outputs such as annual and cumulative cash flows, NPV, IRR, and payback period. An analytical description of the methodology and a similar tool can be found in Ims [5], while a more detailed description of the tool used in this analysis is presented by Olsen [6]. The main principles of the methodology used in the techno-economic tool are illustrated in Figure 16.1.

The analysis of a network deployment project is performed for a certain user-defined study period. The services to be provided are defined, as well as their market penetration and tariff levels over the study period. The revenues for each year are calculated by combining yearly market penetration and tariff information.

Network planning and dimensioning is done mostly outside the TONIC tool. The resulting shopping list indicates the volumes of all network cost components for each

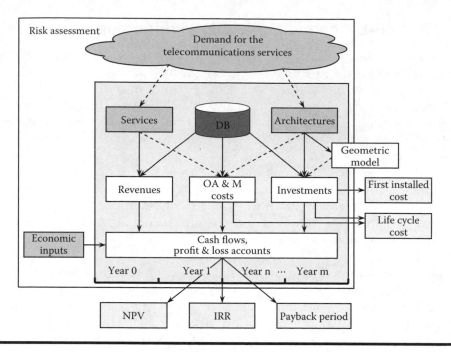

Figure 16.1   Framework for techno-economic analyses [5].

year of the study period. A geometric model provided by the tool can be used to estimate the amount of cable, ducts, and civil works required in the access network.

The costs of the network components are calculated using a cost database integrated in the tool. The network architecture and the shopping list, together with the cost database, give the investments for each year. The investments are usually spread over the study period. The first installed cost is calculated by discounting all the investments to the start of the study period.

The operation, administration, and maintenance (OA&M) costs include repair parts, repair work, and operation and administration costs. The first two of these are automatically calculated by the tool with user-defined parameters, but the last one has to be included in the models manually. The investment costs together with the OA&M costs give the life-cycle costs of the project.

16.2.2   Sensitivity and Risk Analyses

Most of the inputs and assumptions required for techno-economic models are forward-looking and by definition uncertain. However, small variations in the inputs may have a great impact on the outcome of the analysis. Sensitivity and risk analyses are useful in coping with this uncertainty.

Sensitivity analysis is a simple technique used to locate and assess the potential impact of risk on a project's value. The aim is to identify the impact of changes in key assumptions on the profitability (e.g., the NPV) of the project. The results of the sensitivity analysis can be plotted in so-called sensitivity graphs, showing the project NPV as a function of the analyzed parameter.

Risk analysis using Monte Carlo simulation, for example, is a complement to sensitivity analysis. In Monte Carlo simulation, hundreds or thousands of possible combinations of variable values are generated according to predefined probability distributions. Each combination produces an NPV, and the NPVs of all scenarios together produce a probability distribution, which is the outcome of the simulation. In order to be useful, the Monte Carlo simulation requires probability distributions to be specified for all the key variables.

16.3 Technical Performance of Fixed WiMAX Systems

16.3.1 Standards and Interoperability

The IEEE 802.16-2004 standard [7] specifies the air interface for fixed broadband wireless access (BWA) systems. The standard includes separate physical (PHY) layer specifications for single carrier (SC), orthogonal frequency division multiplexing (OFDM), and orthogonal frequency division multiple access (OFDMA) systems, operating on various licensed and license-exempt bands between 2 and 60 GHz. Furthermore, the standard allows manufacturers to choose between frequency divisiion duplex (FDD) and time division duplex (TDD) schemes, various channel bandwidths, and point-to-multipoint (PMP) and mesh network topologies. For a more detailed introduction to the IEEE 802.16 standard, see Eklund et al. [8] and Ghosh et al. [9].

The IEEE 802.16e-2005 standard specifies a system for combined fixed and mobile BWA systems. The standard provides enhancements to the 802.16-2004 standard to support higher layer handovers between base stations (BSs) or sectors and mobility at vehicular speeds. Operation is limited to licensed bands below 6 GHz. Also, the 802.16e-2005 standard includes a number of options for the equipment manufacturers to choose from.

Because of the numerous options, compliance to standards is not sufficient to guarantee interoperability. Therefore the WiMAX Forum was founded in 2001 to promote and certify the compatibility and interoperability of BWA systems from different equipment manufacturers. To drive interoperability, the WiMAX Forum has specified a number of system profiles (i.e., subsets of options in the IEEE standards that have to be implemented by the equipment manufacturers in order to get their systems certified). The certification process for fixed WiMAX systems was launched in mid-2005.

16.3.2 Frequency Bands and Regulations

The selection of frequency bands for WiMAX deployments has an important effect on the achievable range and capacity of the BS cells and sectors. The 802.16-2004 standard allows the use of various licensed and license-exempt frequency bands in the range of 2 to 60 GHz. The first WiMAX-certified products will operate in the licensed 3.5 GHz frequency band, followed by systems for both the 2.5 GHz licensed band as well as the 5.8 GHz license-exempt band. Characteristics of these bands are listed in Table 16.1.

The choice between licensed and unlicensed bands has a major effect on network deployment. The licensed spectrum provides protection from interference, but the operator must deal with the license acquisition process. Use of the unlicensed spectrum has the advantage of faster deployment, but interference from other networks or radio devices that operate on the same frequency band cannot be avoided, and the transmission power levels are typically more limited. Here, only the licensed bands will be considered.

In addition to the initially available bands, a spectrum for fixed BWA systems is available at frequency bands above 10 GHz. In many European countries, large amounts

Table 16.1 Initially Available Frequency Bands for WiMAX-Certified Systems

Frequency Band (GHz)	2.300–2.400	2.500–2.690	3.400–3.600	5.725–5.850
Availability	South Korea	U.S., U.K., Canada, Australia, Central America	Worldwide except U.S.	North America, South America
Licensed/ unlicensed	Licensed	Licensed	Licensed	Unlicensed
Profile for 802.16-2004	No	No	Yes (FDD / TDD)	Yes (TDD)
Profile for 802.16e-2005	Yes (TDD)	Yes (TDD)	Yes (TDD)	No
Channel bandwidth	8.75 MHz	3 MHz, 5.5 MHz	3.5 MHz, 7 MHz	10 MHz
Typical spectrum allocation per operator	27 MHz	3×5.5 MHz + 6 MHz (total 22.5 MHz per license)	Varies, from 2×7 MHz to 2×56 MHz	No licenses

Various sources.

of spectrum are available around the 26 GHz, 28 GHz, and 40 GHz frequency bands, and similar allocations have been made in other parts of the world. The characteristics of these bands are very different from the sub-10-GHz bands, requiring strict line-of-sight (LOS) link conditions and experiencing higher path loss and rain fading. On the other hand, spectrum allocations are typically large, on the order of 100 to 200 MHz per operator, allowing higher data rates to be offered.

The regulatory policy for 2.5 GHz and 3.5 GHz band license allocations has generally favored beauty contests, although some countries (e.g., Canada and Austria) have organized auctions. This has kept the spectrum cost low. Most licenses are regional instead of nationwide: in North America, 100% of licenses are regional and in Europe 77% are [10]. In most cases, multiple licenses per region are allocated to promote competition. It should be noted that 78% of the cumulative 721 licenses globally are limited by the regulator to allow only fixed broadband wireless local loop (WLL) services. This regulatory policy has resulted in a very high number of license holders compared to, for instance, 3G. Not surprisingly, the portfolio of license holders includes, in addition to established fixed and mobile operators, many challengers such as power companies and other local players. Low license costs have attracted less powerful and passive license holders. Thus the WiMAX market is still fragmented and rather unpredictable.

In this chapter, our focus is on the licensed bands of 2.5 GHz and 3.5 GHz, as these are the first ones available for systems based on the 802.16-2004 and 802.16e-2005 standards. In Europe, the 2.5 GHz band is currently reserved for International Mobile Telecommunications (IMT)-2000 systems, although there is an ongoing debate about applying the principle of technology neutrality to the band or parts of it, thus making it available for WiMAX. The 3.5 GHz band is not available in the United States.

16.3.3 Range and Coverage

The range and coverage of WiMAX BS sectors can be estimated by link budget calculations and suitable path loss models. In situations where the detailed geography of the service area is unknown, empirical models based on measurement statistics are useful. Generally the mean value of path loss can be modeled using a log-distance path loss model:

$$L = L_{d_0} + 10n\log_{10}\left(\frac{d}{d_0}\right) = 20\log_{10}\left(\frac{\lambda}{4\pi d_0}\right) + 10n\log_{10}\left(\frac{d}{d_0}\right) \; dB, \qquad (16.1)$$

where d is the distance, d_0 is a reference path loss, L_{d_0} is the path loss at $d = d_0$, and n is the path loss exponent. The reference path loss can be, for example, the free space path loss or based on field measurements. The path loss exponent depends on the area type, is found by field measurements, and usually varies between two and five. Log-distance path loss models are often extended with additional correction factors, making the models applicable to different operating frequencies and antenna heights.

By far, empirical propagation models have been developed mainly for mobile systems below 3 GHz, and the applicability of the models for higher frequency bands is somewhat uncertain. Examples of possible candidates for the 2.5 GHz and 3.5 GHz bands include the Stanford University Interim (SUI) models [11] and the ECC-33 model [12]. Here we have used the SUI path loss models.

As an example, Figure 16.2 shows the predicted mean path losses of SUI path loss models in the 3.5 GHz frequency band, assuming a BS height of 30 m and a customer premises equipment (CPE) antenna height of 6 m. The figure also shows link budgets for fixed and mobile WiMAX systems using different types of CPEs, assuming typical parameters of commercial systems and a 10 dB fade margin. As the figure illustrates, the choice between indoor and outdoor CPEs has a significant effect on the achievable range. An assumed 22 dB decrease in the link budget as a result of a lower gain antenna (18 dBi versus 6 dBi) and building penetration loss (10 dB) drops the link range by a factor of three to four.

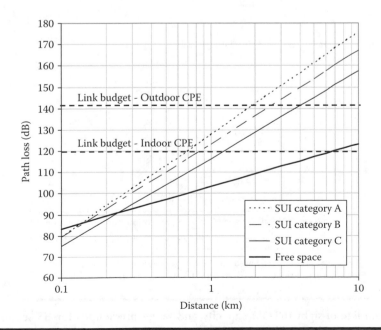

Figure 16.2   Predicted mean path loss curves for 3.5 GHz WiMAX systems.

For systems operating on the 2.5 GHz frequency band, the path loss, as predicted by Equation (16.1), is somewhat lower than for 3.5 GHz systems. Actual range figures are presented in the case study section.

16.3.4 Capacity

When dimensioning BWA networks, a natural unit of analysis is one BS sector, the capacity of which depends on the channel bandwidth and the spectral efficiency of the utilized modulation and coding scheme, as well as on the characteristics of the environment. The 802.16 PHY layer utilizes adaptive burst profiles to maximize the sector capacity, meaning that inside one BS sector each CPE may choose the most suitable modulation and coding type irrespective of other CPEs [8]. On average, CPEs located further away from the BS have to use more robust and less effective modulation types.

Radio access network deployments can be either coverage or capacity limited. In coverage limited cases, the capacity demand of the service area can be fulfilled with a minimum number of BS cells optimized for maximum range. In capacity limited cases, the number of BS cells required to fulfill the capacity demand is greater than what is needed to provide full coverage to the service area.

The capacity of WiMAX BS sectors also depends on the characteristics of the service area. In coverage limited deployments, the most robust modulation types (i.e., binary phase-shift keying [BPSK]) are in use, decreasing the average sector capacity. In capacity limited deployments, the least efficient modulation types are not required, increasing the average sector capacity. Figure 16.3 illustrates this trade-off between capacity and range for different propagation environments and CPE types.

In coverage limited deployments, the average sector capacity is calculated to be 10 Mbps for a 7 MHz channel. In capacity limited deployments, the sector capacity is

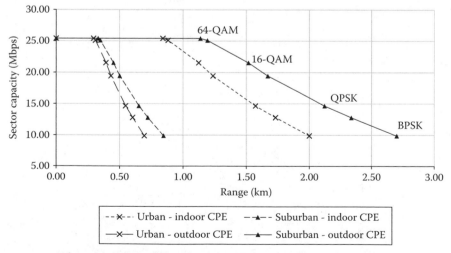

Urban area predictions based on SUI category A path loss model
Suburban area predictions based on SUI category B path loss model

Figure 16.3   Non-line-of-sight (NLOS) capacity and range predictions for BS sectors (frequency band = 3.5 GHz, channel bandwidth = 7 MHz, receiver sensitivity requirements as in IEEE 802.16-2004) [7].

between 10 Mbps and 25 Mbps, the range of a 25 Mbps sector being about 38% of a 10 Mbps sector. Assuming the 22 dB loss, the achievable range of indoor CPEs is found to be shorter than the range of outdoor CPEs utilizing the most efficient 64-QAM modulation. For the 2.5 GHz frequency band, the channel bandwidth is 5.5 MHz, resulting in about 21% lower sector capacity.

Various technologies have been proposed to improve the capacity/coverage performance of WiMAX systems, including adaptive antenna systems, spatial multiplexing, and interference cancelation, as well as hybrid automatic repeat request (HARQ) and adaptive subcarrier/power allocation, introduced in detail in Ghosh et al. [9].

16.4   Case Study: Fixed WiMAX Network Deployment

For the purposes of our analysis, an advanced techno-economic model was created, utilizing an Excel-based tool created by the IST-TONIC project [13]. The model takes into account various market- and technology-related parameters and assumptions, and gives basic profitability measures such as NPV and IRR as outputs. The required inputs and the internal logic of the model are illustrated in Figure 16.4.

The model is used to analyze an 802.16-2004-based fixed WiMAX network deployment, providing fixed broadband services for residential customers and small to medium-size enterprises (SMEs). All the calculations are done over a 5 year study period (2006 to 2010) using a discount rate of 10%. The value of the network at the end of the study period is assumed to be the depreciated value of cumulative investments. In the following sections we will go through the other inputs and assumptions.

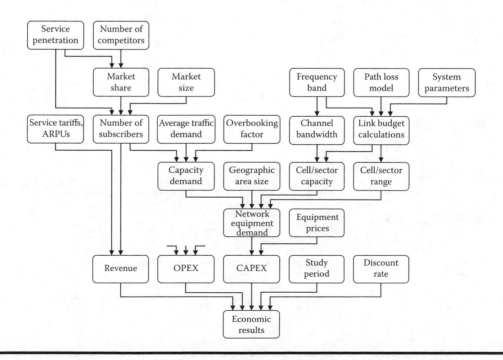

Figure 16.4   Model for techno-economic analysis of WiMAX networks.

16.4.1 Market- and Service-Related Inputs

In addition to the various technical parameters and assumptions, well-formulated forecasts for service penetration, tariffs, and traffic evolution are also required for the techno-economic analysis. As a basis for the forecasts, we have used broadband statistics available from the OECD [14]. Our forecasts and the assumptions behind them are discussed in the following subsections.

16.4.1.1 Service Penetration

WiMAX systems are aimed at providing broadband access services for fixed, portable, and mobile devices as an alternative to a number of existing and emerging technologies. As in all communications services, a complete service offering requires sufficient network coverage and capacity, end-user terminals, as well as overlay applications, services, and content to be available at a price acceptable to consumers.

Systems based on both the 802.16-2004 and 802.16e-2005 standards can be deployed to offer fixed broadband access services as a competitor or complement to other fixed broadband networks. The most important alternatives are currently based on DSL and cable modem systems. The penetration of fixed broadband services varies significantly between different markets, as illustrated in Figure 16.5.

For the purposes of this study, we have forecasted the penetration of broadband services to follow the curves shown in Figure 16.6. The forecast is based on data available from the OECD, and assumes a service availability of 100%.

Different areas inside each country have different characteristics in terms of service availability and penetration. In most countries, large cities are covered by both DSL and cable network infrastructure, whereas in smaller towns and suburban areas, only DSL might be available. In sparsely populated rural areas the situation is often worse, as even DSL availability may be limited.

Accordingly, the expected number of subscribers for WiMAX operators varies between countries and between area types. In urban areas, the population density and

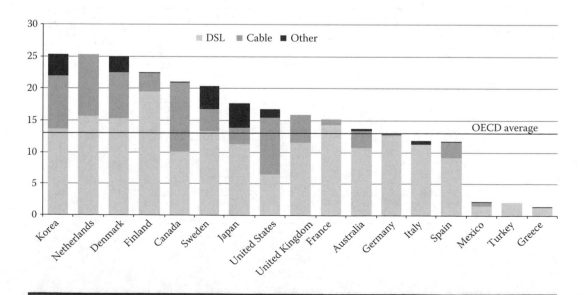

Figure 16.5   Broadband subscribers per 100 inhabitants, by technology, December 2005 [14].

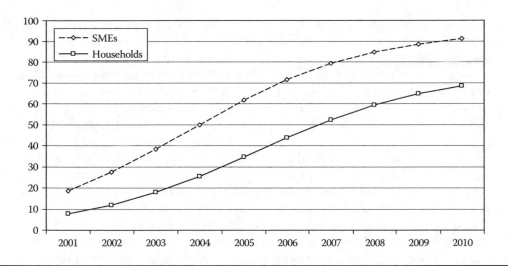

Figure 16.6   Broadband penetration forecasts.

number are higher, and smaller cells are required to match the capacity requirements. Furthermore, the competition is more intense and a new operator must expect a smaller market share and higher sales and marketing costs. In suburban areas, population density is somewhat lower, the required cell radius is larger, and the WiMAX operator can expect less competition since the availability of alternative technologies (DSL or cable) may not be universal. Rural areas are more sparsely populated and there is a pending demand for broadband access, because DSL and cable coverage may be very low. Therefore the competition is less intense.

For this analysis, we separate the areas into urban, suburban, and rural, each having different household and business densities and experiencing different levels of competition (see Figure 16.2). For simplicity and comparability, 50,000 households and 5,000 SMEs are assumed to exist in each area. In the urban areas, both DSL and cable networks are assumed to provide full coverage, whereas in the suburban areas, DSL availability is assumed to be 95% and cable is assumed to cover 50% of the DSL area. In rural areas, DSL availability is assumed to be 75% and no cable services are available.

16.4.1.2   Service Characteristics

For the analysis, we consider the network operator's service portfolio to consist of wholesale bitstream services of various data rates. Typical service classes enabled by current

Table 16.2   Characteristics of Area Types

Area Type	Urban	Suburban	Rural
Area size (km^2)	500	50,000	25,000,000
Household density (1/km^2)	5,000,000	50,000	100
Business density (1/km^2)	50,000	2000	1
Competitors	2	1.5*	1*
DSL availability	100%	95%	75%

*Only in areas with DSL coverage, no competition in residual markets.

DSL and cable modem technologies are asymmetric, having data rates of, for example, 512/1024/2048 kbps in the downlink and 128/256/512 kbps in the uplink. Higher downlink data rates of up to 8 to 12 Mbps are also becoming increasingly available.

For simplicity, we have not made separate forecasts for service classes with different data rates. Instead, we have forecasted the evolution of the average maximum data rates (uplink + downlink) for both households and SMEs. The annual growth rate is assumed to be 20%, from the initial levels of 1 Mbps for households and 2 Mbps for SMEs.

The overbooking factor (or contention ratio) has an important role in the dimensioning of broadband access networks. The maximum line capacities of subscribers are not required constantly and the traffic flows of different users can be statistically multiplexed. In the case of DSL, the traffic is multiplexed from the digital subscriber line access multiplexer (DSLAM) onward, whereas in the case of WiMAX networks, multiplexing happens in the air interface. Typically, overbooking factors of 10 to 20 are used among operators, although higher and lower values are also possible. Often the operators do not inform their customers about the overbooking factors in use.

For the purposes of this study, overbooking factors of 20 for residential users and 4 business users are assumed throughout the study period. For network dimensioning purposes, the traffic patterns of residential and business users are assumed to be overlapping. During the residential user, hours the traffic demand of business users is assumed to be 20% of maximum, and vice versa.

16.4.1.3   Competition Model and WiMAX Operator's Market Share

To forecast the market share evolution of the WiMAX operator, a simple competition model is used, taking into account broadband service penetration, competition level, and DSL availability. In areas with no DSL coverage, the WiMAX operator is assumed to get all of the subscribers desiring broadband services. The WiMAX operator is assumed to satisfy the latent demand of these previously unserved areas during its first year of operation. In areas with existing DSL or cable coverage, the WiMAX network operator has to compete with other network operators. The WiMAX operator's market share is calculated based on the broadband penetration curve and the competition level, which defines how many network operators will be sharing the new subscribers coming to the market. For example, in urban areas with two competing alternatives (DSL and cable), a third of the new subscribers is assumed to choose the WiMAX network. Churn is assumed to have no effect on the WiMAX operator market share (i.e., the amount of incoming and outgoing churning subscribers is assumed to be equal).

16.4.1.4   Service Tariffs

The wholesale tariffs of the WiMAX network operator have to be competitive with other network operators, otherwise it is not able to attract service operators to its network. Accordingly, we assume the WiMAX network operator charges the same bitstream tariffs from the service operators as those being charged by existing DSL network operators. The tariff assumptions were done in a top-down manner, by first forecasting the retail average revenue per unit (ARPU) evolution and then calculating the bitstream service tariffs as a percentage of the retail tariffs.

The retail tariffs of broadband service classes (e.g., 2 Mbps, 1 Mbps, 512 kbps, with different upload speeds) have decreased continuously during the past few years, but at the same time users are migrating to service classes with higher data rates. Therefore, the broadband ARPU (average revenue per user) is not decreasing as fast as the individual service class tariffs. We assume a 15% annual decrease in the retail ARPUs throughout the study period from the 2005 levels of $30 and $200 per month (excluding VAT), for households and businesses, respectively. In order to avoid margin squeeze, the wholesale tariffs have to decrease accordingly. For the purposes of our study, we have assumed the bitstream service tariffs to be 80% of the retail tariffs.

16.4.2 Technology-Related Inputs

16.4.2.1 Selected Frequency Band and System Characteristics

We have limited the scope of our study to systems operating in the licensed 2.5 GHz and 3.5 GHz frequency bands. The bands are available in the United States and Europe and are supported by many equipment manufacturers. The licensed nature of the band allows relatively high transmit powers to be used, and interference can be controlled by the operator owning the exclusive rights to the band.

The analyzed system is based on the IEEE 802.16-2004 standard and utilizes the OFDM PHY layer specification. Channel bandwidth is 5.5 MHz for the 2.5 GHz frequency band and 7 MHz for the 3.5 GHz band. TDD is used to separate uplink and downlink transmissions. The spectrum allocation of the operator is assumed to be large enough to allow six-sector BSs to be deployed without cochannel interference having any significant effect on the capacity.

Simple link budget calculations showing the assumed link gains and losses are shown in Table 16.3, based on typical values reported by WiMAX equipment manufacturers.

16.4.2.2 Network Architecture and Cost Components

In our study, we are mostly interested in the cost of building the network infrastructure and in the revenue levels required to cover these costs. Therefore our analysis is made from the viewpoint of a network operator offering wholesale services to service operators. Costs do not include service operator-specific elements related to, for example, sales and marketing or billing and customer care, reducing the number of uncertain parameters considerably.

In our study, the network operator builds and operates the WiMAX access network and the required transmission links toward a core network access point. Figure 16.7 shows the network architecture considered for the WiMAX deployment, as well as a respective DSL network architecture.

The major capital expenditure (CAPEX) elements of WiMAX network deployments include the costs of BSs, core point-to-point (P2P) transmission links, and CPEs. We

Table 16.3 System Characteristics and Link Budget Calculation

Case	Outdoor CPE	Indoor CPE
Transmit power	23 dBm	23 dBm
BS antenna gain	16 dBi	16 dBi
CPE antenna gain	18 dBi	6 dBi
Receiver sensitivity	−95 dBm	−95 dBm
System gain	152 dB	140 dB
Fade margin	10 dB	10 dB
Building penetration loss	0 dB	10 dB
Link budget	142 dB	120 dB

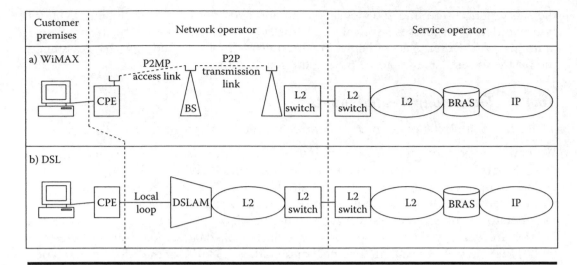

Figure 16.7 WiMAX network deployment architecture.

assume that the CPEs are owned by the network operator and leased to the end users at typical DSL/cable modem rental prices, making the service offering competitive with DSL/cable offerings. For the transmission network part, we assume the connections between the BSs and the core network are built with P2P radio links (based on, for example, another variation of the IEEE 802.16-2004 standard, or proprietary technologies) owned by the WiMAX operator.

Operational expenditure (OPEX) elements related to the network rollout include the maintenance and administration of network elements, leasing of equipment space and antenna sites, and installation costs of BSs and subscriber stations (SSs). An important part of OPEX consists of the costs related to the installation of outdoor CPEs, as a technician is required to visit each new subscriber to install and direct the CPE outdoor antennas.

Table 16.4 lists the CAPEX and OPEX assumptions used in our techno-economic model, together with the expected price evolution over the study period. Price levels of

Table 16.4 CAPEX and OPEX Assumptions for WiMAX Network Deployments

Cost Component	Price in 2006	Price Evolution
Spectrum license fee (e.g., 8×7 MHz)	$0	—
WiMAX 3.5 GHz BS	$10,000	15% per year
WiMAX 3.5 GHz BS sector	$7,000	15% per year
BS installation cost	$5,000 per BS + $500 per sector	—
BS site rental	$1,800 per BS per year + $1,200 per sector per year	—
Transmission link equipment (P2P radio link + port in core switch)	$20,000 per BS	10% per year
P2P radio link site rental	$2,400 per BS per year	—
WiMAX 3.5 GHz indoor CPE	$250	20% per year
WiMAX 3.5 GHz outdoor CPE	$350	20% per year
Outdoor CPE installation cost	$100 per installation	—
Network equipment administration and maintenance costs	15% of cumulative investments	—

Table 16.5   Range of WiMAX Cells

Area Type Frequency Band	Urban		Suburban		Rural	
	2.5 GHz	3.5 GHz	2.5 GHz	3.5 GHz	2.5 GHz	3.5 GHz
Cell range, NLOS, indoor CPE	0.83 km	0.70 km	1.02 km	0.85 km	—	—
Cell range, NLOS, outdoor CPE	2.40 km	2.00 km	3.25 km	2.70 km	—	—
Cell range, obstructed LOS, outdoor CPE	—	—	—	—	12.0 km	10.0 km

2006 represent an industry average, and the price evolution is based on the assumption of WiMAX becoming a mass market technology. The price of the OPEX elements is assumed not to change over the study period.

16.4.2.3   Network Dimensioning

The maximum sector ranges (Table 16.5) were calculated assuming the system characteristics shown in Table 16.3. For urban areas, the SUI Category A path loss model was used, whereas the suburban area sector range was calculated using the SUI Category B model [11].

For rural areas, the assumptions on cell ranges are based on an estimate of the density of existing radio masts (of mobile networks). Because of the relatively high range requirement, the link conditions have to be sufficiently good to reach the subscribers. We assume that with these cell ranges, all the potential subscribers in the service area will be reached and existing BS sites and towers can be used for deployment.

The techno-economic model calculates the number of cells and sectors required to fulfill the coverage and capacity demands of the service area. The model takes into account the capacity/range relationship of the BS sectors, optimizing the rollouts so that the number of BS cells is minimized. In many cases the initial rollout utilizes more robust modulation types for maximum coverage, but in the later years, as the subscriber density increases, it becomes more efficient to build new BS cells and utilize more efficient modulation types for higher capacity.

16.4.3   Results

16.4.3.1   Economic Results with Base Assumptions

Using the market-, service-, and technology-related inputs and assumptions presented in the previous sections, we calculated the economic results for a number of scenarios. Figure 16.8 shows the results for 2.5 GHz and 3.5 GHz systems in urban, suburban, and rural areas with different household densities.

In urban areas, the two more densely populated area scenarios provide clearly positive NPVs and payback in less than 4 years. When household density drops to 500 households (hh)/km^2, the profitability becomes less clear. For urban area deployments, there are no clear differences in the techno-economic performance between 2.5 GHz and 3.5 GHz systems.

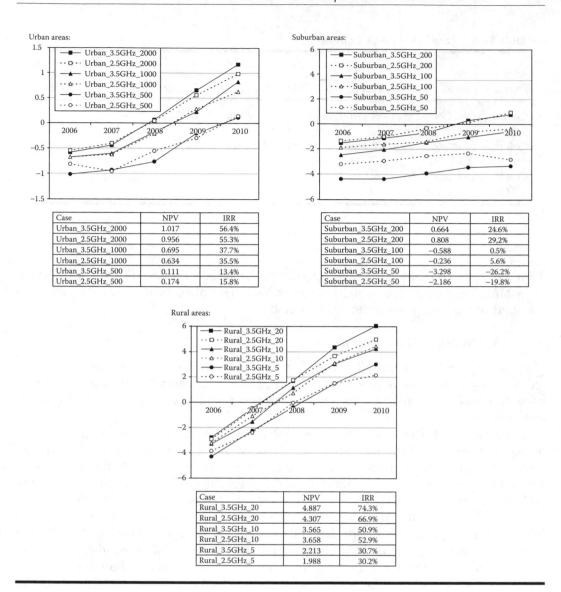

Urban areas:

Case	NPV	IRR
Urban_3.5GHz_2000	1.017	56.4%
Urban_2.5GHz_2000	0.956	55.3%
Urban_3.5GHz_1000	0.695	37.7%
Urban_2.5GHz_1000	0.634	35.5%
Urban_3.5GHz_500	0.111	13.4%
Urban_2.5GHz_500	0.174	15.8%

Suburban areas:

Case	NPV	IRR
Suburban_3.5GHz_200	0.664	24.6%
Suburban_2.5GHz_200	0.808	29,2%
Suburban_3.5GHz_100	−0.588	0.5%
Suburban_2.5GHz_100	−0.236	5.6%
Suburban_3.5GHz_50	−3.298	−26.2%
Suburban_2.5GHz_50	−2.186	−19.8%

Rural areas:

Case	NPV	IRR
Rural_3.5GHz_20	4.887	74.3%
Rural_2.5GHz_20	4.307	66.9%
Rural_3.5GHz_10	3.565	50.9%
Rural_2.5GHz_10	3.658	52.9%
Rural_3.5GHz_5	2.213	30.7%
Rural_2.5GHz_5	1.988	30.2%

Figure 16.8   Results of the techno-economic analysis for urban, suburban, and rural areas.

In suburban areas, only the most densely populated area turns out positive, whereas with a household density of 50 hh/km², the profitability of WiMAX deployment is poor. The 2.5 GHz systems clearly provide better results compared to 3.5 GHz, taking advantage of the higher cell range.

In rural areas, all three area types prove to be profitable, providing fast payback times and high NPV and IRR levels. Differences between 2.5 GHz and 3.5 GHz systems are in general small.

The effect of subscriber density on the economics of network deployments is clearly visible in the different area types. Lower household/SME density leads to less efficient use of network equipment and higher initial investments. In dense areas, more efficient modulation schemes can be utilized in the early years when the subscriber base is low, improving the subscribers-to-BS ratio and decreasing the number of BS sectors and cells

Table 16.6   Number of BS Sectors and Cells Required in the Different Urban Area Types

		2006	*2007*	*2008*	*2009*	*2010*
Urban_3.5GHz_2000	BS cells	3	5	8	11	15
	BS sectors	18	30	40	62	90
	Modulation	BPSK	QPSK	16-QAM	16-QAM	16-QAM
Urban_3.5GHz_1000	BS cells	5	8	9	14	16
	BS sectors	18	30	53	83	90
	Modulation	BPSK	QPSK	QPSK	QPSK	16-QAM
Urban_3.5GHz_500	BS cells	10	10	14	16	20
	BS sectors	30	44	79	83	120
	Modulation	BPSK	QPSK	QPSK	QPSK	QPSK

required. This is illustrated in Table 16.6, which shows the required number of BS sectors and cells for urban area types with different household/SME densities, as well as the most robust modulation required in each.

As Table 16.6 shows, the required investments increase significantly as the household/SME density decreases and area size increases. To reach the same number of subscribers in 2010, the required number of BS cells and sectors is 33% higher in the sparsest urban area compared to the densest one, strongly affecting the economic feasibility of the deployment. In the earlier years the situation is worse, as the networks are more clearly coverage limited.

The effect of differences in the competitive situations in different area types can be seen in the slopes of the cash balance curves in Figure 16.8. In rural areas, satisfying the needs of the significant residual market requires heavy initial investments, but gives the WiMAX operator a significantly larger subscriber base and annual revenues compared to the urban and suburban areas. In urban areas, the investments are smaller, but so are the achievable revenue flows. This is illustrated in Figure 16.9, which shows the investment and OPEX breakdowns for each of the three main area types, together with the annual service revenues.

The revenue flows of the operator start to decline in the fourth and fifth year as the markets are saturating and ARPU continues to decrease. Investments to CPEs leased to the subscribers constitute the most significant part of costs in the early years, but overall the relative importance of OPEX compared to CAPEX increases throughout the years. In 2010, OPEX costs constitute about 70% of overall costs in all area types.

16.4.3.2   Indoor versus Outdoor CPEs

In urban and suburban areas, the use of outdoor CPEs helps to minimize the number of BS cells and sectors, especially in the early years of deployment. On the other hand, outdoor CPEs are more expensive than indoor CPEs and bear an extra installation cost. Furthermore, special permission for installing outdoor antennas may have to be acquired from property owners, placing an extra burden on operators and potential subscribers.

Thus indoor CPEs are in many ways preferable to outdoor CPEs. The drawback, however, is the significantly lower link range resulting from the lower gain, omnidirectional antenna and loss in signal strength when penetrating walls and windows of buildings. Figure 16.10 shows the results for 3.5 GHz and 2.5 GHz systems deployed in urban areas with different household densities, assuming that outdoor CPEs are not used at all.

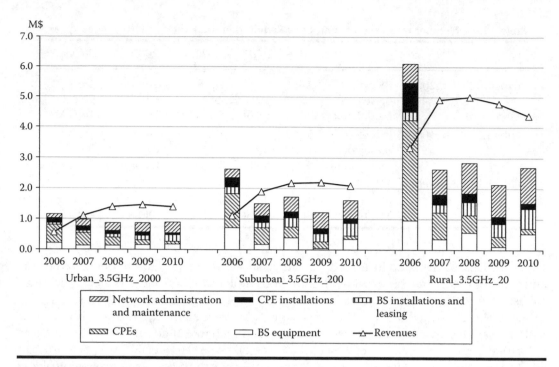

Figure 16.9   Breakdown of OPEX and CAPEX for three area types.

As Figure 16.10 shows, only the 2000 hh/km^2 scenario proves to be profitable, whereas with lower household densities the results are much worse. The better coverage and signal penetration of 2.5 GHz systems brings significant benefits compared to 3.5 GHz systems in these scenarios.

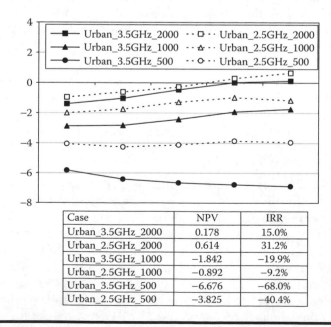

Figure 16.10   Economic results for the urban areas when only indoor CPEs are used.

16.4.4 Sensitivity Analysis

To cope with the uncertainty inherent in many of our assumptions, sensitivity analyses were done on the key input parameters. The parameters included in the sensitivity analyses were BS sector capacity, BS sector range, BS price, CPE price, and the service ARPU level. For each parameter, 50% deviations from the base assumptions were defined as the upper and lower limits for the sensitivity analysis. The parameters were analyzed one by one, changing their values between the minimum and maximum and plotting the respective NPV values into sensitivity graphs.

The NPV sensitivity to ARPU level, BS price, and CPE price assumptions (shown in Figure 16.11) is rather similar in all the area types. Differences exist in the base NPV levels and in the slopes of the sensitivity curves, resulting from the different scales of the

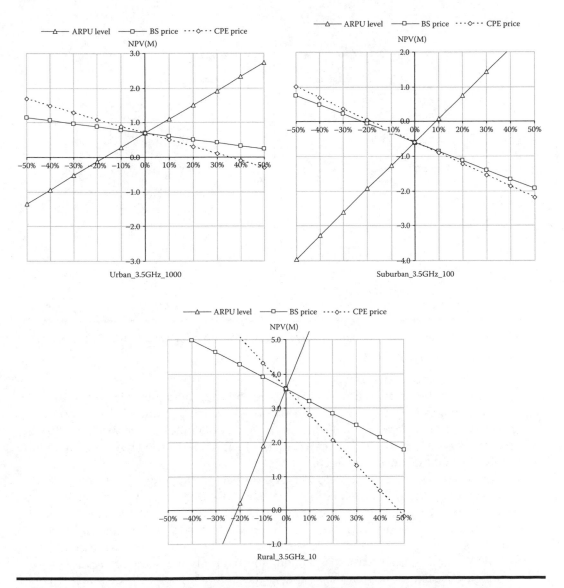

Figure 16.11 NPV sensitivity to changes in ARPU level, BS price, and CPE price assumptions.

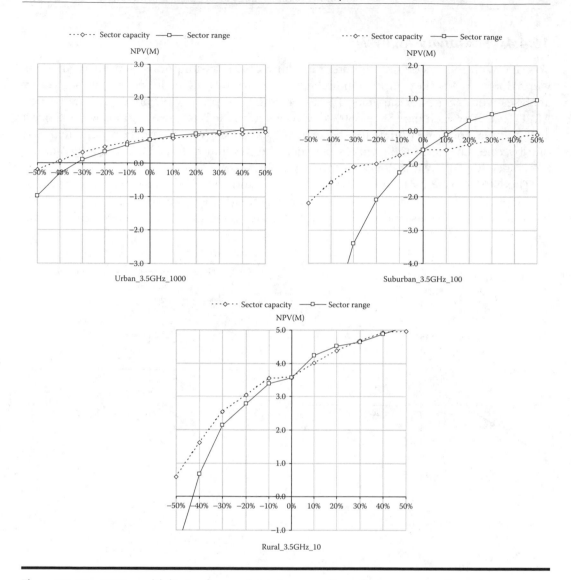

Figure 16.12 NPV sensitivity to changes in sector capacity and range assumptions.

projects regarding both investments and acquired subscriber amounts. The sensitivity graphs show that the profitability of WiMAX network deployments is significantly more sensitive to changes in CPE prices than in BS prices, which is natural as CPEs constitute a larger share of the operator's investments. However, changes in the retail ARPU level (or alternatively in the network operator's share of ARPU) have the largest impact on the profitability of WiMAX deployments.

Sensitivity analysis with regard to WiMAX BS sector capacity and range produces different results for the different area types, as shown in Figure 16.12. The shape of the sensitivity graphs is similar in all the areas, although the slope of the curves and the relative importance of sector capacity and range differs between the scenarios.

The NPV sensitivity to sector range is similar in the urban and suburban areas, growing slowly when the range is increased, and falling more rapidly when the range is decreased below the base assumption. This shows that as the number of BS cells becomes capacity limited rather than coverage limited, the importance of the cell range diminishes.

The importance of BS sector capacity varies between area types. The suburban area results are interesting in that the project NPV is quite indifferent to increases in the sector capacity, suggesting that the network deployment is still coverage limited with the base assumptions. In rural areas, the base network deployment is capacity limited and the NPV is more sensitive to changes in the sector capacity.

16.4.5 Risk Analysis

In our study, risk analysis was performed for the urban, suburban, and rural areas with household densities of 1000, 100, and 10 hh/km², respectively. For each case, sector capacity, sector range, CPE price, and monthly ARPU were identified as the most critical input parameters. Each of these variables was then assumed to follow the beta distribution, the minimum value being 80% and maximum value being 120% of the base assumption. The shape of the distribution was assumed to be such that with a probability of 90% the values would be between 90% and 110% of the base assumption value. These distributions were used to run a Monte Carlo simulation with 1000 samples, using the NPV as the output.

Figure 16.13 shows an example of the results from the Monte Carlo simulation for the urban area. As shown, with our assumptions the risk of the nonprofitable business case in this area type is quite small.

For the urban and rural scenarios, the monthly ARPU and the CPE price were the dominant contributors to the uncertainty in NPV. In suburban areas, cell range replaced the CPE price as the second dominant input parameter, highlighting the coverage limited nature of the network deployment. Table 16.7 shows the statistics for all the risk analysis simulations.

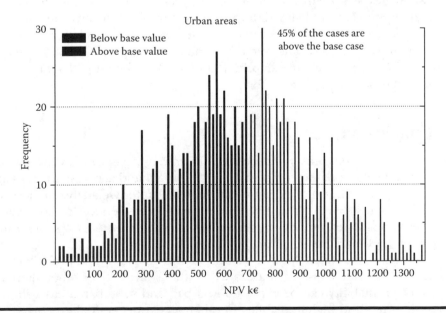

Figure 16.13   Risk simulation using Monte Carlo analysis for urban areas.

Table 16.7   Results From the Risk Analysis Simulations

NPV/Scenario	Urban_3.5GHz_1000	Suburban_3.5GHz_100	Rural_3.5GHz_10
Mean	673,839	−673,569	3,749,081
Standard deviation	287,271	602,746	1,167,983
Minimum	−370,789	−2,585,085	444,826
Maximum	1,576,025	964,410	7,542,988
Range	1,946,814	3,549,495	7,098,162
Percentage of simulations above the base case	45.10%	45.00%	53.60%

16.5   Open Issues

Our analysis has focused on fixed WiMAX deployments, assuming basic broadband Internet access services to households and SMEs to generate the revenues. In the future, WiMAX is foreseen to be added to portable and mobile devices, offering an alternative to WLANs and 3G high-speed packet access (HSPA) networks. Techno-economic analysis of mobile WiMAX is left for future work.

The generally low cost of WiMAX spectrum licenses can be seen as a driver for WiMAX toward mobile services. The gap is significantly reduced due to the lower cost of equipment on lower frequency bands, but still the cost of WiMAX spectrum is essentially lower than that of 3G in most countries. If regulators continue this policy for mobile WiMAX licences, WiMAX operators will get an additional economic advantage over 3G operators.

Overall, regulatory confusion exists between Universal Mobile Telecommunications System (UMTS)/3G and WiMAX spectrum allocations. It is tempting to predict that the regulatory differentiation gradually disappears, allowing both technologies to cover all mobile wireless spectrum. This is because both technologies are likely to converge toward similar OFDM-based technology and have mobility support. The battle between the IEEE and Third-Generation Partnership Project (3GPP) evolution paths will be strongly influenced by the decisions of regulators.

16.6   Conclusion

Fixed wireless access networks based on the IEEE 802.16-2004 standard have been proposed to induce facilities-based competition in the broadband access market, as an alternative technology to DSL and cable networks. In this chapter we analyzed these WiMAX network deployments for providing fixed broadband access services to households and businesses in various area types and market conditions for systems operating in either the 2.5 GHz or 3.5 GHz frequency bands.

Our results show that fixed WiMAX network deployments can be profitable in dense urban areas as well as in those rural areas where the availability of other alternatives is limited. Lower profitability can be expected in urban and suburban areas with medium population densities and high availability of other access network alternatives.

The trade-off between coverage and capacity is evident in our case scenarios; in some area types the networks are coverage limited, whereas in others the sector/cell capacity defines the required investments. Generally, coverage limited network deployments are less profitable. The performance of WiMAX systems appears to be suitable for the broadband traffic demands of today, but the emergence of services requiring data rates greater than 2 to 4 Mbps per subscriber will be problematic for WiMAX operators.

Sensitivity and risk analyses reveal that the CPE price and broadband tariff levels are the most critical parameters defining the profitability of WiMAX network deployments. Decreasing the total cost of CPEs, including both equipment and installation costs, is seen as the main requirement for WiMAX to succeed against DSL and cable networks.

References

1. D. Katsianis, I. Welling, M. Ylönen, D. Varoutas, T. Sphicopoulos, N.K. Elnegaard, B.T. Olsen, and L. Budry, The financial perspective of the mobile networks in Europe, *IEEE Personal Commun.*, 8(6), 58, 2001.
2. T. Monath, N.K. Elnegaard, P. Cadro, D. Katsianis, and D. Varoutas, Economics of fixed broadband access network strategies *IEEE Commun. Mag.*, 41(9), 132, 2003.
3. D. Varoutas, D. Katsianis, T. Sphicopoulos, K. Stordahl, and I. Welling, On the economics of 3G Mobile Virtual Network Operators, *Wireless Personal Commun.*, 36(2), 129, 2006.
4. B.T. Olsen, D. Katsianis, D. Varoutas, K. Stordahl, J. Harno, N.K. Elnegaard, I. Welling, F. Loizillon, T. Monath, and P. Cadro, Technoeconomic evaluation of the major telecommunication investment options for European players, *IEEE Network Mag.*, 20(4), 6, 2006.
5. L. Ims, ed., *Broadband Access Networks—Introduction Strategies and Techno-economic Evaluation*, Chapman & Hall, New York, 1998.
6. B.T. Olsen, OPTIMUM—a techno-economic tool, *Telektronikk*, 95 (2/3), 239–250, 1999.
7. IEEE, IEEE Standard for Local and Metropolitan Area Networks — Part 16: Air Interface for Fixed Broadband Wireless Access Systems, Standard 802.16-2004, IEEE, Washington, DC, 2004.
8. C. Eklund, R.B. Marks, K.L. Stanwood, and S. Wang, IEEE standard 802.16: a technical overview of the wirelessMAN air interface for broadband wireless access, *IEEE Commun. Mag.*, 40(6), 98, 2002.
9. A. Ghosh, D.R. Wolter, J.G. Andrews, and R. Chen, Broadband wireless access with WiMAX/802.16: current performance benchmarks and future potential, *IEEE Commun. Mag.*, 43(2), 129, 2005.
10. Maravedis Inc., Spectrum Analysis: The Critical Factors in BWA/WiMAX versus 3G, Maravedis, Inc., Montreal, Quebec, Canada, 2006.
11. IEEE 802.16.3c-01/29r4, Channel Models for Fixed Wireless Applications, IEEE 802.16 Broadbar wireless Aceess Working group, available at http://www.ieee802.org/16/.
12. Electronic Communication Committee (ECC), European Conference of Postal and Telecommunications Administrations (CEPT), The analysis of the coexistence of FWA cells in the 3.4–3.8 GHz band, ECC Report 33, May 2003.
13. IST-TONIC Project, Techno-economics of IP optimised networks and services, IST-2000-25172, available at http://www-nrc.nokia.com/tonic/.
14. OECD, ICT statistics, available at http://www.oecd.org/.

Chapter 17

Capacity of OFDMA-Based WirelessMAN

Hongxiang Li and Hui Liu

Contents

Orthogonal frequency division multiple access (OFDMA) has emerged as one of the prime multiple-access schemes for broadband wireless networks (e.g., IEEE 802.16e Mobile WiMAX, IEEE 802.20, IEEE 802.22, 3G LTE, etc.). In particular, the OFDMA-based IEEE 802.16e wireless metropolitan area network (MAN) (Mobile WiMAX) is rapidly gaining popularity among wireless service providers because of its open standard, high throughput, and low equipment costs. As a special case of multicarrier multiple-access schemes, OFDMA exclusively assigns each subcarrier to only one user, eliminating intra-cell interference (ICI). For fixed or portable applications where the frequency selective channels are slowly varying, an intrinsic advantage of OFDMA is its capability to exploit the so-called multiuser diversity through subcarrier allocation [1,2]. Furthermore, OFDMA has the advantage of easy decoding at the receiver side due to the absence of ICI. Other advantages of OFDMA include finer granularity and better link budget in uplink communications.

In the case of Mobile WiMAX, the IEEE 802.16e standard actually encompasses a variety of modem and access technologies. There are three basic physical/media access control (PHY/MAC) modes in the 802.16e documents: (i) single-carrier/time division multiple access (TDMA) with two different flavors, (ii) orthogonal frequency division multiplexing (OFDM)/TDMA, and (iii) orthogonal frequency division multiple access (OFDMA). The single-carrier mode and the OFDM mode are both inherited from the IEEE 802.16-2004 fixed broadband wireless access standard. Most of the design and standardization activities in 802.16e are in the OFDMA mode, which provides the only scalable scheme for wide area mobile networks. In fact, Mobile WiMAX only considers the OFDMA mode of 802.16e. Therefore, this chapter only describes OFDMA and related issues.

While the industry has unequivocally converged to OFDMA in the post-code division multiple access (CDMA) era, the theoretical analysis on the fundamental capacity of OFDMA is lacking. For example, there have been statements that the information capacity limit of a multicarrier system can be achieved by OFDMA; however, no rigorous proof can be found in the literature. Until we obtain a thorough understanding of the true potential of OFDMA (e.g., whether OFDMA can deliver the maximum capacity of multicarrier communications), it will be challenging to measure and compare the performance of wireless metropolitan area networks (MANs) in the field. The objective of this chapter, therefore, is to provide a measuring stick against which one can evaluate a practical broadband system. More specifically, we are interested in the channel capacity of OFDMA and its optimality relative to the generic multicarrier communication systems. In the following sections we describe the characteristics of OFDMA and present the notion of "multiuser diversity" in OFDMA networks. A quantitative analysis on OFDMA capacity will follow. In light of the importance of multiple-input multiple-output (MIMO) technologies in broadband wireless communications, we broaden the scope of our analysis to include MIMO/OFDMA systems. The results, both analytical and numerical, provide important insights on the design and implementation of practical OFDMA systems.

17.1   Multicarrier, Multiuser, MIMO Broadband Systems

The OFDMA scheme can be derived from a generic broadband multicarrier framework. The basic idea of multicarrier modulation is to divide the transmitted bitstream into substreams and send these over parallel narrowband subchannels. Typically the subchannels are orthogonal to each other under multipath environments (i.e., finite impulse response

[FIR] channel models). The data rate on each of the subchannels is much less than the total data rate and the corresponding subchannel bandwidth is much less than the total system bandwidth. The number of substreams is chosen to ensure that each subchannel has a bandwidth less than the coherence bandwidth of the multipath channel, resulting in only flat fading on each subchannel. Moreover, multicarrier modulation can be implemented efficiently using digital signal processing. The most popular realization is the fast fourier transform (FFT)-based orthogonal frequency division multiplexing (OFDM) where the intersubcarrier interference (ISI) can be completely eliminated through the use of a cyclic prefix.

In the operations of multicarrier, multiuser communications, conceptually all users can transmit simultaneously through all subchannels, although such a configuration can result in prohibitively expensive receivers. In practical WirelessMANs, additional rules often apply to limit the available subchannels in each user. At the PHY layer, MIMO-enabled multicarrier systems offer better resistance against fading than a traditional single-input single-output (SISO) system. More importantly, the higher dimensional spatial operations can potentially lead to a multiplicative increase in capacity [3–6]. Multiuser communications can be accommodated by sharing the radio resources in both the frequency and the space domains, leading to the so-called multicarrier, multiuser, MIMO (M^3) system. Notice that an unconstrained M^3 system has little practical value due to a large number of overlapping signals on each subcarrier. Nevertheless, the model is theoretically significant, as it presents the capacity upper limit for all multicarrier-based system. More details are provided in the following sections.

17.1.1 OFDMA

Definition 1 *An OFDMA system is defined as one in which each user occupies a subset of subcarriers (termed an OFDMA traffic channel), and each traffic channel is assigned exclusively to one user at any time [7].*

In OFDMA, users are not overlapped in the frequency domain at any given time. However, the frequency bands assigned to a particular user may change over time, as shown in Figure 17.1 (each type of shade represents resources allocated exclusively to a user). In addition, the subcarriers assigned to be a particular user may be either localized (adjacent) or distributed across the entire frequency band.

Example 1

The IEEE 802.16a–e standards have an OFDMA mode with bandwidth options of either 1.25, 5, 10, or 20 MHz. Depending on the bandwidth, the entire spectrum is divided into 128, 512, 1024, or 2048 subcarriers. For example, a 20 MHz band with a 2048-FFT yields a subcarrier spacing of 9.8 kHz [8]. In the time domain, the resource is further divided into frames and subframes that can be allocated to different users.

While there are many ways to partition the radio resources in multiuser communications, OFDMA is fundamentally advantageous over other schemes when it comes to real system operations.

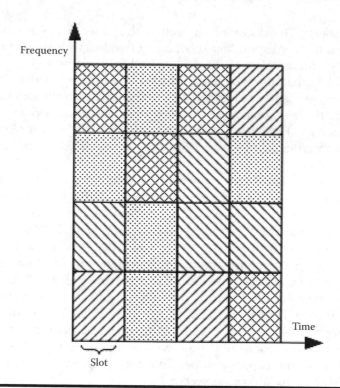

Figure 17.1   OFDMA in the frequency and time domain.

■ Granularity: Early broadband access systems utilize OFDM-TDMA to offer a straight-forward method of multiple access: each user uses a small number of OFDM symbols in a time slot and multiple users share the radio channel through TDMA. The method has two obvious shortcomings. First, every time a user utilizes the channel, it has to burst its data over the entire bandwidth, leading to a high peak power and therefore low radio frequency (RF) efficiency. Second, when the number of sharing users is large, the TDMA access delay can be excessive. OFDMA is a much more flexible and powerful way to achieve multiple-access with an OFDM modem. In OFDMA, the multiple access is not only supported in the time domain, but also in the frequency domain, just like traditional frequency division multiple access (FDMA) minus the guard-band overhead. As a result, an OFDMA system can support more users with much less delay. The finer data rate granularity in OFDMA, as illustrated in Figure 17.2, is paramount to rich media applications with diverse quality of service (QoS) requirements.

■ Link budget: Since each TDMA user must burst its data over the entire bandwidth during the allocated time slots, the instantaneous transmission power (dictated by the peak rate) is the same for all users, regardless of their actual data rates. This inevitably creates a link budget deficit that handicaps low-rate users. Unlike TDMA, an OFDMA system can accommodate a low-rate user by allocating only a small portion of its band (proportional to the requested rate). For example, by reducing the effective transmit bandwidth to 1/64 of the system bandwidth, OFDMA can provide an approximately 18 dB uplink budget advantage over OFDM-TDMA.

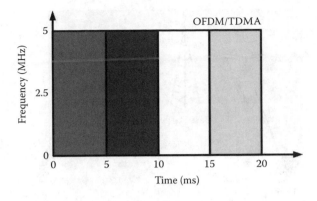

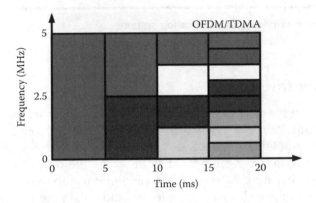

Figure 17.2   Resource partitioning in OFDM/TDMA and OFDMA.

■ Receiver simplicity: OFDMA has the advantage of easy decoding at the receiver side, as it eliminates the intracell interference, avoiding CDMA type of multiuser detection. This is not the case in multicarrier (MC)-CDMA, even if the codes are designed to be orthogonal. Users' signals can only be detected jointly, since the code orthogonality is destroyed by frequency selective fading. The fact that users' channel characteristics must be estimated also favors OFDMA. In MC-CDMA, the users' channel responses must be estimated using complex joint estimation algorithms. Furthermore, OFDMA is the least sensitive multiple access scheme to system imperfections [9]. Because of these intrinsic features, OFDMA has been adopted in several modern wireless systems (e.g., IEEE 802.16a–e [8], 3GPP-LTE, IEEE 802.22, and IEEE 802.20).

■ Multiuser diversity: Since broadband signals experience frequency selective fadings, the frequency response of the channel varies over the whole frequency spectrum. The fact that each user has to transmit its signal over the entire spectrum in OFDM-TDMA/CDMA leads to an averaged-down effect in the presence of deep fading and narrowband interference. On the other hand, OFDMA allows different users to transmit over different portions of the broadband spectrum (traffic channel). Since different users perceive different channel qualities, a deep faded channel for one user may still be favorable to others. Therefore, through judicious channel allocation, the system can potentially outperform interference-averaging techniques.

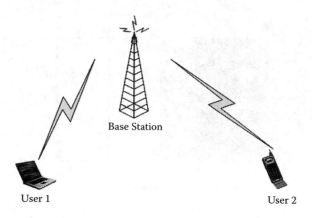

Figure 17.3   A two-user wireless communication system.

17.1.2   *Multiuser Diversity*

Multiuser diversity is a recently identified diversity in multiuser wireless communication systems. In general, multiuser diversity gain arises from the fact that in a wireless system with many users, the utility value (e.g., achievable data rate) of a given resource unit varies from one user to another. Such fluctuations allow the overall system performance to be maximized by allocating each radio resource unit to the user that can best exploit it. To illustrate the multiuser diversity gain, let us study a simple example as follows.

Example 2

Figure 17.3 shows a single-cell OFDMA system with one base station serving two users. In this example, we have the following assumptions:

1. The two users are independent, that is, their channel responses are independent.
2. The users have perfect knowledge of the channel state information.
3. The base station gathers the channel measurements from the two users and allocates channels (i.e., subcarriers) based on these measurement reports.

Since the signal-to-interference plus noise ratios (SINRs) across subcarriers characterize the channel response, we shall use the system average SINR to illustrate the multiuser diversity. Figure 17.4 shows the frequency selective channel responses of two users. Due to interference and noise, some of the subcarriers are in deep fading. However, since the two users are independent, a deep faded subcarrier for one user may be an excellent one for the other user. If we use TDMA, then the SINR value for each subcarrier is the average of the two users, which is indicated in Figure 17.5a. On the other hand, if we use OFDMA with intelligent resource allocation, then each subcarrier can be utilized more efficiently by allocating it to the user that has the highest channel frequency response, which is indicated in Figure 17.5b. In Figure 17.5, the bold curves represent the achieved SINR values on subcarriers.

It is important to note that despite the promise of multiuser diversity, it is unclear whether OFDMA can exploit the maximum potential of multicarrier communications.

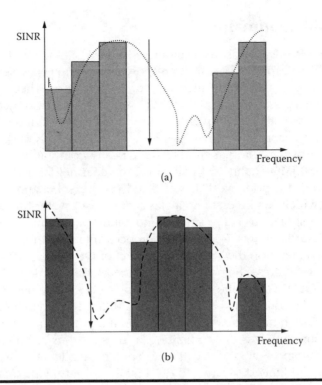

(a)

(b)

Figure 17.4   Channel frequency responses of the two users.

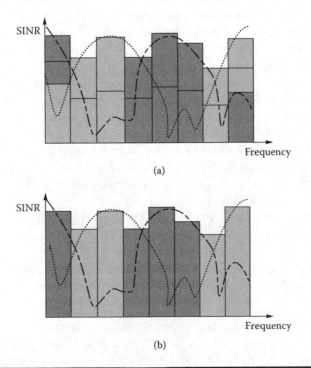

(a)

(b)

Figure 17.5   Achieved SNR levels in (a) TDMA and (b) OFDMA with resource allocation.

17.2 OFDMA Optimality

An important performance measure in multicarrier multiple access is the sum rate capacity. The input sum capacity optimization (power allocation) problem for the general M^3 channels has been studied in the literature for several cases. The capacity region of a Gaussian multiple-access SISO channel with ISI was characterized by Cheng and Verdu [10]. The optimal power allocation over time for independent and identically distributed (i.i.d.) fading uplink channels was investigated by Tse and Hanly [11]. Yu [12] proposed an efficient iterative water-filling method to numerically compute the optimal power allocation for uplink MIMO channels. In all cases, the channel information is assumed to be known. For OFDMA, prior work has focused on the subchannel allocation/power loading problem, which can be cast into a classic multiuser water-filling framework. The topic has been the subject of many studies and a number of implementation schemes can be found in the literature [7–13, and references therein]. In the remainder of this chapter, our interest is not on the algorithmic aspect of OFDMA subchannel loading, but rather, we attempt to address the problem at a more fundamental level: what is the relationship between OFDMA and the optimal multiple-access scheme? We concentrate on whether OFDMA, as a constrained multicarrier multiple-access scheme, is sum rate optimal from an information theoretic prospective. And if not, the objective is to characterize the nature of the sum rate loss due to limiting, at most, one user on each subchannel.

Given the importance of OFDMA, the amount of academic research on OFDMA optimality is still quite limited (relative to 3G/CDMA) to date. Jang and Lee [14] proved the sum rate optimality of OFDMA in the downlink with adaptive quadrature amplitude modulation (QAM) and independent decoding. Li and Liu [15] generalized the results to downlink OFDMA with MIMO, again assuming only independent decoding. Note that the sum rate achieved by independent decoding is always inferior to the optimal sum capacity, which can be achieved by superposition coding and successive interference cancelation [16]. More recently, Michel and Wunder [17] presented results on OFDMA optimality for downlink MIMO channels from an information-theoretic viewpoint. In Cheng and Verdu [10] and Knopp and Humblet [18], it is stated that in order to maximize the uplink sum capacity, only one user should transmit at any given frequency over the entire spectrum. However, this result is only valid for continuous frequency selective channels and cannot be applied to discrete multicarrier multiple-access systems with a finite number of subchannels (as in OFDMA). As a matter of fact, as the main result of this chapter shows, OFDMA is suboptimal with nontrivial probabilities in many scenarios. Other uplink OFDMA optimality studies can be found in the literature [19–21]. However, none of these results provide a quantitative analysis with regard to OFDMA optimality with a finite number of subchannels.

In the rest of this chapter we investigate OFDMA optimality for the following two scenarios:

■ Multicarrier multiuser uplink/downlink channels in SISO systems.
■ Multicarrier multiuser uplink/downlink channels in MIMO systems.

We invoke the following assumptions for the rest of the chapter: (i) slow fading channels (also called the quasi-static scenario) [22]; in other words, the channel coefficients are random, but remain constant long enough for subchannel allocation/power loading; and (ii) perfect knowledge of the channel state information (CSI) for both the transmitter and receiver.

We use the following notation conventions throughout the chapter:

K the total number of users

N the total number of subcarriers

h_{kn} the channel gain associated with subcarrier n and user k in the SISO channel

p_{kn} the transmitting power of user k on subcarrier n in the SISO channel

p_k user k's total transmitting power in the uplink channel

$\mathbf{H}_{kn}$ the channel gain matrix associated with subcarrier n and user k in the MIMO channel

$\mathbf{Q}_{kn}$ the spatial covariance matrix of user k on subcarrier n in the MIMO channel

B the total system bandwidth

B_n the subchannel bandwidth

B_c the coherent bandwidth

P_o the probability of OFDMA being sum rate optimal

$e_i\{\mathbf{A}\}$ the ith eigenvalue of matrix A

$e_{max}\{\mathbf{A}\}$ the maximum eigenvalue of matrix A

$\max_{x}\{\}$ maximize over x

17.3   The Optimality of SISO/OFDMA

For the SISO case, it is known that OFDMA is always the sum rate optimal multiple access scheme in downlink channels [16], thus the discussion in this section focuses on uplink channels. We consider a generic discrete multicarrier multiuser uplink channel and derive the necessary and sufficient conditions under which OFDMA is sum rate optimal. We further derive the probabilities of these conditions for both low and high signal-to-noise ratio (SNR) regions. Another contribution of this section is that we show the number of shared subchannels under the optimal solution should be less than the number of total users. As a result, the performance gap between the OFDMA scheme and the optimal solution becomes negligible when the number of subchannels is large. On the other hand, it is difficult to completely quantify the performance gap between OFDMA and the optimal solution under arbitrary configurations (e.g., the number of subchannels, the number of users, SINR, etc.), and the topic remains an area of future research.

17.3.1   System Model

We consider a generic multicarrier multiple-access system. The noninterfering parallel subchannels can be created using FFT and inverse FFT (IFFT) operations [23]. For the generic multicarrier multiple-access, all the users can transmit on all the subchannels subject to individual peak power constraints $p_k : \sum_{n=1}^{N} p_{kn} \leq p_k$, where $1 \leq k \leq K$. We assume the total bandwidth B is fixed and the subchannel bandwidth B_N decreases linearly with N, that is, $B_N \sim B/N$ [16].

Let x_{kn} be the information-bearing signal from user k on subchannel n. The received signal on subchannel n can be expressed as:

$$y_n = \left(\sum_{k=1}^{K} x_{kn}\sqrt{p_{kn}}h_{kn} \right) + v_n, \tag{17.1}$$

where v_n is the additive white Gaussian noise (AWGN) noise with variance N_0. The capacity region on subchannel n is given in Cover and Thomas [24] (assuming $E\left(\|x_{kn}\|^2\right) = 1$; and x_{jn}, x_{kn} are independent zero-mean random variables):

$$c_n = (R_{1n}, \ldots, R_{Kn}) : \sum_{k \in S} R_{kn} \leq B_N \log_2 \left(1 + \frac{\sum\limits_{k \in S} p_{kn}\|b_{kn}\|^2}{N_0 B_N}\right)$$

$$\forall S \subset \{1, 2, \ldots, K\}. \tag{17.2}$$

Equation (17.2) indicates that the sum of rates on subchannel n for any subset of the K users is upper limited by the capacity of a "superuser" with received power equal to the sum of received powers associated with the particular user subset on this subchannel. For slow fading channels, information is delivered through parallel subchannels in multicarrier, the total capacity region over all subchannels and all users is then [22,24,25]:

$$C([p_{kn}]_{K \times N}) = (R_1, \ldots, R_K) : \sum_{k \in S} R_k \leq B_N \sum_{n=1}^{N} \log_2 \left(1 + \frac{\sum\limits_{k \in S} p_{kn}\|b_{kn}\|^2}{N_0 B_N}\right)$$

$$\forall S \subset \{1, 2, \ldots, K\}. \tag{17.3}$$

The capacity region in Equation (17.3) can be achieved by superposition coding and successive interference cancellation [16]. We choose the sum-of-rate capacity (C_{sum}) as the figure of merit. To maximize the sum capacity, the transmission power needs to be distributed optimally under the individual peak power constraints. The following optimization problem is formulated accordingly [26,27]:

$$\arg\max_{p_{kn}} C_{sum} = B_N \sum_{n=1}^{N} \log_2 \left(1 + \frac{\sum\limits_{k=1}^{K} p_{kn}\|b_{kn}\|^2}{N_0 B_N}\right) \tag{17.4}$$

$$s.t. \sum_{n=1}^{N} p_{kn} \leq p_k, \tag{17.5}$$

$$p_{kn} \geq 0 \tag{17.6}$$

$$\forall\, k, n, \text{ where } 1 \leq k \leq K, 1 \leq n \leq N.$$

The optimization problem of Equation (17.4) is also widely used in frequency-flat fading channels [1,11] and multiuser water filling with informed static MIMO channels [12,28,29].

Remark 1 *Joint decoding is needed to achieve the sum capacity in Equation (17.4). A suboptimal, but simpler scheme is independent decoding. With independent decoding, the objective* *function in Equation (17.4) becomes*

$$C_{sum} = B_N \sum_{n=1}^{N} \sum_{k=1}^{K} \log_2 \left(1 + \frac{p_{kn}\|h_{kn}\|^2}{\sum_{i=1, i\neq k}^{K} p_{in}\|h_{in}\|^2 + N_0 B_N} \right). \tag{17.7}$$

The optimality analyses in the following sections can be applied to independent decoding as well.

In the remainder of this section we refer to the solution to Equation (17.4) as the sum rate optimal multicarrier multiple-access scheme, or simply the optimal solution.

Define S_n as the active user set on subchannel n: $\{i \in S_n$ iff $p_{in} > 0\}$. Clearly, S_n $(S_n \subset \{1, 2, \ldots, K\})$ is a subset of the K users. As a special multiple-access scheme, OFDMA can be defined mathematically as

$$|S_n| \leq 1, \quad \text{for } n = 1, \ldots, N.$$

That is, each subchannel is exclusively assigned to no more than one user. For discrete multicarrier systems, the optimal solution scheme may or may not be OFDMA. The rest of the section focuses on the relationship between OFDMA and the optimal solution. Our primary interest is to determine (i) under what conditions is OFDMA the optimal multiple-access scheme; (ii) what is the probability for these conditions; and (iii) in the case where OFDMA is not optimal, what is the performance gap between OFDMA and the optimal solution.

17.3.2 Analysis of OFDMA Optimality

We first provide a general theorem that asserts the optimal solution of the problem in Equation (17.4) without completely solving it. The results allow us to derive the necessary and sufficient conditions under which OFDMA is optimal. Based on these conditions, we calculate P_o (the probability of OFDMA being optimal) for the small and the large SNR regions, respectively. In the case where OFDMA is not optimal, we further show that the number of shared subchannels should be less than K in the optimal solution, and thus the performance gap between the OFDMA scheme and the optimal solution is negligible when $N >> K$.

Theorem 1 *Referring to Equation (17.4), the sum rate capacity C_{sum} is maximized only if the following conditions are satisfied:*

(i) *Subchannel n is utilized by at least one user if and only if* $\max_k \left\{ \frac{\|h_{kn}\|^2}{\lambda_k} \right\} > N_0$; *and the transmitting users must have the highest weighted power gain* $\max_k \left\{ \frac{\|h_{kn}\|^2}{\lambda_k} \right\}$, *where λ_k is the Lagrange multiplier that enforces the power constraint on the kth user.*

(ii) *If the optimal solution suggests OFDMA, each user allocates its power within its allocated subchannels by water filling with water level at $\frac{1}{\lambda_k}$ [24].*

The proof uses the following lemma.

Lemma 1

Karush-Kuhn-Tucker conditions (first-order necessary conditions) [30, Theorem 12.1]: For a general optimization problem

$$\min_{\mathbf{x} \in R^n} f(x) \tag{17.8}$$

$$s.t. \quad h_i(\mathbf{x}) = 0, \quad i \in \mathcal{E}$$

$$h_i(\mathbf{x}) \geq 0, \quad i \in \mathcal{I},$$

where $f(\mathbf{x})$ and $h_i(\mathbf{x})$ are all smooth, real-valued functions on a subset of R^n, and $\mathcal{I}$ and $\mathcal{E}$ are two finite sets of indices. The Lagrange of Equation (17.8) is defined as

$$L(\mathbf{x}, \lambda) = f(\mathbf{x}) - \sum_{i \in \mathcal{E} \cup \mathcal{I}} \lambda_i h_i(\mathbf{x}).$$

If $\mathbf{x}^*$ is a local solution of Equation (17.8) and the linear independence constraint qualification (LICQ) holds at $\mathbf{x}^*$, then there is a Lagrange multiplier vector λ^* with components $\lambda_i^*, i \in \mathcal{E} \cup \mathcal{I}$, such that the following conditions are satisfied at $(\mathbf{x}^*, \lambda^*)$:

$$\nabla_{\mathbf{x}} L(\mathbf{x}^*, \lambda^*) = 0 \tag{17.9}$$

$$h_i(\mathbf{x}^*) = 0, \quad i \in \mathcal{E} \tag{17.10}$$

$$h_i(\mathbf{x}^*) \geq 0, \quad i \in \mathcal{I} \tag{17.11}$$

$$\lambda^* \geq 0, \quad i \in \mathcal{I} \tag{17.12}$$

$$\lambda_i^* h_i(\mathbf{x}^*) = 0, \quad i \in \mathcal{E} \cup \mathcal{I}. \tag{17.13}$$

Furthermore, if Equation (17.8) is a convex optimization problem (the objective function $f(\mathbf{x})$ is convex and the feasible region is also convex), then conditions from Equation (17.9) to Equation (17.13) are also sufficient conditions to determine the global optimal solution $\mathbf{x}^*$ [31].

We can prove that $-C_{sum}$ in Equation (17.4) is a convex function with respect to p_{kn} (see Appendix A). Since both the equality and inequality constraints are linear and the LICQ condition holds as well, the optimization problem of Equation (17.4) can be cast into a classic convex programming problem [31]. The Karush-Kuhn-Tucker (KKT) conditions can be used to determine the optimal solution. In order to do so, we first construct the Lagrangian function of Equation (17.4) as

$$L(p, \lambda) = -B_N \sum_{n=1}^{N} \log_2 \left(1 + \frac{\sum_{k=1}^{K} p_{kn} \|h_{kn}\|^2}{N_0 B_N} \right)$$

$$+ \sum_{k=1}^{K} \lambda_k \left(\sum_{n=1}^{N} p_{kn} - p_k \right) - \sum_{k=1}^{K} \sum_{n=1}^{N} \lambda_{kn} p_{kn}, \tag{17.14}$$

where λ_k and λ_{kn} are the Lagrange multipliers associated with Equation (17.5) and Equation (17.6), respectively. The KKT conditions state that (p^*, λ^*) is the optimal solution if and only if the pair satisfies the following conditions:

$$\frac{B_N \|b_{kn}\|^2}{\sum_{i=1}^{K} \|b_{in}\|^2 p_{in}^* + N_0 B_N} + \lambda_{kn}^* = \lambda_k^* \tag{17.15}$$

$$\sum_{n=1}^{N} p_{kn}^* - p_k = 0 \tag{17.16}$$

$$p_{kn}^* \geq 0 \tag{17.17}$$

$$\lambda_{kn}^* p_{kn}^* = 0 \tag{17.18}$$

$$\lambda_{kn}^* \geq 0, \tag{17.19}$$

where $1 \leq k \leq K$ and $1 \leq n \leq N$.

Remark 2 *For optimal power allocation with independent decoding, we can arrive at the same optimal conditions as in Equation (17.15) to Equation (17.19). However, these conditions are not sufficient conditions because the objective function of Equation (17.7) is neither convex nor concave.*

Using the above results, the proof of the first claim of Theorem 1 can be found in Appendix B and the second claim is quite straightforward.

From Theorem 1, we conclude that, under the optimal solution, (i) no user should transmit on subchannel n if $\max\{\frac{\|b_{kn}\|^2}{\lambda_k}\} \leq N_0$; (ii) multiple users may transmit on the same subchannel n if they have the same weighted power gain $\max\{\frac{\|b_{kn}\|^2}{\lambda_k}\} > N_0$. In this case, OFDMA is no longer the optimal solution. One may argue that the probability of multiple users having identical weighted gains is zero due to random channels, thus OFDMA is indeed the optimal solution. While intuitively true, such an argument is invalid for finite discrete multiple-access channels. A counter example is provided for illustration.

Example 3

Define $\alpha_1 = \frac{\|b_{11}\|^2}{\|b_{21}\|^2}$, $\alpha_2 = \frac{\|b_{12}\|^2}{\|b_{22}\|^2}$, $n_{kn} = \frac{B_N N_0}{\|b_{kn}\|^2}$ in a two-user, two-subchannel scenario. The closed-form analytical solutions of Equation (17.4) is summarized in Table 17.1 for different channel conditions.

From Table 17.1, we clearly observe that OFDMA may or may not be optimal, depending on channel gains $\{b_{kn}\}$ and user's individual transmitting power $\{p_k\}$. Under the scenario where $\alpha_1 \leq \alpha_2$ and $|\alpha_2 p_1 + n_{22} - n_{21}| < p_2$, the optimal solution yields $\frac{b_{12}}{\lambda_1} = \frac{1}{2}(b_{12}p_1 + b_{22}p_2 + b_{22}n_{21} - N_0) = \frac{b_{22}}{\lambda_2}$. That is, with probability 1, the two users will have the same weighted power gain on subchannel 2. On the other hand, when $\alpha_2 p_1 + n_{22} - n_{21} \geq p_2$ and $\frac{p_2}{\alpha_1} + n_{11} - n_{12} \geq p_1$, we have $\frac{b_{11}}{\lambda_1} \leq \frac{b_{21}}{\lambda_2}$ on subchannel 1 and $\frac{b_{12}}{\lambda_1} \geq \frac{b_{22}}{\lambda_2}$ on subchannel 2. The probability that two users have the same weighted power

Table 17.1   Example 3

Conditions	Power Loading/Subchannel Allocation
$\alpha_1 \leq \alpha_2$ $\lvert \alpha_2 p_1 + n_{22} - n_{21} \rvert < p_2$	$p_{11} = 0,$ $p_{21} = \frac{1}{2}(\alpha_2 p_1 + p_2 + n_{22} - n_{21}),$ $p_{12} = p_1,$ $p_{22} = \frac{1}{2}(-\alpha_2 p_1 + p_2 + n_{21} - n_{22}),$
$\alpha_2 p_1 + n_{22} - n_{21} \geq p_2$ $\frac{p_2}{\alpha_1} + n_{11} - n_{12} \geq p_1$	$p_{11} = 0, \quad p_{12} = p_1,$ $p_{21} = p_2, \quad p_{22} = 0,$
$\alpha_1 \leq \alpha_2$ $\left\lvert \frac{p_2}{\alpha_1} + n_{11} - n_{12} \right\rvert < p_1$	$p_{11} = \frac{1}{2}(p_1 - \frac{p_2}{\alpha_1} + n_{12} - n_{11}),$ $p_{21} = p_2,$ $p_{12} = \frac{1}{2}(p_1 + \frac{p_2}{\alpha_1} + n_{11} - n_{12}),$ $p_{22} = 0,$
$\alpha_1 \geq \alpha_2$ $\lvert \alpha_1 p_1 + n_{21} - n_{22} \rvert < p_2$	$p_{11} = p_1,$ $p_{21} = \frac{1}{2}(-\alpha_1 p_1 + p_2 + n_{22} - n_{21}),$ $p_{12} = 0,$ $p_{22} = \frac{1}{2}(\alpha_1 p_1 + p_2 + n_{21} - n_{22}),$
$\alpha_1 p_1 + n_{21} - n_{22} \geq p_2$ $\frac{p_2}{\alpha_2} + n_{12} - n_{11} \geq p_1$	$p_{11} = p_1, \quad p_{12} = 0,$ $p_{21} = 0, \quad p_{22} = p_2,$
$\alpha_1 \geq \alpha_2$ $\left\lvert \frac{p_2}{\alpha_2} + n_{12} - n_{11} \right\rvert < p_1$	$p_{11} = \frac{1}{2}(p_1 + \frac{p_2}{\alpha_2} + n_{12} - n_{11}),$ $p_{21} = 0,$ $p_{12} = \frac{1}{2}(p_1 - \frac{p_2}{\alpha_2} + n_{11} - n_{12}),$ $p_{22} = p_2,$
$\alpha_2 p_1 + n_{22} - n_{21} \leq -p_2$ $\frac{p_2}{\alpha_2} + n_{12} - n_{11} \leq -p_1$	$p_{11} = 0, \quad p_{12} = p_1,$ $p_{21} = 0, \quad p_{22} = p_2,$
$\frac{p_2}{\alpha_1} + n_{11} - n_{12} \leq -p_1$ $\alpha_1 p_1 + n_{21} - n_{22} \leq -p_2$	$p_{11} = p_1, \quad p_{12} = 0,$ $p_{21} = p_2, \quad p_{22} = 0,$
$\alpha_1 = \alpha_2$ $\frac{p_2}{\alpha_2} + n_{12} - n_{11} > -p_1$	$\forall 0 < p_{11} < p_1, 0 < p_{21} < p_2 \text{ satisfy}$ $2(\alpha_1 p_{11} + p_{21}) = \alpha_1 p_1 + p_2 + n_{22} - n_{21}$

gain is zero. Given random channel realizations, clearly there is a nontrivial probability that OFDMA is not the optimal solution.

17.3.2.1   Necessary and Sufficient Conditions

Next, we derive the necessary and sufficient conditions under which OFDMA is optimal. Let us define a specific OFDMA subchannel allocation scheme as $\Phi = \{\phi_1, \phi_2, \ldots, \phi_K\}$, where ϕ_k is the set of subchannels assigned to user k [7–13]. Denote $I_k = \lvert \phi_k \rvert$ as the number of subchannels assigned to user k. Since $I_k \geq 1$, OFDMA requires $N \geq K$. Under the OFDMA subchannel allocation Φ, we can calculate its power loading $\Psi = (p^*, \lambda^*)$ using single-user water filling [24]. Here, we are interested to know whether this specific OFDMA power allocation scheme is optimal to Equation (17.4).

Using Theorem 1, we arrive at the following results.

Proposition 1

The necessary and sufficient conditions for OFDMA power allocation Ψ to be the optimal solution to Equation (17.4) are

$$\frac{\|b_{kn}\|^2}{\lambda_k} = \max_i \left\{ \frac{\|b_{in}\|^2}{\lambda_i} \right\}, \quad for \ S_n = \{k\}$$

$$\max_i \left\{ \frac{\|b_{in}\|^2}{\lambda_i} \right\} \leq N_0, \quad for \ S_n = \emptyset$$

(17.20)

where $n = 1, 2, \ldots, N$ and the weight $\frac{1}{\lambda_k}$ is user k's single-user water-filling level.

Proof Necessary conditions: Assume the OFDMA power allocation scheme Ψ is the optimal solution to Equation (17.4). In the case $|S_n| = 1$, part (i) of Theorem 1 asserts that the active user k must have the highest weighted power gain on subchannel n for any $n \in \phi_k$; in the case $|S_n| = 0$, part (ii) of Theorem 1 requires $\max_i \left\{ \frac{\|b_{in}\|^2}{\lambda_i} \right\} \leq N_0$. Thus Equation (17.20) is a necessary condition for Ψ to be optimal.

Sufficient condition: With OFDMA power allocation Ψ, we can verify that (p^*, λ^*) satisfies the KKT conditions of Equation (17.15) to Equation (17.19) if Equation (17.20) is true. By Lemma 1, $\Psi = (p^*, \lambda^*)$ is the optimal solution to Equation (17.4). Therefore Equation (17.20) is also a sufficient condition for Ψ to be the optimal allocation.

Now that we know OFDMA is not always the optimal solution, our next interest is in P_ϕ, the probability that OFDMA is indeed sum rate optimal. The probability of conditions in Equation (17.20) defines P_Ψ, that is, the probability that the OFDMA allocation Ψ is the optimal allocation. Note that P_ϕ is the sum of P_Ψ for all possible $\{\Psi\}$. We can numerically determine P_ϕ as long as the probability density function (PDF) of the channel gain b_{kn} is given. However, the computation can be prohibitively expensive when K and N are large. For tractability, we provide analyses on two extreme scenarios: (i) the low SNR case and (ii) the high SNR case. Our numerical results in the next section show that our analyses for low and high SNR regions provide the upper and lower bounds for P_o in the moderate SNR region, respectively.

17.3.2.2 Low SNR Case

In the low SNR region, we assume

$$\sum_{i=1}^{K} \|b_{in}\|^2 p_{in} \ll B_N N_0, \quad for \ \forall n.$$

(17.21)

Then Equation (17.15) becomes

$$\frac{\|b_{kn}\|^2}{N_0} + \lambda_{kn} = \lambda_k, \quad 1 \leq k \leq K, \quad 1 \leq n \leq N.$$

(17.22)

The following theorem provides the solution to Equation (17.4) in the low SNR scenario.

Theorem 2 *(i) C_{sum} is maximized if and only if each user allocates all its power to only one subchannel and the assigned subchannel to user k has the highest channel gain $\|b_{kn}\|$ over all n; (ii) $P_\phi = \frac{N!}{(N-K)! N^K}$.*

Proof We use reduction to absurdity to prove the first claim: assume in the optimal solution that the user k transmits on both subchannel n_1 and n_2, that is, $p_{kn_1} \neq 0$, $p_{kn_2} \neq 0$. From Equation (17.18), we have

$$\lambda_{kn_1} = \lambda_{kn_2} = 0.$$

Plugging the above result into Equation (17.22) yields

$$\frac{\|b_{kn_1}\|^2}{N_0} = \lambda_k = \frac{\|b_{kn_2}\|^2}{N_0}.$$

That is, $\|b_{kn_1}\| = \|b_{kn_2}\|$. However, the probability of $\{\|b_{kn_1}\| = \|b_{kn_2}\|\}$ is zero because of random channel gains. Thus, in the optimal solution, each user should allocate all its power to only one subchannel.

Next, let us assume that user k allocates its power p_k to subchannel n_1 ($p_{kn_1} = p_k$). If $n_1 \neq \arg\max_{n}(\|b_{kn}\|)$, then there exists a subchannel n_2 such that $p_{kn_2} = 0$ and $\|b_{kn_1}\| < \|b_{kn_2}\|$. From Equation (17.18) and Equation (17.19), we have $\lambda_{kn_1} = 0$ and $\lambda_{kn_2} \geq 0$. Plugging $\lambda_{kn_1} = 0$ into Equation (17.22) yields

$$\frac{\|b_{kn_1}\|^2}{N_0} = \lambda_k. \tag{17.23}$$

Plugging $\lambda_{kn_2} \geq 0$ into Equation (17.22) yields

$$\frac{\|b_{kn_2}\|^2}{N_0} + \lambda_{kn_2} = \lambda_k. \tag{17.24}$$

From Equation (17.23) and Equation (17.24), we have $\|b_{kn_1}\| \geq \|b_{kn_2}\|$, which contradicts the fact that $\|b_{kn_1}\| < \|b_{kn_2}\|$. As a result, we have proved that

$$p_{kn}^* = \begin{cases} p_k, & n = \arg\max_{n}(\|b_{kn}\|) \\ 0, & otherwise \end{cases}, \tag{17.25}$$

that is, each user must allocate all its power to the subchannel with the maximum gain.

For part (ii), notice that P_ϕ simply is the probability that the subchannels of highest gains are different for different users, therefore

$$P_\phi = \binom{N}{1} \cdots \binom{N-K+1}{1} \bigg/ \binom{N}{1}^K = \frac{N!}{(N-K)!N^K}. \tag{17.26}$$

Figure 17.6 plots the P_ϕ versus N with different K. From this figure we can see that there is a significant chance that OFDMA is optimal under low SNRs. Further, the P_ϕ increases with N and decreases with K.

17.3.2.3 High SNR Case

For the high SNR case, we assume

$$\sum_{i=1}^{K} \|b_{in}\|^2 p_{in} \gg N_0 B_N, \quad \forall \text{ utilized subchannel } n. \tag{17.27}$$

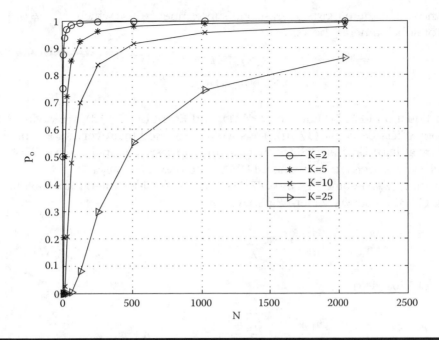

Figure 17.6 P_o for low SNR.

Then Equation (17.15) for the utilized subchannel reduces to

$$\frac{B_N \|b_{kn}\|^2}{\sum_{i=1}^{K} \|b_{in}\|^2 p_{in}} + \lambda_{kn} = \lambda_k. \tag{17.28}$$

The following theorem states the solution to Equation (17.4) in high SNR case.

Theorem 3 *(i)* C_{sum} *is maximized only if every subchannel is utilized, that is, there is no empty subchannel. (ii) Assuming* $\|b_{kn}\|$ *s are i.i.d. R.V. with Rayleigh distribution,* P_ϕ *is given by*

$$P_o = \sum_{I_1=1}^{N-(K-1)} \cdots \sum_{I_{K-1}=1}^{N-(I_1+\cdots+I_{K-2})} \left\{ \binom{N}{I_1} \cdots \binom{N - \sum_{j=1}^{K-2} I_j}{I_{K-1}} P_\Psi \right\} \tag{17.29}$$

where

$$P_\Psi = \prod_{k=1}^{K} \left[\int_0^1 \prod_{j \neq k} \left(1 - x^{\frac{I_j}{I_k} \frac{p_k}{p_j}}\right) dx \right]^{I_k}. \tag{17.30}$$

Proof We prove part (i) using reduction to absurdity. Assuming subchannel n_1 is empty, then Equation (17.15) becomes

$$\frac{B_N \|b_{kn_1}\|^2}{N_0} + \lambda_{kn_1} = \lambda_k, \quad \forall k. \tag{17.31}$$

We know that there exists at least one subchannel n_2 for which $p_{kn_2} \neq 0$. Equation (17.28) for subchannel n_2 becomes

$$\frac{B_N \|b_{kn_2}\|^2}{\sum_{i=1}^{K} \|b_{in_2}\|^2 p_{in_2}} = \lambda_k. \tag{17.32}$$

From Equation (17.27), Equation (17.31), and Equation (17.32), we obtain $\lambda_{kn_1} < 0$, which contradicts Equation (17.19). Thus we have argued that every subchannel should be used by at least one user. Since there is no empty subchannel, each OFDMA power loading Ψ corresponds to only one OFDMA subcarrier allocation Φ.

To calculate P_ϕ, assume that the OFDMA scheme Ψ (Φ) is the optimal solution. From Equation (17.32), user k's water-filling level over ϕ_k is

$$\frac{1}{\lambda_k} = \frac{p_k}{B_N I_k}. \tag{17.33}$$

Equation (17.20) yields

$$\frac{\|b_{1n}\|}{\|b_{in}\|} \geq \sqrt{\frac{I_1}{I_i} \frac{p_i}{p_1}}, \quad \forall i \neq 1, \quad n \in \phi_1$$

$$\vdots \tag{17.34}$$

$$\frac{\|b_{Kn}\|}{\|b_{in}\|} \geq \sqrt{\frac{I_K}{I_i} \frac{p_i}{p_K}}, \quad \forall i \neq K, \quad n \in \phi_K.$$

Appendix C derives the probability of Equation (17.34) for Rayleigh channels:

$$P_\Psi = \left[\int_0^1 \prod_{j \neq 1} \left(1 - x^{\frac{I_j p_1}{I_1 p_j}}\right) dx \right]^{I_1} \cdots \left[\int_0^1 \prod_{j \neq K} \left(1 - x^{\frac{I_j p_K}{I_K p_j}}\right) dx \right]^{I_K}.$$

For each fixed set of $\{I_1, I_2, \ldots, I_K\}$, the number of possible Ψs (Φs) is $\binom{N}{I_1} \cdots$ $\binom{N - \sum_{j=1}^{K-2} I_j}{I_{K-1}}$. Including all possible sets of $\{I_1, I_2, \ldots, I_K\}$, P_o is given by

$$P_\phi = \sum_{I_1=1}^{N-(K-1)} \cdots \sum_{I_{K-1}=1}^{N-(I_1+I_2+\ldots+I_{K-2})} \left\{ \binom{N}{I_1} \cdots \binom{N - \sum_{j=1}^{K-2} I_j}{I_{K-1}} P_\Psi \right\}.$$

Remark 3 *When $K = N$, both Equation (17.26) and Equation (17.29) yield the same $P_o = K!/K^K$.*

Note that we assume i.i.d. channels to simplify the derivation of Equation (17.30). In reality, the adjacent subcarriers have similar channel gains when subcarriers have strong correlation. One solution is to group these adjacent subcarriers as a single subchannel, as in WiMAX IEEE 802.16e. The grouped subchannels are much less correlated and can be approximated by i.i.d. analysis. Our numerical results validate this approximation.

17.3.2.4 Suboptimal Case

In the case when the optimal solution is not OFDMA, we would like to know what the performance gap is between OFDMA and the optimal solution. The following theorem provides the answers.

Theorem 4 *(i) The total capacity C_{sum} is maximized only if the number of shared subchannels is less than K. (ii) The capacity gap between OFDMA and the optimal solution diminishes when $N >> K$.*

The proof of Theorem 4 uses the following definitions and corollaries in Rockafellar [32]:

Definition 2 *(i) A network is called a tree if it is connected and contains no elementary circuits but has at least one arc. (ii) An arc set $F \subset A$ is said to form a forest in the network G if and only if no elementary circuits are included in F. (iii) A maximal forest in G is defined to be a forest that is not strictly contained in any other forest of G. (iv) A spanning tree for G is a tree meeting every node of G.*

Corollary 1 *(i) In a connected network (with at least one arc), a maximal forest is the same thing as a spanning tree. (ii) A connected network is a tree if and only if [the number of arcs] = [the number of nodes] − 1 > 0.*

If the optimal solution requires m ($m > 1$) users to transmit on subchannel n, we define a network g_n with node set $u_n \subset \{1, \ldots, K\}$, where $|u_n| = m$ and each node in u_n represents one of the m users. For any two different nodes (users) $i, j \in u_n$, we obtain from Theorem 1 the following constraint:

$$\frac{\|b_{in}\|^2}{\lambda_i^*} = \frac{\|b_{jn}\|^2}{\lambda_j^*}. \tag{17.35}$$

Introducing an arc a_{ij} between node i and j to denote the above constraint, then g_n is a connected network with $\binom{m}{2}$ different arcs.

By eliminating unnecessary arcs without changing the constraints among all nodes, we can transform g_n into a tree t_n with only $m - 1$ arcs. The idea is illustrated with the following example.

Example 4

For a subchannel n used by three users, Figure 17.7 shows the network g_n. We can remove the arc j_{13} without losing the constraint between node 1 and 3. To see this, we have from Equation (17.35) the constraints $\frac{\|b_{1n}\|^2}{\|b_{2n}\|^2} = \frac{\lambda_1^*}{\lambda_2^*}$ and $\frac{\|b_{2n}\|^2}{\|b_{3n}\|^2} = \frac{\lambda_2^*}{\lambda_3^*}$ for arc j_{12} and j_{23}, respectively. Multiplying these two equations yields $\frac{\|b_{1n}\|^2}{\|b_{3n}\|^2} = \frac{\lambda_1^*}{\lambda_3^*}$, which is the same constraint indicated by arc j_{13}. Thus t_n is equivalent to g_n in Figure 17.7.

If OFDMA is suboptimal, there exists r ($r \geq 1$) trees corresponding to r different subchannels used by multiple users. These r trees together form a new network G. Before the proof of Theorem 4, we first prove the following lemma.

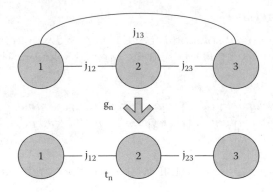

Figure 17.7   Example 4.

Lemma 2
With probability 1, the network G comprising r trees forms a forest.

Proof G forms a forest if and only if no elementary circuits are included in G [32]. Without loss of generosity, we assume there exists a circuit $c : 1 \longleftrightarrow 2 \longleftrightarrow 3 \ldots n \longleftrightarrow 1$. Since each tree contains no elementary circuits, the circuit c contains arcs from at least two different trees, say arc $j_{12} \in t_1$, $j_{23} \in t_2$, and $j_{1n} \in t_x$, where $t_1 \neq t_2$. From Equation (17.35), we have

$$\frac{\|b_{11}\|^2}{\|b_{21}\|^2} = \frac{\lambda_1^*}{\lambda_2^*} = \frac{\|b_{1x}\|^2}{\|b_{nx}\|^2} \cdots \frac{\|b_{32}\|^2}{\|b_{22}\|^2}. \tag{17.36}$$

Since $\|b_{kn}\|$s are random variables with continuous PDFs, Equation (17.36) holds with zero probability. Thus, with probability 1, the network G comprising all possible trees contains no elementary circuits (i.e., G forms a forest).

Based on the above definitions, corollaries, and Lemma 2, we can easily prove Theorem 4.

Proof of Theorem 4   From Lemma 2, we know G is a forest. Furthermore, when the number of trees reaches maximum, by Definition 1 (iii), G becomes a maximal forest. This is true because we can always add an arc (tree) to G if G is not a maximal forest. By Corollary 1 and Definition 1 (iv), the maximal forest G is a spanning tree and [the number of arcs] $= K - 1$. Note that a tree has at least one arc. Therefore we have proved that the maximum number of trees $r_{\max} \leq K - 1$. That is, the number of subchannels that can be assigned to multiple users is less than K.

For the r subchannels used by multiple users, from the second claim of Theorem 1, we have

$$\sum_{k=1}^{K} \|b_{kn}\|^2 p_{kn}^* = B_N \left\{ \max \left\{ \frac{\|b_{kn}\|^2}{\lambda_k} \right\} - N_0 \right\}.$$

From Equation (17.2), the sum capacity on these r subchannels is

$$c_{r,sum} = \frac{B}{N} \sum_{n=1}^{r} \log_2 \left(\frac{\max\left\{ \frac{\|b_{kn}\|^2}{\lambda_k} \right\}}{N_0} \right).$$ (17.37)

Because $\max\{\frac{\|b_{kn}\|^2}{\lambda_k}\}$ and $r < K$ are finite, we know $c_{r,sum}$ is inverse proportional to N. If we assign each subchannel to only one user with the maximum weighted power gain, the capacity gap between OFDMA and the optimal solution is less than $c_{r,sum}$. From Equation (17.37), this gap is negligible when $N >> K$. Essentially OFDMA becomes the optimal solution ($c_{r,sum} \to 0$) when $N \to \infty$ with finite K.

Remark 4 *With some nonzero probabilities, the solution to the KKT conditions of Equation (17.15) to Equation (17.19) requires multiple users to transmit over a finite number of subchannels. However, the capacity gap between OFDMA and the optimal solution vanishes when $N \to \infty$. This is exact the case in Cheng and Verdu [10] and Knopp and Humblet [18] where a discrete multicarrier reduces to continuous frequency selective channels.*

17.3.3 Numerical Results

While our analysis provides insights on OFDMA optimality in both low and high SNR regions, it is difficult to derive explicit analytical results for the general case (arbitrary SNR and N values). In this subsection we first present a fast numerical algorithm to compute the optimal solution to Equation (17.4). We then calculate P_o by examining whether or not the optimal solution in each channel realization is OFDMA. Finally, we evaluate the performance of OFDMA by simulations.

For Gaussian vector multiple-access channels, Yu [12] proposed a fast iterative water-filling method and proved it converges to the optimal point. A similar algorithm was proposed by Tse and Hanly [11], where the optimal power allocation over time was characterized. Based on the same idea, we have the following algorithm for calculating the optimal sum capacity in multicarrier uplink SISO channels:

Algorithm 1: Iterative water filling for multicarrier uplink SISO channels
   initialize p_{kn}, $k = 1, \ldots, K$; $n = 1, \ldots, N$.
   while (the desired accuracy is not reached)
      for $k = 1$ to K
         for $n = 1$ to N

$$Z_{kn} = I + \sum_{j=1, j \neq k}^{K} \|b_{jn}\|^2 p_{jn}$$

         end

$$p_{kn} = \arg\max_{p_{kn}} \sum_{n=1}^{N} \log_2 \left(Z_{kn} + \|b_{kn}\|^2 p_{kn} \right)$$

      end
   end

Corollary 2 *For any initial power assignment, Algorithm 1 converges to the optimal solution in (17.4).*

Since multicarrier subchannels are parallel channels, the proof of Theorem 2.4 in Yu [12] can be applied here by allowing single-user water filling over the frequency domain.

In practice, the parallel frequency selective subchannels are usually correlated. We consider subchannels with different correlation profiles in our simulations. The correlation among subchannels depends on the coherent bandwidth B_c and subcarrier spacing B_n. A general approximation for coherent bandwidth is $B_c \approx 1/T_m$, where T_m is typically taken to be the root mean square (rms) delay spread of the power delay profile. The power delay profile is often modeled as having a one-sided exponential distribution [16]:

$$A_c(\tau) = \frac{1}{T_m}e^{-\tau/T_m}, \quad \tau \geq 0, \tag{17.38}$$

where the average and rms delay spread are the same as T_m.

The coherent bandwidth varies in different applications. For example, the subcarrier spacing in 802.11a/g is 0.3125 MHz for a 20 MHz bandwidth with 64 subcarriers, which corresponds to a large B_c. In WiMAX (IEEE 802.16e), the same bandwidth employs hundreds of subcarriers (from 128 up to 2048 depending on the application) because of a relatively small B_c.

Assuming the power delay profile of Equation (17.38), we consider two scenarios in the simulation: (1) weak correlation among subchannels, with $B_c = B_n$; (2) strong correlation among subchannels, with $B_c = 10B_n$. For both scenarios, we plot P_o and OFDMA capacity as a function of K, N, and SNR. We assume power control is used so that the BS has the same received power for all users. In the uplink scenario, users are usually separated far enough to be considered independent. For each user, we assume correlated Rayleigh fading among subchannels.

Figure 17.8 to Figure 17.11 show P_o versus N for different Ks with different SNRs. We choose SNRs over a large range from -25 dB to 25 dB.

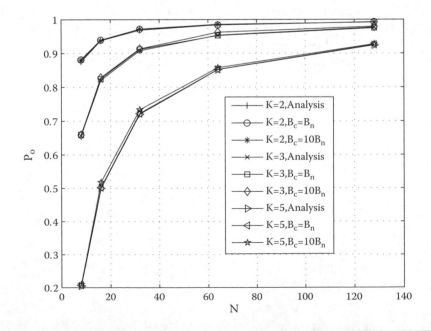

Figure 17.8 P_o **for SNR** $= -25$**dB.**

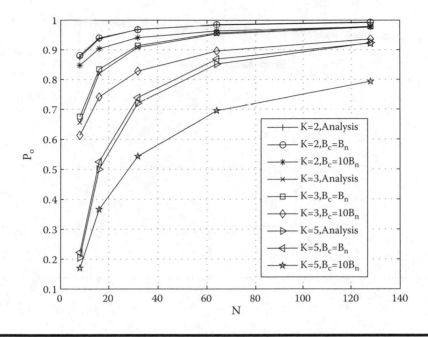

Figure 17.9 P_o for SNR $= -15$dB.

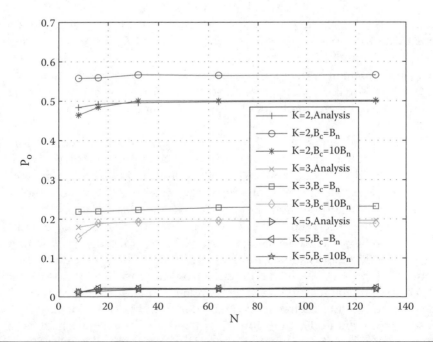

Figure 17.10 P_o for SNR $= 5$ dB.

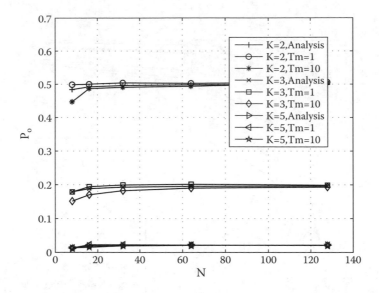

Figure 17.11 P_o for SNR = 15 dB.

We can see from Figure 17.8 that the numerical results match perfectly with our analysis in the low SNR region (Theorem 2) for both weak and strong subchannel correlations. In both cases, P_o increases with N and decreases with K.

In Figure 17.9, the P_o curves for $B_c = B_n$ match the low SNR analysis very well. This is even true for a moderate SNR of 5 dB. However, P_o decreases considerably for $B_c = 10B_n$. The intuitive explanation is that in the low SNR region, each user tries to find the best subchannel over all subcarriers. But the strong correlation ($B_c = 10B_n$) among subchannels makes it difficult because adjacent subchannels have similar channel gains. In other words, bigger coherent bandwidth reduces the number of "independent subchannels."

From Figure 17.10 and Figure 17.11, we observe that the P_o curves for $B_c = B_n$ decrease in the moderate SNR region (5 dB) and match with our analysis (Theorem 3) in the high SNR region (15 dB). We expect these results because subchannels are less correlated for $B_c = B_n$ and can be approximated as i.i.d. These are exactly the assumption we made for the derivation of Equation (17.30).

Interestingly, we observe that the P_o curves for $B_c = 10B_n$ agree with Equation (17.30) for both SNR = 5dB and SNR = 15 dB, which is especially true for large N. Note that in the high SNR region, each user is generally assigned multiple subchannels instead of only one. This presents the key difference from the low SNR region. Actually, even though adjacent subchannels are strongly correlated, a large N in the high SNR region provides sufficient multiuser channel diversity compared to the i.i.d. case.

We observe from Figure 17.8 to Figure 17.11 that simulation results match our analyses in both the low and high SNR regions, regardless of whether subchannels are correlated or not. In particular, our analyses for low and high SNR regions provide the upper and lower bounds for P_o in the moderate SNR region, respectively. In general, P_o increases with N and decreases with K and the SNR. Higher correlation among subchannels reduces P_o in the moderate SNR region.

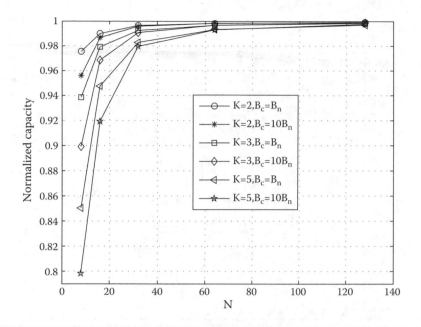

Figure 17.12 Normalized OFDMA capacity, SNR $= -25$ dB.

Next we illustrate the performance gap between OFDMA and the optimal solution. From Theorem 4 we know that OFDMA is optimal when $N \to \infty$. If all users have the same individual power constraints, by symmetry, λ_ks are the same for all users. The optimal subchannel allocation policy is to allow only the user with the best channel gain $\|b_{kn}\|$ to transmit on subchannel n. We adopt this policy as the suboptimal subchannel allocation scheme for finite Ns and numerically evaluate its performance. Figure 17.12 to Figure 17.15 show the average OFDMA capacity versus N for different Ks. For each set of parameters (K, N, and SNR), we normalize OFDMA capacity with respect to the maximum capacity (i.e., the optimal sum capacity is always one).

Similar to P_o, we can see the normalized OFDMA capacity increases with N and decreases with K. Interestingly, even though P_o decreases with the SNR, the normalized OFDMA capacity increases with the SNR. Regardless of the SNR region, higher correlation degrades OFDMA capacity as expected, since correlation reduces multiuser channel diversity. While OFDMA is not always optimal, it captures a majority of the same rate (80%) in all cases. Overall, OFDMA is nearly optimal in most practical situations, especially when N is large.

17.3.4 Conclusion

In this section we investigated the sum rate optimality of OFDMA in uplink multicarrier systems with a finite number of subchannels. We derived the necessary and sufficient conditions under which OFDMA is sum rate optimal and calculated the probabilities of these conditions for both low and high SNR regions. When OFDMA is suboptimal, we showed that the number of shared subchannels under the optimal solution must be less than the number of total users. While OFDMA is not optimal in general, the performance

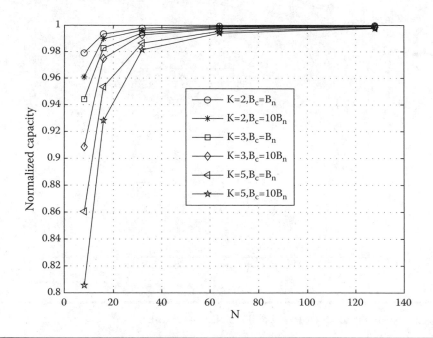

Figure 17.13   Normalized OFDMA capacity, SNR = −15 dB.

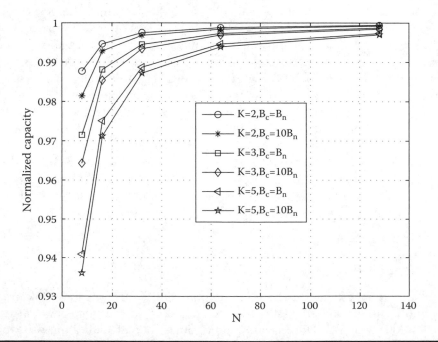

Figure 17.14   Normalized OFDMA capacity, SNR = 5 dB.

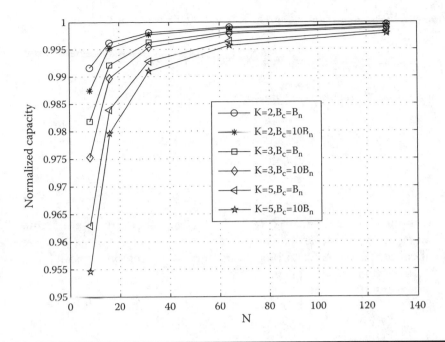

Figure 17.15   Normalized OFDMA capacity, SNR = 15 dB.

gap between the OFDMA scheme and the optimal solution is small in most cases. The difference diminishes when the number of subchannels is sufficiently large. Finally, we give a fast numerical algorithm to calculate the optimal solution and our simulations validate the analyses for both weak and strong subchannel correlations.

17.4   The Optimality of MIMO/OFDMA

In this section we extend our study on the OFDMA optimality to MIMO systems. In particular, we investigate the OFDMA optimality in both uplink and downlink MIMO multicarrier systems with a finite number of subcarriers. Toward this end, we derive the necessary and sufficient conditions under which OFDMA is optimal on any boundary point of the capacity region and derive the probabilities of these conditions for the low SNR region. In addition, for MISO channels, we also prove that OFDMA is always downlink sum rate optimal and provide uplink optimality analysis by applying our SISO results. Because of the added space dimension, it has been speculated that OFDMA is only suboptimum almost always. To answer this question we quantify OFDMA performance for the general MIMO channel and show the that the probability of OFDMA being sum rate optimal is nontrivial regardless of the SNR regions. Finally, we present three suboptimal OFDMA subcarrier allocation schemes and evaluate their performance.

17.4.1   System Model

We model the uplink MIMO channels and formulate the optimization problems for both uplink and downlink using their duality. Let $\boldsymbol{x}_{kn}$ be the input vector from user k on subcarrier n. The received signal on subcarrier n can be expressed as

$$y_n = \left(\sum_{k=1}^{K} \mathbf{H}_{kn} \mathbf{x}_{kn} \right) + \mathbf{v}_n.$$

The capacity region on subcarrier n is given in Goldsmith [16]:

$$C_n = (R_{1n}, \ldots, R_{Kn}) : \sum_{k \in S} R_{kn} \leq B_N \log_2 \left(\det \left(\mathbf{I} + \frac{\sum_{k \in S} \mathbf{H}_{kn} \mathbf{Q}_{kn} \mathbf{H}_{kn}^H}{N_0 B_N} \right) \right)$$

$$\forall S \subset \{1, 2, \ldots K\}, \tag{17.39}$$

where the input covariance matrix $\mathbf{Q}_{kn} = E[\mathbf{x}_{kn} \mathbf{x}_{kn}^H]$ is positive semidefinite because of the nonnegativity condition of the powers. Since information is delivered through parallel subcarriers in the multicarrier, the total capacity region over all subcarriers and all users for the uplink channel is given by

$$C = (R_1, \ldots, R_K) : \sum_{k \in S} R_k \leq B_N \sum_{n=1}^{N} \log_2 \det \left(\mathbf{I} + \frac{\sum_{k \in S} \mathbf{H}_{kn} \mathbf{Q}_{kn} \mathbf{H}_{kn}^H}{N_0 B_N} \right)$$

$$\forall S \subset \{1, 2, \ldots K\}$$

$$s.t. \ \mathbf{Q}_{kn} \succeq 0, \ \sum_{n=1}^{N} tr(\mathbf{Q}_{kn}) \leq p_k$$

$$1 \leq k \leq K, \ 1 \leq n \leq N.$$

To maximize the system capacity, we need to find a set of covariance matrices $\mathbf{Q}_{kn}$ that achieve the extreme (boundary) point of the capacity region. This is equivalent to maximize a weighted sum of individual rates $C_r = \sum_{k=1}^{K} \mu_k R_k$ with $\mu_k \geq 0$ and $\sum_{k=1}^{K} \mu_k = 1$. Without loss of generality, we assume $\mu_1 \geq \mu_2 \geq \ldots \geq \mu_K \geq \mu_{K+1} = 0$. Then the optimization problem can be formulated as follows [12]:

$$\arg \max_{\mathbf{Q}_{kn}} C_r = B_N \sum_{n=1}^{N} \sum_{k=1}^{K} (\mu_k - \mu_{k+1}) \log_2 \det \left(\mathbf{I} + \frac{\sum_{j=1}^{k} \mathbf{H}_{jn} \mathbf{Q}_{jn} \mathbf{H}_{jn}^H}{N_0 B_N} \right)$$

$$\tag{17.40}$$

$$s.t. \ \mathbf{Q}_{kn} \succeq 0, \ \sum_{n=1}^{N} tr(\mathbf{Q}_{kn}) \leq p_k$$

$$1 \leq k \leq K, \ 1 \leq n \leq N.$$

The sum capacity is defined when $\mu_1 = \mu_2 = \ldots = \mu_K$, that is,

$$\arg \max_{\mathbf{Q}_{kn}} C_s = B_N \sum_{n=1}^{N} \log_2 \det \left(\mathbf{I} + \frac{\sum_{k=1}^{K} \mathbf{H}_{jn} \mathbf{Q}_{jn} \mathbf{H}_{jn}^H}{N_0 B_N} \right)$$

$$\tag{17.41}$$

$$s.t. \ \mathbf{Q}_{kn} \succeq 0, \ \sum_{n=1}^{N} tr(\mathbf{Q}_{kn}) \leq p_k.$$

Using the M^3 uplink-downlink duality [16], the corresponding downlink optimization problems are formulated as [17]

$$\arg\max_{\mathbf{Q}_{kn}} C_r = B_N \sum_{n=1}^{N} \sum_{k=1}^{K} (\mu_k - \mu_{k+1}) \log_2 \det \left(\mathbf{I} + \frac{\sum_{j=1}^{k} \mathbf{H}_{jn} \mathbf{Q}_{jn} \mathbf{H}_{jn}^{H}}{N_0 B_N} \right) \tag{17.42}$$

$$s.t. \ \mathbf{Q}_{kn} \succeq 0, \ \sum_{k=1}^{K} \sum_{n=1}^{N} tr(\mathbf{Q}_{kn}) \leq P,$$

and

$$\arg\max_{\mathbf{Q}_{kn}} C_s = B_N \sum_{n=1}^{N} \log_2 \det \left(\mathbf{I} + \frac{\sum_{k=1}^{K} \mathbf{H}_{kn} \mathbf{Q}_{kn} \mathbf{H}_{kn}^{H}}{N_0 B_N} \right) \tag{17.43}$$

$$s.t. \ \mathbf{Q}_{kn} \succeq 0, \ \sum_{n=1}^{N} \sum_{k=1}^{K} tr(\mathbf{Q}_{kn}) \leq P,$$

where the power constraint $P = \sum_{k=1}^{K} p_k$ on the downlink equals the sum of individual power constraints on the uplink. It is shown in Yu [12] that Equation (17.40) to Equation (17.43) are convex programming problems. In the remainder of this section we refer to the solutions to Equation (17.40) through Equation (17.43) as the optimal solutions, which provide a set of covariance matrices $\mathbf{Q}_{kn}$ that achieve the maximum capacity of M^3 system.

The rest of the discussion concerns whether OFDMA can achieve the capacity.

17.4.2 Necessary and Sufficient Conditions of OFDMA Optimality

If the optimal solution requires $\mathbf{Q}_{kn} \succ 0$ ($\mathbf{Q}_{kn}$ is positive definite), then user k is active on subcarrier n. Otherwise, if $\mathbf{Q}_{kn} = 0$, user k cannot transmit (inactive) on subcarrier n. Similar to the SISO case, we define S_n as the active user set on subcarrier n: $\{i \in S_n \text{ if } \mathbf{Q}_{in} \succ 0\}$. Since the capacity-achieving subcarrier allocation scheme may or may not be OFDMA, the maximum capacity in Equation (17.40) through Equation (17.43) upper bounds the achievable rates of OFDMA.

In this subsection we derive the necessary and sufficient conditions under which OFDMA is the optimal solution to Equation (17.40) through Equation (17.43). Normalizing $B_N = 1$ and assuming $N_0 = 1$ for simplicity, we have the following theorem which asserts the optimality of uplink OFDMA:

Theorem 5 *(i) Given an uplink OFDMA subcarrier allocation scheme Φ, the corresponding power loading Ψ is the water-filling solution over frequency and space. (ii) The Uplink OFDMA subcarrier allocation Φ (power allocation Ψ) is optimal to Equation (17.40) if and only if*

$$e_{\max}\left\{\mu_k \mathbf{H}_{kn}^{H}\left(\mathbf{I} + \mathbf{H}_{kn}\mathbf{Q}_{kn}\mathbf{H}_{kn}^{H}\right)^{-1}\mathbf{H}_{kn}\right\}/\lambda_k \geq e_{\max}\{\mathbf{A}_{ln}\}/\lambda_l, \text{ for } S_n = \{k\}, \forall l \neq k$$

$$\max_{l}\left\{\mu_l e_{\max}\left\{\mathbf{H}_{ln}^{H}\mathbf{H}_{ln}\right\}/\lambda_l\right\} \leq 1, \text{ for } S_n = \emptyset$$

where

$$\mathbf{A}_{ln} = \max\{\mu_l - \mu_k, 0\}\mathbf{H}_{ln}^{H}\mathbf{H}_{ln} + \min\{\mu_l, \mu_k\}\mathbf{H}_{ln}^{H}\big(\mathbf{I} + \mathbf{H}_{kn}\mathbf{Q}_{kn}\mathbf{H}_{kn}^{H}\big)^{-1}\mathbf{H}_{ln} \qquad (17.44)$$

where $1/\lambda_k$ is the OFDMA water-filling level for user k over assigned subcarrier ϕ_k, and $e_i\{\mathbf{A}\}$ denotes the ith eigenvalue of $\mathbf{A}$.

The proof is given in Appendix D. For sum capacity ($\mu_1 = \mu_2 = \ldots = \mu_K = u$), Theorem 5 becomes the following corollary.

Corollary 3 *The uplink OFDMA subcarrier allocation scheme Φ is optimal to Equation (17.41) if and only if*

$$e_{\max}\big\{\mathbf{H}_{kn}^{H}\big(\mathbf{I} + \mathbf{H}_{kn}\mathbf{Q}_{kn}\mathbf{H}_{kn}^{H}\big)^{-1}\mathbf{H}_{kn}\big\}/\lambda_k \geq e_{\max}\big\{\mathbf{H}_{ln}^{H}\big(\mathbf{I} + \mathbf{H}_{kn}\mathbf{Q}_{kn}\mathbf{H}_{kn}^{H}\big)^{-1}\mathbf{H}_{ln}\big\}/\lambda_l$$

$$\text{for } S_n = \{k\}, \forall l \neq k$$

$$\max_l \big\{\mu e_{\max}\big\{\mathbf{H}_{ln}^{H}\mathbf{H}_{ln}\big\}/\lambda_l\big\} \leq 1, \text{for } S_n = \emptyset. \qquad (17.45)$$

In the downlink optimization problem, Equation (17.82) and Equation (17.87) in Appendix D are still valid except that $\lambda_k = \lambda_l = \lambda$, which is the lagrangian multiplier associated with the total power constraint. Thus we have the following theorem for downlink channels:

Theorem 6 *(i) Given a downlink OFDMA subcarrier allocation scheme Φ, the corresponding power loading Ψ is a water-filling solution over frequency, space, and users. (ii) The downlink OFDMA subcarrier allocation Φ (power allocation Ψ) is optimal to Equation (17.42) if and only if*

$$e_{\max}\big\{\mu_k\mathbf{H}_{kn}^{H}\big(\mathbf{I} + \mathbf{H}_{kn}\mathbf{Q}_{kn}\mathbf{H}_{kn}^{H}\big)^{-1}\mathbf{H}_{kn}\big\} \geq e_{\max}\{\mathbf{A}_{ln}\}, \text{for } S_n = \{k\}, \forall l \neq k$$

$$\max_l \big\{\mu_l e_{\max}\big\{\mathbf{H}_{ln}^{H}\mathbf{H}_{ln}\big\}/\lambda\big\} \leq 1, \text{for } S_n = \emptyset \qquad (17.46)$$

where $1/\lambda$ is the downlink OFDMA water-filling level.

For sum capacity, Theorem 6 becomes the following corollary.

Corollary 4 *The downlink OFDMA subcarrier allocation scheme Φ is optimal to Equation (17.43) if and only if*

$$e_{\max}\big\{\mathbf{H}_{kn}^{H}\big(\mathbf{I} + \mathbf{H}_{kn}\mathbf{Q}_{kn}\mathbf{H}_{kn}^{H}\big)^{-1}\mathbf{H}_{kn}\big\} \geq e_{\max}\big\{\mathbf{H}_{ln}^{H}\big(\mathbf{I} + \mathbf{H}_{kn}\mathbf{Q}_{kn}\mathbf{H}_{kn}^{H}\big)^{-1}\mathbf{H}_{ln}\big\}$$

$$\text{for } S_n = \{k\}, \forall l \neq k$$

$$\max_l \big\{\mu e_{\max}\big\{\mathbf{H}_{ln}^{H}\mathbf{H}_{ln}\big\}/\lambda\big\} \leq 1, \text{for } S_n = \emptyset. \qquad (17.47)$$

Remark 5 *The uplink-downlink duality suggests some relationship between uplink and downlink optimality conditions. If all users have equal power in the uplink and the number of subcarriers is large enough, by symmetry, we have $\lambda_k = \lambda_l$ for any $l \neq k$, thus the uplink optimality conditions of Equation (17.44) and (Equation 17.45) are the same as the downlink optimality conditions of Equation (17.46) and Equation (17.47).*

For a set of channel matrices $\mathbf{H}_{kn}$s, we can easily determine whether or not a given OFDMA scheme Φ is optimal using Theorems 5 and 6. However, to establish the relationship between OFDMA and the optimal solution, we have to test every Φ to see if it satisfies Equation (17.44) to Equation (17.47) and the computation is inhibitive when K and N are large. Next, we discuss some special cases in which explicit OFDMA optimality results can be derived.

17.4.3 Special Cases

In this subsection we provide OFDMA optimality analysis for two special cases: (a) the low SNR region in MIMO, and (b) the multiple input single output (MISO) channels. In the low SNR region, we extend our SISO analysis to MIMO channels.

Assume

$$e_{\max}\left(\sum_{k=1}^{K} \mathbf{H}_{kn}\mathbf{Q}_{kn}\mathbf{H}_{kn}^{H}\right) \ll 1 \text{ for } \forall n. \tag{17.48}$$

For $s < K$, it is obvious $e_{\max}(\sum_{j=1}^{s} \mathbf{H}_{jn}\mathbf{Q}_{jn}\mathbf{H}_{jn}^{H}) \ll 1$ so that we have $(\mathbf{I} + \sum_{j=1}^{s} \mathbf{H}_{jn}\mathbf{Q}_{jn}\mathbf{H}_{jn}^{H})^{-1} \simeq \mathbf{I}$.

Under this assumption, Equation (17.76) becomes

$$\mu_k \mathbf{H}_{kn}^{H}\mathbf{H}_{kn} + \Lambda_{kn} = \lambda_k \mathbf{I}. \tag{17.49}$$

Denote the ith spatial channel on subcarrier n as subchannel (n, i), thus the following theorem provides the optimal uplink solution to Equation (17.40) in the low SNR scenario.

Theorem 7 *In low SNR uplink MIMO channels, (i) C_r is maximized if and only if each user allocates all its power to only one subchannel and the assigned subchannel to user k has the highest eigenvalue $e_i\{\mathbf{H}_{kn}\mathbf{H}_{kn}^{H}\}$ over all n and i; (ii) $P_\phi = \frac{N!}{(N-K)!N^K}$.*

Proof of Theorem 7 Using reduction to absurdity, assume that user k transmits on both subchannel (n_1, i_1) and (n_2, i_2) in the optimal solution, that is, $e_{i_1}\{\mathbf{Q}_{kn_1}\} > 0$ and $e_{i_2}\{\mathbf{Q}_{kn_2}\} > 0$, then the corresponding eigenvalues of Λ_{kn_1} and Λ_{kn_2} are zeros. From Equation (17.49), we have $e_{i_1}\{\mathbf{H}_{kn_1}^{H}\mathbf{H}_{kn_1}\} = \lambda_k = e_{i_2}\{\mathbf{H}_{kn_2}^{H}\mathbf{H}_{kn_2}\}$, which happens with zero probability because $(n_1, i_1) \neq (n_2, i_2)$. So each user should allocate all its power to only one subchannel. From Equation (17.49), it is trivial to verify that the subchannel (n^*, i^*) assigned to user k has the largest eigenvalue $e_i\{\mathbf{H}_{kn}^{H}\mathbf{H}_{kn}\}$ over all n and i. As a result, we have proved that

$$e_i(Q_{kn}^*) = \begin{cases} p_k, & (n, i) = \arg\max_{n,i}\{e_i\{\mathbf{H}_{kn}^{H}\mathbf{H}_{kn}\}\} \\ 0, & \text{otherwise.} \end{cases} \tag{17.50}$$

Notice that P_ϕ is simply the probability that assigned subcarriers are different for different users; we have the same result as the SISO case:

$$P_o = \frac{N!}{(N-K)!N^K}. \tag{17.51}$$

Figure 17.6 illustrates the P_o versus N with different K. We see the P_ϕ increases with N and decreases with K. When $N \to \infty$, OFDMA is always the optimal solution.

The following theorem states OFDMA optimality in the low SNR region for downlink channels.

Theorem 8 *In the low SNR region, downlink OFDMA is always the optimal solution to Equation (17.42), subcarrier n is assigned to the user with the highest eigenvalue $e_{\max}\{\mathbf{H}_{kn}\mathbf{H}_{kn}^H\}$ over all k.*

Proof Under the assumption of Equation (17.48), Equation (17.76) for the downlink becomes

$$\mu_k \mathbf{H}_{kn}^H \mathbf{H}_{kn} + \Lambda_{kn} = \lambda \mathbf{I}. \tag{17.52}$$

Using reduction to absurdity, assume that both user k_1 and k_2 transmit on subcarrier n in the optimal solution, that is $e_{\max}\{\mathbf{Q}_{k_1 n}\} > 0$ and $e_{\max}\{\mathbf{Q}_{k_2 n}\} > 0$, then the corresponding eigenvalues of $\Lambda_{k_1 n}$ and $\Lambda_{k_2 n}$ are zeros. From Equation (17.52), we have $\mu_{k_1} e_{\max}\{\mathbf{H}_{k_1 n}^H \mathbf{H}_{k_1 n}\} = \lambda = \mu_{k_2} e_{\max}\{\mathbf{H}_{k_2 n}^H \mathbf{H}_{k_2 n}\}$, which happens with zero probability. Thus only the user who has the highest eigenvalue $e_{\max}\{\mathbf{H}_{kn}^H \mathbf{H}_{kn}\}$ is allowed to transmit on subcarrier n.

Note that the high SNR case for MIMO channels is more involved and no conclusion can be drawn.

Although it is generally difficult to characterize OFDMA optimality analytically in MIMO systems in the special case of MISO systems, similar results to SISO can be obtained for sum capacity optimality. For uplink channels, Theorem 4 can be applied directly to the uplink MISO case.

Theorem 9 *(i) The total capacity C_s is maximized only if the number of shared subcarriers is less than K. (ii) The capacity gap between OFDMA and the optimal solution diminishes when $N \gg K$. (iii) In the high SNR region, C_s is maximized only if every subcarrier is utilized, that is, there is no empty subcarrier.*

Proof The proof of parts (i) and (ii) is exactly the same as in the SISO case except that the distribution of power gain b_{kn}^2 is replaced by the distribution of the largest eigenvalue $e_{\max}\{\mathbf{H}_{kn}^H \mathbf{H}_{kn}\}$.

For part (iii), since $\mathbf{H}_{kn}$ is a $1 \times T$ vector in MISO channels, Equation (17.76) becomes

$$\frac{\mathbf{H}_{kn}^H \mathbf{H}_{kn}}{1 + \sum_{j=1}^{K} \mathbf{H}_{jn} \mathbf{Q}_{jn} \mathbf{H}_{jn}^H} + \Lambda_{kn} = \lambda_k \mathbf{I}. \tag{17.53}$$

In the high SNR region, we assume

$$\mathbf{H}_{jn}\mathbf{Q}_{jn}\mathbf{H}_{jn}^{H} \gg 1 \text{ if } \mathbf{Q}_{jn} > 0.$$

We prove part (iii) using reduction to absurdity. Assuming, in the optimal solution, subcarrier n_1 is null, then Equation (17.53) becomes

$$\mathbf{H}_{kn_1}^{H}\mathbf{H}_{kn_1} + \Lambda_{kn_1} = \lambda_k\mathbf{I}.$$

Since both $\mathbf{H}_{kn_1}^{H}\mathbf{H}_{kn_1}$ and Λ_{kn_1} are positive semidefinite, we have

$$e_{\max}\left(\mathbf{H}_{kn_1}^{H}\mathbf{H}_{kn_1}\right) \leq \lambda_k. \tag{17.54}$$

We know that there exists at least one subcarrier n_2 for which $\mathbf{Q}_{kn_2} \neq 0$. Diagonalizing Equation (17.53), we have

$$\frac{e_{\max}\left(\mathbf{H}_{kn_2}^{H}\mathbf{H}_{kn_2}\right)}{1 + \sum_{j=1}^{K} \mathbf{H}_{jn_2}\mathbf{Q}_{jn_2}\mathbf{H}_{n_2}^{H}} = \lambda_k\mathbf{I}. \tag{17.55}$$

Since $1 \ll \sum_{j=1}^{K} \mathbf{H}_{jn_2}\mathbf{Q}_{jn_2}\mathbf{H}_{jn_2}^{H}$, Equation (17.54) and Equation (17.55) contradict each other. Therefore we have proved part (iii).

Note that the SIMO case is more involved and no simple conclusion can be drawn so far.

For downlink MISO channels, we have the following theorem.

Theorem 10 *The downlink OFDMA is always the optimal solution to Equation (17.43) in SIMO channels and subcarrier n is assigned to the user who has the highest eigenvalue* $e_{\max}\{\mathbf{H}_{kn}\mathbf{H}_{kn}^{H}\}$ *over all k.*

The proof is quite straightforward and omitted here. Theorem 10 is expected since downlink OFDMA is the optimal solution in the SISO case.

17.4.4 Numerical Algorithms

For the general MIMO channels, it is difficult to derive the optimal solution analytically. Exhaustive search is needed to determine OFDMA achievable rates. In this subsection we first propose fast numerical algorithms to compute the optimal sum capacity for generic multicarrier MIMO channels. The results provide benchmarks for the OFDMA scheme. The second contribution here is that we evaluate simple suboptimal OFDMA subcarrier allocation schemes and quantify their performance with respect to the optimal OFDMA scheme.

In the moderate and high SNR regions, it is generally difficult to characterize OFDMA optimality analytically (i.e., to calculate P_ϕ and quantify the performance gap between OFDMA and the optimal solution). However, since Equation (17.40) through Equation (17.43) are convex programming problems, we can numerically calculate the optimal solution using standard convex optimization methods. The results serve as the upper bound for the OFDMA rate where its upper limit can be determined by exhaustively searching all possible Φs.

For the single-carrier sum capacity optimization problem of Equation (17.41), the optimal solution can be numerically obtained using the well-known iterative water-filling method proposed in Yu [12]. By slightly extending this result, we have the following algorithm for calculating the optimal sum capacity in multicarrier uplink MIMO channels:

Algorithm 2: Iterative water-filling for multicarrier uplink MIMO channels

   initialize $\mathbf{Q}_{kn}$, $k = 1, \ldots, K$; $n = 1, \ldots, N$.
   while (the desired accuracy is not reached)
      for $k = 1$ to K
         for $n = 1$ to N

$$\mathbf{Z}_{kn} = \mathbf{I} + \sum_{j=1, j \neq k}^{K} \mathbf{H}_{jn} \mathbf{Q}_{jn} \mathbf{H}_{jn}^{H}$$

        end

$$\mathbf{Q}_{kn} = \arg\max_{\mathbf{Q}} \sum_{n=1}^{N} \log_2 \det \left(\mathbf{Z}_{kn} + \mathbf{H}_{kn} \mathbf{Q}_{kn} \mathbf{H}_{kn}^{H} \right)$$

      end
   end

Corollary 5   *Algorithm 2 converges to the optimal solution in (17.41).*

Since multicarrier subchannels are parallel independent channels, the proof of Theorem 2.4 in Yu [12] can be applied here by allowing single-user water filling over both frequency and space.

Michel and Wunder [17] proposed an algorithm to compute the optimal sum capacity of downlink multicarrier MIMO channels based on the iterative water-filling algorithm in Yu [12]. By using Algorithm 2, we arrive at a more efficient and simplified algorithm:

Algorithm 3: Iterative water filling and convex optimization for multicarrier downlink MIMO channels

   initialize p_k, $k = 1, \ldots, K$.
   while (the desired accuracy is not reached)
      1: for given power allocation p_k, $k = 1, \ldots, K$, calculate $\mathbf{Q}_{kn}$ using Algorithm 2
      2: update p_k, $k = 1, \ldots, K$ using convex optimization methods
   end

Corollary 6   *For any initial power assignment, Algorithm 3 converges to the optimal solution in (17.43).*

Since Equation (17.43) is a convex optimization problem, the proof is straightforward and omitted here.

As noted earlier, computation of the OFDMA rate by exhaustive search is expensive when K and N are large. To gain some insight into OFDMA performance with light computation, we consider an example multiuser multicarrier MIMO system with $K = N = R = T = 2$.

For both the uplink and downlink channels, there are only four possible Φs:

Φ_1 User 1 transmits on subchannel 1
 　　 User 2 transmits on subchannel 2
Φ_2 User 1 transmits on subchannel 2
 　　 User 2 transmits on subchannel 1
Φ_3 User 1 transmits on both subchannels
 　　 User 2 is inactive
Φ_4 User 1 is inactive
 　　 User 2 transmits on both subchannels

Under each Φ, the maximum capacity is achieved by single-user water filling. We compute the water-filling solutions for all four Φs to determine the best $\Phi_{\max}$, which is the upper limit of OFDMA performance. To calculate the optimal sum capacity, we choose $\Psi_{\max}$ as the initial point for Algorithms 1 and 2 and the algorithms converge extremely fast to the optimal solution.

Let $p_1 = p_2 = P/2$. Figure 17.16 shows the probability of OFDMA being sum rate optimal with different SNRs. One notable observation is that P_o is nontrivial regardless of the SNR for both the uplink and downlink channels. In the low SNR region, P_ϕ is approximately 0.5 and 1 for the uplink and downlink, respectively, which validates our

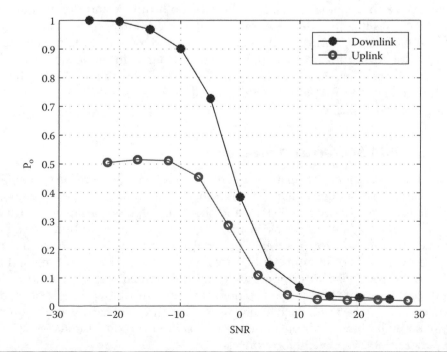

Figure 17.16   The probability of OFDMA optimality.

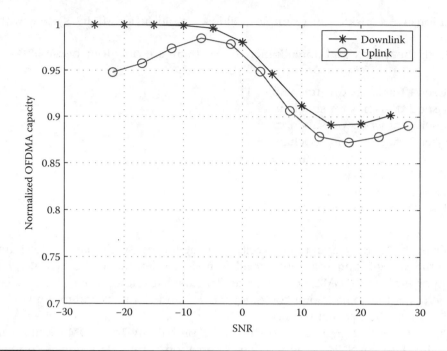

Figure 17.17 Normalized OFDMA capacity.

analysis. We see P_o decreases dramatically in the moderate SNR region and becomes steady in the high SNR region.

We normalize the optimal capacity to one for both uplink and downlink channels. Figure 17.17 shows the OFDMA performance gap with different SNRs. Interestingly, we observe that OFDMA performance degrades in the moderate SNR region and increases in the high SNR region for both the uplink and downlink. In the low SNR region, the OFDMA capacity is optimal for the downlink and increases with the SNR for the uplink, reaching a maximum 98% at about −5 dB. In all SNR cases, OFDMA captures at least 87% of the optimal capacity.

17.4.5 OFDMA Subcarrier Allocation

The above OFDMA optimality analysis asserts that there is a nontrivial probability that OFDMA is indeed the optimal multiple access scheme. Even when OFDMA is suboptimal, the performance gap between OFDMA and the optimal solution is small. Under the framework of OFDMA, an exhaustive search is generally needed to find the optimal Φ. The complexity is $o\left(K^N\right)$ for possible Φs over all subcarriers. For each Φ, the capacity needs to be evaluated to find out the best OFDMA scheme. The cost of the exhaustive search is thus exponential with respect to N and polynomial with respect to K.

To reduce the subcarrier allocation complexity, suboptimal low complexity subcarrier allocation criteria can be used. Specifically, we choose three subcarrier allocation criteria, namely, maximum eigen criterion, sum criterion [17], and product criterion [15], which are independent of the power allocation:

$$S_n = \arg\max_k \left\{ e_{\max}\left(\mathbf{H}_{kn}^H \mathbf{H}_{kn} \right) \right\} \qquad (17.56)$$

$$S_n = \arg\max_k \left\{ \sum_{i=1}^{R_{kn}} e_i \left(\mathbf{H}_{kn}^H \mathbf{H}_{kn} \right) \right\} \tag{17.57}$$

$$S_n = \arg\max_k \left\{ \prod_{i=1}^{R_{kn}} e_i \left(\mathbf{H}_{kn}^H \mathbf{H}_{kn} \right) \right\}, \tag{17.58}$$

where R_{kn} is the rank of $\mathbf{H}_{kn}^H \mathbf{H}_{kn}$.

The computation complexity for the above criteria is only $o(KN)$. To justify such simplifications, we notice from Theorem 8 that Equation (17.56) is actually the optimal subcarrier allocation scheme in the low SNR region of downlink channels. In the high SNR region of downlink channels, we assume $e_i(\mathbf{H}_{kn}^H \mathbf{H}_{kn}) \gg 1$. The optimal water filling allocates the same amount of power q_n / R_{kn} to all the R_{kn} spatial subchannels on subcarrier n, where q_n is the amount of power assigned to subcarrier n. We obtain the following approximations for OFDMA capacity on subcarrier n:

$$\arg\max_k \left\{ \log \left(\det \left(\mathbf{I} + \mathbf{H}_{kn} \mathbf{Q}_{kn} \mathbf{H}_{kn}^H \right) \right) \right\}$$

$$= \arg\max_k \left\{ \log \prod_{i=1}^{R_{kn}} \log \left(1 + e_i \left(\mathbf{H}_{kn}^H \mathbf{H}_{kn} \right) q_n / R_{kn} \right) \right\}$$

$$\approx \arg\max_k \left\{ \log \prod_{i=1}^{R_{kn}} e_i \left(\mathbf{H}_{kn}^H \mathbf{H}_{kn} \right) q_n / R_{kn} \right\}$$

$$= \arg\max_k \left\{ \prod_{i=1}^{R} e_i \left(\mathbf{H}_{kn}^H \mathbf{H}_{kn} \right) \right\}$$

when $R_{1n} = \ldots = R_{Kn} = R$.

Therefore the product criterion tends to be more accurate when the SNR is high.

We now compare the three suboptimal criteria against the optimal criterion using exhaustive search. For downlink channels, Figure 17.18 shows the probability of being optimal OFDMA as a function of the SNR. Normalizing the optimal OFDMA capacity to one, Figure 17.19 shows the achievable sum rate with different SNRs. From these two figures we can see that the maximum eigen criterion and sum criterion have similar performance and are more suitable in the low SNR region. On the other hand, the product criterion tends to be a better approximation when the SNR is high.

Similar results are obtained for uplink channels in Figure 17.20 and Figure 17.21. However, we notice that the product criterion is inferior to the other two criteria regardless of the SNR region. Compared with the downlink case, we see these three criteria are far less efficient in the uplink channels.

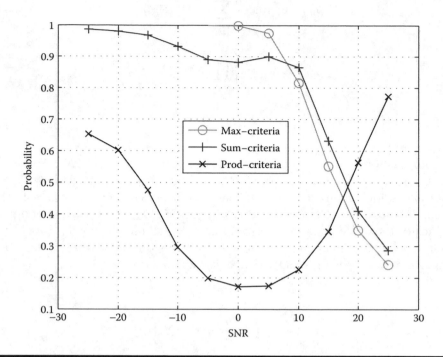

Figure 17.18   Probability of being optimal OFDMA versus SNR in downlink.

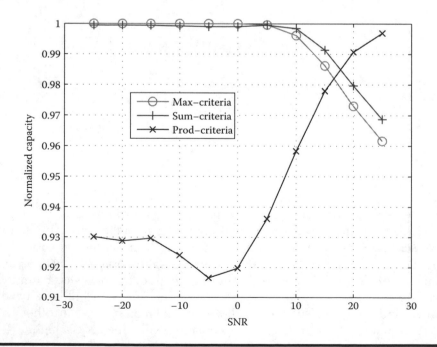

Figure 17.19   Capacity versus SNR in downlink.

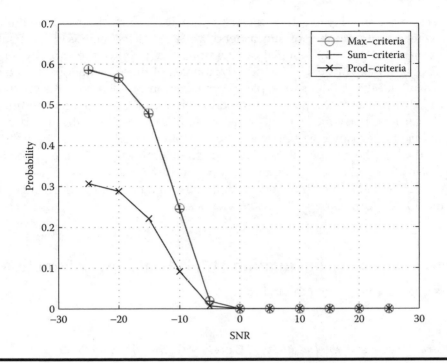

Figure 17.20 Probability of being optimal OFDMA versus SNR in uplink.

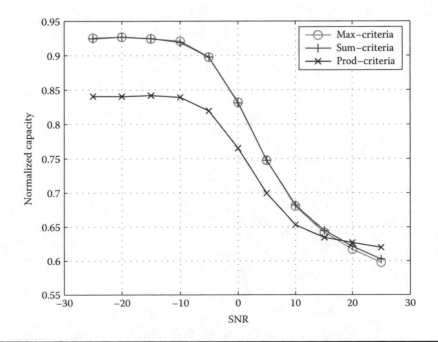

Figure 17.21 Capacity versus SNR in uplink.

17.4.6 Conclusion

In this section we investigated the OFDMA optimality in uplink and downlink MIMO systems with a finite number of subcarriers. We derived the necessary and sufficient conditions for OFDMA being optimal on any boundary point of the capacity region. In the low SNR case, we further derive the probabilities of these conditions. In the special uplink/downlink MISO/SIMO sum capacity optimization problem, our results show that OFDMA is always sum rate optimal in the downlink, and in the uplink the number of shared subcarriers under the optimal solution should be less than the number of total users, thus the performance gap between the OFDMA scheme and the optimal solution is negligible when the number of subcarriers is sufficiently large. For the general MIMO case, we provide algorithms to compute the optimal sum capacity and compare this with OFDMA capacity and show P_o is nontrivial. We also quantify the performance of three suboptimal OFDMA subcarrier allocation schemes.

Appendix A: $-C_{sum}$ in Equation (17.4) Is a Convex Function

We prove the statement using the following lemma [31].

Lemma 3
$f(x)$ is (strictly) convex if the Hessian of f is positive (definite) semidefinite.

We see from Equation (17.4) that C_{sum} is simply the sum of the capacity on each subchannel. In order to prove $-C_{sum}$ is a convex function, we only need to prove the convexity of each summation term, that is,

$$-C_{sn} = -B_N \log_2 \left(1 + \frac{\sum\limits_{k=1}^{K} p_{kn} \|b_{kn}\|^2}{B_N N_0} \right),$$

where $1 \leq n \leq N$.

For any subchannel n, we have

$$-\frac{\partial^2 C_{sn}}{\partial p_{jn} \partial p_{in}} = \frac{B_N \|b_{in}\|^2 \|b_{jn}\|^2}{\left(\sum\limits_{k=1}^{K} \|b_{kn}\|^2 g_{kn} + B_N N_0 \right)^2} > 0,$$

where $1 \leq i, j \leq K$.

Then the Hessian matrix of $-C_{sn}$ can be decomposed as

$$\mathbf{S} = \frac{B_N}{\left(\sum\limits_{k=1}^{K} \|b_{kn}\|^2 g_{kn} + B_N N_0 \right)^2} \mathbf{T}^T * \mathbf{T},$$

where $\mathbf{T} = \left[\|b_{1n}\|^2, \|b_{2n}\|^2, \ldots, \|b_{Kn}\|^2 \right]$. It is obvious that $\mathbf{S}$ is strictly positive definite. With Lemma 3, we state that $-C_{sn}$ is strictly convex and thus $-C_{sum}$ is strictly convex.

Appendix B: Proof of the First Claim of Theorem 1

Proof Note that Equation (17.15) can be rewritten as

$$\sum_{i=1}^{K} \|b_{in}\|^2 \, p_{in}^* = B_N \left(\frac{\|b_{kn}\|^2}{\lambda_k^* - \lambda_{kn}^*} - N_0 \right). \tag{17.59}$$

From Equation (17.19), we have

$$\sum_{i=1}^{K} \|b_{in}\|^2 \, p_{in}^* \geq B_N \left(\frac{\|b_{kn}\|^2}{\lambda_k^*} - N_0 \right). \tag{17.60}$$

If there exists at least one user k whose transmitting power $p_{kn}^* > 0$, we have $\lambda_{kn}^* = 0$ from Equation (17.18). Thus Equation (17.59) becomes

$$\sum_{i=1}^{K} \|b_{in}\|^2 \, p_{in}^* = B_N \left(\frac{\|b_{kn}\|^2}{\lambda_k^*} - N_0 \right). \tag{17.61}$$

Since $\sum_{i=1}^{K} \|b_{in}\|^2 \, p_{in}^* > 0$, we have $\frac{\|b_{kn}\|^2}{\lambda_k^*} > N_0$. Thus $\max_k \left\{ \frac{\|b_{kn}\|^2}{\lambda_k} \right\} > N_0$.

On the other hand, if $\frac{\|b_{kn}\|^2}{\lambda_k^*} > N_0$, from Equation (17.60), we obtain $\sum_{i=1}^{K} \|b_{in}\|^2 \, p_{in}^* > 0$, that is, there exists at least one users with nonzero transmitting power on subchannel n.

Now we use reduction to absurdity to determine which user should transmit on subchannel n when $\max_k \left\{ \frac{\|b_{kn}\|^2}{\lambda_k} \right\} > N_0$. Assume user k_1 transmits ($p_{k_1 n}^* > 0$) with weighted power gain $\frac{\|b_{k_1 n}\|^2}{\lambda_{k_1}^*} < \max_k \left\{ \frac{\|b_{kn}\|^2}{\lambda_k} \right\}$. Equation (17.61) yields

$$\sum_{i=1}^{K} \|b_{in}\|^2 \, p_{in}^* = B_N \left(\frac{\|b_{k_1 n}\|^2}{\lambda_{k_1}^*} - N_0 \right). \tag{17.62}$$

We know, however, there exists a user k_2 with weighted power gain $\frac{\|b_{k_2 n}\|^2}{\lambda_{k_2}^*} > \frac{\|b_{k_1 n}\|^2}{\lambda_{k_1}^*}$. Equation (17.60) yields

$$\sum_{i=1}^{K} \|b_{in}\|^2 \, p_{in}^* \geq B_N \left(\frac{\|b_{k_2 n}\|^2}{\lambda_{k_2}^*} - N_0 \right) > B_N \left(\frac{\|b_{k_1 n}\|^2}{\lambda_{k_1}^*} - N_0 \right). \tag{17.63}$$

Equation (17.62) and Equation (17.63) contradict each other. Thus we have proved that only those users with the maximum weighted power gain are allowed to transmit on subchannel n.

Appendix C: P_Ψ for the High SNR Case

In this appendix we derive P_Ψ, the probability of Equation (17.34), assuming $\|b_{kn}\|$s are i.i.d. Rayleigh RVs.

For different subchannels m and n, since the event

$$A_1 = \left\{ \frac{\|b_{kn}\|}{\|b_{in}\|} \geq \sqrt{\frac{I_k \, p_i}{I_i \, p_k}}, \, \forall i \neq k, \, n \in \phi_k \right\} \tag{17.64}$$

is independent of the event

$$A_2 = \left\{ \frac{\|h_{jm}\|}{\|h_{im}\|} \geq \sqrt{\frac{I_j}{I_i} \frac{p_i}{p_j}}, \, \forall i \neq j, \, n \in \phi_j \right\}, \tag{17.65}$$

all we need is to derive $P\{A_1\}$. Via some random variable transformation, $P\{A_1\}$ becomes

$$P\left\{ \|h_{kn}\| \geq \sqrt{\frac{I_k}{I_i} \frac{p_i}{p_k}} \|h_{in}\|, \, \forall i \neq k \right\}$$

$$= P\left\{ \|h_{kn}\| \geq \max\left\{ \sqrt{\frac{I_k}{I_i} \frac{p_i}{p_k}} \|h_{in}\|, \, \forall i \neq k \right\} \right\}$$

$$= P\left\{ \max\left\{ \sqrt{\frac{I_k}{I_i} \frac{p_i}{p_k}} \|h_{in}\|, \, \forall i \neq k \right\} - \|h_{kn}\| \leq 0 \right\}. \tag{17.66}$$

Let $X = \max\left\{ \sqrt{\frac{I_k}{I_i} \frac{p_i}{p_k}} \|h_{in}\|, \, \forall i \neq k \right\}$, $Y = -\|h_{kn}\|$ and $Z = X + Y$. Since $\|h_{kn}\|$ has Rayleigh distribution, using the monotonic transformation formula, we know immediately that the PDF of Y is

$$f_Y(t) = \begin{cases} -\dfrac{2}{b} t e^{-t^2/b} & t \leq 0 \\ 0, & t > 0, \end{cases} \tag{17.67}$$

where b is a constant. The CDF of $\sqrt{\frac{I_k}{I_i} \frac{p_i}{p_k}} \|h_{in}\|$ is thus

$$F(t) = \begin{cases} 1 - e^{-t^2 \frac{I_i p_i}{b I_k p_k}} & t \geq 0 \\ 0, & t < 0. \end{cases} \tag{17.68}$$

Now we derive the distribution of X,

$$F_X(t) = P\{X \leq t\}$$

$$= P\left\{ \sqrt{\frac{I_k}{I_1} \frac{p_1}{p_k}} \|h_{1n}\| \leq t, \ldots, \sqrt{\frac{I_k}{I_{k-1}} \frac{p_{k-1}}{p_k}} \|h_{k-1,n}\| \leq t, \right.$$

$$\left. \sqrt{\frac{I_k}{I_{k+1}} \frac{p_{k+1}}{p_k}} \|h_{k+1,n}\| \leq t, \ldots, \sqrt{\frac{I_k}{I_K} \frac{p_K}{p_k}} \|h_{Kn}\| \leq t \right\}$$

$$= \begin{cases} \displaystyle\prod_{i \neq k} (1 - e^{-t^2 \frac{I_i p_i}{b I_k p_k}}) & t \geq 0 \\ 0, & t < 0. \end{cases} \tag{17.69}$$

Since X and Y are independent, we have $F_Z(z) = F_X(t) * f_Y(t)$, where " $*$ " means convolution. Then Equation (17.66) becomes

$$F_Z(0) = \int_{-\infty}^{+\infty} F_X(t) f_Y(-t) dt$$

$$= \int_0^{+\infty} \prod_{i \neq k} \left(1 - e^{-t^2 \frac{I_i p_i}{b I_k p_k}}\right) \frac{2}{b} t e^{-t^2/b} dt. \tag{17.70}$$

Let $x = e^{-t^2/b}$, and Equation (17.70) becomes

$$F_Z(0) = \int_0^1 \prod_{i \neq k} \left(1 - x^{\frac{I_i p_k}{I_k p_i}}\right) dx. \tag{17.71}$$

Therefore

$$P\left\{\|h_{kn}\| \geq \sqrt{\frac{I_k}{I_i} \frac{p_i}{p_k}} \|h_{in}\| \text{ for } \forall i \neq k\right\}$$

$$= \int_0^1 \prod_{i \neq k} \left(1 - x^{\frac{I_i p_k}{I_k p_i}}\right) dx. \tag{17.72}$$

Since parallel subchannels are independent, the probability of Equation (17.34) is given by

$$P_\Psi = \left[\int_0^1 \prod_{j \neq 1} \left(1 - x^{\frac{I_j p_1}{I_1 p_j}}\right) dx\right]^{I_1} \cdots \left[\int_0^1 \prod_{j \neq K} \left(1 - x^{\frac{I_j p_K}{I_K p_j}}\right) dx\right]^{I_K}. \tag{17.73}$$

Appendix D: Proof of Theorem 5

Proof For OFDMA subcarrier allocation Φ, power loading for each user in the uplink is independent because information is delivered through parallel subcarriers. The optimization problem of Equation (17.40) becomes

$$\arg\max_{Q_{kn}} C_o = \sum_{k=1}^K \sum_{n_k=1}^{|\phi_k|} \log_2 \det\left(\mathbf{I} + \mathbf{H}_{kn_k} \mathbf{Q}_{kn_k} \mathbf{H}_{kn_k}^H\right)$$

$$s.t. \ \mathbf{Q}_{kn_k} \succeq 0, \ \sum_{n_k=1}^{|\phi_k|} tr(\mathbf{Q}_{kn_k}) \leq p_k, \tag{17.74}$$

$$n_k \subset \phi_k, \ 1 \leq k \leq K.$$

By singular value decomposition (SVD) decomposition of the space channel matrix $\mathbf{H}_{kn}$ on each subcarrier, the multicarrier MIMO channel can be decomposed into $\sum_{n=1}^N R_{kn}$ parallel subchannels, where R_{kn} is the rank of $\mathbf{H}_{kn}$. To maximize C_o, the transmitter must align its transmit direction with the right singular vector of the effective channel and allocate an appropriate amount of energy in each direction in a water-filling fashion [12]. Note that in the uplink, each user k water-fills its individual power over the $\sum_{n_k=1}^{|\phi_k|} R_{kn_k}$ assigned frequency and space subchannel set (subcarrier set ϕ_k) to maximize any boundary point in the capacity region.

For downlink channels, Equation (17.40) becomes

$$\arg\max_{Q_{kn}} C_o = \sum_{n=1}^{N} \log_2 \det\left(\mathbf{I} + \mathbf{H}_{kn}\mathbf{Q}_{kn}\mathbf{H}_{kn}^{H}\right)$$
$$s.t.\ \mathbf{Q}_{kn} \succeq 0,\ \sum_{k=1}^{K}\sum_{n=1}^{N} tr(\mathbf{Q}_{kn}) \leq P. \tag{17.75}$$

Different from the uplink case, the total power is water filled over all $\sum_{n=1}^{N} R_{kn}$ parallel subchannels in the downlink.

For part (ii), if (Φ, Ψ) is the optimal solution to Equation (17.40), we apply the KKT conditions and have

$$\mathbf{H}_{kn}^{H}\left(\sum_{s=k}^{K} \mathbf{F}_{sn}\right)\mathbf{H}_{kn} + \Lambda_{kn} = \lambda_k \mathbf{I} \tag{17.76}$$

$$\sum_{n=1}^{N} tr(\mathbf{Q}_{kn}) - p_k \leq 0 \tag{17.77}$$

$$\lambda_k \geq 0 \tag{17.78}$$

$$\mathbf{Q}_{kn} \succeq 0 \tag{17.79}$$

$$\Lambda_{kn} \succeq 0 \tag{17.80}$$

$$tr(\mathbf{Q}_{kn}\Lambda_{kn}) = 0 \tag{17.81}$$

$$1 \leq k \leq K,\ 1 \leq n \leq N,$$

where $\mathbf{F}_{sn} = (\mu_s - \mu_{s+1})(\mathbf{I} + \sum_{j=1}^{s} \mathbf{H}_{jn}\mathbf{Q}_{jn}\mathbf{H}_{jn}^{H})^{-1}$.

For any subcarrier $n \in \phi_k$, if $S_n = \{k\}$, user k is the only active user on subcarrier n. The KKT condition of Equation (17.76) for user k on subcarrier n, setting $\mathbf{Q}_{ln} = 0$ for all $l \neq k$, becomes

$$\mu_k \mathbf{H}_{kn}^{H}\left(\mathbf{I} + \mathbf{H}_{kn}\mathbf{Q}_{kn}\mathbf{H}_{kn}^{H}\right)^{-1}\mathbf{H}_{kn} + \Lambda_{kn} = \lambda_k \mathbf{I}. \tag{17.82}$$

For any matrix $\mathbf{H}_{kn}$ we can obtain its SVD as $\mathbf{H}_{kn} = \mathbf{U}_{kn}\Sigma_{kn}\mathbf{V}_{kn}^{H}$, where the $R \times R$ matrix $\mathbf{U}_{kn}$ and the $T \times T$ matrix $\mathbf{V}_{kn}$ are unitary matrices and Σ_{kn} is a $R \times T$ diagonal matrix of singular values $\{\sigma_{kn}(i)\}$ of $\mathbf{H}_{kn}$. These singular values have the property that $\sigma_{kn}^2(i)$ is the ith eigenvalue of $\mathbf{H}_{kn}^{H}\mathbf{H}_{kn}$. When only user k transmits on subcarrier n, the optimal $\mathbf{Q}_{kn}$ has the eigenvalue decomposition $\mathbf{Q}_{kn} = \mathbf{V}_{kn}\Delta_{kn}\mathbf{V}_{kn}^{H}$ [12], where Δ_{kn} is a diagonal matrix. Plug $\mathbf{H}_{kn}$ and $\mathbf{Q}_{kn}$ into Equation (17.82), we have

$$\mu_k \Sigma_{kn}^{H}\left(\mathbf{I} + \Sigma_{kn}\Delta_{kn}\Sigma_{kn}^{H}\right)^{-1}\Sigma_{kn} + \widetilde{\Lambda}_{kn} = \lambda_k \mathbf{I} \tag{17.83}$$

where $\widetilde{\Lambda}_{kn} = \mathbf{V}_{kn}^{H}\Lambda_{kn}\mathbf{V}_{kn}$ is diagonal matrix.

Note that Equation (17.83) is a set of scalar equations

$$\frac{\mu_k}{1/e_i\{\Sigma_{kn}^H\Sigma_{kn}\} + e_i\{\Delta_{kn}\}} + e_i\{\widetilde{\Lambda}_{kn}\} = \lambda_k.$$

Since user k is active on subcarrier n, at least the largest eigenvalue of $\mathbf{H}_{kn}^H\mathbf{H}_{kn}$ is used during water filling. Using Equation (17.81), we have

$$\frac{\mu_k}{1/e_{\max}\{\Sigma_{kn}^H\Sigma_{kn}\} + e_{\max}\{\Delta_{kn}\}} = \lambda_k. \qquad (17.84)$$

Note that Equation (17.84) is equivalent to

$$e_{\max}\left\{\mu_k\mathbf{H}_{kn}^H(\mathbf{I} + \mathbf{H}_{kn}\mathbf{Q}_{kn}\mathbf{H}_{kn}^H)^{-1}\mathbf{H}_{kn}\right\} = \lambda_k. \qquad (17.85)$$

On the other hand, the KKT condition for the inactive users l on subcarrier n becomes

$$\max(\mu_l - \mu_k, 0)\mathbf{H}_{ln}^H\mathbf{H}_{ln} \qquad (17.86)$$

$$+ \min(\mu_k, \mu_l)\mathbf{H}_{ln}^H(\mathbf{I} + \mathbf{H}_{kn}\mathbf{Q}_{kn}\mathbf{H}_{kn}^H)_{ln}^{-1}\mathbf{H}_{ln} + \Lambda_{ln} = \lambda_l\mathbf{I}.$$

Let $\mathbf{A} = \max\{\mu_l - \mu_k, 0\}\mathbf{H}_{ln}^H\mathbf{H}_{ln} + \min\{\mu_l, \mu_k\}\mathbf{H}_{ln}^H(\mathbf{I} + \mathbf{H}_{kn}\mathbf{Q}_{kn}\mathbf{H}_{kn}^H)^{-1}\mathbf{H}_{ln}$. Since both $\mathbf{A}_{ln}$ and Λ_{ln} are positive semidefinite, we have

$$e_{\max}\{\mathbf{A}_{ln}\} \le \lambda_l. \qquad (17.87)$$

Combining Equations (17.85) and (17.87), we have the capacity region optimality condition of Equation (17.44).

For any subcarrier $n \in \phi_k$, if $S_n = \emptyset$, there is no active user on subchannel n (user k gets zero power during water filling). The KKT condition of Equation (17.76) for any user l on subcarrier n becomes

$$\mu_l\mathbf{H}_{ln}^H\mathbf{H}_{ln} + \Lambda_{ln} = \lambda_l\mathbf{I}, \qquad (17.88)$$

thus we have $\mu_l e_{\max}\{\mathbf{H}_{ln}^H\mathbf{H}_{ln}\}/\lambda_l \le 1$, that is, $\max_l\left\{\mu_l e_{\max}\{\mathbf{H}_{ln}^H\mathbf{H}_{ln}\}/\lambda_l\right\} \le 1$.

On the other hand, if Equation (17.44) is true, we can verify that Ψ satisfies the KKT conditions of Equations (17.76) to (17.81). Since Equation (17.40) is a convex programming problem, the KKT conditions of Equations (17.76) through (17.81) are also sufficient for Ψ being optimal. Therefore we have proved that Equation (17.44) is necessary and sufficient conditions for Ψ being optimal.

References

1. R. Knopp and P.A. Humblet, Information capacity and power control in single-cell multiuser communications, in *IEEE International Conference on Communications*, vol. 1 p. 331, IEEE, Washington, DC, 1995.
2. D. Tse, Optimal power allocation over parallel Gaussian broadcast channels, *International Symposium on Information*, p. 27, IEEE, Washington, DC, 1997.
3. S.M. Alamouti, A simple transmitter diversity scheme for wireless communications, *IEEE J. Select. Areas Commun.*, 1, 1451, 1998.
4. R.S. Blum, Y. Li, J.H. Winters, and Q. Yan, Improved space-time coding for MIMO-OFDM wireless communications, *IEEE Commun. Lett.*, 48, 1873, 2001.
5. H. Bolckei, D. Gesbert, and A.J. Paulraj, On the capacity of OFDM-based spatial multiplexing systems, *IEEE Trans. Commun.*, 50(2), 225, 2002.
6. A. Goldsmith, S.A. Jafar, N. Jindal, and S. Vishwanath, Capacity limits of MIMO channels, *IEEE Trans. Select. Areas Commun.*, 21, 684, 2003.
7. D. Kivanc, G. Li, and H. Liu, Computationally efficient bandwidth allocation and power control for OFDMA, *IEEE Trans. Wireless Commun.*, 2, 1150, 2003.
8. IEEE, IEEE Standard for Local and Metropolitan Area Networks—Part 16: Air Interface for Fixed Broadband Wireless Access Systems, IEEE, Washington, DC, 2005.
9. M. Moeneclaey, M. Van Bladel, and H. Sari, Sensitivity of multiple-access techniques to narrow-band interference, *IEEE Trans. Commun.*, 49(3), 497, 2001.
10. R.S. Cheng and S. Verdu, Gaussian multiaccess channels with ISI: capacity region and multiuser water-filling, *IEEE Trans. Inform. Theory,* 39(3), 773, 1993.
11. D. Tse and S. Hanly, Multiaccess fading channel—part I: polymatroid structure, optimal resource allocation and throughput capacities, *IEEE Trans. Inform. Theory,* 44(7), 2796, 1998.
12. W. Yu, *Competition and Cooperation in Multiuser Communication Environments*, Ph.D. Dissertation, Stanford University, Stanford, CA, 2002.
13. K. Kim, H. Kim, Y. Han, and S.L. Kim, Iterative and greedy resource allocation in an uplink OFDMA system, *15th International Symposium on Personal, Indoor and Mobile Radio Communications*, vol. 4, p. 2377, IEEE, Washington, DC, 2004.
14. J. Jang and K.B. Lee, Transmit power adaptation for multiuser OFDM systems, *IEEE J. Select. Areas Commun.,* 21(2), 171, 2003.
15. G. Li and H. Liu, On the optimality of downlink OFDMA MIMO system, *IEEE Asilomar Conference*, vol. 1, p. 324, IEEE, Washington, DC, 2004.
16. A. Goldsmith, *Wireless Communications*, Cambridge University Press, New York, 2005.
17. T. Michel and G. Wunder, Optimal and low complex suboptimal transmission schemes for MIMO-OFDM broadcast channels, *IEEE International Conference on Communications*, vol. 1, p. 438, IEEE, Washington, DC, 2005.
18. R. Knopp and P.A. Humblet, Multiple-accessing over frequency-selective fading channels, in *6th International Symposium on Personal, Indoor and Mobil Radio Communications*, vol. 3, p. 1326, IEEE, Washington, DC, 1995.
19. S. Ohno, P.A. Anghel, G.B. Giannakis, and Z.-Q. Luo, Multi-carrier multiple access is sum-rate optimal for block transmission over circulant ISI channels, *IEEE International Conference on Communications*, vol. 3, p. 1656, IEEE, Washington, DC, 2002.
20. Z.-Q. Luo, T.N. Davidson, G.B. Giannkis, and K.M. Wong, Transceiver optimization for block-based multiple access through ISI channels, *IEEE Trans. Inform. Theory*, 52(4), 1037, 2004.
21. J. Lee, Rate and power allocation for multi-carrier communication systems, Ph.D dissertation, Stanford University, Stanford, CA, available at http://www.stanford.edu/jungwon/research/JungwonLeeThesis.pdf.
22. D. Tse and P. Viswanath, *Fundamentals of Wireless Communication*, Cambridge University Press, New York, 2005.

23. H. Liu and G. Li, *OFDM-Based Broadband Wireless Networks*, John Wiley & Sons, New York, 2005.
24. T.M. Cover and J.A. Thomas, *Elements of Information Theory*, John Wiley & Sons, New York, 1991.
25. R.G. Gallager, *Information Theory and Reliable Communication*, John Wiley & Sons, New York, 1968.
26. G. Caire and S. Shamai, On the capacity of some channels with channel side information, *IEEE International Symposium on Information Theory*, p. 42, IEEE, Washington, DC, 1998.
27. W. Hirt and J.L. Massey, Capacity of the discrete-time Gaussian channel with intersymbol interference *IEEE Trans. Inform. Theory*, 34(3), 380, 1988.
28. S. Verdu, Multiple-access channels with memory with and without frame synchronization, *IEEE Trans. Inform. Theory*, 35(3), 605, 1989.
29. E. Telatar, Capacity of multi-antenna Gaussian channels, *Euro. Trans. Telecommun.*, 10(6) 585, 1999.
30. J. Nocedal and S.J. Wright, *Numerical Optimization* Springer-Verlag, New York, 1999.
31. S. Boyd and L. Vandenberghe, *Convex Optimization*, Cambridge University Press, New York, 2004.
32. R.T. Rockafellar, *Network Flows and Monotropic Optimization*, John Wiley & Sons New York, 1984.

Index

moving networks, 274
in PANs, 274
of relay stations (RSs), 58
support in RAN technologies, 273
technology-independent, 274
MONAMI6 (*See* Mobile Nodes and Multiple
Interfaces in IPv6)
Moving networks, 274
Moving Picture Experts Group (MPEG), 12,
152
MPEG (*See* Moving Picture Experts Group)
MPLS (*See* multi-protocol label switching)
MQAM (*See* M-ary quadrature amplitude
modulation)
MRC (*See* maximum ratio combining)
MSCTP (*See* mobile stream control transmission
protocol)
MSH (*See* mesh)
MSs (*See* mobile stations)
MSSs (*See* mesh SSs (MSSs); (*See* mobile subscriber
stations (MSSs)
MUL (*See* mobile uplink)
Multicarrier modulation, 388–389
Multi-carrier spread spectrum (MC-SS), 82
Multihomed devices, 238, 239
Multihoming Mobile IP (M-MIP), 250–251
Multihop relay, 165, 166, 167; (*See also* radio
resource management)
cross-layer design for, 214–215
Multipath fading, 82, 84
Multiple-input multiple-output (MIMO), 20, 318; (*See
also* adaptive MIMO switching); (*See also*
Space-time block codes)
and adaptive antenna systems, 29
capacity, 71, 73–74
defined, 71
for Mobile WiMAX Downlink, link-level
performance
MIMO channel model, 100–101
PHY layer abstraction, 109–113
SM MIMO mode, 104–109
STC MIMO mode, 101–104
for Mobile WiMAX Downlink, system-level
performance
aggregate sector throughput, 123
deployment scenario and link budget,
113–114
interference model and loading, 115–116
network topology and frequency
reuse, 113
OFDMA scheduler and HARQ, 116–117, 118
propagation model, 114–115
simulation results, 118–124
spectral efficiency, 122–123
multidimensional diversity, 82
multi-user system model, 77
and OFDM, 71
versus smart antennas, 98–99
space diversity, 71, 82
space-time signal processing, 71
system model, single user, 72–73

transmission schemes, 71, 78, 324
in WiMAX
capacity, 74–77
relationship with, 69–70
Multi-protocol label switching (MPLS), 91
Multiuser communications, 389
Multiuser diversity, 391, 392
MVNOs, *see* Mobile virtual network operators
(MVNOs)

N

Native clock (CLKN), 281
Near zero intermediate frequency (NZIF), 27
Neighborhood detection, 272
NETLMM (*See* Network-based Localized Mobility
Management)
NetMan (*See* Network Management Task Group)
Network deployment
architecture, 376
capacity limited, 370
coverage limited, 370
expenditures, 376–377
Network dimensioning, 374, 377
Network identification (ID), 275
Network Management Task Group (NetMan), 293
Network selection
access discovery, 247–248
access network selection
analysis and comparison, 253
performance of algorithm, 253–254
state-of-the-art algorithms, 251–253
types of algorithms, 251
always cheapest (AC), 244, 262
architecture design, 241, 242–245
challenges with state-of-the-art, 241–242
consumer utility-based strategy, 254–255
consumer surplus, 260–261, 262
throughput prediction scheme, 255
user utility function, 255–259
decision methodology, 249
decisions, 237
future, 238–239
handover execution process, 240
metrics and user preferences, 241, 245–248
selection triggers, 240
traditional, 238
mobility management functionality, 249–251
performance evaluation, 262–264
Network-assisted association, 286
Network-based Localized Mobility Management
(NETLMM), 288
NLOS (*see* non-line-of-sight)
Non-line-of-sight (NLOS), 2, 7
Non-real-time polling service (nrtPS), 2, 98, 134,
154
bandwidth allocation, 159
Non-real-time traffic, 161–164
Non-real-time variable rate service (NRT-VR), 208,
210
NrtPS (*See* non-real-time polling service)